CARBON NANOTUBES - FROM RESEARCH TO APPLICATIONS

Edited by **Stefano Bianco**

Carbon Nanotubes - From Research to Applications
Edited by Stefano Bianco

Published by InTech
Janeza Trdine 9, 51000 Rijeka, Croatia

Publishing Process Manager Viktorija Zgela

Technical Editor Teodora Smiljanic

Cover Designer InTech Design Team

Additional hard copies can be obtained from orders@intechopen.com

Carbon Nanotubes - From Research to Applications, Edited by Stefano Bianco
p. cm.
ISBN 978-953-307-500-6

For EU product safety concerns:
IN TECH d.o.o., Prolaz Marije Krucifikse Kozulić 3, 51000 Rijeka, Croatia,
info@intechopen.com or visit our website at intechopen.com.

Meet the editor

Dr. Stefano Bianco received his Master's Degree in Physics from the University of Turin, Italy, in 2003 and his Ph.D. in Electronic Devices from Polytechnic University of Turin, Italy, in 2007. He was a postdoctoral fellow (2007 - 2009) at the Physics Department of the Polytechnic University of Turin, working on carbon nanotube synthesis, characterization and application. Currently he is a researcher of the Italian Institute of Technology, working on innovative materials and technologies for energy harvesting applications, mainly investigating novel systems for dye-sensitized solar cells and water splitting devices. He is co-author of more than 20 scientific papers in international peer-reviewed journals and of five book chapters.

Contents

Preface

Since their discovery in 1991, carbon nanotubes have been considered as one of the most promising materials for a wide range of applications, in virtue of their outstanding properties. During the last two decades, both single-walled and multi-walled CNTs probably represented the hottest research topic concerning materials science, equally from a fundamental and from an applicative point of view. Research on CNT synthesis, combined with basic understanding on growth phenomena, contributed to the development of a well controlled process and, thanks to this, the CNT production is now mature and ready for industrialization. Moreover, CNT application and integration in very different systems, from nanoelectronics to optics, from biosensing to reinforcing in nanocomposites, was extensively demonstrated.

There is a prevailing opinion among the research community that CNTs are now ready for application in everyday world. This book provides an (obviously not exhaustive) overview on some of the amazing possible applications of CNT-based materials in the near future. Some interesting topics concerning CNT surface treatment and integration with other materials are covered, both from a theoretical and from an experimental point of view. Particular emphasis is devoted to the application of carbon nanostructures as reinforcements in nanocomposites. In fact, this field currently appears as the most promising for the immediate market application, since the tailoring of material characteristics with CNT integration could open new astonishing possibilities in materials science.

Stefano Bianco
Fondazione Istituto Italiano di Tecnologia,
Italy

Part 1

Theory, Characterization and Application of Carbon Nanotubes

1

Nitrogen-Containing Carbon Nanotubes - A Theoretical Approach

M. Leonor Contreras and Roberto Rozas
University of Santiago de Chile, Usach, Faculty of Chemistry and Biology
Chile

1. Introduction

Carbon nanotubes are important materials for a variety of scientific and technological applications due to their unique properties (Frank et al., 1998; Ganji, 2008; Hone et al., 2000; Marulanda, 2010; Yu et al., 2000). Of course, their properties depend on their structure. For instance, nanotube conductivity depends on chirality, diameter, and length (Alam & Ray, 2007; Hamada et al., 1992; Saito et al., 1992; S.H. Yang et al, 2008). The purity of the nanotubes and the presence of defects also affect their conductivity.

Chirality or helicity refers to the way the nanostructure arises by the folding of a graphene sheet. Nanotube chirality is usually characterised by two integers, n and m, known as Hamada indices, defining three classes of nanotubes. For instance, *armchair* (n,n) nanotubes exhibit metallic behaviour, *zigzag* (n,0) nanotubes are semiconductors, and *chiral* (n,m) nanotubes exhibit metallic behaviour if the difference (n - m) is a multiple of 3 and semiconductor behaviour otherwise (Charlier, 2002). For instance, a (7,1) *chiral* nanotube is a conductor but the *chiral* (7,3) nanotube is not.

Within the numerous potential applications imagined for carbon nanotubes, hydrogen storage represents the most promising application capable of making a safe, efficient and "green" contribution to fuel cells with hydrogen management in the solid state. The principal hydrogen-adsorption mechanisms associated with nanotube hydrogen uptake are the physisorption and chemisorption of hydrogen.

During physisorption, hydrogen interacts with selected sites of a carbon nanotube or substrate. The interaction energy increases as the substrate polarisability increases. Density-functional theory calculations indicate that nanotube-hydrogen interactions are weak and that hydrogen diffusion from the nanotube is facilitated by slightly increasing temperature (Mpourmpakis et al., 2006). The hydrogen-binding energies, calculated using density-functional theory are small and similar for metallic and semiconducting nanotubes, indicating that substantial adsorption is only possible at very low temperatures (Cabria et al., 2006). The same conclusion is reached by studying hydrogen adsorption in carbon-nanotube arrays through molecular dynamic simulation (Kovalev et al., 2011), in which a second adsorption layer is detected at 80 °K. This second layer of hydrogen is not detected at room temperature.

Through chemisorption, hydrogen is covalently bonded to carbon atoms in such a way that a change of sp^2 to sp^3 carbon hybridisation occurs, which is manifested in the C−C bond length values. A typical (sp^3)C−C(sp^3) bond length is 1.54 Å. The calculated C−C bond

lengths for fully hydrogenated nitrogen-containing carbon nanotubes, obtained using density-functional theory, are somewhat longer and range from 1.54 to 1.57 Å (Contreras et al., 2010) depending on the nanotube configuration.
Experimental work by Dillon et al. (Dillon et al., 1997), which reported 10 wt. % of hydrogen uptake by single-walled carbon nanotubes at room temperature, stimulated many theoretical and experimental studies of carbon nanotubes as an ideal hydrogen carrier (Bilic & Gale, 2008; Dinadayalane et al, 2007; Kaczmarek et al, 2007). The hydrogen binding energies in these cases are clearly dependent on chirality, tube diameter, hydrogen occupancy, and endohedral vs. exohedral binding.
For example, based on density-functional theory calculations of atomic hydrogen adsorption on carbon nanotubes at very low occupancies (i.e., 1 or 2 adsorbed hydrogen atoms), F.Y. Yang et al. reported that the binding energies for *zigzag* nanotubes increase as the nanotube diameter increases and are higher than the binding energies for *armchair* nanotubes (F.H. Yang et al., 2006). In contrast, calculations for *armchair* nanotubes by Dinadayalane et al. at the same level of density-functional theory showed that binding energy (or exothermicity) of hydrogen chemisorption decreases as the nanotube diameter increases (Dinadayalane et al., 2007). However, for a single hydrogen atom adsorbed on a single-walled-carbon-nanotube surface, density-functional theory calculations indicate that both the binding energy (chemisorption) and the diffusion barrier for a hydrogen atom decrease as the tube diameter increases (Ni & Zeng, 2010). In this case, the binding energy is not strongly affected by the tube chirality (Ni & Zeng, 2010).
It is clear that exohedral binding is more energetically favourable than endohedral binding because the conversion of sp^2 to sp^3 hybridisation upon hydrogen binding is easier for the carbon atoms of the exterior carbon-nanotube walls (F.H. Yang et al., 2006). It is also apparent that adsorbed hydrogen acts as an autocatalyst for further hydrogenation, as was reported by Bilic and Gale after investigating the chemisorption of molecular hydrogen on small-diameter *armchair* carbon nanotubes using density-functional theory (Bilic & Gale, 2008). Bilic and Gale found that only small-diameter nanotubes (diameters up to 10 Å) have the theoretical potential for a high hydrogen uptake by chemisorption (Bilic & Gale, 2008).
However, from a quantitative viewpoint, some of the experiments performed at room temperature resulted in very low hydrogen-storage capacities, generating debate and much controversy (Baughman et al., 2002; G. Zhang et al., 2006). Density-functional theory calculations of both the energy-barrier and the Gibbs-free-energy changes for hydrogen on a (10,0) single-walled carbon nanotube when changing from a physisorption to a chemisorption state (Han & Lee, 2004) suggest a major obstacle for the practical use of the carbon nanotube as a hydrogen storage medium.
Several research groups' results have indicated that hydrogen uptake depends on factors, such as the nanotube type and purity, the gas temperature and pressure, and the equipment used for the experimental determination, all of which affect reproducibility. A recent detailed discussion by Yao (Yao, 2010) about different experimental and theoretical studies critically analyses the influencing factors that must be considered for a better evaluation of carbon nanotubes as good candidates for hydrogen storage. Two important points mentioned are the nanotube purity after synthesis and the presence of some heteroatoms that could modify the nanotube surface electronic density.
Using a volumetric measurement setup specifically designed for carbon nanotubes, Liu et al. obtained results for different types of nanotubes, showing that the reliable hydrogen storage capacity of carbon nanotubes under a pressure of approximately 12 MPa at room

temperature is less than 1.7 wt. % (Liu et al., 2010). This value is far below the benchmark of 6.5 wt. % set by the US Department of Energy for the on-board application of hydrogen storage systems, suggesting that hydrogen uptake in pure carbon nanotubes is not a good alternative for on-board applications.

Doping is also an important factor to consider (Griadun, 2010). Semiconducting carbon nanotubes doped with 2–10% of nitrogen become metallic (Charlier, 2002; Czerw et al., 2001). Interestingly, nitrogen-doped carbon nanotubes constitute a good metal-free catalyst system for oxygen reduction reactions in alkaline media (Gong et al., 2009). In acidic media, these nitrogen-doped carbon nanotubes show a higher current density and a higher oxygen reduction reaction rate constant compared to conventional Pt-based catalysts (Xiong et al., 2010). Nitrogen-doped carbon nanotubes behave as convenient catalysts for these reactions and have excellent environmental and economical profiles because they are electrochemically more reactive and more durable than Pt-containing materials. Density-functional theory calculations at the B3LYP/6-31G* level indicate that the metal-free nitrogen-doped carbon nanotubes have promising catalytic ability for C-H methane activation (Hu et al., 2011) that is comparable to that of noble-metal catalysts and enzymes.

In addition, the nitrogen doping of carbon nanotubes has a significant effect on hydrogen storage capacity. Density-functional theory calculations for atomic hydrogen adsorption indicate that nitrogen-doping forms an electron-rich six-membered ring structure and decreases the adsorption energies in single-walled carbon nanotubes (Zhou, et al., 2006). Nevertheless, doping nanotubes with nitrogen considerably enhances the hydrogen dissociative adsorption, substantially reducing the hydrogen diffusion barrier according to density-functional theory studies on nitrogen-doped (8,0) nanotubes (Z.Y. Zhang & Cho, 2007).

Hydrogen molecules can diffuse inside nitrogen-doped *zigzag* (10,0), *chiral* (7,5) and *armchair* (6,6) nanotubes (with diameters of approximately 8 Å), as indicated by molecular dynamics simulation (Oh et al., 2008), suggesting that these nitrogen-doped nanostructures could be applied as effective media for the storage of hydrogen molecules. However there has not been any publication estimating the amount (wt. %) of hydrogen uptake achieved by these nitrogen-doped nanotubes. Importantly, most of these studies are conducted only for structures with small nitrogen content. Research on the adsorption of molecular hydrogen on the external surface of single-walled (8,0) nanotubes decorated with atomic nitrogen (approximately 14.6 wt. % of nitrogen content) using both density-functional theory and molecular dynamics found that the system can store up to 9.8 wt. % of hydrogen at 77°K (Rangel et al., 2009) and that 6.0 wt. % of hydrogen remains adsorbed at 300°K at ambient pressure with an average adsorption energy of −80 meV/(H_2). These results suggest that nanotubes with higher nitrogen content could potentially constitute a high-capacity hydrogen storage medium.

Experimental measurements on hydrogen storage associated with other nitrogen-doped carbon structures indicated that nitrogen-doped microporous carbon had both an 18% higher hydrogen-storage capacity and significantly higher heats of hydrogen adsorption than a pure carbon structure with a similar surface area (L.F. Wang & R.T. Yang, 2009). In addition, nitrogen-doped carbon xerogels enhanced hydrogen adsorption at 35°C (K.Y. Kang et al., 2009).

Because the incorporation of nitrogen atoms into carbon nanotubes affords structures with the ability to participate in hydrogen bonding, these nitrogen-doped nanostructures may have additional chemical properties, such as the immobilization of transition metals (Feng et al., 2010), or the coupling of gold nanoparticles (Allen et al., 2008), which could be useful for potential biomedical applications.

Nitrogen-doped nanotubes are less toxic than undoped carbon nanotubes, but some concern about their safe use remains (Pastorin, 2009; Stern & McNeil, 2008). Experimental research involving the analysis of the toxicological effects on both mice and amoeba cell viability caused by nitrogen-doped or undoped carbon nanotubes indicates that nitrogen-doped carbon nanotubes are less harmful and more biocompatible than the undoped nanotubes (Terrones, 2007).

For undoped carbon nanotubes, a recent scientific study (Nayak et al., 2010) investigated a variety of parameters concerning the toxicity of either single- or multi-walled carbon nanotubes, with and without functionalisation, to assess their cytotoxic profile; this assessment was based on several critical parameters, such as tube length, concentration, dispersibility, and purity, using colorimetric assays to measure the activity of mitochondrial reductase. The results of these studies show that the purity and dispersibility of the nanotubes are the most critical parameters to guarantee their safe application in biology and medicine when used in a normal concentration range (10-150 μg/ml). This finding is an important contribution to the field, assuring the safe use of ultrapure nanotubes.

All of the aforementioned features make the study of the properties, stability and hydrogen chemisorption energies of carbon nanotubes with high nitrogen content quite interesting and necessary. As an extreme, nitrogen nanotubes or nitrogen nanoneedles formed by units of N_{2m} (m = 2-6) with hydrogen as the terminal atoms (with almost 100 wt. % of nitrogen) as well as nitrogen nanobundles with a carbon backbone have been studied using the density-functional theory method (J.L. Wang et al., 2006). J.L. Wang et al. reported that the mentioned nitrogen nanostructures and the nitrogen nanobundles have low stability but are proper minima with all real frequencies at the level of B3LYP/6-31G** having electronic properties that might be modulated as a function of the local charge environment.

There are only a few studies on nitrogen-containing carbon nanotubes with high nitrogen content. However, the synthesis of nitrogen-doped carbon nanotubes (Trasobares et al., 2002) can be selectively performed with either sp^2 or sp^3 nitrogen atoms (Zhong et al., 2007), and nitrogen configuration can be controlled during the fabrication of the nitrogen-doped carbon nanotubes to obtain the desired nanotube properties (S.H. Yang et al., 2008). Nitrogen-doped carbon nanotubes with different nitrogen contents synthesised by chemical vapour deposition (CVD) with pyridine as the nitrogen source and acetylene as the carbon source contain pyridinic, pyrrolic and graphitic types of C-N bonds, as revealed by X-ray photoelectron spectroscopy (XPS) (Y. Zhang et al., 2010).

Fully exohydrogenated nitrogen-containing carbon nanotubes with high nitrogen content, having sp^3 nitrogen atoms, have been reported to be stable compounds (Contreras et al, 2010) with promising expected properties that have not yet been fully studied.

Our aim in this work is to theoretically investigate the structural geometry, energetic stabilities, and electronic properties and to calculate the hydrogen-chemisorption energy for a particular family of nitrogen-containing carbon nanotubes using the density-functional theory method at the B3LYP/6-31G* level of theory. These nanotubes have very small diameters (≈0.3 nm) and a C_4N_2 cyclic unit with a pyrimidine-like disposition as the repetitive layer (with 36-37 wt. % nitrogen content). We also would like to clarify whether their structural and electronic properties are affected by the presence of different terminating units at the nanotube ends. The final aim of this work is the evaluation of the possibility that these nitrogen-containing carbon nanotubes behave as hydrogen-storage materials and to determine the influence of both the nanotube configuration and length on their properties.

2. Computational methodology

In this work, chemisorption refers to exhaustive chemisorption with completely saturated products. The full exohydrogenated nanostructures were built using the HyperChem v7.0 program (Hyperchem, release 7.0), starting from layers defined as cyclic units containing four *sp*³ carbon atoms and two *sp*³ nitrogen atoms forming a pyrimidine-like framework (nitrogen atoms are separated by a single carbon atom, -N1-C2-N3-C4-C5-C6-, as shown in Figure 1). The previous visualisation of nitrogen doping for different nanotube configurations was performed using a specially designed *ad hoc* graphical interface (Contreras et al., 2009).
Nanostructures were built by covalently arranging one layer on top of the other in such a way that the carbon and nitrogen atoms were *sp*³-hybridised, thus forming open-ended nanotubes terminated with hydrogen atoms oriented parallel to the nanotube primary axis. Different configurations were characterised according to the rotation angle between adjacent layers, giving *S*, *O*, *M*, and *P* configurations for rotation angles θ of 0°, 60°, 120°, and 180°, respectively (*O*, *M*, and *P* were chosen based on the **o***rtho*, **m***eta*, and **p***ara* positions of disubstituted benzene rings, with carbon atom number 2 taken as a reference, as shown in Figure 2).

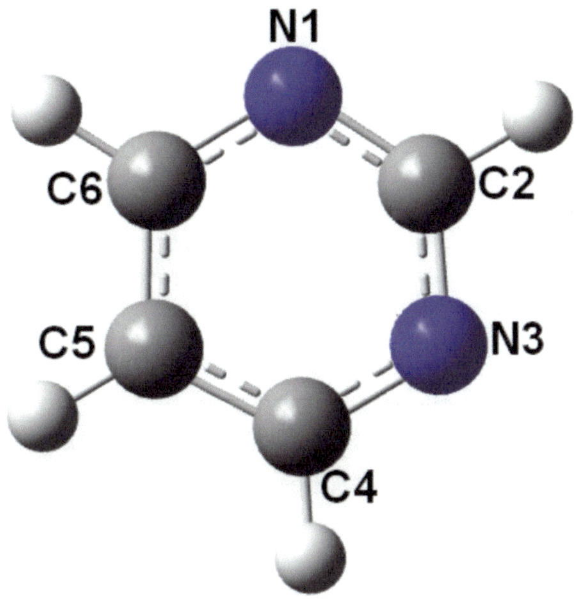

Fig. 1. Representation of the cyclic unit used as framework for building up the nitrogen-containing carbon nanotubes. Carbon atom denoted as C2 is used as reference for defining nanotube configuration.

All nanostructures were optimised by the density-functional theory (DFT) method at the B3LYP/6-31G* level of theory (Becke, 1993; Lee et al., 1988) using the Gaussian 03 suite of programs (Frisch et al., 2004). For the verification and characterisation of energy minima,

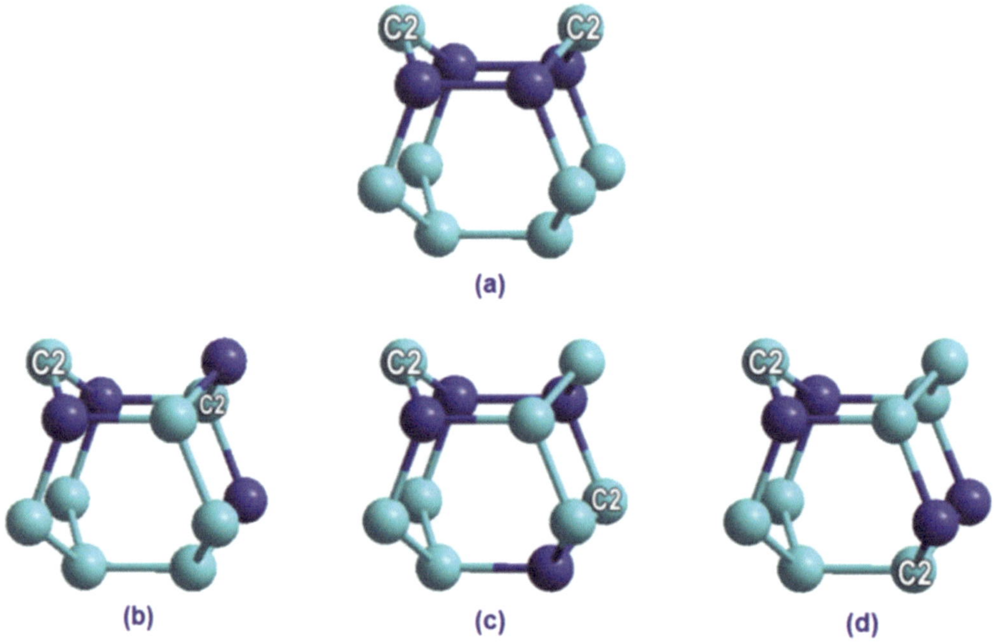

Fig. 2. Union of two consecutive layers defining the configuration of the studied nitrogen-containing carbon nanotubes. The first layer remains fixed, and the second layer is rotated. (a) eclipsed, *S*-type; (b) rotated 60°, *O*-type; (c) rotated 120°, *M*-type; (d) rotated 180°, *P*-type.

harmonic vibrational frequency calculations for optimised geometries at the same level of theory were performed, all of which yielded zero imaginary frequencies. Band gaps were calculated as the difference of $E_{LUMO} - E_{HOMO}$. No symmetry constraints were used. Charges were assigned using the Mulliken population analysis method, which partitions the total charge among the atoms in the molecule (in the present study, the sum of the Mulliken charges = 0.000 for each nanostructure).

2.1 Notation

All of the nanotubes studied in this work are nitrogen-containing carbon nanotubes with high nitrogen content, as was explained previously. Chemisorption is done to 100% hydrogen coverage.

2.1.1 Configuration

S, *O*, *M*, and *P* configurations describe the continuous rotation angle between one layer and the next, which corresponds to 0°, 60°, 120°, and 180°, respectively, as was explained above.

2.1.2 Length of the tube

This term describes the number assigned for determining how many layers or cyclic units of pyrimidinic topology are participating, which is expressed before the configuration character under consideration. For instance, 8M is assigned to a nanotube having 8 layers with a rotation angle between contiguous layers of 120°. Different structures of 4 to 12 layers and nanotubes of up to 20 layers for some configurations have been studied.

2.1.3 Diameter

All nanostructures studied here belong to the (3,0) type with a diameter value of ≈0.3 nm. No variations in diameter have been considered.

2.1.4 Terminal groups

Open nanotubes ended with hydrogen atoms located coaxially to the nanotube were studied. The orientation variation of these hydrogen atoms may affect the total energy of the nanotube up to ≈50 kcal/mol (Contreras, et al., 2010; J.L. Wang et al., 2006). The effect of the terminal groups created by changing the three terminal hydrogen atoms by one unit of nitrogen, phosphorus, NO_3 group or a cycle of five carbon atoms (designated as 5C) at each end of the nanotube was also studied. In this way, the nanotubes remain closed at both extremes.

2.1.5 Chemisorption

A capital letter H is added to the name of the terminal group for nanotubes with adsorbed hydrogen (chemisorbed). The considered nanotubes have both extremes symmetrically bonded. In this way, a nanotube of 8 layers with an M configuration with chemisorbed hydrogen and with nitrogen as the terminal group at both extremes is designated as 8M-N-H. With no hydrogen chemisorption, the notation of 8M-N is used. If there are also no terminal groups that are different from hydrogen, the nanotube is designated simply as 8M. Therefore, in general terms, the notation is

(number of layers) (configuration)-(terminal groups)-(w/out hydrogen adsorption) (1)

Chemisorption, when present, is exhaustively considered with the formation of completely saturated nanostructures. Therefore, a saturated nanotube in this work is a nanotube for which hydrogen chemisorption occurred exhaustively.

2.1.6 Further calculations

Calculations for nanotubes with 20 layers (206 atoms and 1972 basis functions) were conducted using the Jaguar v7.5 (Jaguar, 2008) and the DFT method at the B3LYP/6-31G* level of theory (Becke, 1993; Lee et al., 1988); with the Gaussian 03, it was not possible to optimise these structures after several days of computation.

Formation energies (or substitution energies) were calculated as

$$E_{Form} = E_{NCNT} + 2nE_C + 2nE_H - E_{CNT} - 2nE_N \tag{2}$$

where E_{NCNT} and E_{CNT} are the total energies for saturated carbon nanotubes with the same number of layers with and without nitrogen, respectively; E_C, E_H, and E_N are the total energies of an isolated carbon atom, hydrogen atom, and nitrogen atom, respectively; and n is the total number of layers in the nanotube. E_C was calculated as

$$E_C = \left(E_{CNT} - (6n+6)E_H\right)/6n \tag{3}$$

and E_H and E_N were derived from one half of the calculated total energy for a hydrogen molecule and a nitrogen molecule, respectively, at the same level of theory. For a better understanding, the process considered for a 6-layer nitrogen-containing carbon nanotube can be written as

$$C_{6n}H_{6n+6} + 2nN \longrightarrow C_{4n}H_{4n+6}N_{2n} + 2nC + 2nH \quad (4)$$

Reaction energies for hydrogen chemisorption (Er) on the external surface of nitrogen-containing carbon nanotubes were calculated using the formula below.

$$Er = E_{(\text{H chemisorbed})} - E_{(\text{without chemisorption})} - hE_H \quad (5)$$

where $E_{(\text{H chemisorbed})}$ denotes the total energy of the hydrogen-chemisorbed nanotube; *h* represents the number of chemisorbed hydrogen atoms; $E_{(\text{without chemisorption})}$ and E_H correspond to the energy of sp^2-hybridised nitrogen-containing nanotubes (ended by hydrogen atoms) and of the hydrogen atom, respectively. Expression (5) can also be written as

$$Er = E_{C_{4n}H_{4n+6}N_{2n}} - E_{C_{4n}H_6N_{2n}} - 4nE_H \quad (6)$$

with n representing the number of layers or the length of the nanotube. Er/H, the reaction energy per hydrogen atom, is calculated as

$$Er / H = Er / 4n \quad (7)$$

Some of the $C_{4n}H_6N_{2n}$ nanostructures (after being optimised to proper minima -with entirely real vibrational frequencies), formed small cycles at both ends of the tube in GaussView (graphical interface of Gaussian 03). To calculate Er/H using more stable structures, nanotubes with the first and the last layers completely saturated were considered for chemisorption. Expressions (6) and (7) remain, respectively, as the following (8) and (9) expressions:

$$Er = E_{C_{4n}H_{4n+6}N_{2n}} - E_{C_{4n}H_{14}N_{2n}} - (4n - 8)\, E_H \quad (8)$$

$$Er / H = Er / 4n - 8 \quad (9)$$

3. Results and discussion

Results will be presented in sections following the indicated order. First, the results obtained for completely hydrogenated open nitrogen-containing nanotubes and ended by hydrogen atoms will be shown, which represent 100% chemisorption. Next, the effect of closing the open nanotubes with different terminal groups will be shown. The results for the nanotubes without chemisorption will follow. Finally, the calculated energies of chemisorption will be presented.

3.1 Geometrical structures

The considered fully hydrogenated nitrogen-containing carbon nanotubes formed by $C_4N_2H_4$ units and ended by hydrogen atoms are ≈0.28 nm in diameter and 0.66—3.96 nm in length. The interlayer bond lengths calculated by density-functional theory for N–N, N–C, and C–C bonds are 1.52 Å, 1.48–1.50 Å, and 1.55–1.56 Å, respectively. A typical (sp^3)C–C(sp^3) bond length is 1.54 Å. The C–N bond length for amines is 1.479 Å. The N–N bond length in nitrogen nanotubes calculated at the density-functional theory level range from 1.42 to 1.52 Å (J.L. Wang et al., 2006).

To analyse the bond lengths and bond angles inside each nanotube layer, all nanotube geometries were optimised at the same level of theory, including pyrimidine and its saturated isomer as a reference. Based on the comparison of the bond angles and bond lengths of pyrimidine and its saturated isomer, the saturated structure has smaller bond angles and higher bond lengths than pyrimidine, as was expected for these structures (see Figure 3).

To analyse the same parameters for nanotubes, an 8M (8 layers, with M configuration) nanotube was selected at random and only layer 2 and layer 4 were considered for the analysis of both the nanotube without hydrogen adsorption (8M-H) and the totally saturated (8M-H-H) nanotube (see Figure 4).

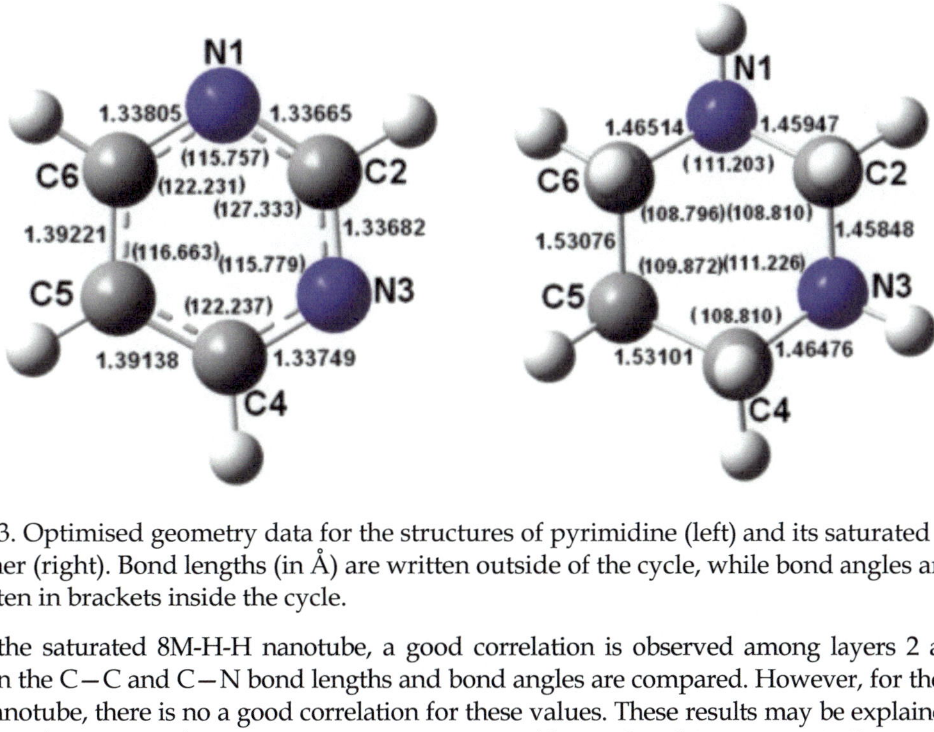

Fig. 3. Optimised geometry data for the structures of pyrimidine (left) and its saturated isomer (right). Bond lengths (in Å) are written outside of the cycle, while bond angles are written in brackets inside the cycle.

For the saturated 8M-H-H nanotube, a good correlation is observed among layers 2 and 4 when the C–C and C–N bond lengths and bond angles are compared. However, for the 8M-H nanotube, there is no a good correlation for these values. These results may be explained by the fact that saturated structure environments around layers 2 and 4 are more similar, and thus the characteristics of the layers along the tube are likely to be maintained. However, for the unsaturated structure, layer 2 is nearest to layer 1, which has three covalent bonds to hydrogen atoms, compared to layer 4, which has no neighbours bonded to hydrogen. Based on the analysis of the intralayer parameters, there is only one pair of C–C bonds in layer 2 that are longer for the saturated structure than those for the unsaturated one. Instead, in layer 4, there are several bonds in the saturated structure that are longer than those in the unsaturated one (except for the N1–C2 and N3–C4 bonds).

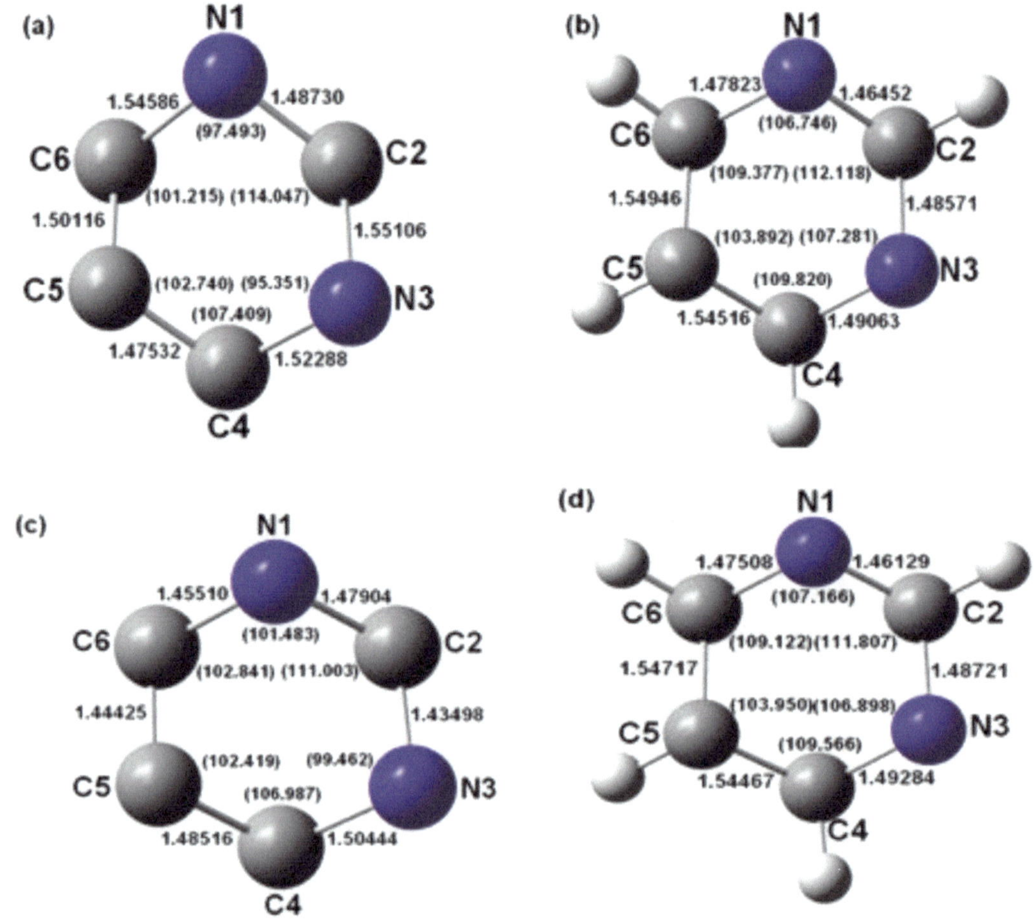

Fig. 4. Structural characteristics for the second (a, b) and fourth (c, d) 8M-nanotube layers: (a and c) without hydrogen chemisorption and (b and d) with hydrogen chemisorption. Bond lengths (in Å) are written outside of the cycle, while bond angles are written in brackets inside the cycle.

3.2 Formation energies

The formation energies of the nitrogen-containing carbon nanotubes were calculated using the density-functional theory method at the B3LYP/6-31-G* level and equation (2). Figure 5 shows the values obtained for nitrogen-containing carbon nanotubes that were totally hydrogenated with different configurations as a function of nanotube length.

A clear difference in stability for different configurations can easily be seen, with an expected regular trend of increasing formation energy as the nanotube length increases. The *O*-configuration is the most stable (with formation energies between 1.95 and 13 eV for nanotubes having between 2 and 10 layers and a formation energy of 30.45 eV for a nanotube of 20 layers -not seen), followed closely by the *P*-configuration. The *S*-configuration is the most unstable (with formation energies between 4 and 24 eV for nanotubes of 2-10 layers long and 53.16 eV for a 20-layer nanotube -not seen), and the *M*-

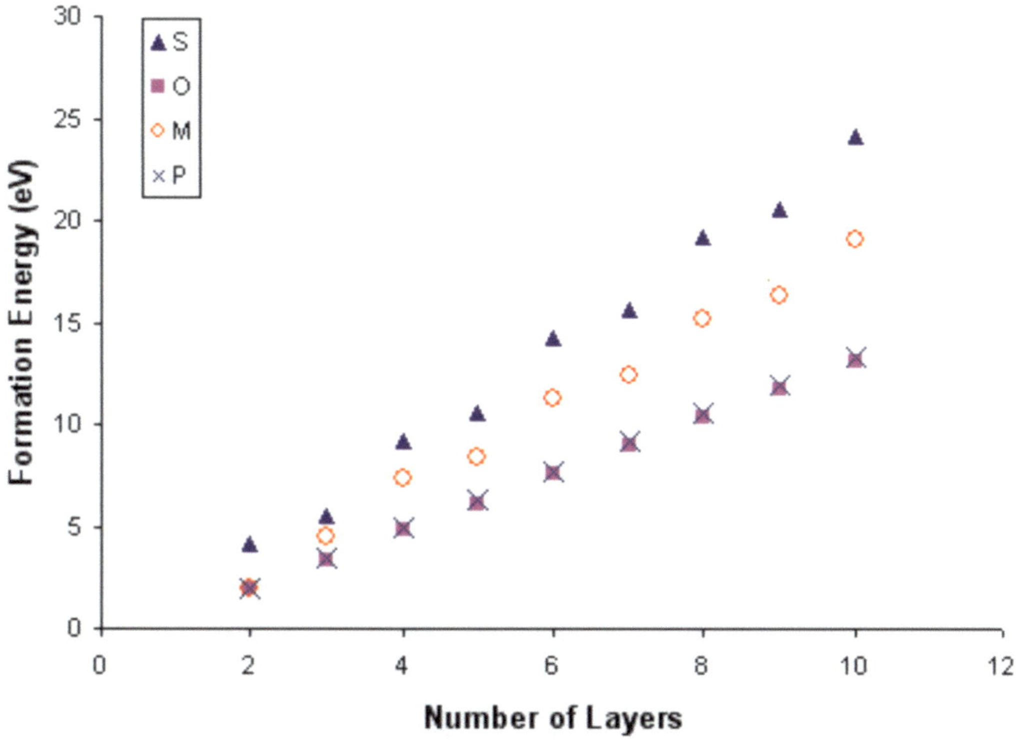

Fig. 5. Formation-energy values for saturated nanotubes of different configurations as a function of length.

configuration was of intermediate stability. For undoped carbon nanostructures of the type $H_3(C_6)_mH_3$ with m=3−6, values equivalent to 8−43 eV have been reported (J.L. Wang, et al., 2006), as calculated using density-functional theory methods. For a nitrogen-doped (5,5) nanotube ($C_{74}N_6$) with a diameter of approximately twice the size of those used in this work, a formation energy of 10.86 eV has been reported (H.S. Kang & Jeong, 2004). A 6*O*-H-H nanotube in this work, has the formula $H_3(C_4H_4N_2)_6H_3$ or $C_{24}H_{30}N_{12}$. Therefore, these systems are not directly comparable.

The formation-energy values have a close relationship to the nanotube structures in this study (see Figure 6). It is clearly observed that the *S*-configuration is the only configuration that has two out of the three interlayer bonds with N−N linkages. Repulsion between lone-pair–lone-pair electron clouds on the nitrogen atoms contributes to the explanation of the larger formation energies for *S*- and *M*-configurations. For *M*-configurations, only one of three interlayer bonds belongs to a N−N bond, which is why *M*-nanotubes have lower formation energies than *S*-nanotubes. For the *O*- and *P*-configurations, the nitrogen atoms are not bonded to each other, and the spatial disposition can be more favourable to the *O*-configuration to better avoid the aforementioned repulsion. Lone-pair–lone-pair repulsive interactions could also explain the observed curvature for the *S*-configuration of the 20-layer totally hydrogenated nanotubes, unlike those in the *O*-configuration (see Figure 7).

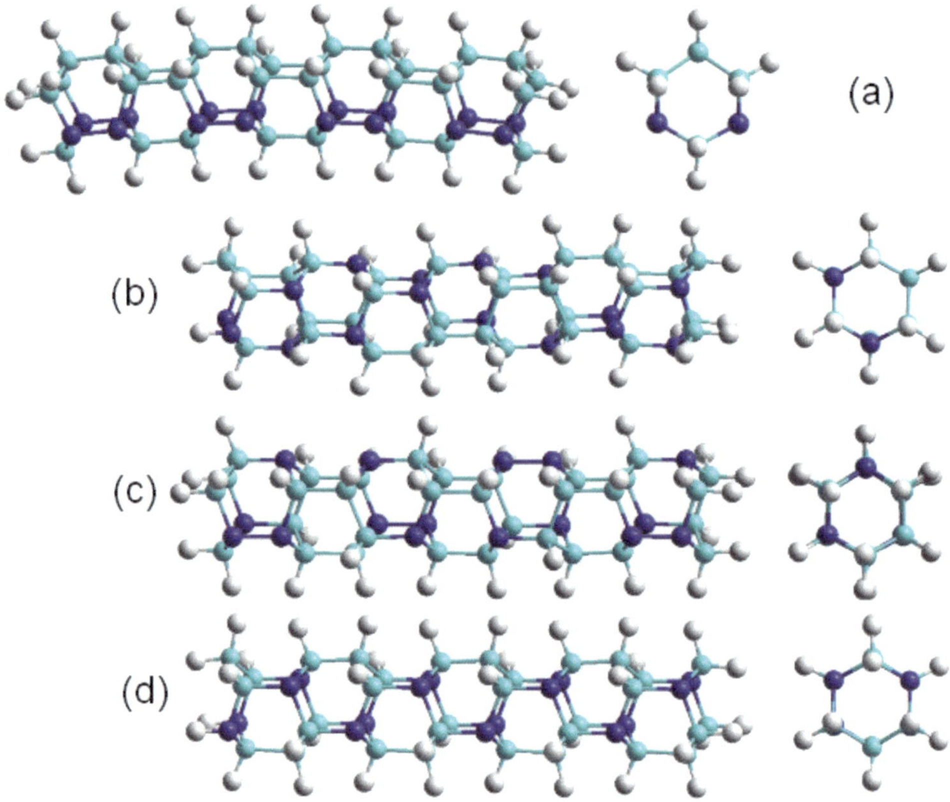

Fig. 6. Fully hydrogenated 8-layered nitrogen-containing carbon nanotubes with different configurations. (a) *S*-type, eclipsed; (b) *O*-type, rotated 60°; (c) *M*-type, rotated 120°; (d) *P*-type, rotated 180°. Top and side views of the optimised structures are shown.

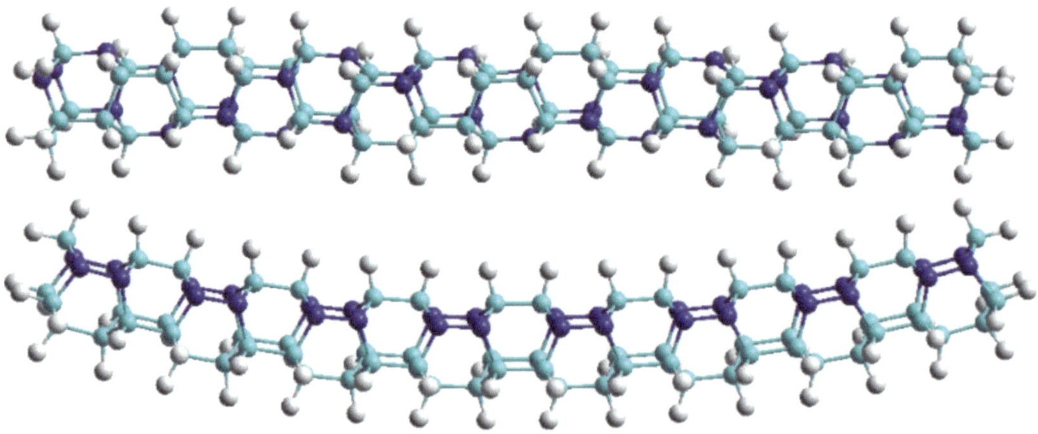

Fig. 7. Optimised geometries for the 20-layer saturated nitrogen-containing carbon nanotubes: (top) the *O*-configuration (the more stable) and (bottom) the *S*-configuration (the more unstable).

3.3 Charge analysis

Charge analysis was conducted for carbon 2 (contiguous to the two nitrogen atoms on each layer) of the most stable configuration of the hydrogen-chemisorbed nanotubes with between 1 and 10 layers. It was found that the charge value on this atom increases from the layer at the extremes of the tube up to the central layers, increasing from -0.031 for layer 1 to 0.297–0.303 for internal layers and then decreasing to 0.259–0.261 for the final layer, while charges for the nitrogen atoms remain constant at -0.481 and -0.486 for N1 and N3, respectively (Contreras et al, 2010).

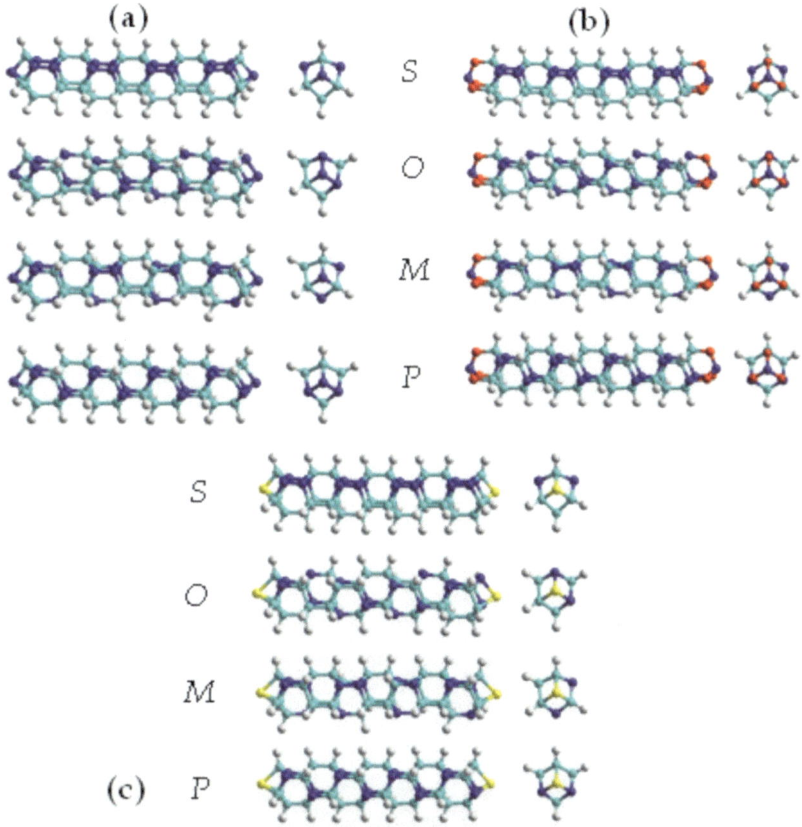

Fig. 8. Eight-layered fully hydrogenated nitrogen-containing carbon nanotubes having different configurations (*S*, *O*, *M*, and *P*) and different terminal groups. (a), (b), and (c) have the terminal groups N, NO_3, and P, respectively. Top and side views of the optimised structures are also shown.

Charge analysis clearly shows that, in the four configurations studied, the charge on C2 has a maximum value at the centre of the nanotube and that the greatest charge difference between the extremes and between one extreme and the maximum value corresponds to the *O*-configuration. Calculations indicate that structures with an *O*-configuration generate the greatest positive charge on C2, possibly because this configuration has a large number of C2 atoms linked to three nitrogen atoms, which does not occur in the other configurations. This fact could be a useful guide for oxygen reduction catalytic properties (Gong et al., 2009).

3.4 Terminal group effect

Figure 8 shows the optimised structures with all real vibrational frequencies obtained for the nanotubes containing terminal groups that close the nanotubes at both extremes as observed from the top and side views for each one.

Figure 9 shows the effect of terminal groups on the band gaps. Here, the band-gap values for nitrogen-containing carbon nanotubes are presented. Those nanotubes with chemisorbed hydrogen correspond to the group of curves above 4 eV, and the group below 4 eV is the corresponding nitrogen-containing carbon nanotubes without chemisorbed hydrogen. The nanotubes of this last group, designed by a single character H, only have hydrogen atoms in the first and in the last layers, with three hydrogen atoms on both end sides, and the nanotube is open.

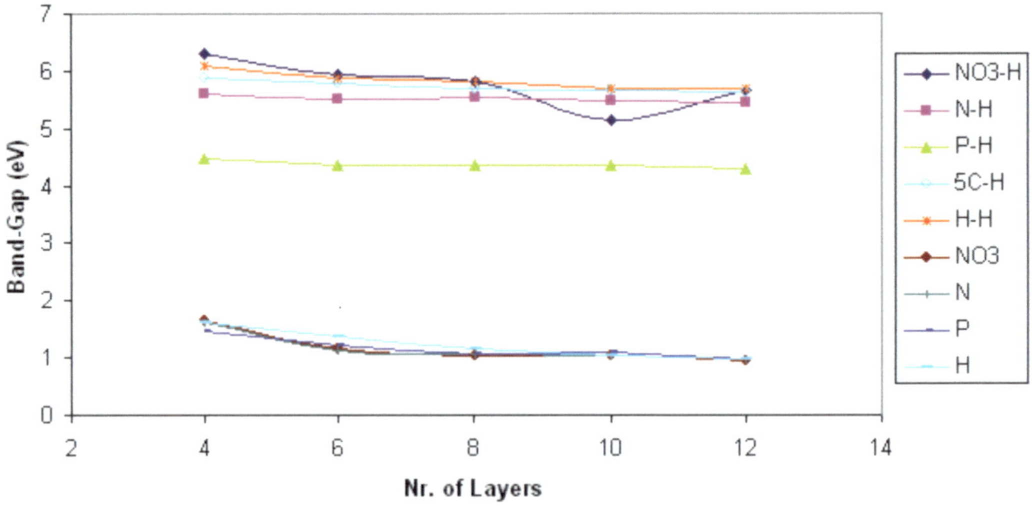

Fig. 9. Band gaps for the *M*-configuration of nitrogen-containing carbon nanotubes having different terminal groups with and without hydrogen chemisorption as a function of length. Greater values (designated by NO_3-H, N-H, P-H, 5C-H; H-H) refer to nanotubes with hydrogen adsorption. Lower values (designated as NO_3, N, P, H) do not have hydrogen chemisorbed. 5C refers to the C5-capped nanotube.

It can clearly be observed from Figure 9 that band-gap values increase for hydrogen-adsorbed nitrogen-containing carbon nanotubes having the *M*-configuration when compared with the band-gap values of the unsaturated nitrogen-containing carbon nanotubes of the same configuration. Band gaps for this last group of nanotubes with no hydrogen adsorbed are insensitive both to the terminal group and to the tube length, especially at and above 8 layers. A different situation is observed for hydrogen-adsorbed nanotubes where the band-gap values decrease by 0.4 eV and 1.6 eV, respectively, for nitrogen and phosphorus terminal groups when compared to an open nanotube with no terminal group. In general, the band gap for this group of nanotubes does not depend on the length, except for the 10-layered nanotube with a NO_3 terminal group, which has a band gap value of $\approx$0.5 eV below the value expected from the general trend.

Thus, when the unsaturated nanotube is closed by terminal groups, its semiconducting property is not strongly affected. However, results suggest that, due to hydrogen chemisorption, the conducting property of nitrogen-containing carbon nanotubes is lost; nanotubes ending with phosphorus atoms are the least affected. This pattern was presented for the *M*-configuration but is also observed for the other studied configurations.

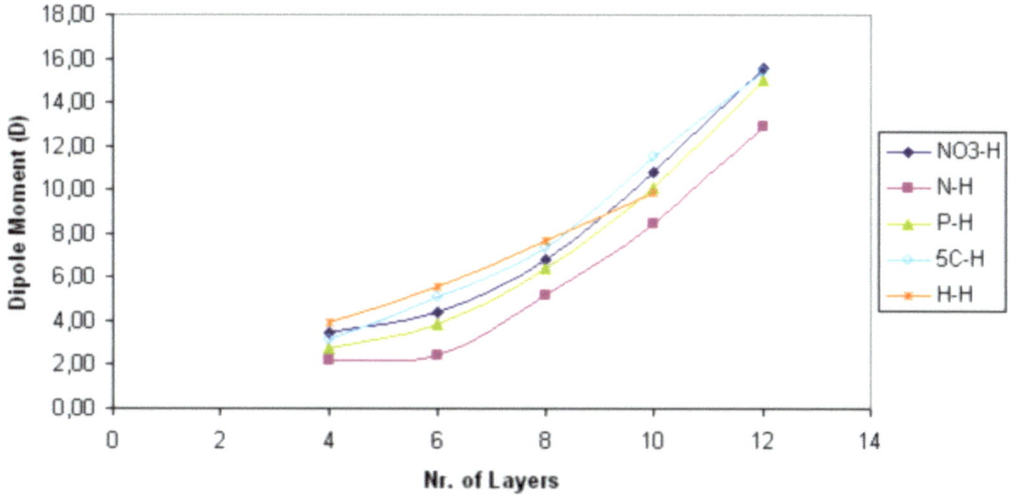

Fig. 10. Dipole moment for the *O*-configuration of hydrogen-chemisorbed nitrogen-containing carbon nanotubes with different end-groups vs. tube length.

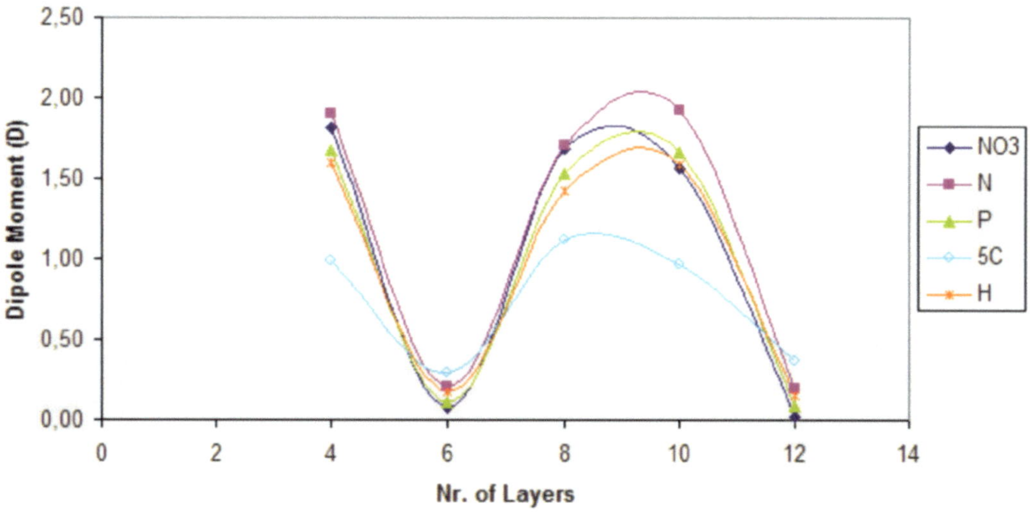

Fig. 11. Dipole moment for the *M*-configuration of unsaturated nitrogen-containing carbon nanotubes with different end-groups vs. tube length.

The terminal group effect on the dipole moment is shown for saturated nitrogen-containing carbon nanotubes with *O*-configuration at different tube lengths in Figure 10 and for unsaturated *M*-nanotubes in Figure 11. The calculated values indicate that the dipole moment clearly depends more on the tube length than on the type of terminal group, with a variation for the saturated nanotubes of within 2—3 D for the different end-groups at the

lengths considered. The respective variation for unsaturated nanotubes is less than 1 D. The dipole moment for the saturated nitrogen-containing carbon nanotubes increases with the number of layers. This trend is also true for saturated nanotubes terminated by hydrogen atoms of different configurations (Contreras et al., 2010).
Regarding the amount of hydrogen uptake by chemisorption, each terminal group replaces three hydrogen atoms at each terminus of the nanotube, and hydrogen is adsorbed at a rate of 4 atoms per layer. In this condition, nanotubes with end-groups will have a lower hydrogen uptake capacity than nanotubes without terminal groups, with values of 4.8, 4.4, and 3.7 wt. % hydrogen uptake for nanotubes ending in N, P, and NO_3, respectively, in comparison with a 7.2 wt. % hydrogen uptake capacity for a nanotube without end-groups, with all the values being for a 4-layer nitrogen-containing carbon nanotube. Therefore, the influence of the end groups is best considered as an anchorage centre for other groups, likely being useful for further functionalisation of the nanotube with specific applications (de Jonge et al., 2005) and also facilitating physisorption because interaction with hydrogen is favoured by the donor hydrogen bond capacity of N, P, and NO_3 groups and by the increase of nanotubes polarisability in the presence of these groups. A single nitrogen atom can bind up to 6 H_2 molecules (Rangel et al., 2009).

3.5 Configuration effect

Previous analyses indicate that configuration strongly affects nanotube stability, with the *O*-configuration, in which each nitrogen atom is linked to three carbon atoms without any N—N bonds, being the most stable. It can also be observed from density-functional theory calculations that, for *S*, *M*, and *P*-configurations in general, the dipole moment increases with the nanotube length, except for the *M*-configuration, in which the dipole moment stays below 2 D independent of nanotube length and the presence of terminal groups in the structure.
The effect of nanotube configuration on the band gap for 8-layered nitrogen-containing carbon nanotubes with different terminal groups coming from hydrogen chemisorption is shown in Figure 12. The same behaviour is observed for nanotubes with and without terminal groups: the band gap depends very little on the configuration, except for *S*-configuration, the more unstable one, where band gap is somewhat lower than for the other configurations.

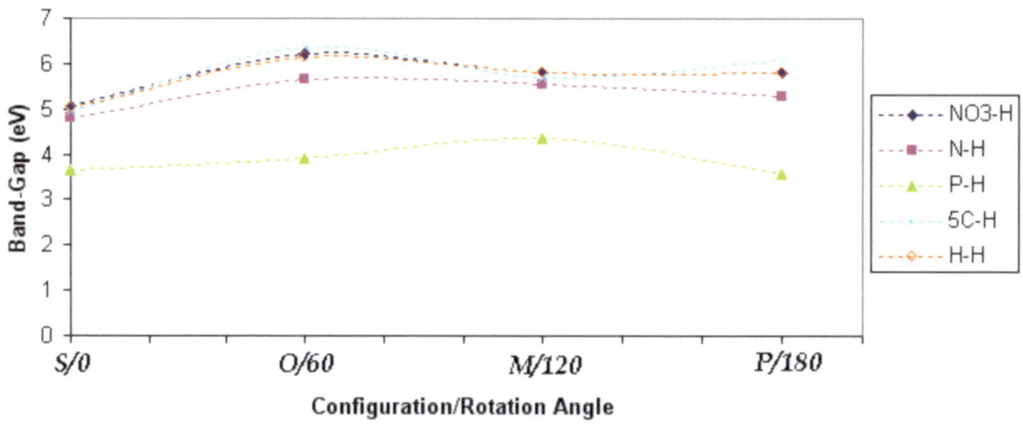

Fig. 12. Band-gap values for hydrogen-chemisorbed 8-layer nitrogen-containing carbon nanotubes with different terminal groups as a function of nanotube configuration.

The dipole moments of 8-layered nitrogen-containing carbon nanotubes for different configuration and terminal groups arising from hydrogen chemisorption are presented in Figure 13.

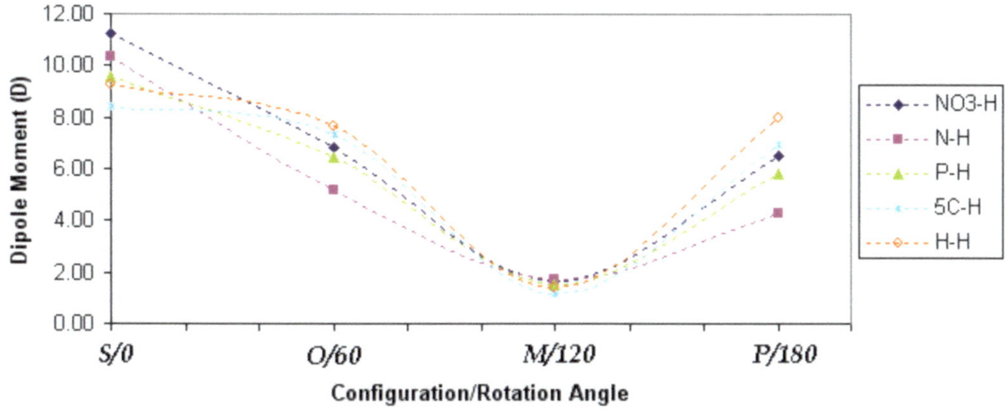

Fig. 13. Dipole moment for hydrogen-chemisorbed 8-layer nitrogen-containing carbon nanotubes with different terminal groups as a function of configuration.

Calculations using density-functional theory methods afford values indicating a configuration effect on the dipole moment for the studied nanotubes, indicating a structural relationship. For *O*- and *P*-configurations, which lack a N−N bond in their structure, the dipole moments have similar values with small differences according to the type of terminal group, decreasing from nanotubes having no terminal group in the order H > 5C > NO_3 > P > N. For the *S*-configurations, where two of the three interlayer bonds are N−N bonds, higher dipole moments are observed for all cases with or without terminal groups. For the case of *M*-configurations, with an interlayer rotational angle of 120°, the dipole moment does not significantly change regardless of the existence or type of functional group, being below a value of 2 D.

3.6 Hydrogen chemisorption energies

Values of hydrogen chemisorption energies for 100% saturation, calculated according to equations (6) and (7) above, for nitrogen-containing carbon nanotubes of different configurations and for carbon nanotubes of different lengths, with no terminal group, are shown in Table 1.

The energies were calculated considering the optimised geometries of the exhaustively chemisorbed nitrogen-containing nanostructures, meaning complete saturation ($C_{4n}H_{4n+6}N_{2n}$). Regular tubular structures with all real vibrational frequencies for all studied nanotube configurations and lengths were obtained. However, the optimised structures of unsaturated molecules ($C_{4n}H_6N_{2n}$), over which hydrogen adsorption takes place, showed three-membered cycles at both extremes of the nanotube with a C−C bond length of 1.51-1.53 Å and C−N bond lengths of 1.45-1.46 Å and 1.47-1.48 Å (Figure 14). These structures were optimised at the same level of theory as the previous structures, up to proper minima characterised by positive vibrational frequencies.

The values shown in Table 1 indicate that the configuration has a distinctive effect, especially on the 4-layered nanotubes. For nanotubes longer than 8 layers, there is no significant variation of the hydrogen chemisorption energies per hydrogen atom, with configuration or length.

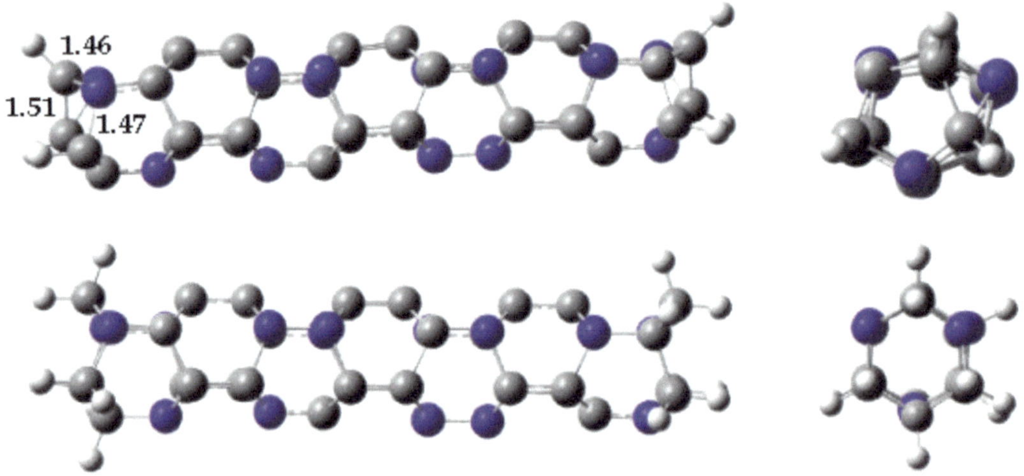

Fig. 14. Optimised structures for a 8M-nitrogen-containing carbon nanotube with no terminal group. $C_{32}H_6N_{16}$ (top) and $C_{32}H_{14}N_{16}$ (bottom). Front and side views.

configuration	Nr. of layers (n)	E_C4nH4n+6N2n (hartrees)	EC4nH6N2n (hartrees)	Er (hartrees)	Er/H (hartrees)	Er/H (kcal/mol)
S	4	-1060.67372179	-1050.525874	-0.74398379	-0.046498987	-29.2
S	6	-1589.19755145	-1573.933203	-1.15855245	-0.048273019	-30.3
S	8	-2117.72125246	-2097.291591	-1.62193362	-0.050685426	-31.8
O	4	-1060.83574369	-1050.533544	-0.89833569	-0.056145981	-35.2
O	6	-1589.44157351	-1574.098557	-1.23722051	-0.051550855	-32.3
O	8	-2118.04670023	-2097.482879	-1.75609323	-0.054877913	-34.4
M	4	-1060.74319832	-1050.450297	-0.88903732	-0.055564832	-34.9
M	6	-1589.30657159	-1574.048135	-1.15264059	-0.048026691	-30.1
M	8	-2117.87029700	-2097.356125	-1.70644400	-0.053326375	-33.5
M	10	-2646.43344985	-2620.807194	-2.11659585	-0.052914896	-33.2
M	12	-3174.92435700	-3144.258419	-2.45434600	-0.051132208	-32.1
P	4	-1060.83275220	-1050.556415	-0.87247320	-0.054529575	-34.2
P	6	-1589.43566275	-1574.113497	-1.21636975	-0.050682073	-31.8
P	10	-2646.64170736	-2621.000194	-2.13185336	-0.053296334	-33.4

Table 1. Chemisorption energy values for different nitrogen-containing carbon nanotubes, with no terminal group.

Values of the hydrogen chemisorption energy for nitrogen-containing carbon nanotube structures with totally saturated first and the last layers and having the general formula $C_{4n}H_{14}N_{2n}$ are given in Table 2. In this case, the optimised regular geometries were obtained without small cycles in their structures (Figure 14).

The Er/H values for partially saturated $C_{4n}H_{14}N_{2n}$ nanotubes are less exothermic than those obtained from less stable geometries, which was expected because less energy per hydrogen atom is to be eliminated by the exhaustive hydrogen adsorption to more thermodynamically stable molecules.

configuration	Nr. of layers (n)	E_C4nH4n+6N2n (hartrees)	E_C4nH14N2n (hartrees)	Er (hartrees)	Er/H (hartrees)	Er/H (kcal/mol)
S	4	-1060.67372179	-1055.5268440	-0.44494579	-0.02780911	-17.5
S	6	-1589.19755145	-1578.9368800	-0.85680745	-0.03570031	-22.4
S	8	-2117.72125246	-2102.3690610	-1.24639546	-0.03894986	-24.4
S	10	-2646.24491823	-2625.7817510	-1.65543923	-0.04138598	-26.0
O	4	-1060.83574369	-1055.6944090	-0.43940269	-0.02746267	-17.2
O	6	-1589.44157351	-1578.9663560	-1.07135351	-0.04463973	-28.0
O	10	-2646.65363361	-2626.2815826	-1.56432303	-0.03910808	-24.5
O	12	-3175.19965990	-3149.6779433	-2.01205660	-0.04191785	-26.3
M	4	-1060.74319832	-1055.6600322	-0.38123412	-0.02382713	-15.0
M	6	-1589.30657159	-1579.1350557	-0.76765189	-0.03198550	-20.1
M	8	-2117.87029700	-2102.6004170	-1.16408400	-0.03637762	-22.8
M	10	-2646.43344985	-2626.0716442	-1.55407765	-0.03885194	-24.4
P	4	-1060.83275220	-1055.7081300	-0.42269020	-0.02641814	-16.6
P	6	-1589.43566275	-1579.2049760	-0.82682275	-0.03445095	-21.6
P	8	-2118.03866452	-2102.7269624	-1.20590612	-0.03768457	-23.6

Table 2. Chemisorption energy values for partially saturated $C_{4n}H_{14}N_{2n}$ nanotubes (nitrogen-containing carbon nanotubes having both the first and the last layer saturated).

The data in Table 2 indicate that, as the nanotube length increases, the hydrogen chemisorption energy increases, favouring chemisorption. The obtained values are comparable with the atomic adsorption energy of hydrogen for nitrogen-doped (8,0) carbon nanotubes with a value of -28.4 kcal/mol (Zhou et al., 2006). A strong effect of the nanotube configuration on the energy values is not observed.

Values of Er/H for hydrogen adsorption over the *M*-configuration of nitrogen-containing carbon nanotubes having N, P, and NO_3 terminal groups at both nanotube extremes are given in Figure 15. Calculations done by density-functional theory at the B3LYP/6-31G* level show no significant variation with the type of terminal group or nanotube length. Theoretically, only 4 hydrogen atoms per layer can be adsorbed by chemisorption.

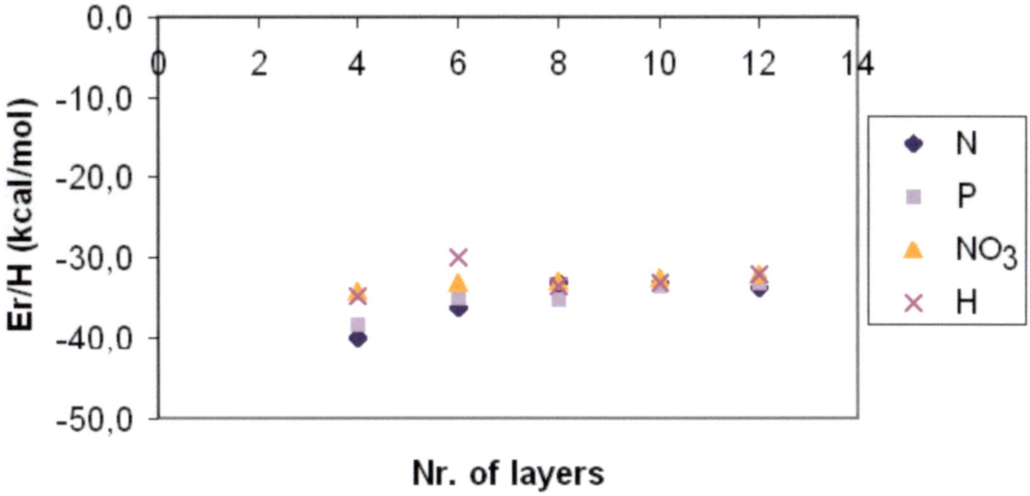

Fig. 15. Er/H values for hydrogen adsorption over nitrogen-containing carbon nanotubes of *M*-configuration with different end-groups vs. length.

The chemisorption process is somewhat more exothermic in the presence of terminal groups than for open nanotubes without terminal groups. The calculated molecular hydrogen absorption energy for a single-walled (8,0) carbon nanotube decorated with atomic nitrogen is also exothermic, with a value of -80 meV/(H_2), equivalent to -1.84 kcal/mol/(H2) (Rangel et al., 2009), as determined by density-functional theory and molecular dynamics. We have not found both experimental neither theoretical studies for a more direct comparison with our results.

4. Conclusions

The structural and energy aspects of hydrogen atom chemisorption on small-diameter nitrogen-containing carbon nanotubes having high nitrogen content have been investigated. The adsorption of hydrogen was examined at full coverage. Stable nanotube structures were fully optimised to proper minima before and after hydrogen chemisorption using density-functional theory methods at the level of B3LYP/6-31G* with all real vibrational frequencies.

The stability was strongly dependant on the configuration of the saturated nitrogen-containing carbon nanotubes, the *O*-configuration being the most stable, probably because the nitrogen atoms are all bonded to carbon atoms, avoiding strong lone-pair—lone-pair repulsions. At the same time, this configuration allows for the development of a positive charge on C2, which theoretically favours nanotube catalytic properties for oxygen reduction reactions.

Hydrogen chemisorption energies for open nitrogen-containing carbon nanotubes ended by hydrogen atoms and similar nanotubes closed at both ends with different units were found to be exothermic and independent of both configuration and length, except for shorter nanotubes.

Chemisorption increases the band gaps ($E_{LUMO\ -\ HOMO}$) of the studied nanostructures (from 1–1.6 eV to 4.5–6.5 eV). These band-gap values decrease by approximately 0.4 eV and 1.6 eV, respectively, for nitrogen and phosphorus terminal groups when compared with open saturated nanotubes without terminal groups. Unsaturated nanotube band gaps are insensitive to terminal groups and length.

The dipole moments of saturated nitrogen-containing carbon nanotubes depend more on the tube length than on the type of terminal group, with a variation within 2—3 D for the different end-groups at the considered lengths. In general, the dipole moment increases as the number of layers increases. This finding is also true for open saturated nitrogen-containing carbon nanotubes of different configurations ended by hydrogen atoms. *S*-configuration nanotubes behave as the most polar, and *M*-configuration nanotubes behave as the most unpolar, among all the studied nanotubes, regardless of terminal group.

Nitrogen-containing carbon nanotubes with small diameters have the capacity to store a full monolayer of hydrogen via chemisorption. Shorter nanotubes and nanotubes without end-groups have higher capacities for hydrogen storage. Hydrogen physisorption studies on these nitrogen-containing carbon nanotubes and the effect of increasing the nanotube diameter constitute an important next step.

Important remaining goals are related to the molecular modelling methods and tools necessary for predicting the properties a particular nanostructure will have —as a hydrogen-storage material, a conductive material, a catalyst, or a further functionalisation centre— which continues to be a scientifically interesting and challenging task.

5. Acknowledgements

This work was partially supported by the Direction of Scientific and Technological Research DICYT-USACH project Nr 061042CF and by the SDT-USACH project Nr CIA 2981. Additionally, the central cluster of the Faculty of Chemistry and Biology and the VRID of the University of Santiago de Chile, Usach, are acknowledged for allocating computational resources.

6. References

Alam, K.M. & Ray, A.K. (2007). A hybrid density functional study of zigzag SiC nanotubes. *Nanotechnology*, Vol.18, No.49 (November 2007) 4957061

Allen, B.L.; Kichambare, P.D. & Star, A. (2008). Synthesis, characterization, and manipulation of nitrogen-doped carbon nanotube cups. *ACS Nano*, Vol.2, No.9, (September 2008), pp. 1914-1920, ISSN 1936-0851

Baughman, R.H.; Zakhidov, A.A. & de Heer, W.A. (2002). Carbon nanotubes - the route toward applications. *Science*, Vol.297, No.5582, (August 2002), pp. 787-792, ISSN 0036-8075

Becke, A.D.J. (1993). Density-functional thermochemistry. 3. The role of exact exchange. *Journal of Chemical Physics*, Vol.98, No.7, pp. 5648-5652, ISSN 0021-9606

Bilic, A. & Gale, J.D. (2008). Chemisorption of molecular hydrogen on carbon nanotubes: a route to effective hydrogen storage? *Journal of Physical Chemistry C*, Vol.112, No.32, (July 2008), pp. 12568-12575, ISSN 1932-7447

Cabria, I.; Lopez, M.J. & Alonso, J.A. (2006). Density functional study of molecular hydrogen coverage on carbon nanotubes. *Computational Materials Science*, Vol.35, No.3, (March 2006), pp. 238-242, ISSN 0927-0256

Charlier, J.C. (2002). Defects in carbon nanotubes. *Accounts of Chemical Research*, Vol.35, No.12, (December 2002), pp. 1063-1069, ISSN 0001-4842

Contreras, M.L.; Avila, D., Alvarez, J. & Rozas, R. (2010). Exploring the structural and electronic properties of nitrogen-containing exohydrogenated carbon nanotubes: a quantum chemistry study. *Structural Chemistry*, Vol.21, No.3, (June 2010), pp. 573-581, ISSN 1040-0400

Contreras, M.L.; Benítez, E.; Alvarez, J. & Rozas, R. (2009). Algorithm for nanotubes computer generation with different configurations. *Algorithms*, Vol.2, No.1, (February 2009), pp 108-120, EISSN 1999-4893

Czerw, R.; Terrones, M.; Charlier, J.C.; Blase, X.; Foley. B.; Kamalakaran, R.; Grobert, N.; Terrones, H.; Ajayan, P.M.; Blau, W.; Tekleab, D.; Rühle, M. & Carroll, D.L. (2001). Identification of electron donor states in N-doped carbon nanotubes. *Nano Letters*, Vol.1, No.9, (September 2001), pp. 457-460, ISSN 1530-6984

De Jonge, N.; Doytcheva, M.; Allioux, M.; Kaiser, M.; Mentink, S.A.M.; Teo, K.B.K.; Lacerda, R. G. & Milne, W.I. (2005). Cap closing of thin carbon nanotubes. *Advanced Materials*, Vol.17, No.4, (February 2005), pp. 451-455, ISSN 0935-9648

Dillon, A.C.; Jones, K.M.; Bekkedahl, T.A.; Kiang, C.H.; Bethune, D.S. & Heben, M.J. (1997), Storage of hydrogen in single-walled carbon nanotubes. *Nature*, Vol.386, No.6623, (March 1997), pp. 377-379, ISSN 0028-0836

Dinadayalane, T.C.; Kaczmarek, A.; Łukaszewicz, J. & Leszczynski, J. (2007). Chemisorption of hydrogen atoms on the sidewalls of armchair single-walled carbon nanotubes.

Journal of Physical Chemistry C, Vol.111, No.20, (May 2007), pp. 7376-7383, ISSN 1932-7447

Feng, H.; Ma, J. & Hu, Z. (2010). Nitrogen-doped carbon nanotubes functionalized by transition metal atoms: a density functional study. *Journal of Materials Chemistry*, Vol.20, No.9, (January 2010), pp. 1702-1708, ISSN 0959-9428

Frank, S.; Poncharal, P.; Wang, Z.L. & de Heer, W.A. (1998). Carbon nanotube quantum resistors. *Science*, Vol.280, No.5370, (June 1998), pp. 1744-1746, ISSN 0036-8075

Frisch, M.J.; Trucks, G.W.; Schlegel, H.B.; Scuseria, G.E.; Robb, M.A.; Cheeseman, J.R.; Montgomery, J.A. Jr.; Vreven, T.; Kudin, K.N.; Burant, J.C.; Millam, J.M.; Iyengar, S.S.; Tomasi, J.; Barone, V.; Mennucci, B.; Cossi, M.; Scalmani, G.; Rega, N.; Petersson, G.A.; Nakatsuji, H.; Hada, M.; Ehara, M.; Toyota, K.; Fukuda, R.; Hasegawa, J.; Ishida, M.; Nakajima, T.; Honda, Y.; Kitao, O.; Nakai, H.; Klene, M.; Li, X.; Knox, J.E.; Hratchian, H.P.; Cross, J.B.; Bakken, V.; Adamo, C.; Jaramillo, J.; Gomperts, R.; Stratmann, R.E.; Yazyev, O.; Austin, A.J.; Cammi, R.; Pomelli, C.; Ochterski, J.W.; Ayala, P.Y.; Morokuma, K.; Voth, G.A.; Salvador, P.; Dannenberg, J.J.; Zakrzewski, V.G.; Dapprich, S.; Daniels, A.D.; Strain, M.C.; Farkas, O.; Malick, D.K.; Rabuck, A.D.; Raghavachari, K.; Foresman, J.B.; Ortiz, J.V.; Cui, Q.; Baboul, A.G.; Clifford, S.; Cioslowski, J.; Stefanov, B.B.; Liu, G.; Liashenko, A.; Piskorz, P.; Komaromi, I.; Martin, R.L.; Fox, D.J.; Keith, T.; Al-Laham, M.A.; Peng, C.Y.; Nanayakkara, A.; Challacombe, M.; Gill, P.M.W.; Johnson, B.; Chen, W.; Wong, M.W.; Gonzalez, C. & Pople, J.A (2004). Gaussian 03, Gaussian Inc, Wallingford CT, Revision D01

Ganji, M.D. (2008). Behavior of a single nitrogen molecule on the pentagon at a carbon nanotube tip: a first-principles study. *Nanotechnology*, Vol.19, No.2, (December 2007), 025709

Gong, K.P.; Du, Z.H.; Xia, Z.H.; Durstock, M. & Dai, L.M. (2009), Nitrogen-doped carbon nanotube arrays with high electrocatalytic activity for oxygen reduction. *Science*, Vol.323, No.5915, (February 2009), pp. 760-764, ISSN 0036-8075

Griadun, V. (2010). Doped Carbon Nanotubes Properties, In: *Carbon Nanotubes*, Marulanda, J.M. (Ed), pp. 147–168, In-Teh, ISBN 978-953-307-054-4, India

Hamada, N.; Sawada, S. & Oshiyama, A. (1992). New one-dimensional conductors-graphitic microtubules, *Physical Review Letters*, Vol.68, No.10, (March 1992), pp. 1579-1581, ISSN 0031-9007

Han, S.S. & Lee, H.M. (2004). Adsorption properties of hydrogen on (10,0) single-walled carbon nanotube through density functional theory. *Carbon*, Vol.42, No.11, (May 2004) pp. 2169-2177, ISSN 0008-6223

Hone, J.; Batlogg, B.; Benes, Z.; Johnson, A.T. & Fisher, J.E. (2000). Quantized phonon spectrum of single-wall carbon nanotubes. *Science*, Vol.289, No.5485, (September 2000), pp. 1730-1733, ISSN 0036-8075

Hu, X.; Zhou, Z.; Lin, Q.; Wu, Y. & Zhang, Z. (2011). High reactivity of metal-free nitrogen-doped carbon nanotube for the C-H activation. *Chemical Physics Letters*, Vol.503, No.4-6, (February 2011), pp. 287-291, ISSN 0009-2614 Hyperchem release 7.0, Hypercube Inc, 1115 NW 4th Street, Gainesville, Florida 32601, USA

Jaguar version 7.5 (2008) Schrödinger, LLC, New York, NY

Kang, H.S. & Jeong, S. (2004). Nitrogen doping and chirality of carbon nanotubes, *Physical Review B*, Vol.70, No.23 (December 2004), pp. 233411-1-233411-4. ISSN 1098-0121

Kang, K.Y.; Lee, B.I. & Lee, J.S. (2009). Hydrogen adsorption on nitrogen-doped carbon xerogels. *Carbon*, Vol.47, No.4, (April 2009), pp. 1171-1180, ISSN 0008-6223

Kaczmarek, A.; Dinadayalane, T.C.; Łukaszewicz, J. & Leszczynski, J. (2007). Effect of tube length on the chemisorptions of one and two hydrogen atoms on the sidewalls of (3,3) and (4,4) single-walled carbon nanotubes: a theoretical study. *International Journal of Quantum Chemistry*, Vol.107, No.12, (October 2007), pp. 2211–2219, ISSN 0020-7608

Kovalev, V.; Yakunchikov, A. & Li, F. (2011), Simulation of hydrogen adsorption in carbon nanotube arrays. *Acta Astronautica*, Vol.68, No.7-8, (April-May 2011), pp. 681-685

Lee, C.; Yang, W. & Parr, R.G. (1988), Development of the colle-salvetti correlation-energy formula into a functional of the electron-density. *Physical Review B*, Vol.37, No.2 (January 1988), pp. 785-789, ISSN 0163-1829

Liu, C.; Chen, Y.; Wu, C.Z.; Xu, S.T. & Cheng, H.M. (2010). Hydrogen storage in carbon nanostructures revisited. *Carbon*, Vol.48, No.2, (February 2010), pp. 452-455, ISSN 0008-6223

Marulanda, J.M. (Ed.) (2010). *Carbon Nanotubes*, In-Teh, ISBN 978-953-307-054-4. India

Mpourmpakis, G.; Tylianakis, E. & Froudakis, G. (2006). Hydrogen storage in carbon nanotubes. A multi-scale theoretical study. *Journal of Nanoscience and Nanotechnology*, Vol.6, No.1, (January 2006), pp. 87-90, ISSN 1533-4880

Nayak, T.R.; Leow, P.C.; Ee, P.L.R.; Arockiadoss, T.; Ramaprabhu, S. & Pastorin, G. (2010). Crucial parameters responsible for carbon nanotubes toxicity. *Current Nanoscience*, Vol.6, No.2, (April 2010), pp. 141-154 ISSN 1573-4137

Ni, M.Y. & Zeng, Z. (2010). Chemisorption and diffusion of hydrogen atoms on single-walled carbon nanotubes. *Journal of Nanoscience and Nanotechnology*, Vol.10, No.8, (August 2010), pp. 5408-5412, ISSN 1533-4880

Oh, K.S.; Kim, D.H.; Park, S.; Lee, J.S.; Kwon, O. & Choi, Y.K. (2008). Movement of hydrogen molecules in pristine, hydrogenated and nitrogen-doped single-walled carbon nanotubes. *Molecular Simulation*, Vol.34, No.10-15, (September-December 2008), pp. 1245-1252, ISSN 0892-7022

Pastorin, G. (2009). Crucial Functionalizations of Carbon Nanotubes for Improved Drug Delivery: A Valuable Option? *Pharmaceutical Research*, Vol.26, No.4, (April 2009), pp. 746-769, ISSN 0724-8741

Rangel, E.; Ruiz-Chavarria, G.; Magana, L.F. & Arellano, J.S. (2009). Hydrogen adsorption on N-decorated single wall carbon nanotubes. *Physics Letters A*, Vol.373, No.30, (July 2009), pp. 2588-2591, ISSN 0375-9601

Saito, R.; Fujita, M.; Dresselhaus, G. & Dresselhaus, M.S. (1992). Electronic-structure of chiral graphene tubules. *Applied Physics Letters*, Vol.60, No.18 (May 1992), pp. 2204-2206, ISSN 0003-6951

Terrones, M. (2007). Synthesis, toxicity, and applications of doped carbon nanotubes. *Acta Microscopica*, Vol.16, No.1-2 (Suppl. 2), pp. 33-34

Stern, S.T. & McNeil, S.E. (2008), Nanotechnology safety concerns revisited. *Toxicological Sciences*, Vol.101, No.1, (June 2007), pp. 4-21

Trasobares, S.; Stephan, O.; Colliex, C.; Hsu, W.K.; Kroto, H.W. & Walton, D.R.M. (2002). Compartmentalized CNx nanotubes: chemistry, morphology, and growth. *Journal of Chemical Physics*, Vol.116, No.20 (May 2002) pp. 8966-8972, ISSN 0021-9606

Wang, J.L.; Lushington, G.H. & Mezey, P.G. (2006). Stability and electronic properties of nitrogen nanoneedles and nanotubes. *Journal of Chemical Information and Modeling*, Vol.46, No.5 (September-October 2006), pp.1965-1971 ISSN 1549-9596

Wang, L.F. & Yang, R.T. (2009). Hydrogen storage properties of N-doped microporous carbon. *Journal of Physical Chemistry C*, Vol.113, No.52, (December 2009), pp. 21883-21888, ISSN 1932-7447

Xiong, W.; Du, F.; Liu, Y.; Perez, A., Jr.; Supp, M.; Ramakrishnan,T.S.; Dai, L. & Jiang, L. (2010). 3-D carbon nanotube structures used as high performance catalyst for oxygen reduction reaction. *Journal of the American Chemical Society*, Vol.132, No.45, (October 2010), pp.15839-15841, ISSN 0002-7863

Yang, F.H.; Lachawiec, J., Jr. & Yang, R.T. (2006), Adsorption of spillover hydrogen atoms on single-wall carbon nanotubes. *Journal of Physical Chemistry B*, Vol.110, No.12, (March 2006), pp. 6236-6244, ISSN 1520-6106

Yang, S.H.; Shin, W.H. & Kang, J.K. (2008). The nature of graphite- and pyridinelike nitrogen configurations in carbon nitride nanotubes: dependence on diameter and helicity. *Small*, Vol.4, No.4, (April 2008), pp. 437–441, ISSN 1613-6810

Yao, Y. (2010). Hydrogen storage using carbon nanotubes, In: *Carbon Nanotubes*, Marulanda, J.M. (Ed), pp. 543–562, In-Teh, ISBN 978-953-307-054-4, India

Yu, M.F.; Files, B.S.; Arepalli, S. & Ruoff, R.S. (2000). Tensile loading of ropes of single wall carbon nanotubes and their mechanical properties. *Physical Review Letters*, Vol.84, No.24, (June 2000), pp. 5552-5555, ISSN 0031-9007

Zhang, G.; Qi, P.; Wang, X.; Lu, Y.; Mann, D.; Li, X. & Dai, H. (2006). Hydrogenation and hydrocarbonation and etching of single-walled carbon nanotubes. *Journal of the American Chemical Soc*iety, Vol.128, No.18, (May 2006), pp. 6026-6027, ISSN 0002-7863

Zhang, Y.; Wen, B.; Song, X.Y. & Li, T.J. (2010). Synthesis and bonding properties of carbon nanotubes with different nitrogen contents. *Acta Physica Sinica*, Vol.59, No.5, (May 2010), pp. 3583-3588, ISSN 1000-3290

Zhang, Z. Y. & Cho, K. (2007). Ab initio study of hydrogen interaction with pure and nitrogen-doped carbon nanotubes. *Physical Review B*, Vol.75, No.7, (February 2007), Art. No. 075420, 6 pp., ISSN 1098-0121

Zhong, Z.; Lee, G.I.; Mo, C.B.; Hong, S.H. & Kang, J.K. (2007). Tailored field-emission property of patterned carbon nitride nanotubes by a selective doping of substitutional N(sN) and pyridine-like N(pN) atoms. *Chemistry and Materials*, Vol.19, No.12 (June 2007), pp. 2918-2920, ISSN: 0897-4756

Zhou, Z.; Gao, X.P.; Yan, J. & Song, D.Y. (2006). Doping effects of B and N on hydrogen adsorption in single-walled carbon nanotubes through density functional calculations. *Carbon*, Vol.44, No.5, (April 2006), pp. 939-947, ISSN 008-6223

2

Dioxygen Adsorption and Dissociation on Nitrogen Doped Carbon Nanotubes from First Principles Simulation

Shizhong Yang[1], Guang-Lin Zhao[2] and Ebrahim Khosravi[1]
[1]*LONI Institute and Department of Computer Science, Southern University and A & M College*
[2]*Physcs Department, Southern University and A & M College*
USA

1. Introduction

The electronic and materials properties of carbon nanotubes (CNTs), like those of recent discovered graphene and earlier found fullerenes, attracted lots of research interests due to their appealing applications[1] in the field of molecular electronics or for the refinement of materials, such as antistatic paints and shieldings, sensor, and catalytic functionality in fuel cells, etc. Like its two/three dimensional siblings graphene and graphite, CNTs consist sp^2 hybridized carbon atoms and are either semiconducting or metallic depending on their helicity. CNTs can be considered in structure as rolling up a single sheet of graphene into a cylinder. Polymer electrolyte membrane fuel cells (PEMFCs) are the best candidates for automobile propulsion owing to their zero emissions, low temperature, and high efficiency. Precious platinum (Pt) catalyst is a key ingredient in fuel cells, which produce electricity and water as the only byproduct from hydrogen fuel.[2] However, platinum is rare and expensive. Reducing the amount of Pt loading by identifying new catalysts is one of the major targets in the current research for the large-scale commercialization of fuel cells. Specifically, developing alternative catalysts to substitute platinum for the oxygen reduction reaction (ORR) in the fuel cell cathodes is essential because the slow kinetics of this reaction causes significant efficiency losses in the fuel cells. Recent intensive research efforts in reducing or replacing Pt-based electrode in fuel cells have led to the development of new ORR electrocatalysts, including carbon nanotube–supported metal particles.[3,4] In 2006, Ozkan and coworkers reported that nitrogen-containing nanostructured carbons and nanotubes have promising catalytic activity towards ORR.[5,6] In a 2008 report, Yang *et al.* at Argonne National Laboratory showed that the vertically-aligned CNT arrays, which are functionalized through nitrogen and iron doping by a chemical vapor deposition (CVD) process, can be electrocatalytically active toward ORR.[7] The functionalized CNTs show promise properties as an alternative non-Pt electrocatalyst with a unique nano-architecture and advantageous material properties for the cathode of PEMFC. They further identified FeN4 sites, which are incorporated into the graphene layers of aligned carbon nanotubes, being electrocatalytic active towards ORR, by their X-ray absorption spectroscopy and other characterization techniques.

In a 2009 publication in *Science,* Gong *et al.*[8] reported that vertically aligned nitrogen-containing carbon nanotubes (VA-NCNTs) can act as a metal-free electrode with a much better electrocatalytic activity, long-term operation stability, and tolerance to crossover effect than platinum for oxygen reduction in alkaline fuel cells. They excluded the effect of metal contaminants and concluded that purely nitrogen doped CNT (metal free) as the active site for ORR. They also performed hybrid density functional theory (DFT) calculations for the hydrogen edge-saturated (5, 5) CNT, in which a nitrogen atom doped in the middle of the nanotube. Based on their experimental and DFT simulation results, they put forward an O_2 reduction mechanism as a four-electron pathway for the ORR on VA-NCNTs with a superb performance. Hu *et al.*[9] studied the triplet O_2 adsorption and activation on the side wall of defect free NCNTs using the DFT based B3LYP method with different NCNT size. From application point of view, however, the defect and edge effects may play an important role in the real reaction environment.

To further understand the properties of nitrogen doped carbon nanotubes as a metal-free electrocatalytic electrode for dioxygen reduction, Yang *et al.*[10] performed first principles DFT calculations for the nitrogen doped (10, 0) CNTs. The stable structure of nitrogen doped CNTs and the properties of the N-doped CNTs for the dioxygen adsorption and reduction were studied. The results show that the nitrogen doping on the open-edges of carbon nanotubes is the most stable structure, that is consistent with the previous experimental results of Ozkan *et al.* [4, 5] To study the dioxygen dissociation, the minimum energy path searching based nudged elastic band (NEB)[11] method was used to calculate the dioxygen dissociation energy barrier. In the following section of this review paper, the detailed method is described. Section 3 presents the optimized open end CNT (10, 0) and one nitrogen edge-doped CNT(10, 0) structures. Section 4 gives the results of dioxygen adsorption. In section 5, recent advance in dioxygen dissociation study is introduced. A brief conclusion is provided in Section 6.

2. Computational method

The first principles DFT calculations with the projector augmented wave (PAW) method[12,13] were performed. The relativistic effect was included in the calculations. The Vienna Ab-initio Simulation Package (VASP)[14,15,16,17] was used in the simulations. The exchange-correlation interaction potentials of the many electron system both in local density approximation (LDA) and in the generalized gradient *approximation* (GGA) with the same model and same parameters were employed. Two sets of data show that they give the consistent results in the stability studies of nitrogen doped (10, 0) CNTs. Thus for the O_2 adsorption simulations, only the LDA results are presented. In these calculations, the 2s and 2p electrons of C, N, and O atoms were included in the valence states. The 1s electrons of the atoms are considered as the core states in a frozen core approximation. Short pieces of (10, 0) CNTs with 16 atomic rings (10 carbon atoms per ring) are included in the simulations. A total of up to 160 carbon atoms are included in the full self-consistent DFT calculations. The nitrogen substitutionally doped CNTs via replacing a carbon atom by a nitrogen atom were simulated. In all of the calculations, the plane wave energy cutoff is fixed at 500 eV. The self consistent energy converged to less than 0.001 meV. All of the atomic coordinates are fully relaxed in all of the DFT calculations and the residue forces are less than 0.05 eV/Å on all the atoms. The CNTs are set in a 17.0 Å × 17.0 Å × 24.9 Å super-cell with the vacuum space separations between CNTs on the sides and along the tube axis direction larger than 9 Å in the simulations, which is large enough to ignore the periodic boundary condition effect. We also performed some test computations and found that by changing the vacuum space separation from 9 Å to 20 Å, only a change of about 0.03 meV/atom in the calculated total energy and 0.0002 μ_B/atom in the calculated magnetic moment are observed. Since a relatively large super-cell is used, only the gamma point is enough in the k-space sampling. Both spin polarized and non-spin polarized DFT calculations are considered in the simulations. The transition state calculations for dioxygen dissociation adsorption were investigated by using the NEB method.[11]

3. Optimized carbon nanotube structures

The synthesized nitrogen-containing carbon nanotubes exhibit a bamboo-like structure.[4,5,7] Distinctively, a great part of the bamboo-like nano-structures consists of open ends of relatively short CNTs.[18,19,20]Individual short CNTs are weakly stacked one on top of the other to create a long nano-fiber. The observed bamboo-like structure along the nano-fiber length can be attributed to the integration of nitrogen into the graphitic structure, altering the nanotube surface from straight cylinder geometry. In order to simulate the open structure of the short carbon nanotubes, a short piece (10, 0) CNT with 16 atomic rings that have 10 carbon atoms per ring was used. There are a total of 160 carbon atoms in the short piece CNT that are included in the full self-consistent DFT calculations. The CNT has a diameter of 7.84 Å, through the center of the carbon atoms, and a total length of about 15.5 Å after the optimization of the structure. Different from the previous calculations, which utilized hydrogen atoms to artificially edge-saturate those CNTs to facility the convergence of the DFT computations,[7] the open-edge of the short piece CNT was not saturated using hydrogen or other atoms. Wei *et al.* simulated the effects of nitrogen substitutional doping in Stone-Wales defects on the transport properties of single-walled nanotubes. They eliminated the open-edges of the short CNTs by matching to two pieces of perfect CNT unit cells.[21]

In this work, nitrogen-doped CNTs either at open-edge sites or in CNT walls was simulated. In addition, the open-edges of the short piece CNTs play an important role in the nitrogen doping, as demonstrated by the previous experimental work, [4,5] as well as by the first-principles calculations as discussed late. In the first-principles self-consistent DFT calculations, all atomic positions are allowed to be fully relaxed. The calculated results show that the carbon atoms on the open-edge of the short CNT have a substantial relaxation, due to the existence of some dangling bonds of the carbon atoms at the open-edge sites. The calculated C-C bond length of the carbon atoms on the open-edge to the atoms of the second atomic ring away from the open-edge of the CNT is 1.39 Å, which is shorter than the C-C bond length of 1.41 to 1.42 Å in the inner atomic rings of the short CNT. This reduction of the C-C bond length for the carbon atoms at the open-edge sites is mainly due to the absence of atoms on the empty side. The calculated bond length of the carbon atoms from the second to third atomic ring is 1.44 Å, which is slightly larger than those in the inner rings (such as those from the third to fourth atomic rings), away from the open-edge of the short CNT. The results indicate that the effect of the relaxation of the third atomic ring from the open-edge of the CNTs and other inner rings would be insignificant.

The spin-polarized DFT calculations for the short piece CNT were also performed. It is interesting to see that the carbon atoms on the open-edge of the CNT possess a magnetic moment of about 0.59 μ_B/atom. Similar spin polarization effect was also shown on a recent study of Möbius graphene nanoribbons.[22] Other carbon atoms away from the open-edge and in the inner wall of the CNT do not have a noticeable magnetic moment from our computational results. This property of the carbon atoms on the open-edge with a magnetic moment is attributed to the existence of some dangling bonds. In the next sub-sections, it is demonstrate that the absorption of other atoms to the open-edge of the CNT will reduce the magnetic moment of the carbon atoms, because the dangling bond is reduced.

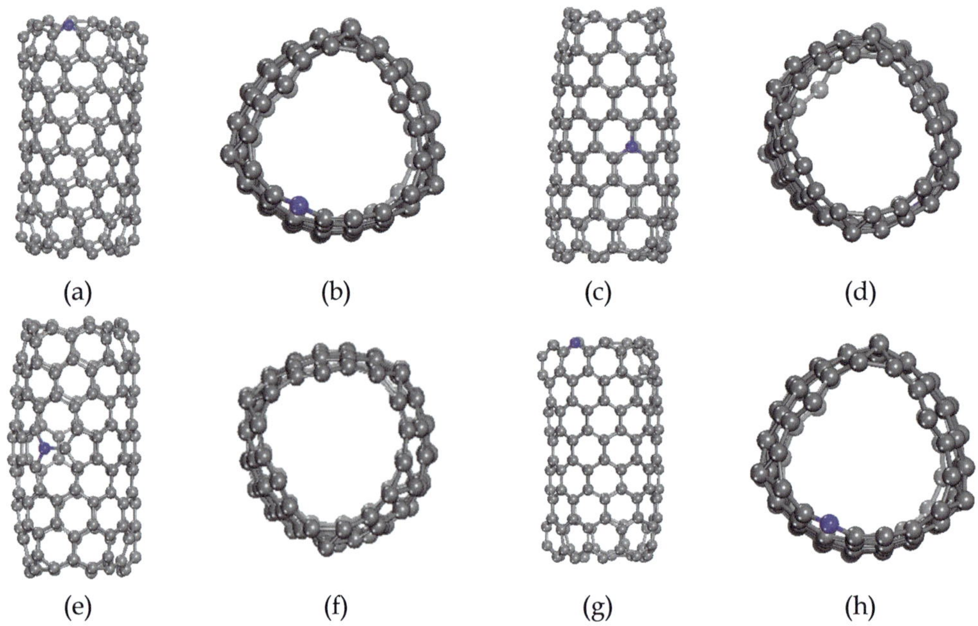

Fig. 1. (a) and (b) Side and top views of a N doped short CNT with N atom on the edge of the CNT; (c) and (d) Side and top views of a N doped short CNT with N atom in the middle side wall of the CNT; (e) and (f) Side and top views of a N doping in the side wall near a (5-7-7-5) Stone-Wales defect in the CNT; (g) and (h) Side and top views of a N doped CNT at the open-edge and a (5-7-7-5) Stone-Wales defect in the CNT wall. The blue ball stands for N atom; the grey balls represent C atoms. The supercell (17.0 Å × 17.0 Å × 24.9 Å) contains 159 carbon atoms and one N atom.

The nitrogen substitutionally doped carbon nanotubes, i. e., a nitrogen atom replacing a carbon atom in the short CNTs were then calculated. Four types of doping configurations were considered. They are, *(i)* a doping nitrogen atom substituting a carbon atom at the open-edge (Figure 1 (a) and (b)); *(ii)* a N atom substituting a C atom at the middle side wall (Figure 1(c) and (d)); *(iii)* a N atom doping at the middle side wall with a (5-7-7-5) Stone-Wales(SW) defect (Figure 1(e) and (f); and *(iv)* a doping N atom at the open-edge with a (5-7-7-5) SW defect in the middle side wall of the short CNT (Figure 1(g) and (h)). We performed full self-consistent first-principles DFT calculations again. All atomic positions are fully relaxed. The calculated total energies of the four types of doping cases are -1534.58 eV, -1531.91 eV, -1530.17 eV, and -1532.54 eV, respectively. The first doping configuration (i.e., a nitrogen atom substituting a carbon atom at the open-edge of the short CNT) has the lowest total energy among the four cases and thus is the most stable configuration within the configurations studied. The total energy of a N atom substituting a C atom on the open-edge (*case 1*) is lower than that of N atom substituting a C atom in the middle side wall of the CNT (*case 2*), by about 2.67 eV. The total energy of the N atom doped CNT at the middle side wall with a (5-7-7-5) Stone-Wales(SW) defect (case 3) is higher than that of case 2 by 1.74 eV and higher than that of case 1 by 4.41 eV. In the optimized structure of Case 3, the nitrogen atom in the side wall with the 5-7-7-5 SW defect is sitting between 5-5 and 7-7 defects, as shown in revised Fig. 1(e). Case 4, i. e., one N atom doped the CNT at the open-edge with a (5-7-7-5) SW defect in the middle side wall has a total energy higher than other three cases. The calculated results show that nitrogen atoms may prefer to stay at the open-edge sites of short CNTs. Both of the LDA and GGA calculations confirmed the same properties. This conclusion is consistent with the previous results of the first-principles DFT calculations of N-doped short carbon nanobells.[23] It agrees with the previous experimental results of Ozkan *et al.*[4,5] The above results are also consistent with the fact that at elevated temperature, the 5-7-7-5 SW defect in single wall CNTs is not stable and will eventually diffuse to the edges of CNTs and be healed by catalytic atoms,[24] although it is possible that the 5-7-7-5 SW defect in single wall CNT could be meta-stable under weak stresses and in catalyst free syntheses environment.[25,26] The finding is in contrast to the simulated results in a recent report for the hydrogen edge-saturated (5, 5) CNT, in which nitrogen atom is doped at the middle (5, 5) CNT sidewall.[7] It should be noticed here that in the real multi-walled CNT application, the nitrogen side wall doped (5-7-7-5) SW defect may be relatively stable considering the existence of wall-defect and wall-wall interaction. To the best of our knowledge, no multi-walled CNT defect stability study report is found due to the relative large size and expensive computation cost. On the other end, the open edge may be considered as an enlarged defect thus at least partially showing the atom adsorption and catalytic property.

In the following discussions, the properties of the nitrogen doping on the open-edges of the short CNTs are introduced. The nitrogen atom on the open-edge of the CNT bonds to two carbon atoms of the second atomic ring. The two N-C bond lengths are 1.31 Å and 1.34 Å, respectively. Because we only included one nitrogen atom substituting one carbon atom on the open-edge, the short CNT in the simulation did not have a cylindrical symmetry. The cross-section of the CNT is distorted and formed to nearly triangular shape by the N doping that can be seen from Figure 1(b). The distortion of the CNT cross-section is also partly attributed to the formation of new C-C bond of about 1.49 to 1.55 Å on the open-edge of the CNT, reducing the dangling bonds of the carbon atoms on the open-edge and forming in more stable sp^2 hybridization.

The effective charge (Bader charge) and the charge transfer of the N and C atoms in the doped CNT were also calculated. The nitrogen atom on the open-edge of the CNT obtained about 3.0 electrons from the nearest neighboring carbon atoms, i.e., turning to $N^{-3.0}$. The C atom with a N-C bond length of 1.31 Å loses about 1.6 electrons to N atom; and the C atom with a N-C bond length of 1.34 Å loses about 1.3 electrons to the N atom. The calculated effective charge may have a computational uncertainty within an estimated range of about 0.1 electrons. The remaining of the charge transfer of about 0.1 electrons to the N atom can be attributed to the second nearest neighboring C atoms or to the computational uncertainty.

The spin-polarized DFT calculation for the short piece CNT with a doping N atom on the open-edge was then calculated. The results show that the nitrogen atom did not have a noticeable magnetic moment. In addition, some of the carbon atoms around the distorted locations of the open-edge of the short CNT have a strong relaxation and lose their magnetic moments. Only the carbon atoms on the open-edges of CNTs, which still have a dangling bond, possess a magnetic moment from 0.1 to 0.5 μ_B/atom, depending on their local environments.

4. Dioxygen adsorption on one nitrogen open edge doped CNT(10, 0)

The first-principles DFT calculations to study dioxygen O_2 adsorption and reduction on the nitrogen edge-doped carbon nanotubes were performed. The calculated results show that the O_2 can chemisorbed on a site close to the nitrogen-carbon complex with a tilted slant away from the nitrogen atom and more close to a carbon atom (Fig. 2a and b, Pauling adsorption model). The O-O bond length of the dioxygen on the end-on Pauling site increased to 1.33 Å, which is longer than the calculated bond-length of 1.22 Å in free gas O_2 molecule state, see Table 1. The O(1)-C and O(2)-N distances for the Pauling adsorption are 1.31 Å and 2.64 Å, respectively. The dioxygen absorbed on the Pauling site obtained partial electrons from carbon-nitrogen complex. The calculated effective charges of two oxygen are $O(1)^{-0.99}$ and $O(2)^{-0.24}$, respectively. The effective charge of the nitrogen atom becomes $N^{-2.9}$ after the dioxygen adsorption, which is not much different from that before the dioxygen adsorption. The electrons obtained by the dioxygen are mainly transferred from the neighboring carbon atoms. The adsorption energy of the dioxygen on the Pauling site is about 1.53 eV/atom. The magnetic momentum of the dioxygen at the Pauling site is greatly reduced from the gas state magnetic momentum of 2 μ_B to 0.027 μ_B. The carbon atom that bonded to O(1) loses its magnetic moment to nearly zero, from about 0.5 μ_B before the dioxygen adsorption. The nitrogen atom still has no noticeable magnetic moment.

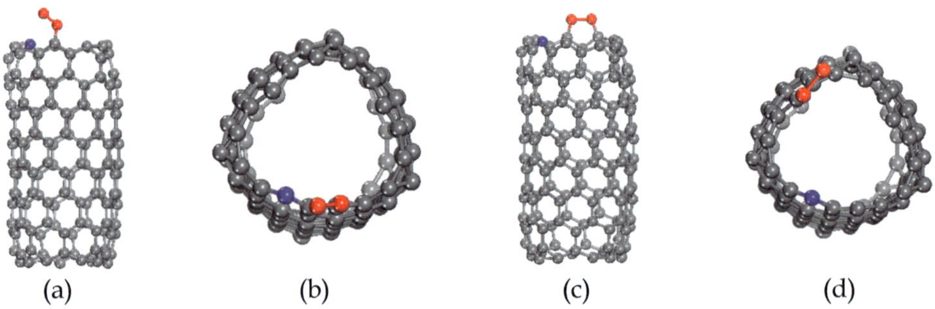

Fig. 2. (a) Side and (b) top views of the dioxygen O_2 adsorbed onto a nitrogen-carbon complex site of the short CNT (Pauling model). The red balls stand for oxygen atoms. (c) Side and (d) top views of the dioxygen O_2 adsorbed on the carbon-carbon long bridge sites away from the N-doping site of the short CNT.

The dioxygen adsorption and reduction at two other locations on the open-edge of the nitrogen doped carbon nanotubes was studied. The first case is the dioxygen adsorption on the opposite side of the CNT open-section away from nitrogen, which is illustrated in Figure 2 (c) and (d). The dioxygen is stabilized in a long bridge site on the open-edge of the CNT. The O-O bond length of the dioxygen on the long bridge site is 1.57 Å, which is even larger than the O-O bond length of the dioxygen at the end-on Pauling site as discussed above. The O(1)-C(1) and O(2)-C(2) distances for the bridge site adsorption are 1.34 Å and 1.38 Å, respectively. The calculated effective charges of two oxygen are $O(1)^{-0.91}$ and $O(2)^{-0.85}$, respectively. The electrons are mainly transferred from the neighboring carbon atoms. The effective charge of nitrogen did not change. The chemisorption energy of the dioxygen on this bridge site is 2.57 eV/atom. The results of the spin-polarized DFT calculations showed that the dioxygen loses its magnetic moment after the chemisorption on the long bridge site to nearly zero. The carbon atoms that adsorb the dioxygen also do not have noticeable magnetic moments, which are different from those of the carbon atoms on the open-edge of CNT with dangling bonds.

The second case of dioxygen adsorption in a short bridge-site on the open-edge of the CNT was also studied. In this case, near the dioxygen adsorption site, the open-edge of the CNT has a noticeable distortion that is attributed to the formation of new C-C bond. The dioxygen adsorption energy at this bridge site is about 0.015 eV/atom, which is much lower than that of the first case of the dioxygen adsorption discussed above. The calculated charges transfer to the two oxygen atom are $O(1)^{-0.03}$ and $O(2)^{-0.01}$, which is much smaller than that of the fist case of the bridge-site adsorption discussed earlier. The O-O bond length is about 1.23 Å, which is nearly the same as that in the free O_2 molecule state. The O(1)-C(1) and O(2)-C(2) distances for this short bridge site adsorption are 3.65 Å and 3.76 Å, respectively. The dioxygen on this site maintained a magnetic momentum of 2 μ_B, which is the same as that of the free gas state. All the data clearly show that the dioxygen adsorption on this short bridge-site is a physisorption, which is attributed to the loss of the dangling bonds of the carbon atoms on the open-edge of the CNT. To save space, this typical physisorption site graph is omitted here.

From Figure 2 (c) and the Bader charge analysis in Table I, it can be seen that the oxygen-carbon form a weak chemical bond (sp hybridization) at long bridge site, since the two carbon atoms at long bridge site would otherwise possess some dangling bonds, while there is strong C-C bond at the short bridge site on the open-edge of the short CNT, thus the interaction between O_2 and carbon atoms at short bridge site is weak physical adsorption.

O_2 Site	O-C (Å)	O-O (Å)	MM(μ_B)	Bader charge change	E_{ad}(eV)
Pauling site (End-on C-N complex, Chemisorption)	1.31	1.33	0.027	$O(1)^{-0.99}$, $O(2)^{-0.24}$	1.53
Long Bridge site (Chemisorption)	1.34	1.57	0.001	$O(1)^{-0.91}$, $O(2)^{-0.85}$	2.57
Short Bridge site (Physisorption)	3.71	1.23	2.0	$O(1)^{-0.03}$, $O(2)^{-0.01}$	0.015

Table 1. The calculated properties of dioxygen O_2 adsorption and reduction on the N-doped short CNTs. In the table, O-C is the shortest O-C distance; O-O is the O-O bond length, and MM is the dioxygen magnetic moment in μ_B. In the gas state, O_2 has a magnetic moment of $2\mu_B$ and a bond length of 1.22 Å from our first-principles DFT calculation. E_{ad} is the low coverage adsorption energy (in eV) per atom.

5. Dioxygen dissociation on one nitrogen open edge doped CNT(10, 0)

To study the dioxygen dissociation energy barrier, the NEB method was used to study the minimum energy path from the Pauling site to the dissociated N-O and C-O state. This section will introduce the major results of the simulations and the detailed results will be organized and publish late.[27] To save the computation time, only top ring atoms are relaxed since the bottom carbon atoms are relatively stable and the relaxation could be safely ignored. It should be mentioned that dioxygen dissociative adsorption minimum energy path from the NEB method is not necessary the only possible reaction path for the ORR or the most important step. However the relative energy barrier comparison can be used as a measure of reactivity.

The calculated initial Pauling site is shown in Figure 3 (a), the stable dissociated N-O(1) and C-O(2) state is shown in Figure 3 (c), while the transition state is shown in Figure 3 (b). The calculated minimum energy barrier is 0.55 eV which is reduced tremendously than the gas phase dissociation energy 5.1 eV. In the transition state (Figure 3(b)), the dioxygen bond length is elongated and the dangling O(1) is moved toward the nitrogen atom. The top carbon ring atoms have a slight re-adjustment to balance the residue forces. It should be emphasized here that in real PEMFC application, the Pauling site dioxygen may be dissociated through other channels even before it is dissociated into N-O(1) and C-O(2) considering the existence of the hydrogen atoms and OH. At the transition state, there is significant charge redistribution on nitrogen, O(1)-O(2), and bonded carbon atoms. It is the local electric field that weakens the O(1)-O(2) bond and dramatically decreases the transition barrier energy.

(a) (b) (c)

Fig. 3. (a) Side view of the initial state of dioxygen O_2 adsorbed onto a nitrogen-carbon complex site of the short CNT (Pauling model, initial state). The red balls stand for oxygen atoms, green ball stands for N atom, while black be C atoms. (b) Side view of the transition state of dioxygen/NCNT complex. (c). Side view of the dissociated two O atom adsorption on the edge carbon and N atoms of the short CNT.

6. Conclusions

In this review report, first principles spin polarized DFT simulations of nitrogen substitutionally doped (10, 0) carbon nanotube (CNT) for dioxygen adsorption and dissociation are performed. The calculated results show that nitrogen prefers to stay at the open-edge of the short CNTs. Two O_2 chemisorption sites are found, the carbon-nitrogen complex (Pauling site) and carbon-carbon long bridge (long bridge) sites. Dioxygen O_2 can be chemisorbed on and reduced on the carbon-nitrogen complex at the open-edge of the

CNT and on the open carbon-carbon sites. The carbon atoms on the open-edge of the short CNT can possess a magnetic moment of 0.59 μB/atom, which is due to the existence of the dangling bonds of these C atoms. The chemisorption of dioxygen O_2 onto both Pauling site and long bridge sites at the open-edge of the short CNTs will reduce the magnetic moments of the carbon atoms to nearly zero. Further spin polarized NEB method minimum energy barrier simulations show that the Pauling site is the possible O_2 dissociation site with a reaction barrier 0.55 eV. The unique open-edge structure and charge redistribution are crucial to the novel doped CNT catalyst design.

7. Acknowledgements

The work is funded in part by the NSF-LASiGMA program (grant number #EPS-1003897), the NASA/LEQSF (2009-2012)-Phase3-03, the LaSPACE/NASA (grant number NNG05GH22H), DOE award No. DE-FE0004734 and DE-FE0003693, and the LONI institute. GLZ acknowledged the funding supports of the National Science Foundation (Award No CBET-0754821) and the Air Force Office of Scientific Research (Award No FA9550-09-1-0367).

8. References

[1] Hirsch, A. *Nature Materials* 2010, *9*, 868.
[2] Subramani, V.; Gangwal, S. K. *Energy & Fuels* 2008, *22*, 814.
[3] Che, G.; Lakshmi, B. B.; Fisher, E. R.; Martin, C. R. *Nature* 1998, *393*, 346.
[4] Kongkanand, A.; Kuwabata, S.; Girishkumar, G.; Kamat, P. *Langmuir* 2006, *22*, 2392.
[5] Matter, P. H.; Ozkan, U. S. *Catal. Lett.* 2006, *109*, 115.
[6] Matter, P. H.; Zhang, L.; Ozkan, U. S. *J. Catal.* 2006, *239*, 83.
[7] Yang, J.; Liu, D. J.; Kariuki, N. N.; Chen, L. X. *Chem. Commun.* 2008, *3*, 329.
[8] Gong, K.; Du, F.; Xia, Z.; Durstock, M.; Dai, L. *Science* 2009, *323*, 760.
[9] Hu, X.; Wu, Y.; Li, H.; Zhang, Z. *J. Phys. Chem. C* 2010, *114*, 9603.
[10] Yang, S.; Zhao, G.; Khosravi E. *J. Phys. Chem. C* 2010, *114*, 3371.
[11] Henkelman, G.; Uberuaga, B. P.; Jonsson, H. *J. Chem. Phys.* 2000, *113*, 9901.
[12] Blöchl, P.E. *Phys. Rev. B* 1994, *50*, 17953.
[13] Kreese, G.; Hafner, J. *Phys. Rev. B* 1999, *59*, 1758.
[14] Kresse, G.; Hafner, J. *Phys. Rev. B* 1993, *47*, 558.
[15] Kresse, G.; Furthmüller, J. *Comp. Mater. Sci.* 1996, *6*, 15.
[16] Kresse, G.; Furthmüller, J. *Phys. Rev. B* 1996, *54*, 11169.
[17] VASP 2010 manual, see website: http://cms.mpi.univie.ac.at/vasp/.
[18] Ma, X.; Wang, E.; Zhou, W.; Jefferson, D. A.; Chen, J.; Deng, S.; Xu, N.; Yuan, J. *Appl. Phys. Lett.* 1999, *75*, 3105.
[19] Ma, X.; Wang, E. G.; Tilley, R. D.; Jefferson, D. A.; Zhou, W. *Appl. Phys. Lett.* 2000, *77*, 4136.
[20] Ma, X.; Wang, E. G. *Appl. Phys. Lett.* 2001, *78*, 978.
[21] Wei, J.; Hu, H.; Zeng, H.; Wang, Z.; Wang, L. *Appl. Phys. Lett.* 2007, *91*, 92121.
[22] Jiang, D.; Dai, S. *J. Phys. Chem. C* 2008, *112*, 5348.
[23] Zhao, G. L.; Bagayoko, D.; Wang, E. G. *Modern Physics Letters* 2003, *9*, 375.
[24] Ding, F. *Phys. Rev. B* 2005, *72*, 245409.

[25] Fahlman, B. D. Materials Chemistry, Springer, Dordrecht, the Netherlands, 2007.
[26] Yakabson, B. I. *Appl. Phys. Lett.* 1998, *72*, 918.
[27] Newell, T.; Yang, S.; Zhao, G.; Khosravi, E. to be submitted.

Hydrogen Adsorptivity of Bundle-Structure Controlled Single-Wall Carbon Nanotubes

Shigenori Utsumi[1] and Katsumi Kaneko[2]
[1]*Department of Mechanical Systems Engineering, Tokyo University of Science, Suwa*
[2]*Research Center for Exotic Nanocarbons, Shinshu University*
Japan

1. Introduction

Hydrogen (H_2) gas is an ideal clean fuel, because H_2 emits only water on burning and the energy content per unit mass is much greater than that of hydrocarbon fuels (Gregory & Oerlemans, 1998). Using H_2 as a fuel has been expected to prevent global warming. To achieve the effective utilization of H_2 energy, the development of its efficient storage method is necessary. H_2 is supercritical gas at room temperature; the critical temperature of H_2 is 33 K. Thus, it is difficult to store large amount of H_2 at room temperature because the supercritical gas does not liquefy even under high pressures. Efficient adsorbents for H_2 storage have been actively studied to overcome this problem.

Single-wall carbon nanotube (SWCNT) is considered to be the most promising material which can contribute to construct a new sustainable chemistry (Iijima, 1991; Iijima & Ichihashi, 1993; Hirsch, 2002; Saito et al., 1998) and particularly a H_2 storage system, because SWCNT bundles have both of internal and interstitial nanospaces which strongly interact even with supercritical H_2 (Liu et al., 1999; Wang & Johnson, 2000; Seung & Young, 2000; Xu et al., 2007; Kim et al., 2007). One SWCNT consists of one graphene sheet rolling up. Thus, SWCNT is a special material referred to as "bi-surface nature material" because the whole carbon atoms are exposed to the both internal and external surfaces, each with different nanoscale curvatures of the SWCNT wall (Noguchi et al., 2007; Fujimori et al., 2010). A SWCNT has a huge geometrical surface area of 2630 m^2 g^{-1}, the same as graphene. The effective surface area of SWCNTs for molecules varies with its tube diameter and the target molecular size. In addition to the large surface area, the differences between surfaces with positive and negative curvature can be exploited to establish unique material science and technology. Ordinary SWCNTs associate to form an ordered bundle structure through dispersion interaction, providing interstitial pore spaces surrounded by carbon walls with positive curvature, which are the strongest molecular sites. Therefore, bundled SWCNTs have considerable potential for application to gas storage, the stabilization of unstable molecules, quantum molecular sieving (Noguchi et al., 2010), specific reaction fields, gas sensing, electrochemical energy storage and so on (Banerjee et al., 2003; Arai et al., 2007).

However, when the interstitial pore width is just comparable to the size of a small molecule, the molecules preadsorbed in the interstitial nanospaces often block further adsorption, or the capacity of the interstitial pore spaces is too small compared with the internal nanospace capacity. Thus, it is necessary to establish a means for tuning the bundle structure for providing the larger capacity of internal and interstitial nanospaces with an optimum size for the target function, as the volume of the interstitial nanospaces at the strongest sites is too small.

Pillaring an SWCNT bundle is the best approach to control interstitial nanoporosity, realizing enhanced adsorptivity for supercritical gases such as H_2, and strengthening the specificity of the molecular recognition function (Abrams et al., 2007; Zhao et al., 2007). Here we report the simple preparation of fullerene (Kroto et al., 1985) (C_{60})-pillared SWCNT bundles by sonication of SWCNTs in a C_{60} toluene solution and the consequent enhancement of the supercritical H_2 adsorptivity of the SWCNTs (Arai et al., 2009). As C_{60} molecules have a conjugated π-electron structure similar to that of SWCNTs, the C_{60}-pillared SWCNT system can be regarded as a new nanocarbon. In fact, naphthalene-pillared SWCNT have pseudo-metallic property (Gotovac-Atlagić et al., 2010).

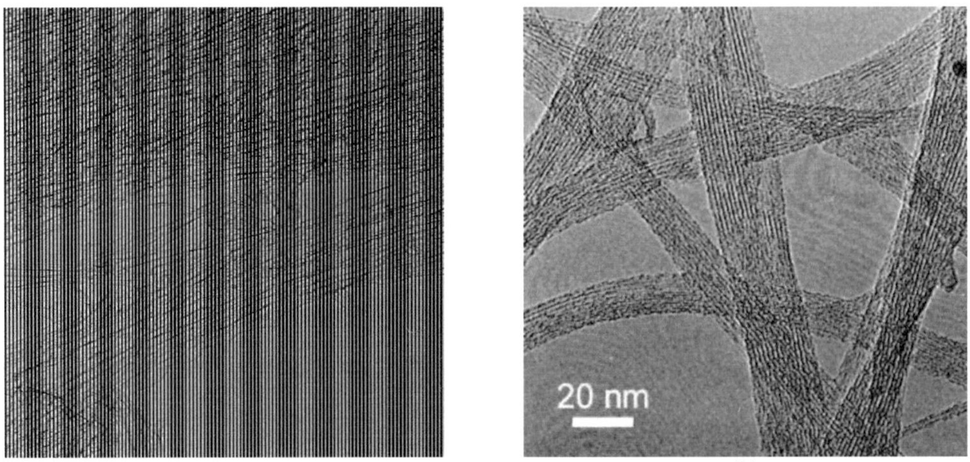

(a) (b)

Fig. 1. TEM images of SWCNT samples used. (a) Mutually isolated SG SWCNT prepared by the CVD method. (b) Well-bundled SWCNTs prepared by the laser ablation method.

Another approach to control the structure of SWCNT bundles is building up of the designed bundles from the isolated SWCNTs (Yamamoto et al., In press). Hata et al. succeeded to prepare mutually isolated SWCNTs of high purity using CVD method, stimulating interfacial researches on SWCNT (Hata et al., 2004). The transmission electron microscopy (TEM) image of SWCNT called as supergrowth SWCNT (SG SWCNT) is shown in Fig. 1a. SG SWCNT has the average diameter of 2.8 nm and the length of 1 mm order. Very recently, the authors evidenced that the monolayer of N_2 molecules adsorbed on the internal wall of the negative curvature of SWCNT is more ordered than that on the external wall of the positive curvature (Ohba et al., 2007). Thus, SWCNT has an explicit bi-surface nature for molecules, which should be applicable to develop intriguing and novel materials of multi-interfacial functions. If we control the bundle structure formation of the isolated SWCNTs induced by drying the SWCNTs dispersed in the solvent, many interstitial sites are formed enough to adsorb supercritical H_2. The effect of surface tension of solvents, which are used to disperse SWCNTs, is focused on to control the bundle structure formation, since the SWCNTs in the bundle are bound by van der Waals force, that is, 27.9×10^{-3} and 22.1×10^{-3} N/m for toluene and methanol at 273 K, respectively.

In this chapter, we report the preparation of C_{60}-pillared SWCNT bundles and SWCNT bundles induced with capillary force-aided drying method, whose the supercritical H_2 adsorptivities are enhanced by C_{60}-pillaring and by the bundle formation.

2. Experimental section

2.1 Preparation and characterization of C_{60}-pillared SWCNT bundle

We used SWCNT samples prepared by the laser ablation of a graphite rod in the presence of Ni and Co (@ Institute of Research and Innovation: IRI) (Yudasaka et al., 1999; Kokai et al., 2000). The produced SWCNT was purified by the following method: SWCNT (200 mg) was added to a 15% hydrogen peroxide solution, and this solution was refluxed with a water bath at 373 K for 5 h to remove amorphous carbons. The residual catalysts of Ni and Co were removed by a 1 M hydrogen chloride solution. Then, SWCNT was filtrated, washed with doubly distilled water, and left at room temperature overnight. The TEM image of the purified SWCNT is shown in Fig. 1b. Characterization data for the purified SWCNT are shown in Fig. 2. Figure 2a shows thermogravimetry (TG) and differential thermogravimetry (DTG) curves measured in the N_2/O_2 flow. The estimated content of Co-Ni catalyst is about 8 wt%. X-ray diffraction (XRD) pattern shown in Fig. 2b measured using CuK_α exhibits a clear peak due to their well-ordered hexagonal bundle structure. The peak at 2θ=6.12° corresponds to 1.44 nm of interlayer distance d of SWCNT. Raman spectra in the radial breathing mode (RBM) band and G- and D-bands regions are shown in Fig. 2c. Very small peak at D-band indicates high-quality of the purified SWCNT. The tube diameter (d_{SWCNT}) is 1.37 nm, determined by the relation of $d_{SWCNT} = 248/w$, where w is the wavenumber of the RBM (Kataura et al., 1999). Closed SWCNT samples were used to clearly show the effect of C_{60}-pillaring. Figure 2d shows the N_2 adsorption isotherms of the purified SWCNT at 77K. The BET specific surface area is 337 $m^2\ g^{-1}$, indicating that the purified SWCNTs were closed. For C_{60}-pillaring, we applied the methods used for the adsorption of organic substances on SWCNTs (Gotovac et al., 2007) and the preparation of peapod SWCNTs (Yudasaka et al., 2003). C_{60}-pillared SWCNTs were prepared by a simple sonication of SWCNT in C_{60} toluene solution with different concentrations. Purified SWCNTs (10 mg) were ultrasonically treated at 28 Hz for 6 h in C_{60}-dissolved toluene solutions of different concentrations up to 2.8 g $L^{-1}_{Toluene}$ in an ice storage. Then, the samples were stood for 24 h, filtrated, and dried in a vacuum at 333 K for 24 h. The amount of C_{60} on the SWCNT bundles was determined by the weight change of the samples before and after C_{60}-pillaring treatment. The C_{60}-pillared SWCNTs are designated as SWCNT-C_{60}(x), where x is the amount in gram of C_{60} doped to 1 g of SWCNTs. Here, SWCNT-C_{60}(0), which was ultrasonically treated in toluene without C_{60}, was also prepared for comparison.

The SWCNT-C_{60}(x) samples were characterized with N_2 adsorption at 77 K, XRD, X-ray photoelectron spectroscopy (XPS), Raman spectroscopy, thermogravimetric analysis (TGA), and high resolution transmission electron microscopy (HR-TEM). The H_2 adsorptivity of the SWCNT-C_{60}(x) samples was examined at 77 K by volumetric method. Samples were pre-evacuated at 423 K and 1 mPa for 2 h for the adsorption measurements of N_2 and H_2 at 77 K. TGA experiments were carried out in N_2 flow (100 ml/min) from ambient temperature to 1273 K at a rate of 5 K/min.

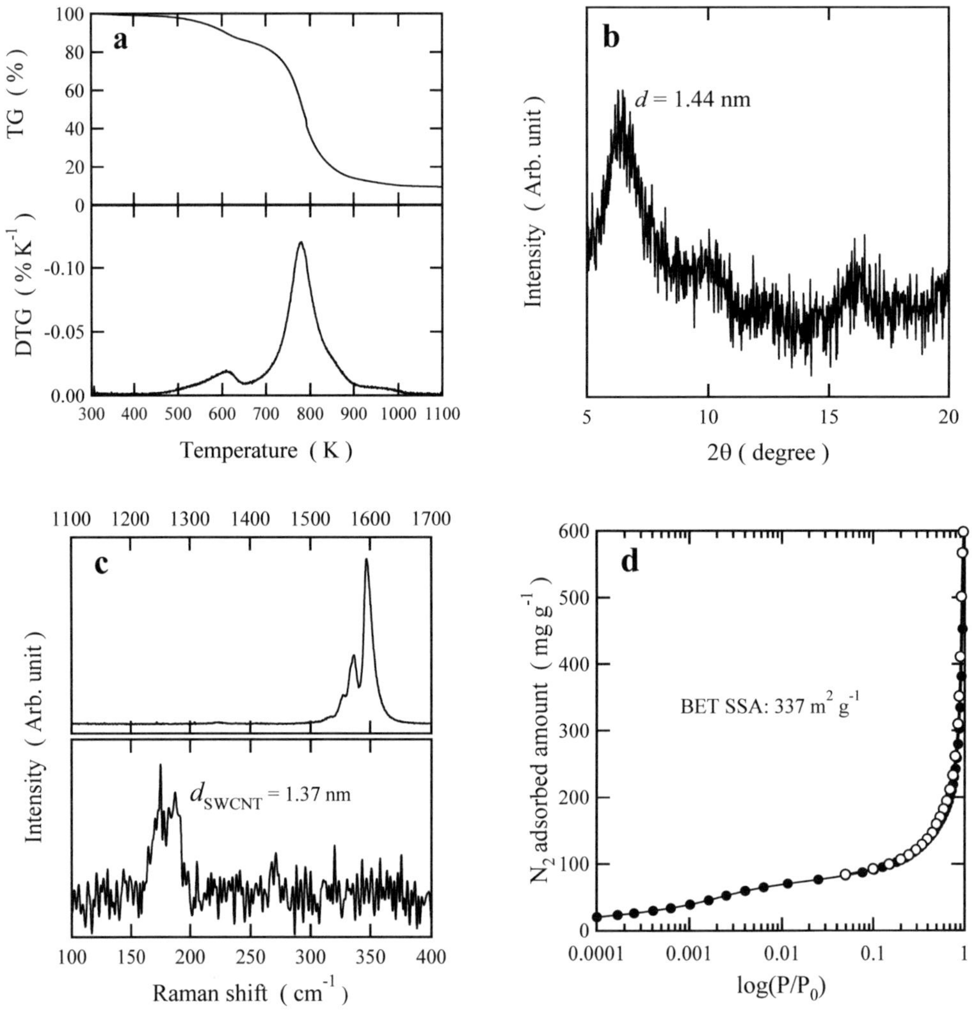

Fig. 2. Characterization data of purified SWCNT bundles prepared by laser ablation. (a) TG (upper)-DTG (bottom) curves. (b) XRD pattern of the superlattice of hexagonal SWCNT bundles measured using CuK_{α}. (c) G- and D-bands (upper) and RBM band (bottom) of Raman spectra. (d) N_2 adsorption isotherm at 77 K in terms of $\log(P/P_0)$.

2.2 Preparation and characterization of predominant bundle formation of isolated SWCNTs

The high purity isolated SWCNTs (SG SWCNTs) were produced by the CVD method (Hata et al., 2004). The SG SWCNTs were sonicated in toluene and methanol around 273 K for 12 h. Then, the SG SWCNTs in each solvent were filtrated and dried at 333 K for formation of the bundle structure by using the capillary force. The obtained SWCNT samples using toluene and methanol are denoted SWCNT/Tol and SWCNT/Met, respectively. The removal condition of residual toluene or methanol on SWCNT samples was determined by TGA before adsorption measurements. The nanopore structures of SWCNT/Tol and

SWCNT/Met were evaluated by N_2 adsorption measurements at 77 K after pretreatment at 423 K and 10^{-4} Pa for 2 h. The high pressure adsorption isotherms of supercritical H_2 were measured at 77 K by gravimetrical method after the pretreatment same as the N_2 adsorption measurement.

3. Results and discussion

3.1 C_{60}-pillared SWCNT bundle formation and its H_2 adsorptivity

The C_{60}-doped amount against the C_{60} concentration of the toluene solution (g $L^{-1}_{Toluene}$) is shown in Figure 3. The C_{60} uptake versus the C_{60} concentration curve has a step near 0.7 g g^{-1}_{SWCNT} of uptake and 0.5 g $L^{-1}_{Toluene}$ of the C_{60} concentration; the step indicates the formation of a stable structure between C_{60} and the SWCNTs. The uptake at the step closely corresponds to amount required for perfect filling of the interstitial spaces by C_{60} molecules, as estimated from the interstitial spaces in the model structure of an SWCNT bundle and the uptake of C_{60} for a trigonal arrangement as shown in Fig. 4 (Williams & Eklund, 2000).

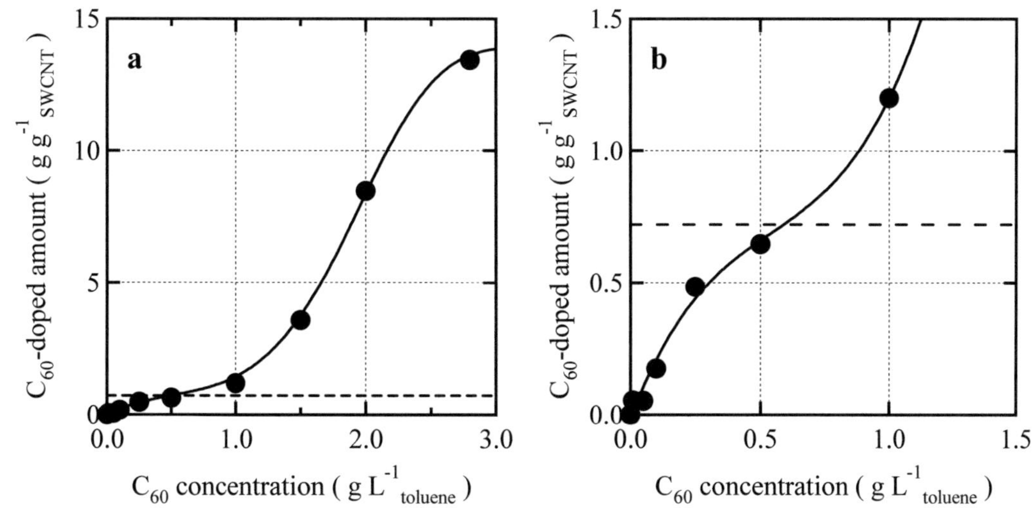

Fig. 3. C_{60}-doped amount on SWCNT bundles against the concentration of C_{60}-toluene solutions. (a) Relationship between the C_{60}-doped amount on 1 g of SWCNT samples (g g^{-1}_{SWCNT}) and the concentration of C_{60}-dissolved toluene. (b) Magnified view of Fig. 3a. The broken lines indicate the expected C_{60}-doped amount corresponding to perfect filling of the interstitial nanospaces with C_{60} in the model structure of C_{60}-pillared SWCNT bundles with hexagonal symmetry.

Figure 5 shows the N_2 adsorption isotherms of C_{60}-pillared SWCNT bundles. The N_2 adsorption amount was dramatically changed by C_{60}-pillaring treatment. The amount adsorbed on SWCNT-C_{60}(0.646) is the greatest. The adsorption isotherms of SWCNT-C_{60}(0.646) and SWCNT-C_{60}(1.68) have an extremely great uptake below P/P_0=0.1; the marked low pressure adsorption is more evidently observed in the isotherms of which abscissa is expressed by the log P/P_0. Accordingly the C_{60}-pillaring treatment increases the nanoporosity having a very strong interaction potential.

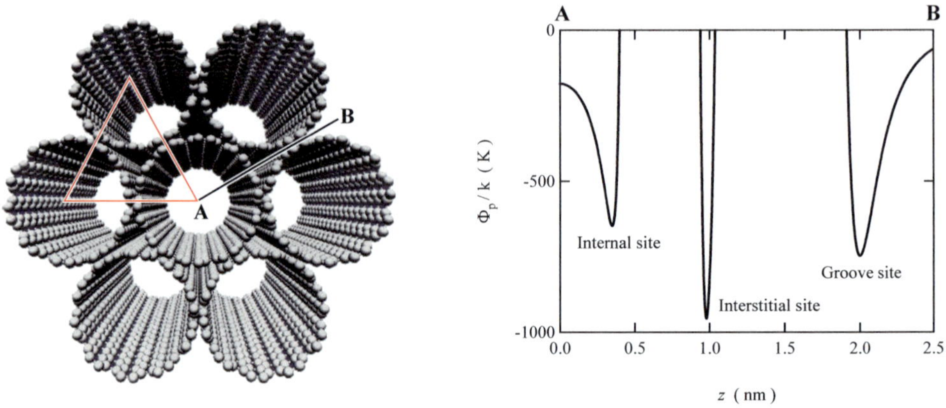

(a) (b)

Fig. 4. Model structure and molecular potential field for hydrogen of an SWCNT bundle. (a) Typical SWCNT bundle composed of (10 10) SWCNTs having interstitial spaces for trigonal arrangement. (b) The potential field on the line connecting A and B in Fig. 4a is shown.

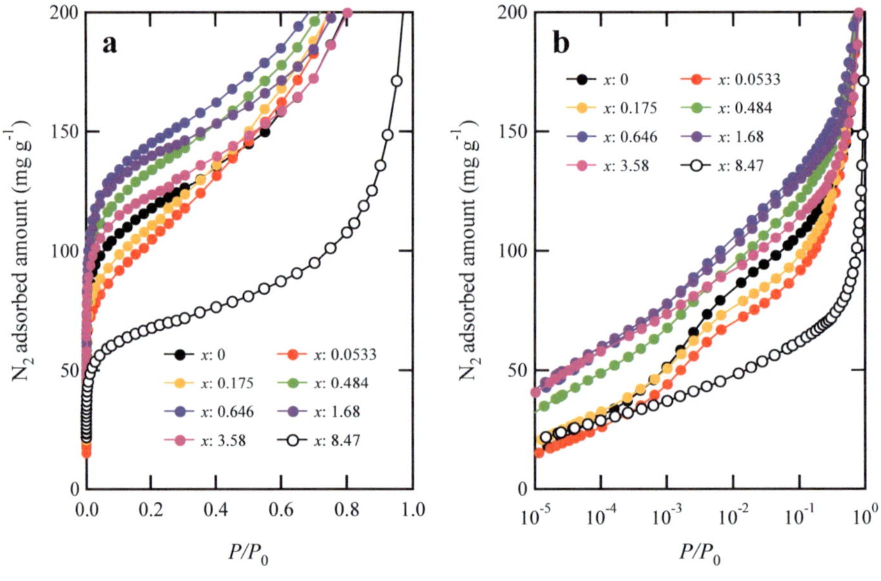

Fig. 5. N_2 adsorption isotherms of SWCNT-C_{60}(x). The abscissas of **a** and **b** are expressed by N_2 relative pressure and the logarithm of N_2 relative pressure, respectively.

The stable structure, that is, SWCNT-C_{60}(0.646), provides the maximum nanopore volume in N_2 adsorption measurements at 77 K. Figure 6 shows the relation between the nanospace volume and the C_{60}-doped amounts x (See Table 1 for detail). The nanospace volume of SWCNT-C_{60}(0.646) is 0.15 cm^3 g^{-1}, which is 1.36 times higher than that of SWCNT-C_{60}(0). Hence, SWCNT-C_{60}(0.646) should have the optimum C_{60}-pillared structure for the acceptance of molecules in the interstitial nanospaces expanded by C_{60}-pillaring. A C_{60} concentration higher than 1.0 g $L^{-1}{}_{Toluene}$ should induce further C_{60}-pillaring and coating of the external surface of the SWCNT bundle, thus reducing the nanopore volume.

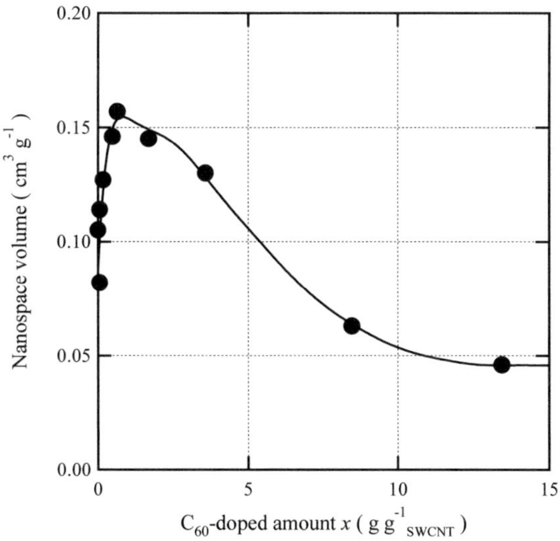

Fig. 6. Nanospace volume of C_{60}-pillared SWCNT as a function of the C_{60}-doped amounts x.

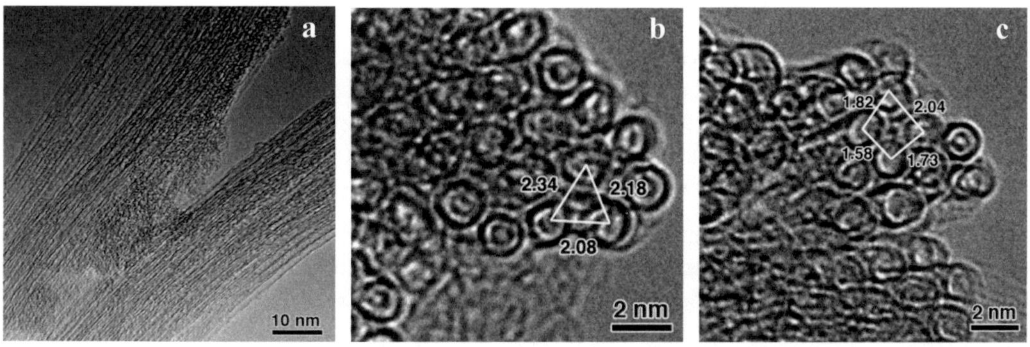

Fig. 7. TEM images of SWCNT-C_{60}(0.646) which has the maximum nanopore volume. (a) Side view of a C_{60}-pillared SWCNT bundle. (b) Cross-section of an expanded hexagonal SWCNT bundle. (d) Cross-section of a distorted-tetragonal array SWCNT bundle.

Figure 7 shows HR-TEM images of the bundle structure of SWCNT-C_{60}(0.646), which should have the optimum structure to adsorb molecules. The wide-range observation (Fig. 7a) shows a well-aligned bundle sheet even after ultrasonic C_{60}-pillaring treatment in the toluene solution. The side-view observation (Fig. 7a) indicates that C_{60} molecules are present on the SWCNT surfaces and there are no peapod SWCNTs (i.e., C_{60} in the SWCNTs). Figure 7b,c shows cross sections of the SWCNT-C_{60}(0.646) bundle having expanded hexagonal and tetragonal arrays, respectively. The intertube distance of the expanded bundle is estimated to be ~2.2 nm, leading to an interlayer distance d'=1.9 nm for hexagonal symmetry and d''=1.8 nm for tetragonal symmetry, under the assumption of uniform bundle structure for each symmetry. The interlayer distances d' and d'' of the bundles of hexagonal and tetragonal superlattices of the C_{60}-pillared SWCNT bundle are 2.03 and 1.92 nm, respectively, according to geometrical evaluation, which were close to the values determined from TEM, as shown in Fig. 8. Thus, SWCNT-C_{60}(0.646) should have a mixed C_{60}-pillared structure with both hexagonal and tetragonal symmetries.

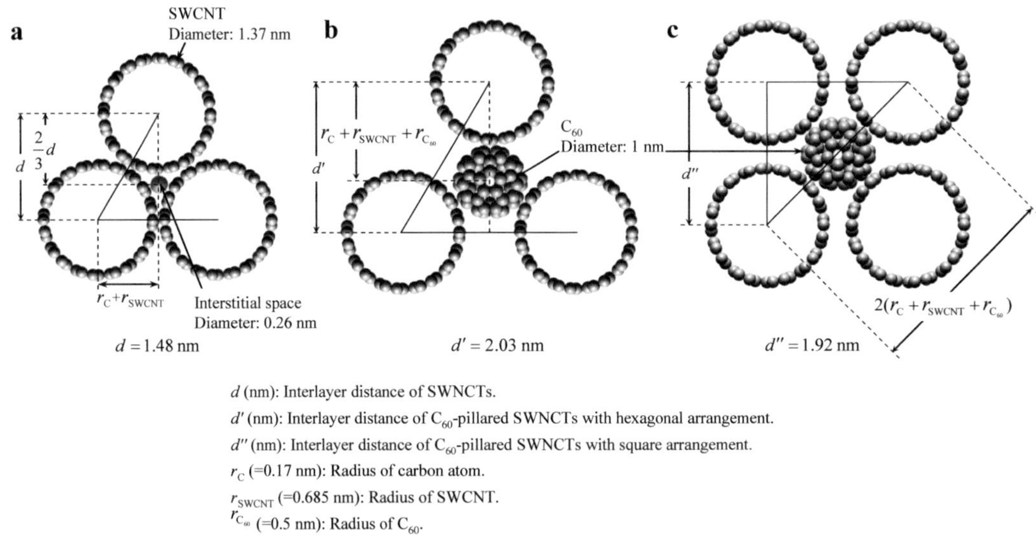

Fig. 8. Structure models and evaluation of interlayer distances for plausible C_{60}-pillared SWCNT bundles. Geometrical derivation of the interlayer distance *d*=1.48 nm of an SWCNT bundle with hexagonal arrangement (a), the interlayer distance *d'*=2.03 nm of an expanded hexagonal C_{60}-pillared SWCNT bundle (b), and the interlayer distance *d''*=1.92 nm of a C_{60}-pillared SWCNT bundle with tetragonal array (c).

Figure 9 shows XRD patterns which support the above-mentioned C_{60}-pillared SWCNT structure. SWCNT-C_{60}(0), which was ultrasonically treated in toluene without C_{60}, gives an explicit peak at 2.81° (X-ray source: MoK_α), corresponding to the interlayer distance (*d*=1.44 nm (experimental)) of a hexagonal lattice of SWCNT arrays (see Fig. 8a), whereas individual SWCNTs have no diffraction peak in the concerned diffraction angle-range. This peak is weakened by the C_{60}-pillaring treatment and a broad peak appears around 2θ=2.0°. The new peak corresponds to an interlayer distance of ~2.0 nm, which is the average of 2.03 and 1.92 nm, derived from the TEM-derived two-structure models. Thus, XRD clearly indicates the formation of C_{60}-pillared SWCNT bundles. However, the pillared structure is not necessarily regular; hexagonal and tetragonal structures coexist, and therefore a broad superlattice peak is observed.

TGA data as shown in Fig. 10 revealed pillaring of C_{60} in the SWCNT bundles. Even SWCNT-C_{60}(0), which has no C_{60} exhibits a remarkable weight decrease (≈ 10%) at 1200 K, which is caused by the elimination of amorphous carbon and oxygen functional groups. The weight decrease at high temperatures increases with increase in the C_{60}-doped amount x. The sample weight decrease is much less than the C_{60}-doped amounts, except in the region of low C_{60}-doped amount. The considerable weight decrease in the low C_{60} region is caused by the elimination of amorphous carbon and oxygen functional groups. On the other hand, only a part of the C_{60}-doped amounts adsorbed on the external surface of the SWCNT bundles can be eliminated in the high C_{60} region. This indicates that major C_{60} molecules inserted in the strong potential sites of interstitial pores cannot be eliminated.

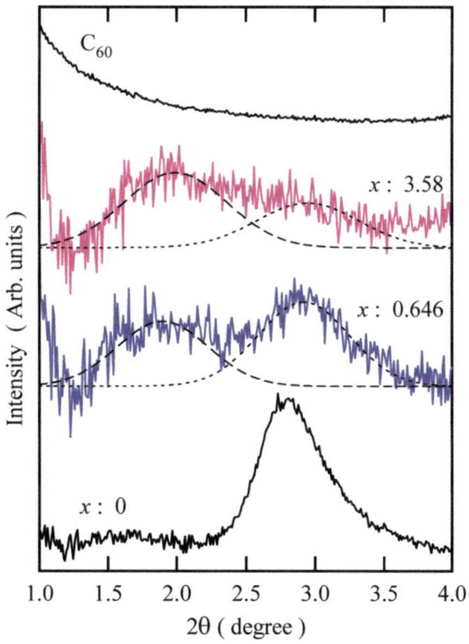

Fig. 9. XRD patterns of SWCNT-$C_{60}(x)$ and C_{60}. The explicit peak of SWCNT-$C_{60}(0)$ at 2.81° evidences an ordered hexagonal SWCNT bundle. The peak position of broken lines almost corresponds to the positions of the superlattice peaks of expanded hexagonal and tetragonal array models of the C_{60}-pillared SWCNT bundle.

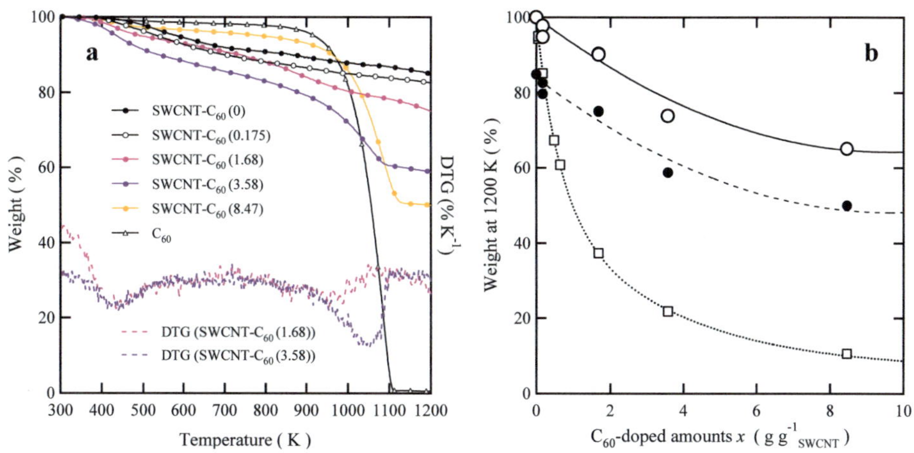

Fig. 10. TGA of SWCNT-$C_{60}(x)$ samples and C_{60}. (a) TG and typical DTG (x: 1.68 and 3.58) curves of SWCNT-$C_{60}(x)$ and C_{60}. (b) Closed circles (●) indicate the weight percents ($TG_{1200\,K}$) of the sample weight at 1200 K against the initial sample weight with C_{60}-doped amount. Open circles (○) indicate the weight change at 1200 K of SWCNT-C_{60}, after correction of the weight decrease of SWCNT itself. Open squares (□) correspond to the theoretical weight decrease in percent of SWCNT-$C_{60}(x)$ under the assumption that doped-C_{60} is completely sublimated.

Surface composition analysis with XPS also supports pillaring of C_{60} molecules in SWCNT bundles. Figure 11 shows the XPS spectra of C_{60}-pillared SWCNT bundles. XPS analysis of the C1s spectrum of SWCNT-C_{60}(0.646) by fitting with the C1s spectra of SWCNT and C_{60} shows the presence of 10% of C_{60} on the bundle surface, much less than the bulk content (39%) (Utsumi et al., 2007). XPS detects electrons only from surface layers in the order of 1 nm; predominant C_{60} molecules are not on the external surface of the SWCNT bundle, but inside of the SWCNT bundle as pillars. The absence of C_{60} on the bundle surface of SWCNT-C_{60}(0.646) is revealed by Raman spectroscopy shown in Fig. 12, which is also surface sensitive. The Raman peak for C_{60} at 1467 cm^{-1} appears only for the SWCNT-C_{60}(x) samples whose x is larger than 1.68 g g^{-1}_{SWCNT} (Dresselhaus et al., 1996; Rao et al., 1997). Thus, all characterization results confirm that SWCNT-C_{60}(0.646) has a promising C_{60}-pillared structure with adequate nanoporosity.

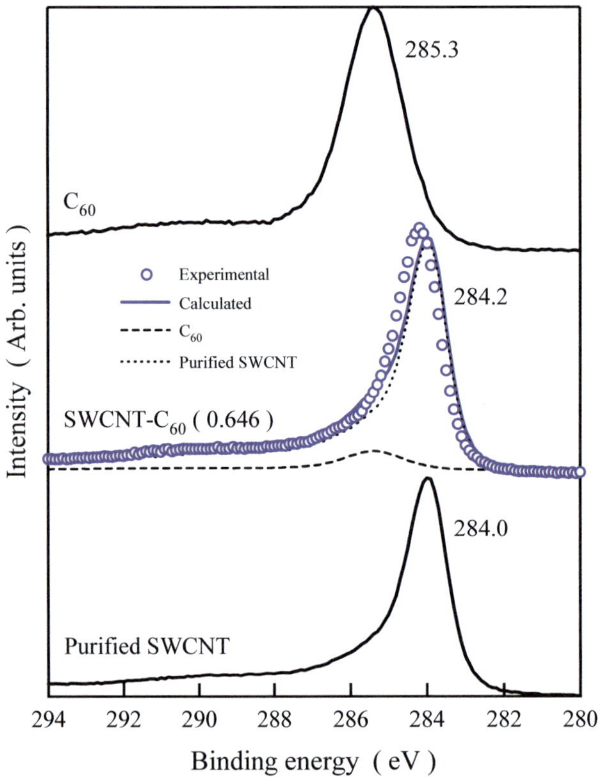

Fig. 11. C1s XPS spectra of SWCNT-C_{60}(x), SWCNT and C_{60}. The results of curve fitting (dotted line: purified SWCNT and broken line: C_{60}) for SWCNT-C_{60}(0.646) are shown. Estimated C_{60} amounts on the surface of SWCNT bundles from the XPS results are 10%.

The change in the interaction strength of the adsorption sites can be sensitively detected by supercritical H_2 adsorption. H_2 adsorption isotherms of SWCNT-C_{60}(x) at 77 K are shown in Fig. 13. For comparison, the isotherm of SWCNT-C_{60}(0) at 40 K is also shown. As the critical temperature of H_2 is 33 K, H_2 at 77 K and 40 K is supercritical, and thereby the adsorption of H_2 needs intensive assistance with interaction potential from solid nanospaces (Kaneko &

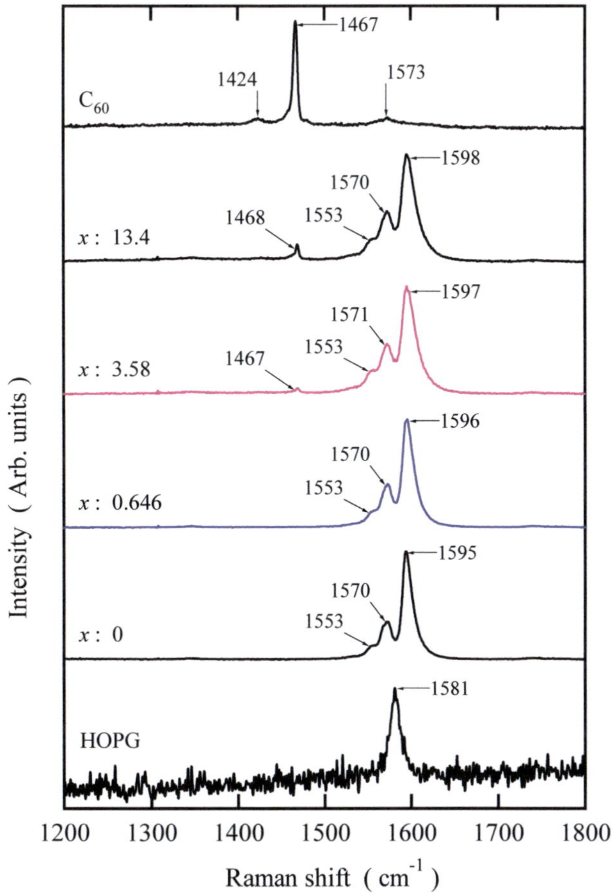

Fig. 12. Raman spectra of SWCNT-C_{60}(x), SWCNT and C_{60}. Selective insertion of C_{60} in the bundle is confirmed by the absence of the Raman peaks of C_{60}.

Murata, 1997; Xu et al., 2007). As shown in Fig. 13, a large amount of H_2 can be adsorbed at enough low temperature such as 40 K (≈ 1 wt% at 0.1 MPa), even though H_2 is supercritical. However, the H_2 adsorption amount at 77 K was less than the half of the amount at 40 K. H_2 adsorptivity of C_{60}-pillared SWCNT bundle was enhanced when the pillaring structure wasoptimum, while excessive C_{60}-doping reduced the H_2 adsorption amounts. Upward concave H_2 adsorption isotherms at relatively low pressure stems from the presence of strong adsorption sites, even for supercritical H_2. The adsorbed amounts per the weight and per the nanopore volume of SWCNT-C_{60}(x) of supercritical H_2 at 77 K are plotted against the C_{60}-doped amount x, as shown in Fig. 14. H_2 adsorptivity of SWCNT-C_{60}(x) per the sample weight markedly enhances the adsorption of H_2, providing almost twice the adsorption amount of SWCNTs in the low-pressure region, and 1.3 times higher in the ambient pressure region. On the other hand, H_2 adsorptivity of SWCNT-C_{60}(x) per the nanopore volume remarkably decreased by C_{60}-pillaring. These results indicate that H_2 adsorptivity of SWCNT-C_{60}(x) enhanced due to the increase in the nanopore volume, even though the interaction between the nanopore and supercritical H_2 were weakened.

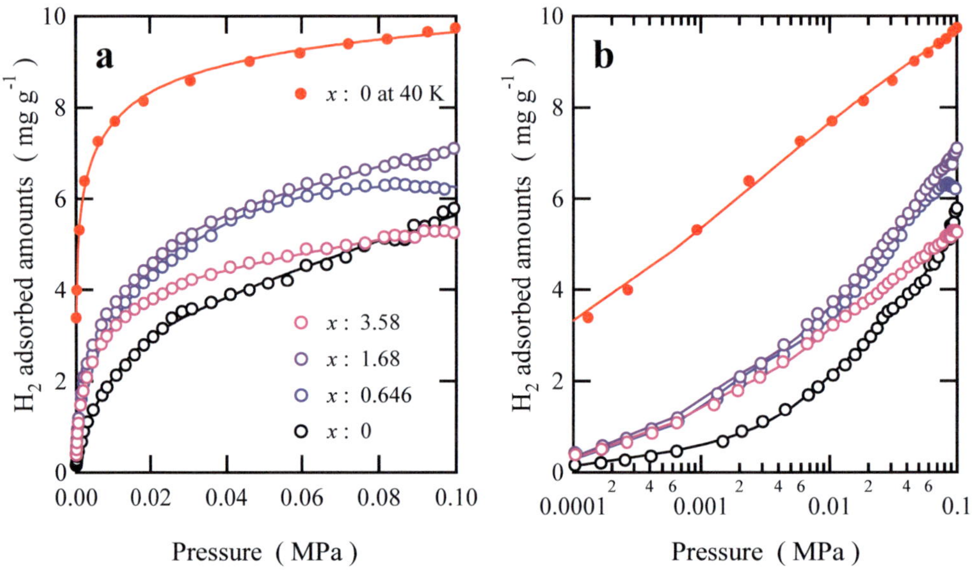

Fig. 13. H_2 adsorption isotherms of SWCNT-C_{60}(x) at 77 K. For comparison, the isotherm of SWCNT-C_{60}(0) measured at 40 K is also shown. The abscissas of **a** and **b** are expressed by the H_2 pressure and the logarithm of the H_2 pressure, respectively.

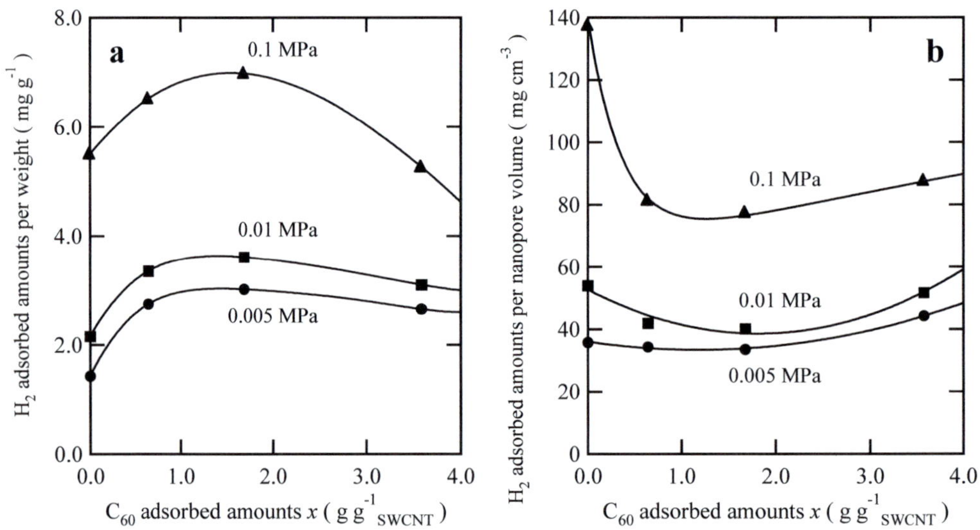

Fig. 14. H_2 adsorption amounts per the weight (a) and the nanopore volume (b) of SWCNT-C_{60}(x) at 0.005 MPa (●), 0.01 MPa (■), and 0.1 MPa (▲) as a function of C_{60} adsorbed amount.

Supercritical gaseous molecules are concentrated in nanopores by the strong molecule-pore interaction and a supercritical gas adsorbed in nanopores is transformed into a quasi-vapor. The Dubinin-Radushkevich (DR) equation extended for a quasi-vapor supercritical gas is expressed in the terms of the quasi-saturated vapor pressure P_{0q} and the inherent nanopore volume V_L by

$$\left[\ln\left(\frac{V_L}{V}\right)\right]^{\frac{1}{2}} = \left(\frac{RT}{\beta E_0}\right)\left(\ln P_{0q} - \ln P\right), \tag{1}$$

where V is the nanopore volume at pressure P, E_0 the characteristic adsorption energy, and β the affinity coefficient. The isosteric heat of adsorption $q_{st,(\theta=1/e)}$ at the fractional filling θ of $1/e$ can be calculated using βE_0 and the enthalpy of vaporization ΔH_v at the boiling point from the relation of

$$q_{st,(\theta=1/e)} = \Delta H_v + \beta E_0 . \tag{2}$$

The DR plot, that is, the plot of $[\ln(V_L/V)]^{1/2}$ versus $\ln P$ gives both values of P_{0q} under the nanopore field and $q_{st,(\theta=1/e)}$ (Kaneko et al., 1992; Kaneko & Murata, 1997). The typical DR plot for SWCNT-C_{60}(1.68) is shown in Fig. 15a, which exhibits a linear relationship. The isosteric heat of H_2 adsorption evaluated from the supercritical DR plot is shown in Fig. 15b. The isosteric heat is in the range of 9.5 to 9.7 kJ mol^{-1}, being almost constant regardless of C_{60}-pillaring. Thus, enhancement of the H_2 adsorptivity of C_{60}-pillared SWCNTs results from the increase in the nanospace volume. This value is much greater than the condensation enthalpy of H_2 molecules at the boiling point (~20 K; 0.22 kJ mol^{-1}) and the isosteric heat on the interstitial sites of (12, 12) SWCNT bundles (~9 kJ mol^{-1}) or on the slit pores of carbon (~8 kJ mol^{-1}) evaluated from Grand Canonical Monte Carlo simulation, indicating strong H_2 molecule-interstitial pore interaction in the C_{60}-pillared SWCNT bundles.

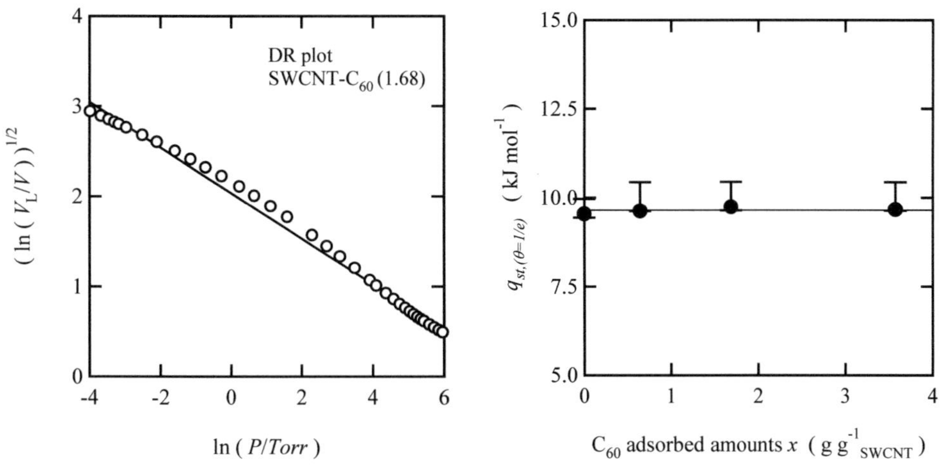

Fig. 15. Representative supercritical DR plot for SWCNT-C_{60}(1.68) and H_2 adsorption heat of SWCNT-C_{60}(x) derived from the supercritical DR plot as a function of C_{60} adsorbed amount.

3.2 Predominant bundle formation from isolated SG SWCNTs

Figure 16 shows the N_2 adsorption isotherms of SG SWCNT, SWCNT/Tol, and SWCNT/Met at 77 K. The N_2 adsorption isotherm of SG SWCNT is of IUPAC type II, showing that SG SWCNT is mutually isolated and the caps are closed. On the other hand, the isotherms of SWCNT/Tol and SWCNT/Met are close to IUPAC Type I, indicating the presence of predominant micropores. The specific surface area (SSA) determined by the α_s method (Kaneko et al., 1998) is shown in Table 2. The SSAs of SWCNT/Tol and SWCNT/Met (750 m^2 g^{-1} and 760 m^2 g^{-1}) are considerably decreased by the sonication treatment, indicating the formation of the bundle structure with interstitial sites which are not accessible by N_2 molecules. SWCNT/Tol and SWCNT/Met should have enough bundle structure. Figure 17a shows the high pressure H_2 adsorption isotherms at 77 K, which are Langmuirian suggesting the presence of considerably strong interaction between H_2 and each SWCNT sample. Figure 17b shows adsorption isotherms expressed by the absolute H_2 amounts per the sample weight, which clearly show the formation of efficient surface for H_2 adsorption. Consequently, interstitial sites were efficiently introduced in the SWCNT bundle by the compression on drying with the aid of capillary force.

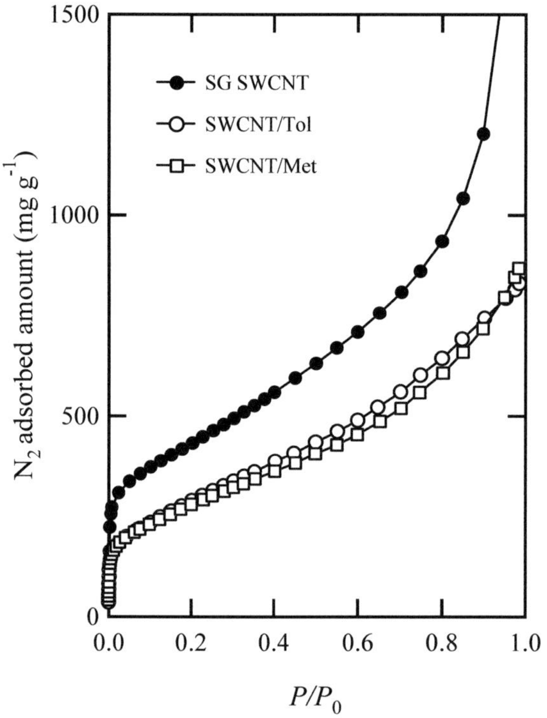

Fig. 16. N_2 adsorption isotherms of SG SWCNT (•), SWCNT/Tol (○), and SWCNT/Met (□) at 77 K.

Sample donation	SSA α_s	H_2 saturated adsorption amount	Quasi-saturated pressure	Heat of adsorption
	m^2/g	mg/g	MPa	kJ/mol
SG SWCNT	1,230	30	15	10.7
SWCNT/Tol	750	33	15	8.9
SWCNT/Met	760	29	13	9.0

Table 2. Parameters of bundle-structure controlled SG SWCNTs.

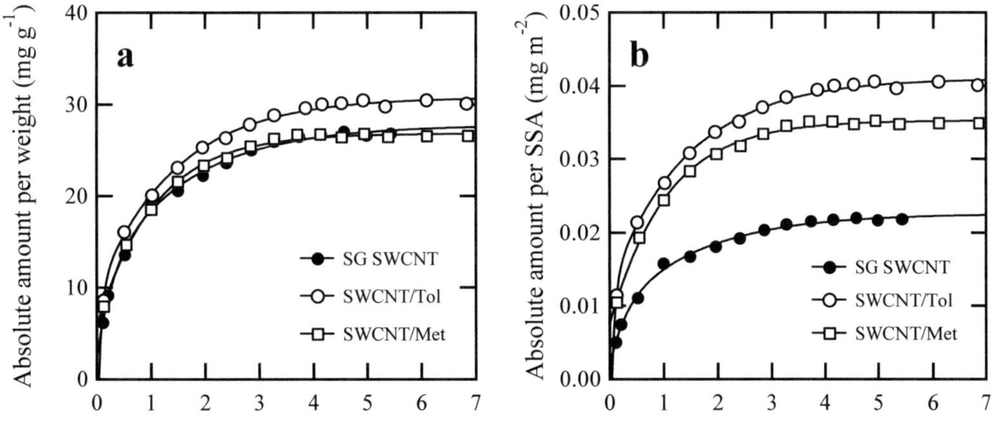

Fig. 17. High pressure H_2 adsorption isotherms of SG SWCNT (•), SWCNT/Tol (○), and SWCNT/Met (□). The vertical axes of **a** and **b** are expressed by the absolute H_2 amounts per the weight and per the specific surface area of the samples, respectively.

4. Conclusion

Two methods for tuning the bundle structure of SWCNT to enhance its H_2 adsorptivity were proposed; a one-step method for C_{60}-pillaring in SWCNT bundles by the cosonication of C_{60} and SWCNT in toluene and predominant bundle formation of the isolated SWCNTs by drying SWCNTs dispersed in toluene or methanol. C_{60}-pillared SWCNT with expanded hexagonal and distorted tetragonal arrays has enhanced H_2 adsorptivity, providing almost twice the adsorption amount of SWCNTs in the low-pressure region, and 1.3 times higher in the ambient pressure region. Isolated SWCNTs treated with toluene and methanol should make enough bundle structure by the compression on drying with the aid of capillary force, forming efficient surface for H_2 adsorption. These results indicate simple and promising tuning routes for SWCNT bundle structures, allowing the utilization of interstitial nanopore spaces for various fields, such as electrochemical, adsorption, sensor, and separation technologies.

5. Acknowledgment

We are grateful to Dr. Kunimitsu Takahashi at Institute of Research and Innovation (IRI) for providing SWCNT samples. We also thank Mr. Daisuke Noguchi for the H_2 adsorption measurements at 40 K. This research was founded with Grant-in-Aids for Scientific Research (S) from Japanese Government. K.K. was supported by Exotic Nanocarbons, Japan Regional Innovation Strategy Program by the Excellence, Japan Science and Technology Agency.

6. References

Abrams, Z. R.; Ioffe, Z.; Tsukernik, A.; Cheshnovsky, O.; & Hanein, Y. (2007). A Complete Scheme for Creating Predefined Networks of Individual Carbon Nanotubes. *Nano Lett.*, Vol. 7, 2666-2671, ISSN 1530-6984.

Arai, M.; Utsumi, S.; Kanamaru, M.; Urita, K.; Fujimori, T.; Yoshizawa, N.; Noguchi, D.; Nishiyama, K.; Hattori, Y.; Okino, F.; Ohba, T.; Tanaka, H.; Kanoh, H.; & Kaneko, K. (2009). Enhanced Hydrogen Adsorptivity of Single-Wall Carbon Nanotube Bundles by One-Step C_{60}-Pillaring Method. *Nano Lett.*, Vol. 9, 3694-3698, ISSN 1530-6984.

Arai, M.; Kanamaru, M.; Matsumura, T.; Hattori, Y.; Utsumi, S.; Ohba, T.; Tanaka, H.; Yang, C.-M.; Kanoh, H.; Okino, F.; Touhara, H.; & Kaneko, K.; (2007). Pore characterization of assembly-structure controlled single wall carbon nanotube. *Adsorption*, Vol. 13, 509-514, ISSN 0929-5607.

Banerjee, S.; Kahn, M. G. C.; & Wong, S. S. (2003). Rational Chemical Strategies for Carbon Nanotube Functionalization. *Chem.-Eur. J.*, Vol. 9, 1898–1908, ISSN 0947-6539.

Dresselhaus, M. S.; Dresselhaus, G.; & Eklund, P. C. (1996). Raman Scattering in Fullerenes. *J. Raman Spectrosc.*, Vol. 27, 351-371, ISSN 0377-0486.

Fujimori, T.; Urita, K.; Ohba, T.; Kanoh, H.; & Kaneko, K. (2010). Evidence of Dynamic Pentagon-Heptagon Pairs in Single-Wall Carbon Nanotubes using Surface-Enhanced Raman Scattering. *J. Am. Chem. Soc.*, Vol. 132, pp. 6764–6767, ISSN 0002-7863.

Gotovac, S.; Honda, H.; Hattori, Y.; Takahashi, K.; Kanoh, H.; & Kaneko, K. (2007). Effect of Nanoscale Curvature of Single-Walled Carbon Nanotubes on Adsorption of Polycyclic Aromatic Hydrocarbons. *Nano Lett.*, Vol. 7, 583-587, ISSN 1530-6984.

Gotovac-Atlagić, S.; Hosokai, T.; Ohba, T.; Ochiai, Y.; Kanoh, H.; Ueno, N.; & Kaneko, K. (2010). Pseudometallization of single wall carbon nanotube bundles with intercalation of naphthalene. *Phys. Rev. B*, Vol. 82, 075136, ISSN 1098-0121.

Gregory, J. M. & Oerlemans, J. (1998). Simulated future sea-level rise due to glacier melt based on regionally and seasonally resolved temperature changes. *Nature*, Vol. 391, 474-476, ISSN 0028-0836.

Hata, K.; Futaba, D. N.; Mizuno, K.; Namai, T.; Yumura, M.; & Iijima, S. (2004). Water-Assisted Highly Efficient Synthesis of Impurity-Free Single-Walled Carbon Nanotubes. *Science*, Vol. 306, 1362-1364, ISSN 0036-8075.

Hirsch, A. (2002). Functionalization of Single-Walled Carbon Nanotubes. *Angew. Chem., Int. Ed.*, Vol. 41, 1853–1859, ISSN 1433-7851.

Iijima, S. (1991). Helical microtubules of graphitic carbon. *Nature*, Vol. 354, 56-58, ISSN 0028-0836.

Iijima, S. & Ichihashi, T. (1993). Single-shell carbon nanotubes of 1-nm diameter. *Nature*, Vol. 363, 603-605, ISSN 0028-0836.

Kaneko, K.; Ishii, C.; Kanoh, H.; Hanzawa, Y.; Setoyama, N.; & Suzuki, T. (1998). Characterization of porous carbons with high resolution α_s-analysis and low temperature magnetic susceptibility. *Adv. Colloid Interface Sci.*, Vol. 76-77, 295-320, ISSN 0001-8686.

Kaneko, K. & Murata, K. (1997). An analytical method of micropore filling of a supercritical gas. *Adsorption*, Vol. 3, 197-208, ISSN 0929-5607.

Kaneko, K.; Shimizu, K.; & Suzuki, T. (1992). Intrapore field-dependent micropore filling of supercritical N_2 in slit-shaped micropores. *J. Chem. Phys.*, Vol. 97, 8705-8711, ISSN 0021-9606.

Kataura, H.; Kumazawa, Y.; Maniwa, Y.; Umezu, I.; Suzuki, S.; Ohtsuka, Y.; & Achiba, Y. (1999). Optical Properties of Single-Wall Carbon Nanotubes. *Synth. Met.*, Vol. 103, 2555-2558, ISSN 0379-6779.

Kim, D. Y.; Yang. C.-M.; Yamamoto, M.; Lee, D. H.; Hattori, Y.; Takahashi, K.; Kanoh, H.; & Kaneko K. (2007). Supercritical Hydrogen Adsorption of Ultramicropore-Enriched Single-Wall Carbon Nanotube Sheet. *J. Phys. Chem. C*, Vol. 111, pp. 17448–17450, ISSN 1932-7447.

Kokai, F.; Takahashi, K.; Yudasaka, M.; & Iijima, S. (2000). Laser Ablation of Graphite–Co/Ni and Growth of Single-Wall Carbon Nanotubes in Vortexes Formed in an Ar Atmosphere. *J. Phys. Chem. B*, Vol. 104, 6777- 6784, ISSN 1520-6106.

Kroto, H. W.; Heath, J. R.; O'Brien, S. C.; Curl, R. F.; & Smalley, R. E. (1985). C_{60}: Buckminsterfullerene. *Nature*, Vol. 318, 162-163, ISSN 0028-0836.

Liu, C.; Fan, Y. Y.; Liu, M.; Cong, H. T.; Cheng, H. M.; & Dresselhaus, M. S. (1999). Hydrogen Storage in Single-Walled Carbon Nanotubes at Room Temperature. *Science*, Vol. 286, 1127-1129, ISSN 0036-8075.

Noguchi, H.; Kondo, A.; Noguchi, D.; Kim, D. Y.; Ohba, T.; Yang, C.-M.; Kanoh, H.; & Kaneko, K.; (2007). Adsorptive Properties of Novel Nanoporous Materials. *J. Chem. Eng. Jpn.*, Vol. 40, 1159-1165, ISSN 0021-9592.

Noguchi, D.; Tanaka, H.; Fujimori, T.; Kagita, T.; Hattori, Y.; Honda, H.; Urita, K.; Utsumi, S.; Wang, Z.-M.; Ohba, T.; Kanoh, H.; Hata, K.; & Kaneko, K. (2010). Selective D_2 Adsorption Enhanced by Quantum Sieving Effect on Entangled Single-Wall Carbon Nanotubes. *J. Phys.: Condens. Matter*, Vol. 22, 334027, ISSN 0953-8984.

Ohba, T.; Matsumura, T.; Hata, K.; Yumura, M.; Iijima, S.; Kanoh, H.; & Kaneko, K. (2007). Nanoscale Curvature Effect on Ordering of N_2 Molecules Adsorbed on Single Wall Carbon Nanotube. *J. Phys. Chem. C*, Vol. 111, 15660-15663, ISSN 1932-7447.

Rao, A. M.; Eklund, P. C.; Hodeau, J.-L.; Marques, L.; & Nunez-Regueiro, M. (1997). Infrared and Raman studies of pressure-polymerized C_{60}s. *Phys. Rev. B*, Vol. 55, 4766-4773, ISSN 1098-0121.

Saito, R.; Takeya, T.; Kimura, T.; Dresselhaus, G.; & Dresselhaus, M. S. (1998). Raman intensity of single-wall carbon nanotubes. *Phys. Rev. B*, Vol. 57, 4145–4153, ISSN 1098-0121.

Seung, M. L.; & Young, H. L. (2000). Hydrogen storage in single-walled carbon nanotubes. *Appl. Phys. Lett.*, Vol. 76, 2877-2879, ISSN 0003-6951.

Utsumi, S.; Honda, H.; Hattori, Y.; Kanoh, H.; Takahashi, K.; Sakai, H.; Abe, M.; Yudasaka, M.; Iijima, S.; & Kaneko, K. (2007). Direct Evidence on C–C Single Bonding in Single-Wall Carbon Nanohorn Aggregates. *J. Phys. Chem. C*, Vol. 111, 5572-5575, ISSN 1932-7447.

Wang, Q. & Johnson, J. K. (2000). Optimization of Carbon Nanotube Arrays for Hydrogen Adsorption. *J. Phys. Chem. B*, Vol. 103, 4809-4813, ISSN 1520-6106.

Williams, K. A. & Eklund, P. C. (2000). Monte Carlo simulations of H_2 physisorption in finite-diameter carbon nanotube ropes. *Chem. Phys. Lett.*, Vol. 320, 325-358, ISSN 0009-2614.

Xu, W.-C.; Takahashi, K.; Matsuo, Y.; Hattori, Y.; Kumagai, M.; Ishiyama, S.; Kaneko, K.; & Iijima, S. (2007). Investigation of hydrogen storage capacity of various carbon materials. *Int. J. Hydrogen Energy*, Vol. 32, 2504-2512, ISSN 0360-3199.

Yamamoto, M.; Itoh, T.; Sakamoto, H.; Fujimori, T.; Urita, K.; Hattori, Y.; Ohba, T.; Kagita, H.; Kanoh, H.; Niimura, S.; Hata, K.; Takeuchi, K.; Endo, M.; Rodrígues-Reinoso, F.; & Kaneko, K. (In press). Effect of nanoscale curvature sign and bundle structure on supercritical H_2 and CH_4 adsorptivity of single wall carbon nanotube. *Adsorption*, ISSN 0929-5607.

Yudasaka, M.; Ajima, K.; Suenaga, K.; Ichihashi, T.; Hashimoto, A.; & Iijima, S. (2003). Nano-extraction and nano-condensation for C_{60} incorporation into single-wall carbon nanotubes in liquid phases. *Chem. Phys. Lett.*, Vol. 380, 42-46, ISSN 0009-2614.

Yudasaka, M.; Kokai, F.; Takahashi, K.; Yamada, R.; Sensui, N.; Ichihashi, T.; & Iijima, S. (1999). Formation of Single-Wall Carbon Nanotubes: Comparison of CO_2 Laser Ablation and Nd: YAG Laser Ablation. *J. Phys. Chem. B*, Vol. 103, 3576- 3581, ISSN 1520-6106.

Zhao, Y.; Sugai, T.; Shinohara, H.; & Saito, Y. (2007). Controlling growth and Raman spectra of individual suspended single-walled carbon nanotubes. *J. Phys. Chem. Solids*, Vol. 68, 284-289, ISSN: 0022-3697.

4

Nanoadhesion and Nanopeeling Forces of Carbon Nanotube on Substrate

Kouji Miura[1], Makoto Ishikawa[1] and Naruo Sasaki[2]
[1]*Aichi University of Education,*
[2]*Seikei University*
Japan

1. Introduction

Detachment and peeling experiments are expected to provide information on adhesion forces and adhesion energies of solid surfaces in contact. Such experiments are important in powder technology, in the formation of adhesive films and in understanding how cracks propagate in solid and how fracture occurs (Israelachvili, 1992). Recently, nanoscale peeling has been studied by extending biological polymer chains such as proteins using atomic force microscopy to clarify the mechanical mechanism of unfolding of the polymer chain. Such experiments have attracted much attention worldwide as a new method of spectroscopy for the structural analysis of biological macromolecules (Evans et al., 1991, Rief et al., 1997). On the other hand, it has been more recently reported that carbon nanotube arrays with a curly entangled top show a macroscopic adhesive force of approximately 100 newtons per square centimeter, almost 10 times that of a gecko foot, and a shear adhesion force much stronger than the normal adhesion force (Qu et al., 2008).

Here, we report the elementary processes of adhesion and peeling of a nanotube on a substrate and have performed adhesion and peeling experiments using a multiwalled carbon nanotube (MWCNT) (Ishikawa et al., 2008, 2009). An MWCNT has been attached parallel to the cantilever to easily peel off the substrate to elucidate the elementary process of adhesion and peeling mechanisms.

2. Experimental

We use a self-detecting cantilever (NPX1CTP003, SII) as a force sensor (Fig.1(a)) (Ishikawa, 2008). The system has a 3D inertial-driven actuator (UNISOKU co., ltd.) as a sample stage (Fig. 1(b)). Thus, the base plate can be set on the conventional stage of a scanning electron microscope (SEM) (S-3000N, Hitachi, base pressure of 2×10^{-3} Pa), instead of the normal sample holder. The sample stage is controlled by a computer via an A/D, D/A compatible board. SEM images can be recorded as a movie using a video recorder. Because the resolution of A/D is 10 mV, the maximum resolution of the force sensor is approximately 0.1 nN. The fabrication of an MWCNT attached probe is performed in the SEM chamber. First, using the chemically etched metal probe that is fixed on the actuator, an MWCNT is pulled out from as-prepared MWCNT powder on the opposite support. Then, the MWCNT on the metal probe is moved toward the tip of a cantilever using an actuator. When an electron beam is irradiated around the contact junction between the MWCNT and the cantilever tip, the hydrocarbon is piled there and it works as glue. Finally, when the metal probe is quickly withdrawn, the MWCNT is attached parallel to the cantilever. The handling of an MWCNT has been developed through improvements in the instruments described in previous reports (Ishikawa et al., 2002, 2008). Once the MWCNT has been sufficiently approached to the substrate surface using a coarse motion, the control was changed from coarse motion to fine motion. The adhesion and peeling experiments were repeated using fine motion.

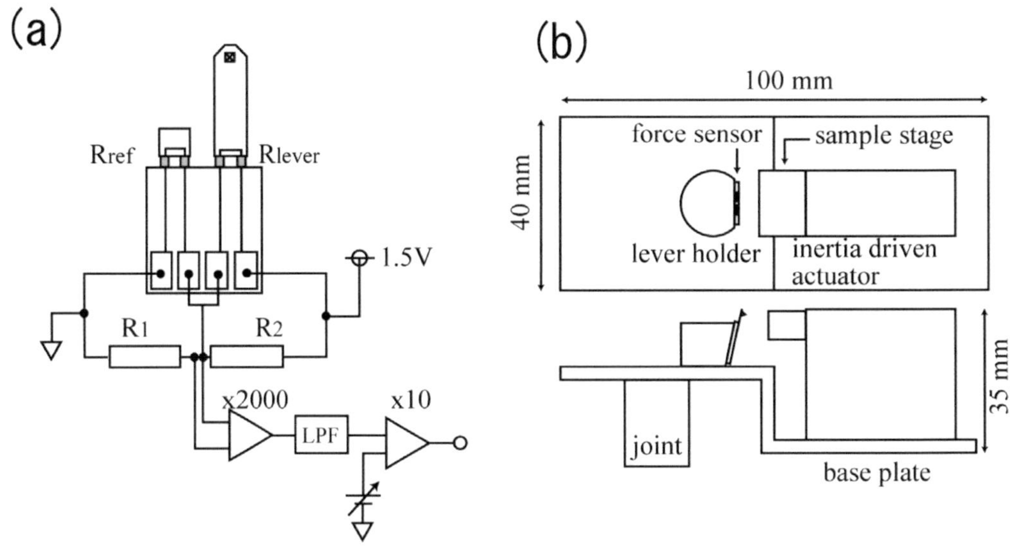

Fig. 1. (a) Schematics of a force detection system with self-detecting cantilever. (b) Overview of the force detection system. Force sensor and inertial driven actuator are fixed on the base plate. The signal from the sensor and the control signal of sample stage are handled by a computer via A/D, D/A compatible board. Thus, the signal is related to the motion of sample stage. SEM images can be recorded by using a video recorder as a movie.

3. Results and discussion

Figure 2(a) shows the vertical force-distance curve using a 400-nm-long MWCNT, where the black and red lines represent adhesion and peeling, respectively. Several frames in the movie recorded during the adhesion and peeling are shown in Fig. 2(b), which are also visible in the movie (Ishikawa, 2011). Figure 2(c) shows the illustrations of the conformational configuration of the MWCNT during the peeling shown in Fig 2(b). The white reversed triangle and white bar on the upper side and the horizon on bottom of each picture in Fig. 2(b) represent the tip of the cantilever, the MWCNT attached to the tip apex and the graphite surface, respectively. At first, the MWCNT takes an initial structure parallel to the graphite substrate surface. When the tip further approaches the graphite surface, the MWCNT attached to the tip apex comes into contact with the graphite surface (point A in Fig. 2(a)). When the tip presses furthermore the graphite surface, the latter deforms downwards with increasing repulsive force. When the peeling begins (red line), the

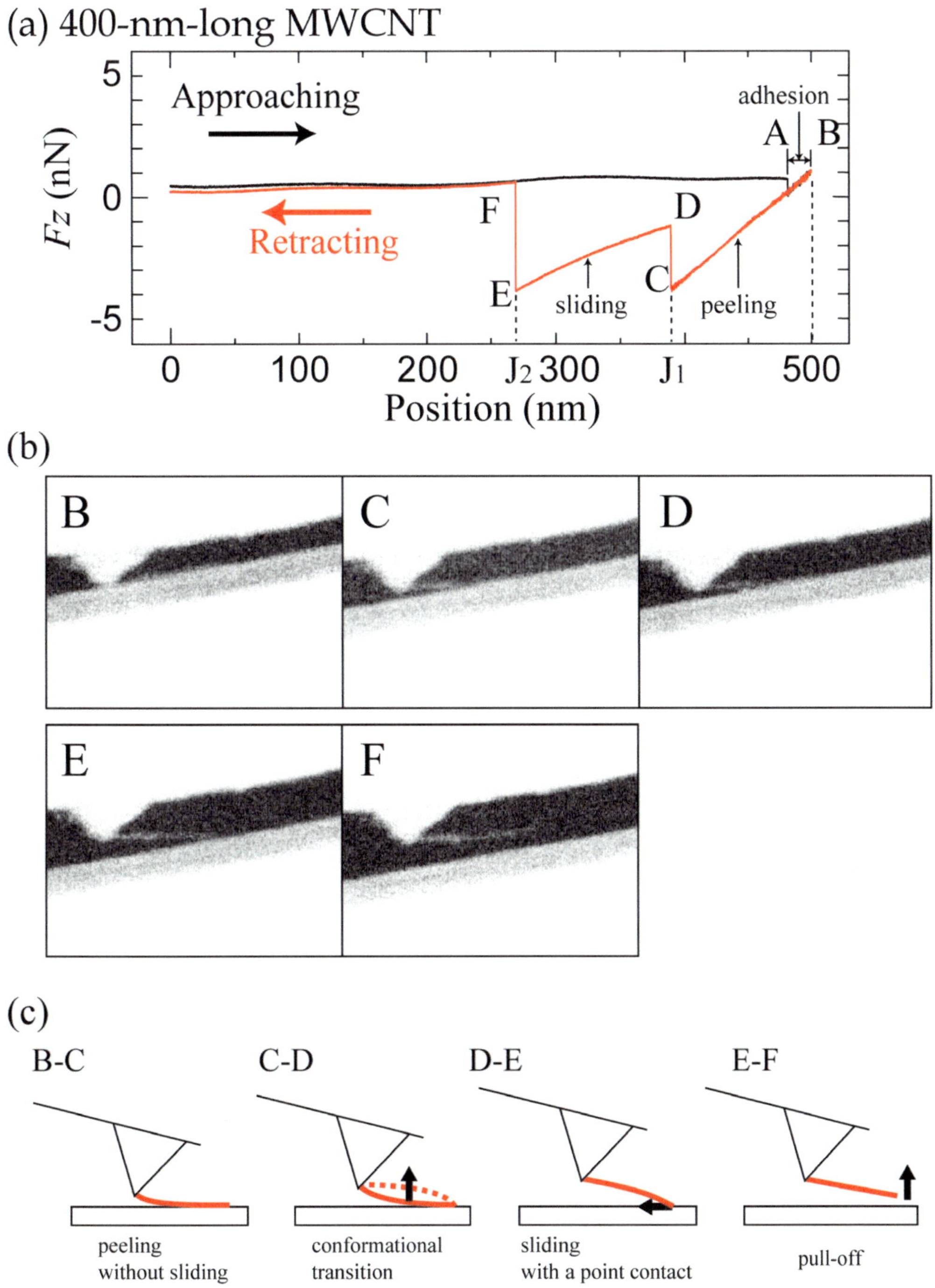

Fig. 2. (a) Vertical force-distance curve measured for the probe of a 400-nm-long MWCNT. (b) Several frames in the movie recorded during adhesion and peeling. (c) Schematic of (b).

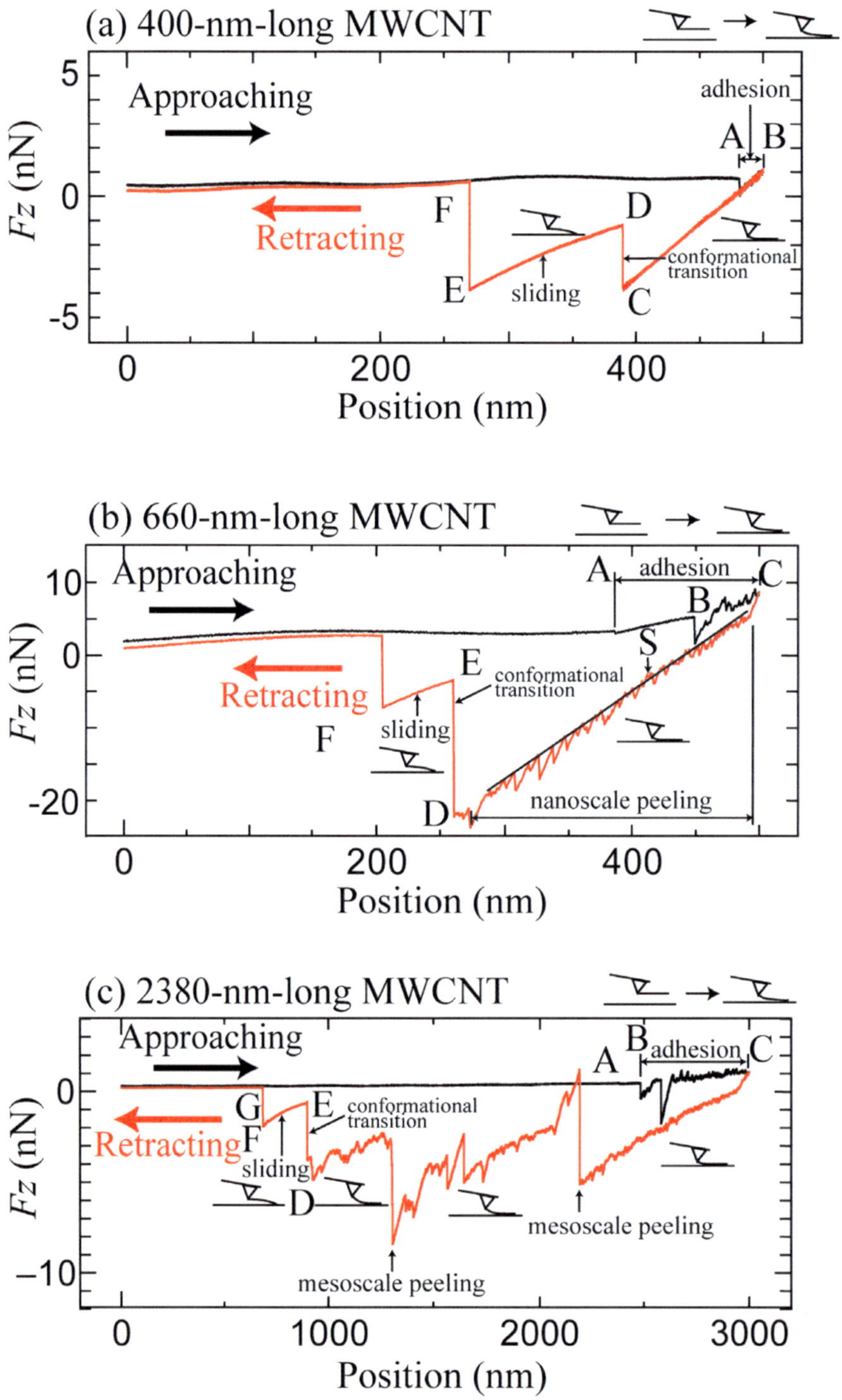

Fig. 3. Vertical force-distance curves measured using (a) 400-nm-long, (b) 660- nm-long, and (c) 2380-nm-long MWCNTs.

vertical force F_z rapidly decreases while holding the line contact between the MWCNT and the graphite surface. As the MWCNT is peeled further from the graphite surface, the transition of the MWCNT shape occurs, which causes the first discontinuous jump (J_1) in the F_z curve. Just after the discontinuous jump J_1, the point contact between the MWCNT and the graphite surface is formed. Frames D and E show the point contact formed between the free edge of the MWCNT and the graphite surface. Here the free edge is pushed onto the graphite surface and atoms on the free edge receive repulsive interaction forces (Sasaki et al., 2006, 2008, 2009a, 2009b). As the MWCNT is peeled further, the free edge of the MWCNT slides on the graphite surface with increasing MWCNT bending in the period between J_1 and J_2. The further retraction of the MWCNT from the surface decreases the repulsive force acting on atoms on the free edge and a relative increase in the effect of the attractive interaction force as shown in Fig. 2(a) (Sasaki et al., 2006, 2008, 2009a, 2009b). Now, when the bending of the MWCNT becomes larger than a certain range, the point contact breaks and the MWCNT is completely peeled from the surface, which makes the last discontinuous jump in the force curve (J_2). When the second discontinuous jump (J_2) occurs, the edge of the MWCNT is completely retracted from the graphite surface. Frames E and F show the images before and after the second discontinuous jump (J_2), respectively. After the MWCNT is moved upward further, the MWCNT takes an original line shape parallel to the graphite surface because the effect of van der Waals interaction from the surface becomes negligibly small and F_z gradually becomes zero.

Figure 3 shows how the vertical force-distance curve depends on the length of the MWCNT. Figure 3(a) is the vertical force-distance curve obtained with the 400-nm-long MWCNT, which is the same as the curve shown in Fig. 2(a). Figures 3(b) and 3(c) show the vertical force-distance curves of the MWCNTs with lengths of 660 nm and 2380 nm, respectively. The black and red lines in each figure represent the force during the adhesion and peeling, respectively. In the approach shown in Fig. 3(b), small stick-slip behaviors appear at the region between B and C after the sudden line contact at A during the adhesion. Small stick-slip behaviors are also observed in the slope between C and D during the peeling, which exhibits a nanoscale intermittent adhesion and peeling without sliding. The discontinuous jump D-E in the peeling shown in Fig. 3(b) shows a conformational transition of the MWCNT. In the slope between E and F, a point contact between the MWCNT and the graphite surface is formed. For a 2380-nm-long MWCNT (Fig. 3(c)), the vertical force-distance curve becomes more complex and the number of discontinuous jumps increases, which exhibits a mesoscale intermittent peeling. Small stick-slip behaviors are also observed at the slopes, which exhibit a nanoscale intermittent peeling without sliding. At the discontinuous jump E-D, the MWCNT bends upwards and performs a conformational transition. In the slope between E and F, the point contact between an MWCNT edge and graphite is formed with sliding.

Here, it should be noted that both the meso- and nanoscale adhesions appear in the vertical force-distance curve during the adhesion. This feature resembles that of the force-distance curve during the peeling, which indicates that a part of the MWCNT comes into sudden line contact with the graphite surface during the approach. It is also revealed in this experiment that as the length of the MWCNT attached to the tip apex increases, the number of discontinuous jumps due to the mesoscale intermittent adhesion and peeling also increases; moreover, the nanoscale intermittent adhesion and peeling appear. This is because the long nanotube behaves like a soft spring, as shown in the simulation (Sasaki et al., 2006, 2008,

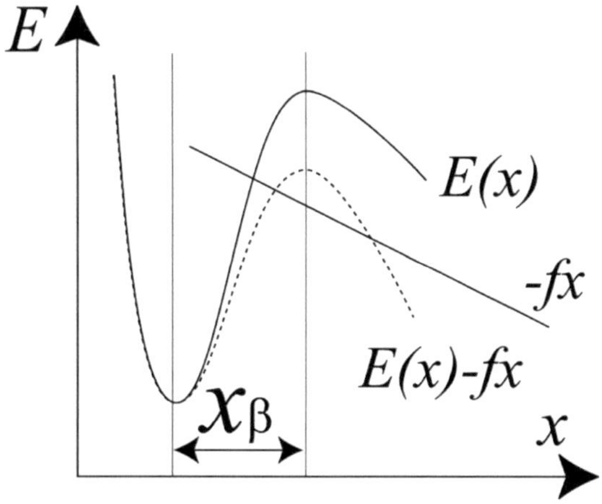

Fig. 4. Conceptual energy landscapes E for bonds are illustrated. The energy barrier height for bonds formed between the MWCNT and substrate is lowered by fx_β with the external force f .

2009a, 2009b). Namely, as the nanotube length increases, the contact time between a nanotube and a substrate during peeling also increases, and thus, a large bending induces a large peeling. Interestingly, it is found that the elastic bending feature of the MWCNT as a nanospring appears in the vertical force-distance curve. Thus, the use of a much softer nanotube is expected to result in a much smaller intermittent adhesion and peeling than those of the nanoscale one.

Furthermore, effect of the peeling velocity on the peeling force is investigated. In the force-distance measurement, the external force f is added to the MWCNT to peel it off the substrate. Thus, the energy barrier height for bonds formed between the MWCNT and substrate is lowered by fx_β with the external force f , where x_β means the effective distance of the potential barrier from the minimum point needed to break the bond, as illustrated in Fig. 4. Then the lifetime for bond is expressed as follows (Evans et al., 1991, Rief et al., 1997),

$$\tau(f) = \tau(0)\exp\left(\frac{-fx_\beta}{k_\beta T}\right), \tag{1}$$

where k_B, T and $\tau(0)$ mean the Boltzmann constant, the temperature and the natural lifetime, respectively. If the external force f is applied with a constant loading rate v_f as a function of time t in the force-distance measurement, then $f(t) = v_f t$. Since the lifetime $\tau(f)$ is the inverse of the rate of dissociation, the probability of the existing bonds, S is given as follows,

$$\frac{dS}{dt} = -\frac{S}{\tau(f)}. \tag{2}$$

$S(t)$ is give as follows using $S(0)=1$,

$$S(t)=\exp\left(-\int_0^t \frac{1}{\tau(f)}dt\right). \tag{3}$$

The probability $P(f)$ that the bond breaks if the external force f acts on it is given by $-dS/df$,

$$P(f)=\frac{\exp\left(-\int_0^t \frac{1}{\tau(f)}dt\right)}{v_f\tau(f)}. \tag{4}$$

If $dP/df=0$, the most probable peeling force f^* is given as follows by Evans et al. (Evans et al., 1991).,

$$f^*=\frac{k_BT}{x_\beta}\ln\left(v_f\right)+\frac{k_BT}{x_\beta}\ln\left(\frac{\tau(0)\cdot x_\beta}{k_BT}\right). \tag{5}$$

Now the peeling velocities are set as 10 to 1000 nm/s. The nanoscale intermittent behaviors for the MWCNTs with lengths of 660nm and 2380 nm, depend strongly on the peeling velocities as shown in Fig. 5(e). Since the peeling is a stochastic process, it is possible that the most probable nanoscale intermittent peeling force, f^*, is given as a function of the loading rate, v_f, obtained from the peak of the histogram of the peeling force f^* specified by the Gaussian distribution (Evans et al., 1991, Rief et al., 1997), which is taken from peak heights from the straight line (see the line S in Fig.3(b)) keeping a constant loading rate. Since it is clearly shown that f^* is linearly related to the logarithm of the loading rate, v_f, the obtained v_f - f^* relation can be fitted to the logarithmic relation of equation (5) obtained by Evans et al. (Evans et al., 1991). x_β and $\tau(0)$ have been estimated to be 0.21 nm and 0.49 sec. from the slope and the *x*-intersection of the fitting line as shown in Fig. 5(e), respectively. The order of magnitude of the position of the potential barrier, x_β, is comparable to that of the atomic-scale distance between the neighboring stable AB-stacking positions within *x-y* plane. Here the AB-stacking registry formed between a graphite flake and a graphite substrate surface is considered (Miura et al., 2003, 2004). This indicates that the peeling of the MWCNT starts when the outermost graphene sheet of the MWCNT goes over the potential barrier of the AB-stacking registry with the graphite substrate surface, as depicted in the lower part of Fig. 5(f). Thus the nanoscale intermittent behaviors occur due to the atomic-scale stochastic sliding of the MWCNT activated thermally within the *x-y* plane on the graphite substrate, which appears in the vertical force curve as shown in Figs.5(b) and 5(c).

However, it was found that the pull-off forces in the vertical force-distance curve in all the MWCNTs depend weakly on the peeling velocities. Figure 5(c) shows x_β = 0.13 nm and $\tau(0)=$ 1.8×10^{10} s for the 400-nm-long MWCNT. Similarly Fig. 5(i) exhibits x_β = 0.52 nm and $\tau(0)=$ 2.2×10^{5} s for the 660-nm-long MWCNT. This indicates that the shorter MWCNT requires a larger force to pull off the graphite substrate than the longer MWCNT.

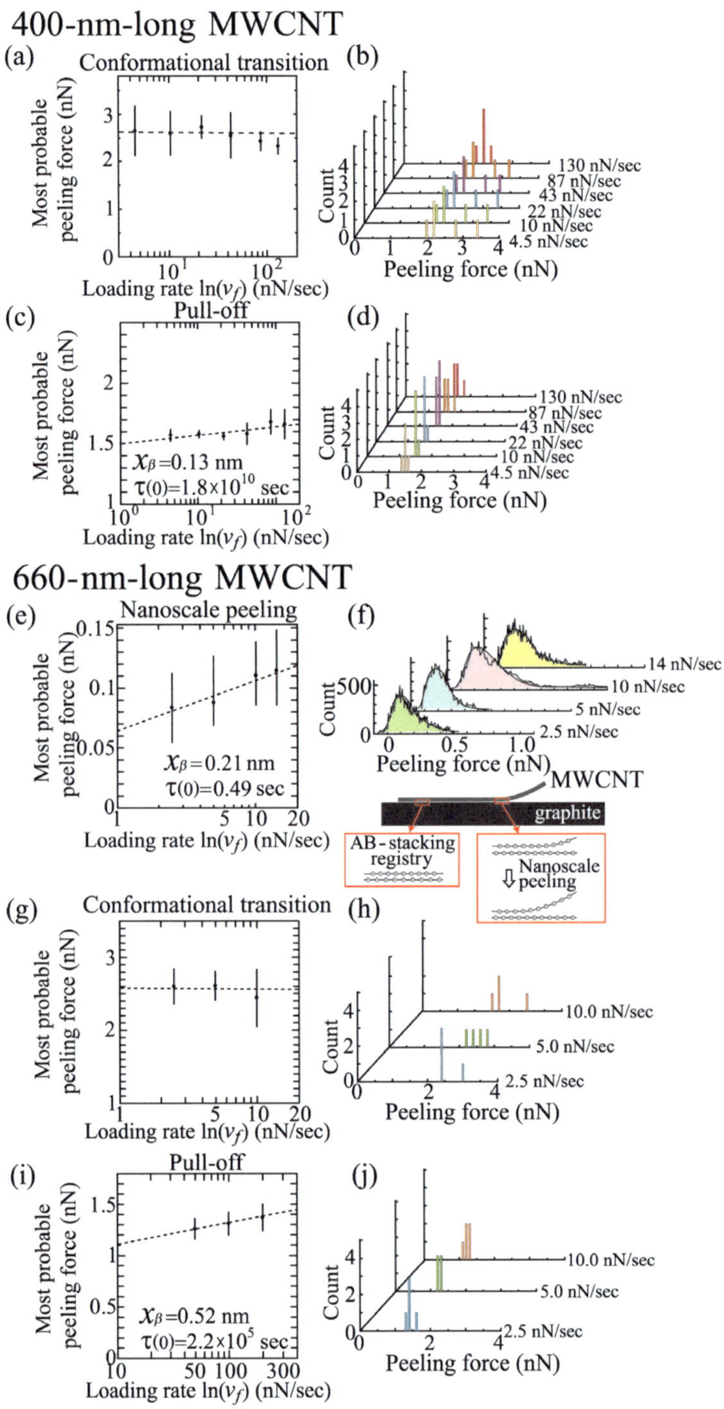

Fig. 5. The histogram of the peeling force, and the most probable peeling force f^* at various force loading rate, v_f for all the processes of the nanoscale peeling, conformational transition, and pull-off for 400-nm-long MWCNT and 660-nm-long MWCNT.

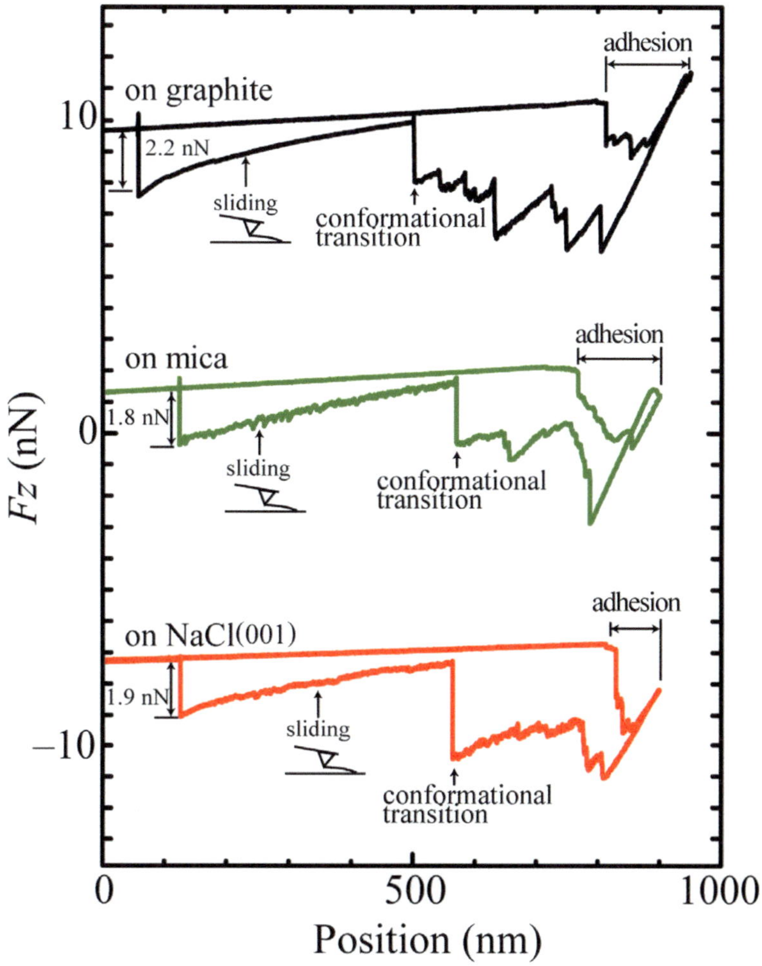

Fig. 6. Vertical force-distance curves measured on graphite, mica and NaCl(001) surfaces.

In either case, it is difficult to pull MWCNTs off the graphite substrate without a lifting force because a natural lifetime $\tau(0)$ is infinitely large. On the other hand, the mesoscale intermittent forces of the conformational transition for 400-nm-long MWCNT, 660-nm-long MWCNT and 2380-nm-long MWCNT, are almost constant (Figs. 5(a) and 5(g)), or slightly decrease with an increase of the loading rate, which indicates that the barrier position x_β of the conformational transition is at least on the order of a MWCNT length and then the slope $k_B T / x_\beta$ in equation (5) nearly goes to zero. Here it should be also noted that the conformational transition occurs within the x-z plane.

Furthermore, to investigate the effect of adsorbates on the adhesion and peeling mechanisms, we have performed experiments on the vertical force-distance curve under ambient conditions. The vertical force-distance curve with the characteristic hysteresis loop obtained under ambient conditions was almost the same as that obtained using an SEM, which indicates that humidity and ambient gases do not strongly affect the main feature of the vertical force-distance curve.

To investigate the effect of substrate on the adhesion and peeling behaviors of an MWCNT, the vertical force-distance curve was measured for graphite, mica and NaCl(001) surfaces, as shown in Fig. 6. First, it was found that the pull-off forces of an MWCNT edge on the substrate corresponding to the final jumps are 2.2, 1.8, and 1.9 nN for graphite, mica and NaCl(001), respectively, which shows that the interaction strength between an MWCNT edge and a substrate is strongest for the graphite surface and comparable for the mica and NaCl(001) surfaces. In the vertical force-distance curve of the graphite surface, the number of discontinuous jumps corresponding to the mesoscale intermittent peeling increases, which is not observed for the mica and NaCl(001) surfaces. This is because, as the interaction strength between the MWCNT and the substrate surface increases, the contact time between them during the peeling becomes longer, and thus, a large bending induces a large peeling. It is thus interesting to note that the features of the vertical force-distance curve in case of the strong interaction between an MWCNT and a substrate resemble those in the case of the soft MWCNT.

4. Conclusions

Here, adhesion and peeling experiments on the MWCNT have been performed. We have experimentally obtained the vertical force-distance curve with the characteristic hysteresis loop, which exhibits multistable states between line contact and point contact of the MWCNT shape during the adhesion and peeling. The line and point contacts are clearly divided by the discrete jump that appeared in the vertical force-distance curve, which shows the nanoscale elastic property of the MWCNT. The adhesion and peeling behaviors of the MWCNT reveal hierarchical structures (or fractal structures) from the nanoscale intermittent to the mesoscale intermittent adhesion and peeling. The nanoscale intermittent behaviors depend strongly on the adhesion and peeling velocities, which reveals that the peeling occurs when a MWCNT goes over the potential barrier of the neighboring AB-stacking registry with the graphite substrate surface. On the other hand, the most probable pull-off forces depend weakly on the peeling velocities, which indicates that it is difficult to pull MWCNTs off the graphite substrate without a finite lifting force. Furthermore the mesoscale intermittent forces of the conformational transition indicate that the barrier position of the conformational transition is at least on the order of a MWCNT length.

This technique will be applicable not only to material science but also to molecular biology because this system makes it possible to analyze the physical properties of a cell, protein molecules and DNA. These results could also provide information on the mechanisms of how to make an adhesion and how a gecko performs, and could be used to propose a guiding principle for designing the artificial superadhesive system beyond a gecko foot.

5. References

Evans, E., Berk, D. & Leung, A., (1991). Detachment of agglutinin-bonded red blood cells. I. Forces to rupture molecular-point attachments. *Biophys. J.*, Vol. 59, No.4 (April, 1991) 838-848, ISSN 1542-0086

Ishikawa, M., Yoshimura, M. & Ueda, K., (2002). A study of friction by carbon nanotube tip. *Appl. Surf. Sci.*, Vol. 188, No. 3-4 (March 2002) 456-459, ISSN 0169-4332.

Ishikawa, M., Harada, R., Sasaki, N. & Miura, K., (2008). Visualization of nanoscale peeling of carbon nanotube on graphite. *Appl. Phys. Lett.*, Vol. 93, No. 8 (August, 2008) 0831221-0831223, ISSN 0003-6951(print), 1077-3118(online)

Ishikawa, M., Harada, R., Sasaki, N. & Miura K., (2009). Adhesion and peeling forces of carbon nanotubes on a substrate. *Phys. Rev. B*, Vol. 80, No. 19 (November 2009) 1934061-1934064, ISSN 1098-0121(print), 1550-235X(online)

Ishikawa, M., (2011). The movie of peeling process of multi-walled carbon nanotube from graphite surface, In: World Wide Web Site, (February, 2011), Available from: http://miuralab.com/box/movie/movie.html

Israelachvili J. N., (January 6, 1992). *Intermolecular and Surface Forces*, Academic Press, ISBN 0123751810, London

Miura, K., Kamiya, S. & Sasaki N., (2003). C_{60} Molecular Bearings. *Phys.Rev.Lett.*, Vol. 90, No. 5 (February, 2003) 0555091-0555094, ISSN 0031-9007(print), 1079-7114 (online)

Miura, K., Sasaki, N. & Kamiya, S., (2004). Friction mechanisms of graphite from a single-atomic tip to a large-area flake tip, *Phys.Rev.B* Vol. 69, No. 7 (February, 2004) 0754201-0754209, ISSN 1098-0121(print), 1550-235X(online)

Qu, L., Dai, L., Stone, M., Xia, Z. & Wang, Z. L., (2008). Carbon Nanotube Arrays with Strong Shear Binding-On and Easy Normal Lifting-Off. *Science*, Vol. 322, No.5899 (October, 2008) 238-242, ISSN 0036-8075(print), 1095-9203(online)

Rief, M., Oesterhelt, F., Heymann, B. & Gaub, H. E, (1997). Single Molecule Force Spectroscopy on Polysaccharides by Atomic Force Microscopy. *Science*, Vol. 275, No. 5304 (28 February 1997) 1295-1297, ISSN 0036-8075(print), 1095-9203 (online)

Sasaki, N., Toyoda, A., Saitoh, H., Itamura, N., Ohyama M. & Miura, K., (2006). Theoretical Simulation of Atomic-Scale Peeling of Single-Walled Carbon Nanotube from Graphite Surface. *e-J. Surf. Sci. Nanotech.*, Vol. 4, (January, 2006) 133-137, ISSN 1348-0391

Sasaki, N., Toyoda, A., Itamura, N. & Miura K., (2008). Simulation of Nanoscale Peeling and Adhesion of Single-Walled Carbon Nanotube on Graphite Surface. *e-J. Surf. Sci. Nanotech.*, Vol. 6, (February, 2008) 72-78, ISSN 1348-0391

Sasaki, N., Saitoh, H., Itamura, N. & Miura, K., (2009a). Analysis of Lateral Orientation of Single-Walled Carbon Nanotube on Graphite. *e-J. Surf. Sci. Nanotech.*, Vol. 7, (January, 2009) 48-52, ISSN 1348-0391.

Sasaki, N., Okamoto, H., Itamura, N. & Miura K., (2009b). Peeling of Graphene Sheet – Simulation Study. *e-J. Surf. Sci. Nanotech.*, Vol. 7, (July, 2009) 783-786, ISSN 1348-0391

5

Evaluation of Histidine Functionalized Multiwalled Carbon Nanotubes for Improvement in the Sensitivity of Cadmium Ions Determination in Flow Analysis

Giovana de Fátima Lima, Fernanda Midori de Oliveira, Maikow de Oliveira Ohara, Mariana Gava Segatelli and César Ricardo Teixeira Tarley
Departamento de Química, Universidade Estadual de Londrina
Brazil

1. Introduction

The determination of heavy metal ions in environmental, clinical and food samples has been a common concern for several researchers and in special case of aquatic environment, there has been a more stringent regulation standards pertaining to the discharge of heavy metals. Some efforts have been carried out to develop many treatment processes to reduce the disposal of heavy metal ions mainly in water bodies. The different treatment processes include precipitation, membrane filtration, ion exchange and adsorption (Tokalıoğlu et al. 2009), however, the levels of heavy metals disposal into water bodies are still rising, mainly in aqueous effluents discharged into aquatic bodies (Mello et al. 2005). As example, lead is considered as the major environmental pollutant and has become the most widely scattered toxic heavy metal in the world as a result of anthropogenic action. It is very well known that lead enters the environment through several industrial activities, such as industrial ammunition, paper board mill, battery manufacturing and coal burning and pigments (Zhan & Zhao, 2003). Lead can cause several effects on central nervous system, blood pressure, kidneys and reproductive system (Needleman, 2004). Similarly to lead, cadmium is also considered as toxic for animals and humans, and it is classified by the International Agency for Research on Cancer as a human carcinogen (IARC, 1993). The *National Council* for the *Environment in Brazil* (CONAMA, 2005) established 0.2 mg L^{-1} as the maximum level of cadmium for effluent discharge in aquatic bodies, while 5.0 and 3.0 μg L^{-1} are the maximum contaminant levels in potable water allowed by EPA(EPA, 2003) and ANVISA (ANVISA, 2004), respectively. Even at low concentration levels (≤ μg L^{-1}), there is bioaccumulation of heavy metals in aquatic organisms, so there is a growing need for their monitoring in different types of water samples. In addition, the heavy metal ions monitoring can provide both important aspects in water quality and controlling the industrial processes (Martinotti et al. 1995). Thus, the development of analytical methods based on preconcentration systems has commonly been the aim of choice for the determination of trace metals in water and biological samples. Moreover, the development of new analytical protocols play an important role in solving instrumental problems commonly associated with low selectivity and sensitivity. Undoubtedly, solid phase preconcentration has been the most technique used for this purpose, owing to its attractive features such as high reproducibly and preconcentration factor, ease of automation and regeneration of solid phase, low cost, simplicity, easy coupling to flow injection analysis (FIA) and high sample throughput (Barbosa et al. 2007).

Basically, preconcentration consists on passing a large quantity of sample through a column containing a solid phase, at which, due to the very low concentration of the analyte in the raw sample, can not be analyzed by simple techniques without this preconcentration step. Moreover, adsorbent extraction can be used for selective retention of some particular chemical forms of a metal, thereby enabling speciation (Campanella et al. 1996, Pyrzyńska et al. 1999). Currently, several solid phase extractors can be found for metal ions preconcentration, such as modified silica (C_{18}), chelating resins, polyurethane foam, activated carbon, fullerenes, modified silica, polymers loaded with chelating reagents and natural adsorbents (Souza & Tarley, 2009). Despite the available solid phase extractors, the development of new solid phase preconcentration systems has still been growing and target of analytical research, where the choice of appropriate extractor is a critical factor to obtain attractive features in preconcentration systems. The new materials developed should present large specific surface area, chemical resistance, high adsorptive capacity, life-time, high clean-up capacity, reversible adsorption and large accessibility of analyte towards the sites of adsorbents. There are numerous attempts on the development of nanomaterials, in which present such characteristics mentioned, but undoubtedly, in the last 20 years, no other nanomaterials have desperted so much attention as carbon nanotubes (CNTs) in the development of solid phase extraction procedures. CNTs can be described as a graphite sheet rolled up into a nanoscale-tube [which are single-wall carbon nanotubes (SWCNTs), or with additional graphene tubes around the core of an SWCNT (which are multi-wall CNTs, MWCNTs]. CNTs have diameters in the range between fractions of nanometers and tens of nanometers and lengths up to several centimeters with both their ends normally capped by fullerene-like structures, having usually a length-to-diameter ratio $\leq 10^5$. The combination of size, structure and topology provides nanotubes with important mechanical and surface properties. However, CNTs are insoluble in organic solvents and inorganic aqueous medium limiting their manipulation in real samples and their walls are not reactive, but their tips can be more reactive. So, several studies have been performed to functionalize the CNTs surface to generate functional groups, such as carbonyl, hydroxyl and carboxylic groups and naturally enhance their dispersion property and adsorption capacity. Studies have demonstrated the excellent performance of these materials as an analytical adsorbent for metal ions, pesticides and phenolic compounds (Liang et al. 2004, Al-Degs et al. 2008, Salam & Burk et al. 2008). Recently, the ability of MWCNTs as a solid adsorbent for the on-line preconcentration and speciation of V^{4+} and V^{5+} coupled to electrothermal atomic absorption spectrometry (ETAAS) has been described (Gil et al. 2007). The application of oxidized MWCNTs with nitric acid for cadmium, lead and cobalt preconcentration with posterior determination by atomic absorption spectrometry has been reported (Barbosa et al. 2007, Tarley et al. 2006, Souza & Tarley, 2009). The pH plays a very important role on extraction of metals ions onto oxidized MWCNTs (Liang et al. 2005). It can be explained by functionalization that has occurred on the surface of the carbon structures producing oxygen-containing groups when submitted to chemical oxidation. This oxidation leads to a reduction of the isoeletric point (IEP) of the MWCNTs to lower pH values. Under these conditions, the sample pH needs to be higher than the IEP of the oxidized MWCNTs (El-Sheikh 2008, Stafiej & Pyrzyńska 2007). The selectivity of raw and oxidized CNTs towards metal ions can be improved by introducing polymers, inorganic oxides and organic compounds onto surface material. The stability, accessibility and selectivity can be achieved through covalent binding between CNTs surface and modifier.

In a recent work, Amais et al. 2007 synthesized a nanocomposite based on alumina (Al_2O_3) supported on MWCNTs to be applied as a sorbent for nickel preconcentration prior to flame atomic absorption spectrometry (FAAS). Significant results mainly on the recovery studies (higher than 95%), stability of the column for more than 200 cycles without loss of adsorption capacity and a high analytical frequency due to the high flow used in the preconcentration step were achieved.

Zang et al. 2009, published an application concerning the use of ethylenediamine-MWCNTs as solid phase extractor for the preconcentration of Cr^{3+}, Fe^{3+} and Pb^{2+} at pH 4.0. A substantial increase on selectivity and sensitivity was observed for these metal ions determination by inductively coupled plasma optical emission spectrometry.

Pacheco et al. 2009 reported the immobilization of L-tyrosine on CNTs as a new adsorbent for cobalt ions preconcentration. The adsorption of other bivalent metal ions (Cu, Ni and Zn) was also investigated. It was observed that bigger cations are more adsorbed on L-tyrosine-CNTs, taking into account the formation of $CuOH^+$ and $NiOH^+$ at pH 9.0.

The preconcentration of Au(II) and Mn(II) in aqueous medium using MWCNTs modified with N,N´-bis(2-hydroxybenzylidene)-2,2´(aminophenylthio)ethane with determination by flame atomic absorption spectrometry was described by Shamspur & Mostafavi, 2009. These metal ions were quantitatively recovered at pH 6.0. The proposed method was useful for Au^{2+} and Mn^{2+} determination with high sensitivity and low limits of detection, 0.01 and 0.03 μg L^{-1}, respectively.

The selectivity of MWCNTs towards Cd^{2+} ions has also been reported by functionalization with L-cysteine (Liu et al. 2008). The preconcentration factor of Cd^{2+} ions by using MWCNTs-cysteine packed into mini-column was found to be 33. Other chemical modifiers can be found in the literature, such as 1-(2-pyridylazo)-2-naphtol (PAN), TiO_2 and di-(2-ethylhexylphosphoricacid)(D2EHPA) and tri-n-octylphosphineoxide(TOPO) (Afzali & Mostafavi 2008, Zhao et al. 2010, Vellaichamy & Palanively, 2011).

The goal of this chapter comprises the modification of MWCNTs with histidine and demonstrates new analytical data obtained in our research group involving the development of a solid phase extraction system for the determination of cadmium ions by FAAS.

2. Materials and methods

2.1 Reagents and solutions

Analytical grade chemical reagents were used throughout and the water used was obtained from a Milli-Q purification system (Millipore, Bedford, MA, USA). Before using, all laboratory glassware were kept overnight in a 10% (v/v) HNO_3 solution, in order to avoid any metal contamination. After that, they were rinsed with deionized water and submitted to a drying step. A standard stock solution of cadmium 1000.0 mg L^{-1} (Merck, Darmstadt, Germany) was used to prepare standard solutions of concentration 30.0 μg L^{-1}. The latter was properly diluted from an intermediate solution of 10.0 mg L^{-1} Cd^{2+}. Concentrated HNO_3

(Merck) and 30% (v/v) H_2O_2 (Merck) were used for the decomposition of the certified reference material in a block digestor. Multi-wall carbon nanotubes (MWCNTs) were supplied by CNT Co., Ltd. (Yeonsu-Gu, Incheon, Korea) with purity > 93%, diameters between 10-40 nm and lengths of 5-20 μm. Histidine (98%) was purchased from Sigma-Aldrich (St. Louis, MO). Buffer solutions including Tris-HCl, ammoniacal, borate and phosphate (Merck) were prepared from their respective salts without further purification.

2.2 Apparatus

All measurements were carried out by using a Shimadzu AA-6601 flame atomic absorption spectrometer (Shimadzu, Tokyo, Japan), equipped with a hollow cathode lamp for cadmium and with a deuterium lamp for background correction. The hollow cathode lamp was operated at 8.0 mA and the wavelength was set at 228.8 nm. The flame composition was operated with an acetylene flow rate of 1.8 L min^{-1} and air flow rate of 10.0 L min^{-1}. An Ismatec Model IPC peristaltic pump (Ismatec IPC-08, Glattzbrugg, Switzerland) furnished with Tygon® tubes was used to propel sample and eluent solutions. The preconcentration/elution steps were selected by using a home-made injector-commutator made of Teflon® (PTFE, polytetrafluoroethylene). The pH of samples were measured by a digital pHmeter 826 pH mobile Metrohm (Herisau, Switzerland). The infrared spectra were obtained by using the conventional KBr pellet technique in a FT-IR Shimadzu 8300 spectrometer, operating in the transmission mode between 4000 and 400 cm^{-1}. The digestion of certified reference material was carried out by using block digestor (Marconi, Piracicaba, Brazil).

2.3 Preparation of MWCNTs functionalized with histidine

Oxidation MWCNTs procedure followed the literature (Tarley et al. 2006). Around 500 mg of MWCNTs were dispersed in 30 ml of concentrated HNO_3, following reflux for 1 h at 120°C. After this step, the sample was successively washed with deionized water in order to eliminate excess acid (pH of washing water was near 7). Finally, the material was dried at 100°C and stored at room temperature. This procedure opens the ends of carbon nanotubes and creates carboxyl group. Next, around 100 mg of oxidized MWCNTs were pulverized and dispersed in 50 ml of 1.3% (w/v) Triton X-100. The suspension was kept in ultrasound batch (40 KHz) during 10 min and stored for 12 h. After this procedure, the suspension was centrifuged (3600 rpm) during 10 min and the material was further dried and stored at room temperature (Liu et al. 2010, Rastogi et al. 2008). The functionalization of MWCNTs was performed in accordance to literature (Pacheco et al. 2009). Around 100 mg of oxidized and dispersed MWCNTs were added to 50 ml of phosphate buffer solution at pH 7.0 and 100 mg of histidine were further added. The suspension was heated at 45 °C in oil batch during 48 h. Finally, the resulting suspension was centrifuged and the fuctionalized MWCNTs was dried at room temperature. A polyethylene mini-column with 6 cm length and 1 cm internal diameter was filled with 15 mg functionalized MWCNTs and glass wool was introduced into the extremities of the mini-column in order to prevent possible losses of adsorbent during system operation.

2.4 On-line preconcentration system

A diagram of the on-line preconcentration system is shown in Figure 1. During the preconcentration step, a volume of 15.0 ml of sample buffered at pH 8.50 with 0.01 mol L^{-1}

ammoniacal buffer percolates through a mini-column filled with the adsorbent material (15 mg), at a flow rate of 4.0 ml min^{-1}. Then, by switching the central part of the injector, the elution step is accomplished at a flow rate of 5.0 ml min^{-1} with 0.8 mol L^{-1} HNO_3, at which Cd^{2+} is released towards FAAS detector. All absorbance signals were taken as peak height.

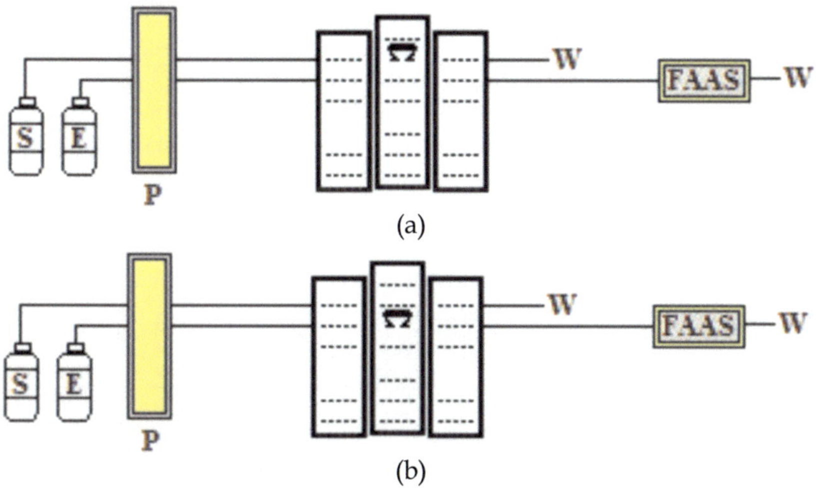

Fig. 1. Diagram of on-line preconcentration system. (a) preconcentration step, (b) elution step. P= peristaltic pump, W= waste, S= sample, E= eluent 0.80 mol L^{-1} HNO_3.

3. Results

3.1 Infrared spectra of raw and histidine functionalized MWCNTs

Fig. 2 shows the infrared spectra of raw and histidine functionalized MWCNTs and histidine functionalized MWCNTs. The band observed at 3445 cm^{-1} corresponds to the

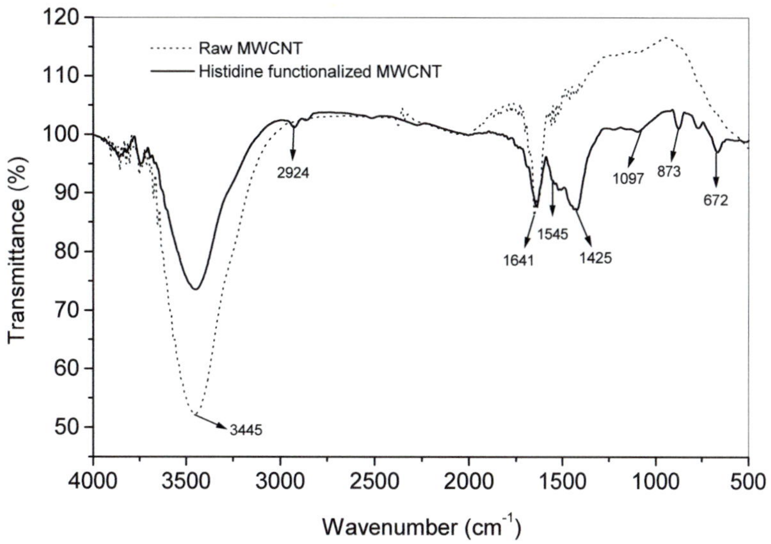

Fig. 2. Infrared spectra of the raw and histidine functionalized MWCNTs.

stretching vibrations of ν(OH), indicating the existence of carboxyl groups on the external surface of the MWCNTs due to oxidation with HNO_3. In addition, the band can be attributed to hydroxyl groups present in water physically adsorbed. The signal at 1097cm^{-1} is assigned to the ν(C-O) stretching vibration (Gaoa, et al. 2009, Vukovic et al. 2009). The band at 2924 cm^{-1} is assigned to asymmetric stretching vibrations of ν(CH_2). The functionalization with histidine can be confirmed from the formation of secondary amide on the MWCNTs at 1545 cm^{-1} (Liu et al. 2008). In addition, the band at 1641 cm^{-1} is attributed to stretching of amide carbonyl (C=O). Another indication of the functionalization is the band at 1425 cm^{-1} corresponding to N-H bending of secondary amide (Pompeo & Resasco, 2002). for The stretching mode of aromatic amine group is observed at 873 cm^{-1} and the stretching mode of C-OH is assigned at 672 cm^{-1} (Shen et al. 2007, Jin et al. 2007).

3.2 Influence of sample pH

The influence of the sample pH was investigated in the range 3.0-9.5 (Figure 3). The concentration of cadmium used in this study was 40.0 μg L^{-1}.

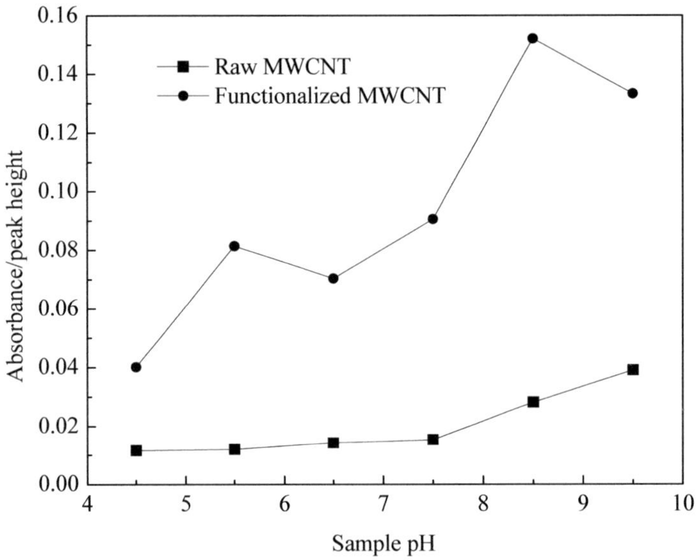

Fig. 3. Influence of sample pH on the adsorption of cadmium ions onto raw and functionalized MWCNTs. Conditions: HNO_3, as eluent, at 0.5 mol L^{-1} concentration and preconcentration flow rate of 4.0 ml min^{-1}.

As observed, the analytical signal increases with increasing of the pH from 3.0 up to 8.5. The very low adsorption of cadmium ions in acid medium is due to protonation of nitrogen atoms of histidine leading to an electrostatic repulsion with cadmium ions. A slight decrease on analytical signal probably occurs above pH 8.5, due to the cadmium precipitation as hydroxides. The pH dependence was also evaluated for raw MWCNTs. As verified, a highly improved adsorption of cadmium ions towards functionalized MWCNTs was observed after comparison with raw MWCNTs, confirming the chemical immobilization of histidine onto MWCNTs surface. Thus, all further works were carried out at pH 8.5.

3.3 Influence of buffer concentration

The sample pH at 8.5 was buffered with ammoniacal buffer. In order to evaluate the effect of sample buffer concentration, the concentration of ammoniacal buffer was studied from 0.01 up to 0.25 mol L^{-1} (Figure 4). Higher responses were clearly noted when low buffer concentrations are employed Such behavior shows that the use of high buffer concentrations, probably by a competitive effect for the adsorbent sites of functionalized MWCNTs, significantly decreases the analytical signal. Despite these findings, the use of low buffer concentrations results in the reduction of the buffering capacity. So, as a compromise between sensitivity and buffering capacity, we have chosen the buffer concentration of 0.05 mol L^{-1}.

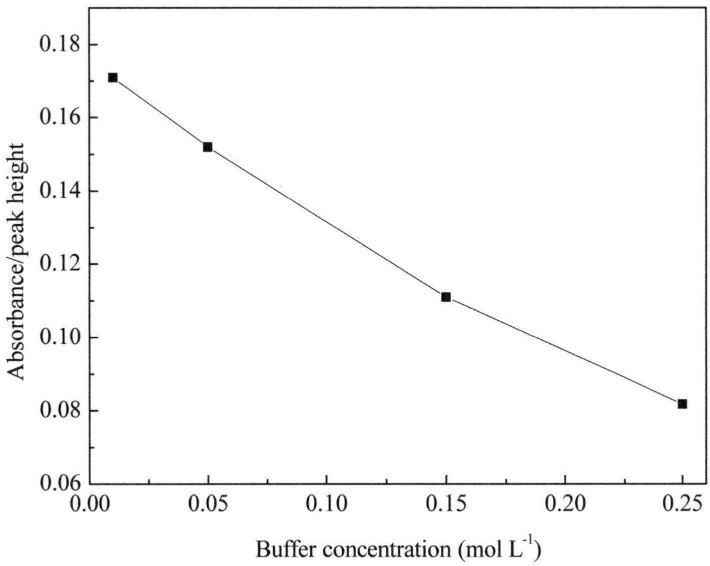

Fig. 4. Influence of buffer concentration on the adsorption of cadmium ions onto functionalized MWCNTs. Conditions: pH 8.5, HNO_3 as eluent at 0.5 mol L^{-1} concentration and preconcentration flow rate of 4.0 ml min^{-1}.

3.4 Influence of eluent concentration

It is well known that eluent concentration plays an important role on the efficiency of solid phase preconcentration systems. Thus, the influence of HNO_3 concentration, as eluent, was investigated in the range 0.5–1.0 mol L^{-1}. An improvement at *ca.* 15% on the analytical response was observed by changing the eluent concentration from 0.5 to 0.8 mol L^{-1}. On the other hand, no significant difference was observed by increasing of eluent concentration from 0.8 up to 1.0 mol L^{-1}. Thus, due to absence of memory effect after each preconcentration/elution step and, in order to preserve the life-time of functionalized MWCNTs, an eluent concentration of 0.8 mol L^{-1} was chosen for subsequent studies.

3.5 Influence of preconcentration flow rate

The preconcentration flow rate in solid phase extraction is an important factor, particularly when high flow rates are used and usually accompanied by decreasing the analytical responses. As expected, the analytical signal decreases with increasing flow rate in range 3.0-5.0 ml min^{-1}. This behavior indicates that the kinetics of mass transfer of cadmium ions towards functionalized MWCNTs surface depends on the flow rate. So, in order to improve the sample throughput of the proposed method, the preconcentration flow rate of 4.0 ml min^{-1} was adopted for further experiments. Under this condition, the required time to preconcentrate 15.0 ml of sample was found to be 3.75 min, which is a usual and satisfactory condition in a flow system.

3.6 Influence of buffer solution

The preconcentration of cadmium ions using different buffers (Tris-HCl, ammoniacal, borate and phosphate) at pH 8.5 and concentration of 0.05 mol L^{-1} was performed. The investigation of influence of the different buffers solutions pointed out to phosphate as the worst buffer. This result was somewhat expected, once even at low cadmium concentrations, phosphate can form insoluble salts with cadmium. Thus, the subsequent experiments were carried out using ammoniacal buffer, in which the signal was slightly higher.

3.7 Interference studies

The preconcentration of cadmium ions was carried out in the presence of co-existing ions in order to assess the feasibility of the proposed method for cadmium monitoring in real and complex samples. Binary solutions containing cadmium ions with Ni^{2+}, U^{4+}, Ca^{2+}, Mn^{2+} or Ba^{2+} in the following proportion (cadmium:co-existing ion) 1:1, 1:10 and 1:100 (w/v) were preconcentrated under optimized conditions. Generally, an error of ± 10%, by comparing the analytical signal of cadmium alone with that obtained by preconcentrating a cadmium solution in the presence of co-existing ions, is tolerable. The results obtained indicated that the method can tolerate the co-existing ions in the following amount: Ni^{2+} (2 mg L^{-1}), U^{4+} (2 mg L^{-1}), Ca^{2+} (4 mg L^{-1}), Mn^{2+} (2 mg L^{-1}) and Ba^{2+} (2 mg L^{-1}). Therefore, the proposed method is not vulnerable to these co-existing ions at high concentrations. Undoubtedly, the analysis of real samples can be susceptive to the interferent ions, however, the determination of cadmium ions by proposed method in water and biological samples is free-interferent, as will be demonstrated later, thus attesting the good selectivity of preconcentration system.

3.8 Breakthrough curve

The maximum adsorption capacity of functionalized MWCNTs towards cadmium was determined from breakthrough experiment. The assay was performed using a mini-column packed with 15 mg of functionalized MWCNTs, where a solution of 0.25 mg L^{-1} Cd^{2+} ions at pH 8.5 in 0.01 mol L^{-1} ammoniacal buffer was continuously percolated at 4.0 ml min^{-1} flow rate. Aliquots of 5 ml of the mini-column effluent were further collected and the amount of Cd^{2+} ions was determined by FAAS. A graph of C/C_o as a function of effluent volume (ml) was built, as shown in Figure 5. The dynamic adsorption capacity of the material was found to be 2.34 mg g^{-1} after loading through mini-column with 250 ml of sample solution.

3.9 Characteristics of the analytical method, application and validation

The analytical characteristic data under optimized conditions were considered very satisfactory taking into account the application of the proposed method for the

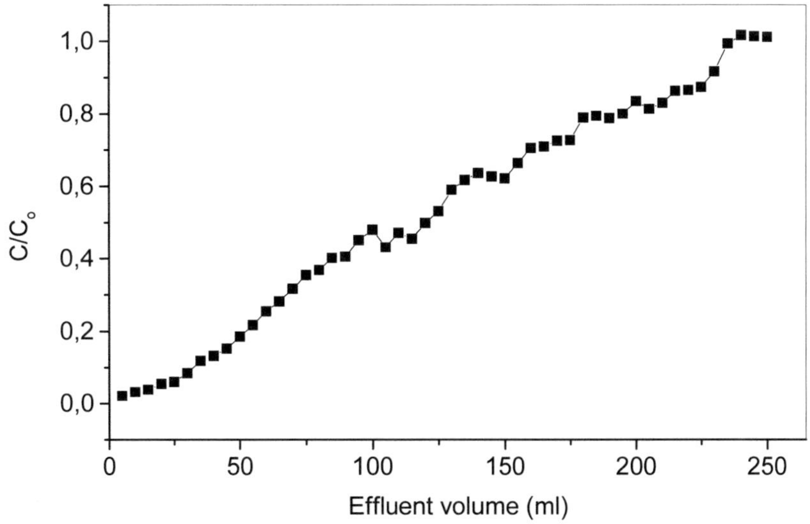

Fig. 5. Breakthrought curve of functionalized MWCNTs using 0.25 mg L^{-1} Cd^{2+} ions at a flow rate of 4.0 ml min^{-1}. C = cadmium concentration in the column effluent and C_o = initial cadmium concentration (0.25 mg L^{-1}).

determination of cadmium in water and biological samples. The analytical curve presented a linear range from 2.0 up to 140 µg L^{-1}. The repeatability of the method, expressed as the relative standard deviation, for the determination of 40 µg L^{-1} Cd^{2+}, was 3.11% (n =10). By preconcentrating 15 ml of sample, a preconcentration factor of 17.7 fold, defined as the ratio of slopes of linear section of the analytical curves before and after the preconcentration, was obtained. So, low limits of detection and quantification, assessed as the preconcentration of analyte giving an equivalent signal to three and ten times the S.D. of the blank signal plus the net blank intensity (Long and Winefordner 1983), were found to be 0.20 and 0.63 µg L^{-1}, respectively. The sample throughput, consumptive index and concentration efficient were 16 samples per hour, 0.85 ml and 4.70 min^{-1}, respectively. The favorable characteristics observed by the proposed preconcentration method allowed its application for the determination of cadmium in different types of water samples and biological materials using external calibration. Mineral and tap waters were spiked with 30 µg L^{-1} Cd^{2+} ions and, as observed from Table 1, recoveries ranging from 95.8 to 104.1% indicate that Cd^{2+} ions are preconcentrated without matrix effect. Igapó lake water was collected in polypropylene bottles from a lake located at Londrina city. The sample was filtered through 0.45 µm cellulose acetate membranes under vacuum and analyzed as soon as possible after pH adjusting with 0.05 mol L^{-1} ammoniacal buffer solution. Certified reference material weighting 150 mg [Tort-2 (Lobster Hepatopancreas)] was decomposed in block digestor with 15.0 ml of concentrated HNO_3 and 5.0 ml 30% (v/v) of H_2O_2 during 4 h. After digestion, the sample was heated on a hot plate to near dryness and then cooled at room temperature with further buffering at pH 8.5, using ammoniacal buffer. As shown in Table 2, the amount of cadmium determined in this certified reference material (26.7 ± 0.6 mg Kg^{-1}), employing external calibration, is in good agreement with certified value (27.7 ± 0.7 mg Kg^{-1}) with confidence interval of 95% (test t-student).

Samples	Cadmium concentration added ($\mu g\ L^{-1}$)	Cadmium concentration found ($\mu g\ L^{-1}$)	Recovery (%)
Mineral water	0	ND	-
	30	31.22 ± 0.40	104.1
Tap water	0	ND	-
	30	29.01 ± 0.26	96.7
Igapó lake	0	ND	-
	30	28.73 ± 0.52	95.8

The results are expressed as mean ± standard deviation. (n = 3)

Table 1. Determination of cadmium ions in spiked water samples with 30 $\mu g\ L^{-1}$

Certified reference material	Certified value (mg Kg^{-1})	Found value (mg Kg^{-1})
Tort 2 (Lobster Hepatopancreas)	26.7 ± 0.6	27.7 ± 0.7

The results are expressed as mean ± standard deviation. (n = 3)

Table 2. Determination of cadmium in certified reference material

4. Conclusions

The present work demonstrates the functionalization of MWCNTs with histidine as a simple, rapid and reliable approach for improving the selectivity and adsorption capacity of MWCNTs towards Cd^{2+} ions. Moreover, the use of histidine as substrate associated with the inherent features of MWCNTs avoids the incompatibility of problems between MWCNTs and aqueous medium in flow injection analysis. The developed method provides satisfactory figures of merit in comparison with previously published methods. The successful application of method was attested by analysis of water samples of different sources as well as certified reference material with high power of detection. The efficient functionalization with histidine was confirmed from 100 preconcentration/elution steps without losses of adsorption capacity towards Cd^{2+} ions.

5. Acknowledgment

The authors would like to acknowledge the assistance of CNPq (Conselho Nacional de Desenvolvimento Científico e Tecnológico), FAPEMIG (Fundação de Amparo à Pesquisa do Estado de Minas Gerais), CAPES (Coordenação de Aperfeiçoamento de Pessoal de Nível

Superior) and INCT of Bioanalytical (Instituto Nacional de Ciência e Tecnologia de Bioanalítica) for the financial support and fellowships.

6. References

Afzali, D. & Mostafavi, A. (2008). Potential of Modified Multiwalled Carbon Nanotubes with 1-(2-Pyridylazo)-2-naphtol as a New Solid Sorbent for the Preconcentration of Trace Amounts of Cobalt(II) Ion, *Analytical Sciences*, Vol. 24 (September 2008), pp. 1135-1139, ISSN 0910-6340.

Al-Degs, Y.S. & Al-Ghouti, M.A. (2008). Preconcentration and determination of high leachable pesticides residues in water using solid-phase extraction coupled with high-performance liquid chromatography, *International Journal of Environmental Analytical Chemistry*, Vol. 88 (June 2008), pp. 487-498, ISSN 0306-7319.

Amais, R.S.; Ribeiro, J.S.; Segatelli, M.G.; Yoshida, I.V.P.; Luccas, P.O. & Tarley, C.R.T. (2007). Assessment of nanocomposite alumina supported on multi-wall carbon nanotubes as sorbent for on-line nickel preconcentration in water samples, *Separation and Purification Technology*, Vol. 58 (December 2007), pp. 122-128, ISSN 1383-5866.

ANVISA (2004). The Brazilian National Agency for Sanitary Vigilance. Public consultation n° 89, December 13.

Barbosa, A.F.; Segatelli, M.G.; Pereira, A.C.; Santos, S.A.; Kubota, L. T.; Luccas, P. O. & Tarley, C.R.T. (2007). Solid-phase Extraction System for Pb (II) Ions Enrichment Based on Multiwall Carbon Nanotubes Coupled On-line to Flame Atomic Absorption Spectrometry, *Talanta*, Vol. 71, (August 2007), pp. 1512–1519, ISSN 0039-9140.

Campanella, L.; Pyrzyńska, K. & Trojanowicz, M. (1996). Chemical speciation by flow-injection analysis. A review, *Talanta*, Vol. 43, No. 6, (December 1996), pp.825-838, ISSN 0039-9140.

CONAMA (2005). National Council for the Environment. Resolution number 357, March 17.

El-Sheikh, A.H. (2008). Effect of oxidation of activated carbon on its enrichment efficiency of metal ions: Comparison with oxidized and non-oxidized multi-walled carbon nanotubes, *Talanta*, Vol. 75 (December 2008), pp. 127-134, ISSN 0039-9140.

Gaoa, Z.; Bandoszc, T.J.; Zhao, Z.; Hand, M. & Qiu, J. (2009). Investigation of factors affecting adsorption of transition metals on oxidized carbon nanotubes, *Journal of Hazardous Materials*, Vol. 167 (January 2009), 34 pp. 357–365, ISSN 030483894.

Gil, R.A.; Goyanes, S.N.; Polla, G.; Smichowski, P.; Olsina, R.A. & Martinez, L.D. (2007). Application of multi-walled carbon nanotubes as substrate for the on-line preconcentration, speciation and determination of vanadium by ETAAS, *Journal of Analytical Atomic Spectrometry*, Vol. 22 (June 2007), pp. 1290-1295, ISSN 0267-9477.

International Agency for Research on Cancer (IARC). (1993). Proceedings of the Meeting of the IARC Working Group on Beryllium, Cadmium, Mercury and Exposures in the Glass Manufacturing Industry, In: *Scandinavian Journal of Work, Environment & Health, Vol. 58, (February 1993), pp. 229-235, ISBN 928321-2584.*

Jin, S.H.; Park, Y. & Yoon. K.H. (2007). Rheological and mechanical properties of surface modified multi-walled carbon nanotube-filled PET composite, *Composites Science and Technology*, Vol. 67 (March 2007), pp. 3434–3441, ISSN 0266-3538.

Liang, P.; Liu, Y.; Guo, L.; Zeng, J. & Lu, H. (2004). Multiwalled carbon nanotubes as solid-phase extraction adsorbent for the preconcentration of trace metal ions and their determination by inductively coupled plasma atomic emission spectrometry, *Journal of Analytical Atomic Spectrometry*, Vol. 19 (October 2004), pp. 1489-1492, ISSN 0267-9477.

Liu, Y.; Li, Y. & Yan, X-P. (2008). Preparation, Characterization, and Application of L-Cysteine Functionalized Multiwalled Carbon Nanotubes as a Selective Sorbent for Separation and Preconcentration of HeavyMetals, *Advanced Functional Materials*, Vol. 18 (May 2008), pp. 1536–1543, ISSN 1616-301X.

Liu, Y.; Yu, L.; Zhang, S.; Yuan, J.; Shi, L. & Zheng, L. (2010). Dispersion of multiwalled carbon nanotubes by ionic liquid-type Gemini imidazolium surfactants in aqueous solution, *Colloids and Surfaces A: Physicochemical and Engineering Aspects*, Vol. 359 (February 2010), pp. 66–70, ISSN 0927-7757.

Long, G.L.; Wineforder, J.D. (1983). Limit of detection. A closer look at the IUPAC definition, *Analytical Chemistry*, Vol. *55* (June 1983), pp. *712A*-724A, ISSN 0003-2700.

Martinotti, W.; Queirazza, G.; Guarinoni, A. & Mori, G. (1995). In-flow Speciation of Copper, Zinc, Lead and Cadmium in Fresh Waters by Square Wave Anodic Stripping Voltammetry Part II. Optimization of Measurement Step, *Analytica Chimica Acta*, Vol. 305, No. (1-3), (December 1995), pp. 183-191, ISSN 0003-2670.

Mello, L.C.; Claudino, A.; Rizzatti, I.; Bortoluzzi, R.L. & Zanetteb, D.R. (2005). Analysis of Trace Metals Cu^{2+}, Pb^{2+} and Zn^{2+} in Coastal Marine Water Samples from Florianópolis, Santa Catarina State, Brazil, *Journal of the Brazilian Chemical Society*, Vol. 16, No.3, (May 2005), pp. 308-315, ISSN 0103-5053.

Needleman, H. (2004). Lead poisoning, *Annual Review of Medicine*, Vol. 55, (February 2004), pp. 209–222, ISSN 0066-4219.

Pacheco, P.H.; Smichowski, P.; Polla, G. & Martinez, L.D. (2009). Solid phase extraction of Co ions using L-tyrosine immobilized on multiwall carbon nanotubes, *Talanta*, Vol. 79 (April 2009), pp. 249–253, ISSN 0039-9140.

Pompeo, F. & Resasco, D.E. (2002). Water Solubilization of Single-Walled Carbon Nanotubes by Functionalization with Glucosamine, *Nano Letters*, Vol. 2 (January 2002), No. 4, pp. 369-373, ISSN 1530-6984.

Pyrzyńska, K. & Trojanowicz, M. (1999). Functionalized cellulose sorbents for on-line preconcentration of trace metals for environmental analysis, *Critical Reviews in Analytical Chemistry*, Vol. 29, No.4, (December 1999), pp. 313-321, ISSN 1040-8347.

Rastogi, R.; Kaushal, R.; Tripathi, S.K.; Sharma, A.L.; Kaur, I. & Bharadwaj, L.M. (2008). Comparative study of carbon nanotube dispersion using surfactants, *Journal of Colloid and Interface Science*, Vol. 328 (October 2008), pp. 421–428, ISSN 0021-9797.

Salam, M.A. & Burk, R. (2008). Novel application of modified multiwalled carbon nanotubes as a solid phase extraction adsorbent for the determination of polyhalogenated organic pollutants in aqueous solution, *Analytical and Bioanalytical Chemistry*, Vol. 390 (April 2008), pp. 2159-2170, ISSN 1618-2642.

Shamspur, T. & Mostafavi, A. (2009). Application of modified multiwalled carbon nanotubes as a sorbent for simultaneous separation and preconcentration trace amounts of Au(III) and Mn(II), *Journal of Hazardous Materials*, Vol. 168 (March 2009), pp. 1548–1553, ISSN 0304-3894.

Shen, J.; Huanga, W.; Wu, L.; Hu, Y. & Ye, M. (2007). Study on amino-functionalized multiwalled carbon nanotubes, *Materials Science and Engineering A*, Vol. 464 (February 2007), pp. 151–156, ISSN 0921-5093.

Souza, J.M.O. & Tarley, C.R.T. (2009). Sorbent separation and enrichment method for cobalt ions determination by graphite furnace atomic absorption spectrometry in water and urine samples using multiwall carbon nanotubes, *International Journal of Environmental Analytical Chemistry*, Vol. 89 (January 2009), pp. 489-502, ISSN 0306-7319.

Stafiej, A., Pyrzyńska, K. (2008). Solid phase extraction of metal ions using carbon nanotubes, *Microchemical Journal*, Vol. 89 (June 2008), pp. 29-33, ISSN 1095-9149.

Tarley, C.R.T.; Barbosa, A.F.; Segatelli, M.G.; Figueiredo, E.C. & Luccas, P.O. (2006). Highly improved sensitivity of TS-FF-AAS for Cd(II) determination at ng L^{-1} levels using a simple flow injection minicolumn preconcentration system with multiwall carbon nanotubes, *Journal of Analytical Atomic Spectrometry*, Vol. 21 (October 2006), pp. 1305-1313, ISSN 0267-9477.

Tokalıoğlu, S.; Vedat, Y.; Şenol, K.; Ali, D.; & Cengiz, S. (2009). Synthesis of a Novel Chelating Resin and its Use for Selective Separation and Preconcentration of Some Trace Metals in Water Samples, *Journal of Hazardous Materials*, Vol. 169, No.(1-3), (September 2009), pp. 593-598, ISSN 03043-894.

US EPA (2003). National Primary Drinking Water Standards, Maximum Contaminant Level, United States Environmental Protection Agency Office of Water.

Vellaichamy, S. & Palanivelu, K. (2011). Preconcentration and separation of copper, nickel and zinc in aqueous samples by flame atomic absorption spectrometry after column solid-phase extraction onto MWCNTs impregnated with D2EHPA-TOPO mixture, *Journal of Hazardous Materials*, Vol. 185 (October 2011), pp. 1131–1139, ISSN 0304-3894.

Vukovic, G.; Marinkovic, A.; Obradovic, M.; Radmilovic, V.; Colic, M.; Aleksic, R. & Uskokovic, P.S. (2009). Synthesis, characterization and cytotoxicity of surface amino-functionalized water-dispersible multi-walled carbon nanotubes, *Applied Surface Science*, Vol. 255 (May 2009), pp. 8067–8075, ISSN 0169-4332.

Zang, Z.; Hu, Z.; Li, Z.; He, Q. & Chang, X. (2009). Synthesis, characterization and application of ethylenediamine-modified Multiwalled carbon nanotubes for selective solid-phase extraction and preconcentration of metal ions, *Journal of Hazardous Materials*, Vol. 172 (July 2009), pp. 958–963, ISSN 0304-3894.

Zhan, X.M. & Zhao, X. (2003). Mechanism of Lead Adsorption from Aqueous Solutions Using an Adsorbent Synthesized from Natural Condensed tannin, *Water Research*, Vol. 37, No. 16, (September 2003), pp. 3905-3912, ISSN 0043-1354.

Zhao, X.W.; Jia, Q.; Song, N.Z.; Zhou, W.H. & Li, Y.S. (2010). Adsorption of Pb(II) from an Aqueous Solution by Titanium Dioxide/Carbon Nanotube Nanocomposites: Kinetics,Thermodynamics, and Isotherms, *Journal of Chemical and Engineering Data*, Vol. 55 (October 2010), pp. 4428–4433, ISSN 0021-9568.

6

Research and Application of CNT Composite Electroplating

Guifu Ding, Yan Wang, Min Deng, Xuemei Cui,
Huiqing Wu and Lida Zhu
National Key Laboratory of Nano/Micro Fabrication Technology
Key laboratory for thin film and microfabrication of Ministry of Education
Institute of Micro and Nano Science and Technology
Shanghai Jiao Tong University
P. R. China

1. Introduction

Carbon nanotubes (CNTs)[1] are one of the most exciting new materials, which possess excellent mechanical properties as well as thermal and electrical characteristics. The Young's modules of CNTs have an order of 1TPa [2, 3], which is much higher than that of the glass fibers (70GP) and most of the other material. Direct measurement on individual nanotubes using atomic force microscopy shows that CNTs can accommodate extreme deformation without fracture [4]. Besides, CNTs are excellent conductors and possess high thermal conductivity.

Many of the remarkable properties of CNTs are established well now [5-8], and their exploitation in a wide range of applications forms a major part of current research and development efforts[9,10]. One of the challenges is to tackle the problem of manipulating CNTs, individually or collectively, to produce a particular arrangement needed for a given application. Moreover, if CNTs are be combined with metal materials to form composites, it is generally important to develop processing methods that disperses the CNTs homogeneously in the metallic matrix

However, it is difficult to make CNTs dispersed uniformly in the metal-matrix. According to Dujardin's report [11], the pure metals would not easily wet the surface of CNTs for their surface tension out of a cut-off limit between 100 and 200mN/m. Therefore, traditional techniques such as powder metallurgic process would bring formidable technical hurdles to prepare high quality metal matrix CNTs composites, which ascribed to the low strength interfacial adhesion [12, 13].

CNT composite electroplating (CCE) is one of the most important techniques for preparing CNT composite films, because no melting and solidification of metal matrix are induced.

The interest in the CCE is driven not only by its applicability to a great variety of metal materials but also by its simplicity. CCE is a cost-effective method usually requiring simple equipment as well as being amenable to scaling-up to large dimensions. In addition, the CCE can be integrated with the micro-electro-mechanical systems (MEMS) [14]. The metal matrix CNTs composite films can be processed to be various micro-scale structures for MEMS devices such as field emission cathodes and electronic contacts [15].

In this study, we consider the following contents including preparation of metal/CNT composite using CCE (Section 2), Fabrication of Ni-matrix CNT field emitters using CCE (Section3), Cu/ CNT contact material prepared by CCE (Section 4) and CNT field emission micro-cathode arrays fabrication (Section 5). Finally, the potential applications of the resulting CNT structures and the scope for future work are highlighted in Sections6 and 7.

2. Preparation of metal/CNT composite using CCE

The most critical problem in preparation of excellent metal/CNT composite film is achievement of a homogenous stable dispersion of CNTs in the plating bath [16].
The influences of cathode-current density plating PH value CNTs concentration in plating bath and agitation mode on the dispersion of CNTs in the metal matrix are discussed in this chapter.
As-produced CNTs are intrinsically inert, often aggregated or entangled, and may contain impurities such as amorphous carbon or catalytic metal particles [17-19]. The amorphous carbon or catalytic metal and all other impurity particles coated on the CNTs will influence the dispersion interface bonding and other electrical, mechanical properties of CNTs in the composite.
Fig.1 shows the schematic diagram of CCE.In the CCE process, adequate CNTs must be conveyed to the cathode surface and embedded in metals through diffusion layer in order to obtain composite coating with moderate CNT content. In the meantime, if CNTs are embedded in the form of aggregation, which will hamper the complement of metal cationic and metal deposition process because of the small clearances formed among CNTs. These will also cause the organization osteoporosis of coating structure and deteriorate the performances. Therefore, adequate amounts of CNTs must be added in the plating for good dispersion degree.

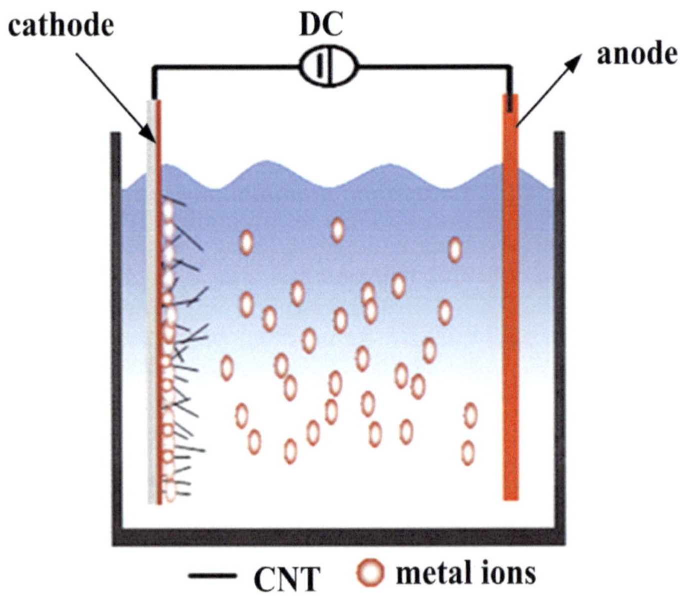

Fig. 1. Schematic diagram of CCE.

In this study, it mainly utilize the mechanical ball mill oxidation processing adding additive and ultrasonic agitation processing to improve suspension stability of CNTs in the electroplating bath.

Long CNTs are more easily aggregated or entangled than shorter CNTs. Mechanical ball mill are often utilized to cut and disperse aggregated CNTs. Fig.2 shows the SEM images of the CNTs composite coatings with the CNTs before (Fig.2A) and after (Fig.2B) ball milling treatment . As can be seen from the images, the mechanical ball milling can effectively improve the dispersibility of the CNTs in the matrix.

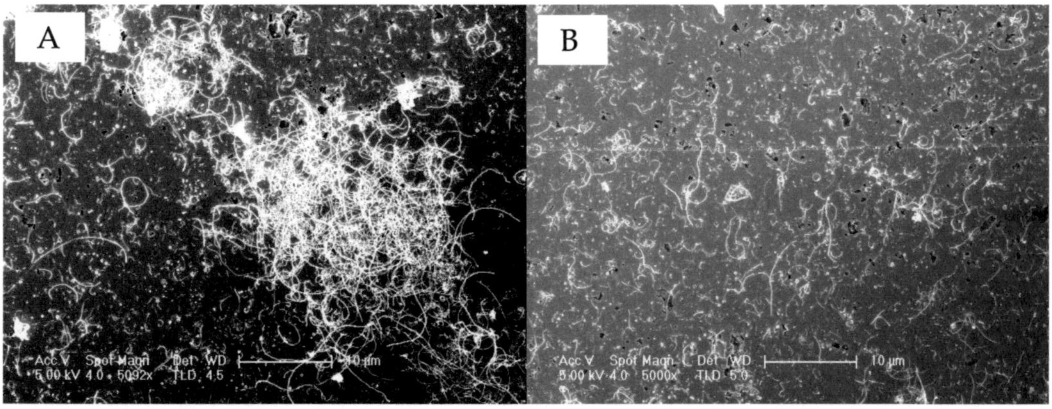

Fig. 2. SEM image of the CNTs composite coatings with the CNTs before (Fig.2A) and after (Fig.2B) ball milling treatment

The purification processes that have been investigated usually utilise the differences in the aspect ratio [20] and oxidation rate between CNTs and impurities [21, 22]. A typical strategy is to use thermal oxidation, then acid reflux in concentrated HCl, and finally a thermal annealing treatment to purify the raw material; a surfactant can subsequently be used to disperse CNTs using sonication [23]. In this study, the CNTs were used after being boiled in potassium hydroxide molten for 5 h and concentrated sulfuric acid for 21h. The CNTs were oxidated in concentrated acid in purification, which reduces the aggregation of the CNTs and enhances their chemical stability in solution. The chemical processing after the CNTs surface defective parts are grafted the hydroxyl and carboxyl group. These ion dissociation increases the negative charge of the CNT surface and strengthens the stability of CNTs suspension.

Fig.3 shows the images of the CNTs before and after being treated by concentrated sulfuric acid. As can be seen from the image, Severe aggregation and tangling exist in the original CNTs (Fig.3A,B) compared with the treated CNTs (Fig.3C,D). In order to validate the existence of the functional groups on the CNTs after being handled by concentrated sulfuric acid, we tried to graft Fe_3O_4 particles on the CTNs by the covalent bond linked. In high resolution electron microscopy, we can see that Fe_3O_4 particles are successfully grafted and uniform disperse on the CNTs (Fig3.D). It is confirmed that there are abundant covalent bond functional groups on the pretreated CNTs.

Besides the chemical pretreatment, adding additive such as ethanol[24], Polyacrylic acid(PA5000), n-pentanol [25], Dodecyl sulfonic acid sodium, tetrahydrofuran (THF) [26-28], dimethly formamide (DMF) [29,30] and deionised water with pyrrole [46,47] is an

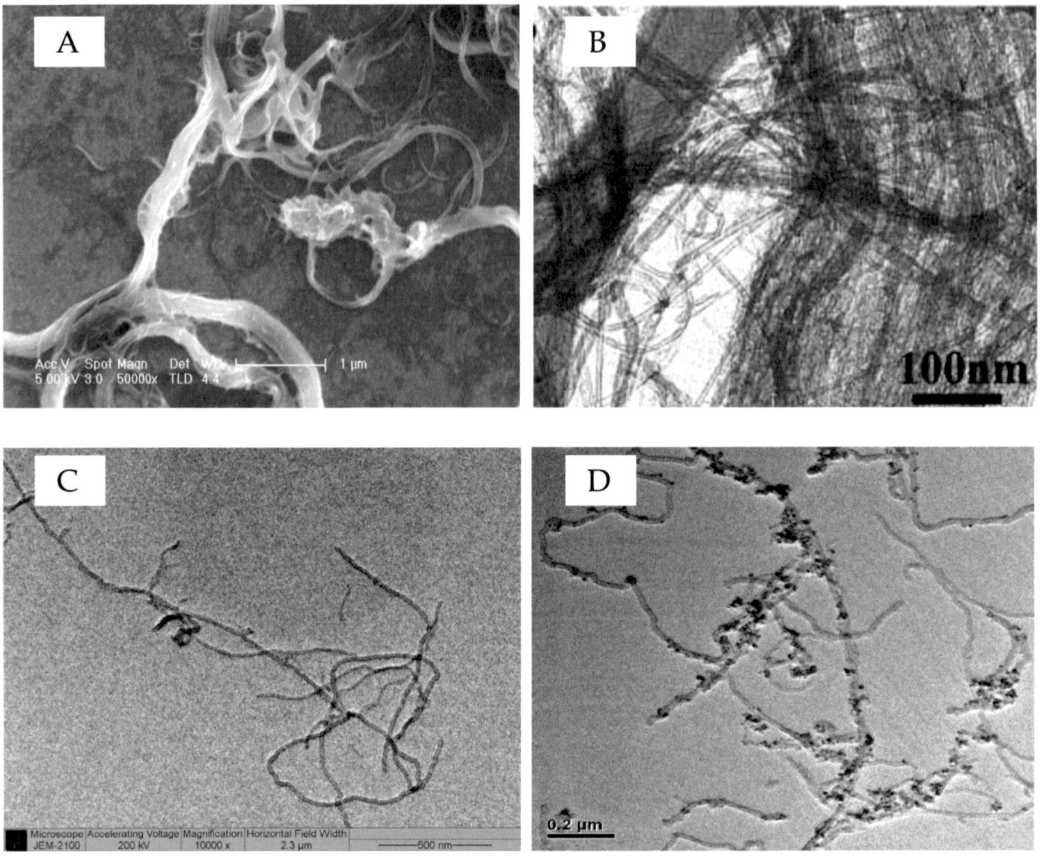

Fig. 3. The SEM (A) and TEM (B) images of the original CNTs and the SEM images of the CNTs after treatment of purification (C) and grafting of Fe_3O_4 (D)

another effective way to raise the CNT dispersion degree in plating bath. In this study, As a conventional bright additive, the molecular weight of 5,000 polyacrylic acid (PA) was added to disperse the CNTs. Though polycyclic acid is a high-performance polymer surfactant, dispersion effect of this low molecular surfactant is small because of its simple molecular structure and small molecular weight. Therefore, in order to achieve the ideal dispersion, increasing the dosage of the low molecular surfactant would be an effective way. The dispersiveability polymer surfactant is better than the low molecular surfactant due to long high molecular chain big molecular weightmore kiss water-based and dredging water-based.

The dispersity of the CNTs in water-based solution is determined using the HPPS nanometer particle size analyzer. Figure 4 shows that the untreated CNTs disperse poorly in the water-based solution. The sizes of the reunion things are in the range from 990nm to 5800nm. After being treated, the sizes of the reunion things fell to the range from 360nm to 1750nm. The dispersity of the CNTs in the water-based solution was impoved.

In addition, ultrasonic agitation processing was utilized in the CCE progress to improve the dispersity of the CNTs in the composite film. Figure 5 shows the growth processes of the metal/CNT composite films prepared by CCE with ultrasonic agitation. The growth mechanism of the metal/CNT film will be changed when the ultrasonic agitation is applied.

Results

Z-Average size (nm):	**895.6**	Peak 1 mean:	990.11	% by Volume:	58.4
Polydispersity index:	**0.461**	Peak 2 mean:	5885.33	% by Volume:	41.6
		Peak 3 mean:	-	% by Volume:	

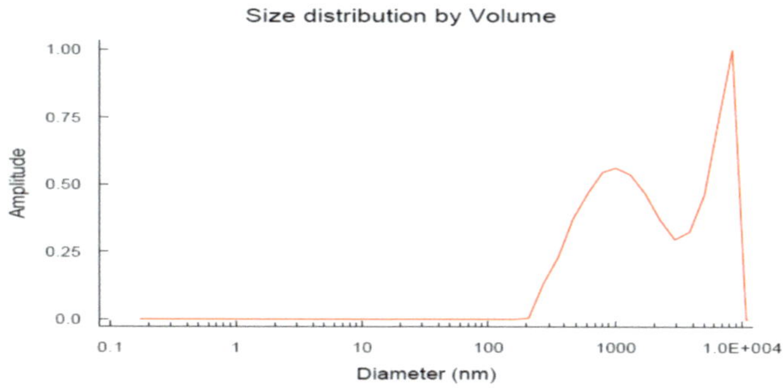

Results

			Diam. (nm)	% Intensity	Width (nm)
Z-Average (d.nm):	617	Peak 1:	1750	61.1	997
PdI:	0.448	Peak 2:	364	38.9	135
Intercept:	0.479	Peak 3:	0.00	0.0	0.00

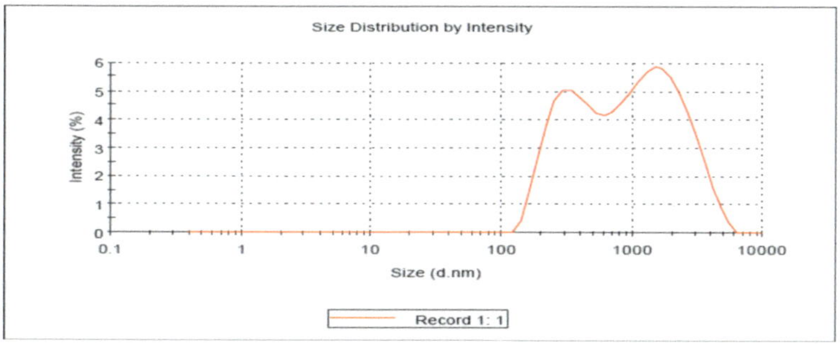

Results

			Diam. (nm)	% Intensity	Width (nm)
Z-Average (d.nm):	242	Peak 1:	273	94.8	115
PdI:	0.285	Peak 2:	4690	5.2	784
Intercept:	0.843	Peak 3:	0.00	0.0	0.00

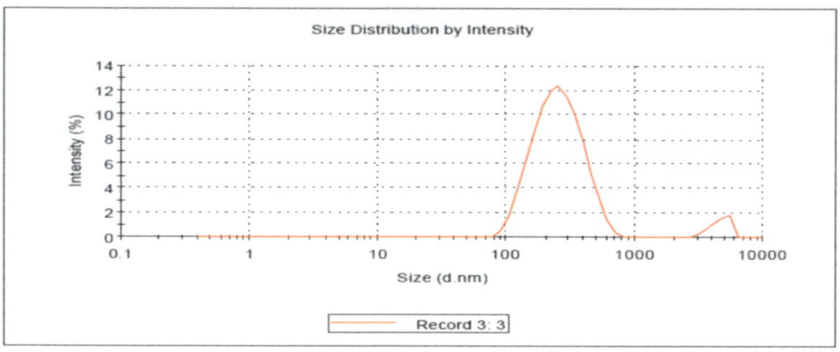

Fig. 4. The granularity dispersion curves of the CNTs in the water-based solution, (a) original CNT; (b) CNT after being treated; (b) after being added bright additive.

Initially, because the CNT aggregates with bigger size will suffer larger impulse force from the electroplate liquid; the CNT aggregates with bigger size are more difficult to be absorbed on the cathode compared with the single CNTs. When the CNT aggregates absorbed on the cathode are washed off, the single CNTs which combine firmly with the substrate would be left on the surface of the composite film. As illustrated in Fig.5(b), the diffusion layer is compressed under the effect of ultrasonic agitation, the agitation further prevent the aggregates from being captured by the cathode and embedded into the composite coating. Moreover, the relatively long single CNTs which metal grains deposit on them may also be pulled out from the metal matrix because of their bigger size and heavier weight compared with other CNTs. For the same reason, the bigger metal grains will be pulled off from the surface of the CNTs too when the combination between the metal grains and the CNT are not enough firm. Finally, the single CNTs are embedded into the metal matrix with the deposit of metal on the cathode and a compact and smooth metal/CNT composite is obtained as a result.

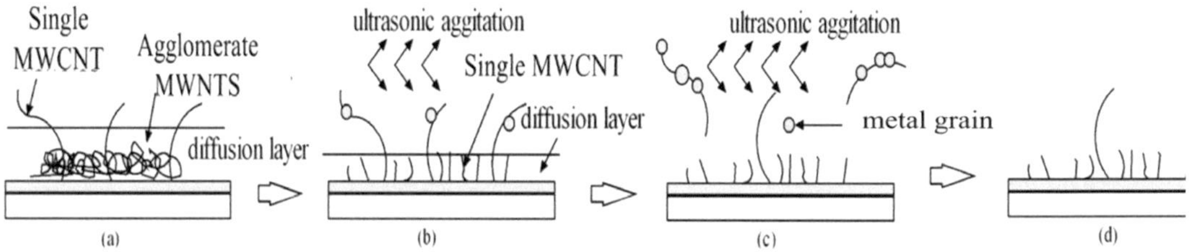

Fig. 5. The growth process of metal/CNT composite films prepared by CCE with ultrasonic agitation. (a) CNT aggregates are absorbed on the substrate. (b) Desorption of the CNT aggregates under ultrasonic agitation. (c) CNTs with metal particles deposited on them are pulled off from the composite film. (d) Single CNT is embedded into the CCE.

Figure 6 shows the SEM images of Ni/CNT composite films prepared by CCE and CCE with ultrasonic agitation. Figure 6 (a) shows the SEM image of Ni/CNT composite films prepared by CCE with electromagnetic stirring and their higher magnification are shown in Fig.6 (b) and(c). It is easy to find that there are a lot of spherical Ni grains are deposited on the CNTs with a diameter from several nanometers to about 700nm.The CNTs accumulate on the cathode, and most of them incorporated into the composite films in the form of aggregates. The CNT aggregates are wrapped by numberless Ni grains. A rough and uneven composite film is formed. Figure 6 (d) shows the SEM image of Ni/CNT composite films prepared by CCE with ultrasonic agitation and their higher magnification SEM images are shown in Fig.6 (e) and (f). From these SEM images, we see that a compact and smooth Ni/CNT composite film has been obtained by CCE with ultrasonic agitation. The CNTs are dispersed homogeneously in the Ni matrix, and most of CNTs incorporated into the composite film in the form of single CNTs without beads-shaped structure is observed.

3. Fabrication of Ni-matrix CNT field emitters by CCE

Electron field emitters based on CNTs are currently being investigated as next-generation cold cathode materials [33–35]. Compared with other field emitters such as spindt-type and silicon field emitter arrays, CNTs possess the advantages of very high aspect ratio, small radius of curvature, lack of vacuum-arcing, low sputter yield, chemical inertness, thermal

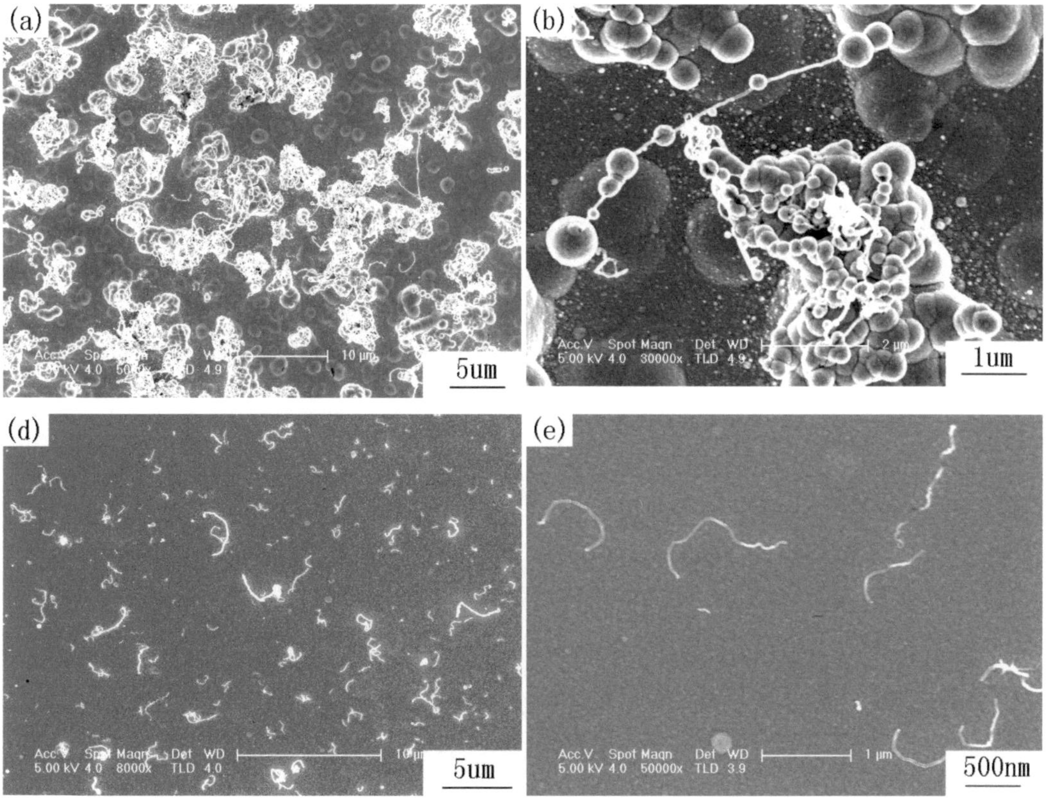

Fig. 6. The SEM images of Ni/CNT composite films prepared with CCE: (a) Ni/CNT composite film prepared without using ultrasonic agitation, (b) is the enlarge image corresponding to (a);(d) Ni/CNT composite film prepared with ultrasonic agitation ;(c)are the enlarge images corresponding to (d).

stability and low work function of electron tunneling [36]. CNT field emission cold cathodes have a potential to be applied to emission devices which include flat panel field emission displays, cathode ray tubes, backlights for liquid crystal displays, outdoor displays and traffic signals [37–40].In order to function as good field emitters, CNTs should have good crystallinity, a clean surface, and good electrical contact with the substrate. Considerable effort has been made to develop various methods in fabricating CNT field emitters [41–47].These methods can be categorized into two fundamental types - direct growth and the screen-printing. Each method has certain advantages and disadvantages. For example, the direct growth is of high efficiency for controlling the CNT alignment, density, and length. However, the preparing method has the following three disadvantages: (1) Due to the self-assembling nature of the CNT synthesis process, it is still difficult to control the length of CNT. Therefore, the distance between the CNT tips and the gate may not be accurately controlled; (2) The adhesion of the CNTs which are catalytically grown on the substrates is often not strong enough to survive the mechanical shaking involved in the fabrication processes;(3) The high temperature condition (800~1000 °C) which may damage some glass substrates is obligatory for this method. In the screen printing, large area field emitters can be fabricated in lower cost. However, for such emitters, the density and heights of CNTs

Table 1 shows several succefully plating bath composition and process conditions for CCE in our present research.

Metal matixs	Solution components	Content	Electroplating conditions
Cu	$CuSO_4 \cdot 5H_2O$	30g/L	Operating temperature：20±1℃; PH:4.5 Cathode-current density：0.1～1A/dm2； Ultrasonic agitation； Anode: pure Cu plate
	H_2SO_4(98%)	76g/L	
	MHT	1ml/L	
	PA5000	3ml/L	
	MWCNTs	4g/L	
Zn	ZnSO4 ·7H2O	110g/L	Operating temperature：20±1; PH：5.0； Cathode-current density：0.9～1.2A/dm2； Ultrasonic agitation； Anode: pure Zn plate
	Na2 SO4	21g/L	
	CH3COONa	20mL/L	
	MWCNTs	2g/L	
Ag	KAg(CN)2:H	80g/L	Operating temperature：25±1℃； PH：4.8 Cathode-current density :0.5～1A/dm2； Ultrasonic agitation； Anode: Pt plate
	30KCNS	150～250g/L	
	KCl	25 g/ L	
	MWCNTs	4g/L	
Ni	Amino-sulfonic nickel	400 g/L	Operating temperature：25±1℃； PH：4.46； Cathode-current density :1A/dm2； Ultrasonic agitation； Anode: pure Ni plate
	Chlorinated nickel	10 g/L	
	Boric acid	15 g/L	
	Dodecyl sulfonic acid sodium	0.05 g/L	
	MHT	4 ml/L	
	Dodecyl sulfonic acid sodium	0.05 g/L	
	Polyacrylic acid (PA5000)	6～8 ml/L	
	MWCNTs	2g/L	

Table 1. Several succefully plating bath composition for CCE in our present research

were random and the turn-on voltage may not be a characteristic of average CNTs but of early igniting CNTs instead. In addition, direct growth and screen printing may cause the tubes be contaminated with some impurities, like metallic catalyst particles, amorphous carbon or organic residues which can introduce further defects into the CNTs during their removal [48, 49]. If all the advantages in both methods are combined and all disadvantages are eliminated, a CNT field emitter with good properties can be fabricated.

In this study, we propose a novel method of fabricating a Ni-matrix CNT (NMCNT) field emitters using CCE and micromachining (CEMM) at room temperature. Pretreated multi-walled CNT and Ni are deposited onto a Cr/Cu conducting layer by CCE; subsequently, protruding tips of CNTs are obtained by etching away a layer of Ni as emitters, followed by emitter pixels being formed by micromachining. This method is not only relatively simple in fabrication process, but also combine advantages of direct growth and screen printing. The NMCNT field emitters we fabricated shows relatively good field-emission properties and stability. It is believed that this is the first report of NMCNT field emitters fabricated by CEMM and the investigation may be helpful to CNT cathodes wide application for industry.

3.1 Experiment

Referring to Fig.7, the fabrication process is described as follows:

1. Cr/Cu (30/50 nm) is deposited on a glass substrate as the conducting layer by sputtering.
2. Photoresist spin coating and photolithography are per-formed to form emitter pixels.
3. The CNT and Ni composite film is deposited on Cr/Cu conducting layer by CCE (the reason for choosing Ni as the basement is its resistance to corrosion). Initially, the multi-walled CNTs with tube diameters ranging from 20 to 40 nm are boiled in potassium hydroxide molten for 5 h and concentrated sulfuric acid for 21 h in order to obtain pure and dispersed emitting materials. Then, the pretreated CNTs are added into Ni electroplating solution, and the solution is sonicated at 21 kHz for 3 h to produce a homogeneous suspension. Finally, the CNT and Ni composite films are deposited on the Cr/Cu conducting layer by composite electroplating using the CNT suspension.
4. The CNT and Ni composite film is polished by polisher to form a flat surface (roughness Ra < 0.2 lm).
5. A layer of Ni is etched away from the CNT and Ni composite film with a flat surface to obtain protruding tips of CNTs as emitters by wet chemical etching, the etching depth is carefully controlled by etching time in order to keep the roots of CNTs still remain in the metal matrix.
6. Photoresist spin coating and photolithography are performed to protect those CNT arrays used as field emitters in the process of reactive ion etching (RIE).
7. Removing the photoresist after RIE of CNTs not used as field emitters.

All experimental steps stated above in (1)-(7) are completed at room temperature. A scanning electron microscope (SEM, JSM-6700F of JEOL Company) is used to analyze the surface morphology of NMCNT field emitters.

3.2 Results and discussion

Figure 8 shows a series of SEM images of NMCNT emitters with different amplified factor. It can be seen that the cathodes with a flat surface were covered with dispersed CNTs (see

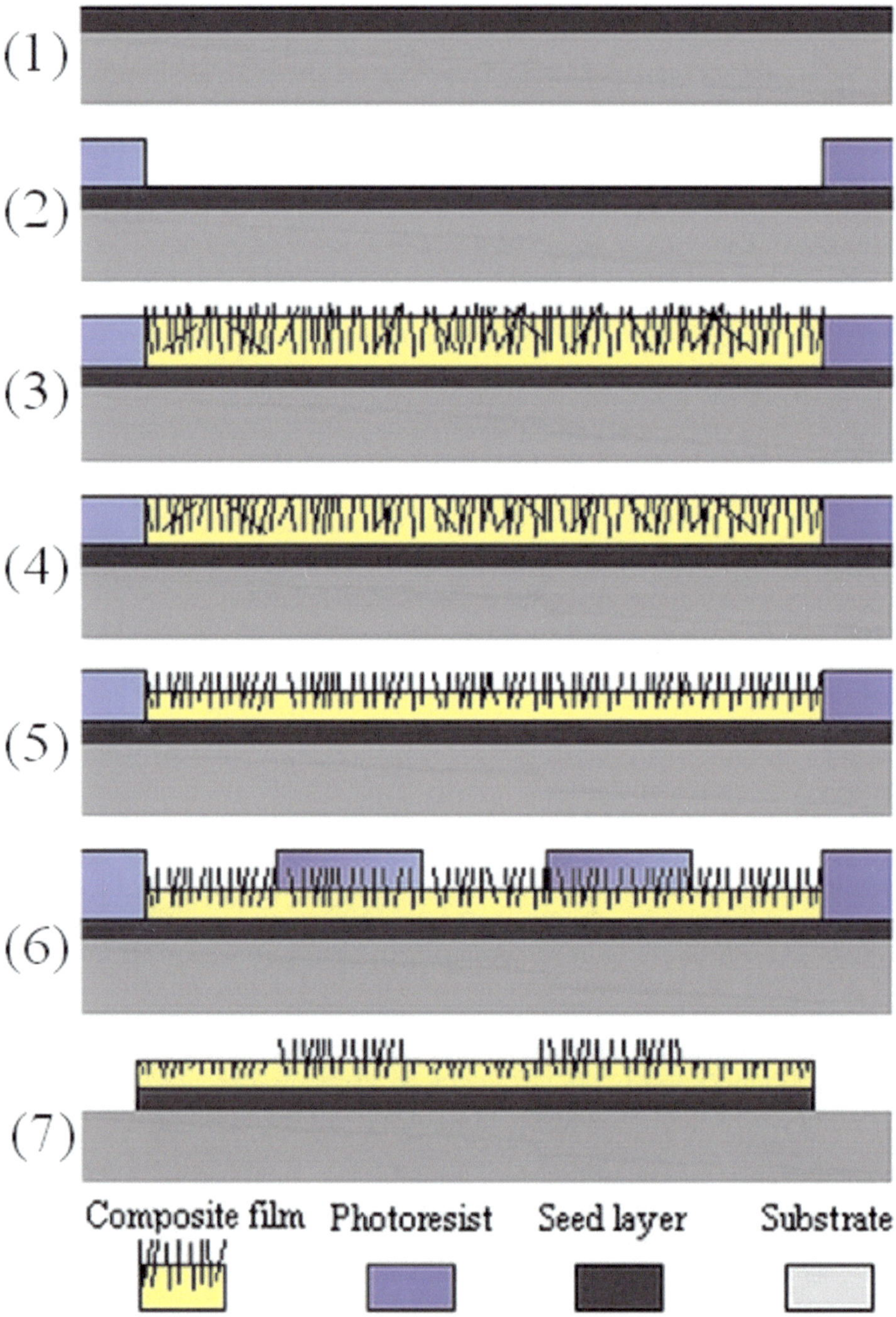

Fig. 7. The fabrication process of NMCNT emitters:(1) sputtering Cr/Cu (30/50nm) conducting layer on a glass substrate, (2) photolithography of NMCNT emitter pixels, (3) CCE of the CNT and Ni composite film, (4) polishing of the surface of CNT and Ni composite film, (5) wet chemical etching of Ni layer, (6) photolithography of NMCNT emitter arrays, (7) removing the photoresist after RIE of CNTs not used as field emitters.

Fig.8a and b), the roots of CNTs were firmly embedded in the cathode and no contamination was induced (see Fig.8c and d). In addition, the distribution and heights of CNTs have been well controlled by the etching time and operation condition of CCE. In conventional CNT emitters, both by screen printed and chemical vapor deposition synthesized, the distribution and height of CNTs were random. In such emitters, the turn-on voltage may not be characteristic of average CNTs but of early igniting CNTs. The early emission from some particular tubes can reveal the apparently low turn-on voltage, which is usually defined by the voltage at a given reference current intensity.

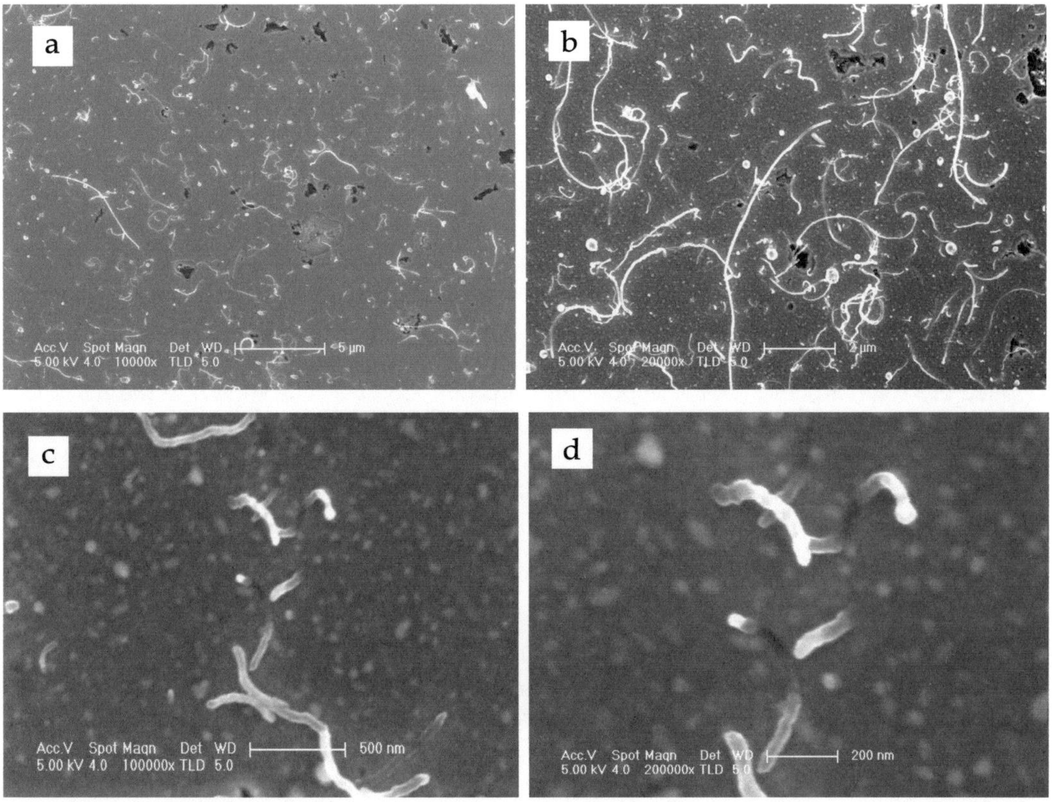

Fig. 8. The SEM images of NMCNT field emitters with different amplified times:(a)10000X, (b)20000X,(c)100000X,(d)200000X.

The SEM image of NMCNT emitter pixels fabricated by CEMM is shown in Fig.9 (a) and their higher magnification SEM image is shown in Fig.9 (b). The spacing between pixels shouldn't be too small, small emitter pixel spacing would induce the field screening effect between adjacent emitters. However, large emitter pixel spacing would decrease the display effect of field emitters. In the fabrication process of NMCNT emitters, it is convenient to adjust the size of emitter pixels by masking and photolithography. The optimized distance between NMCNT emitter pixels was 60um and the area of emitter pixel was 625μm^2 in the present study.

The electron field-emission properties of NMCNT field emitters with a diode structure were measured in a vacuum chamber at a pressure of 1.7×10^{-5}Pa. The distance between the cathode and the anode was 170μm, and the electric field referring to the value of the applied

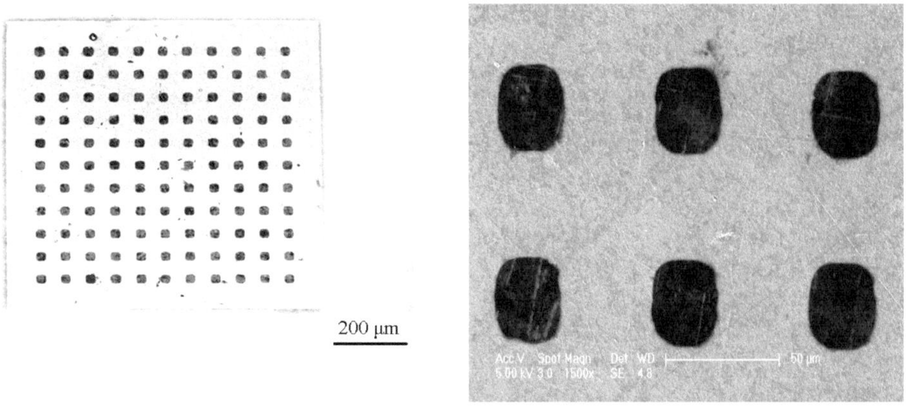

Fig. 9. The SEM images of (a) CNT emitter pixels, (b) a magnified view of pixels.

voltage mentioned in the following descriptions is divided by the electrode distance. In contrast to the distance between electrodes, the surface of the test sample is very flat (roughness Ra < 0.2μm), we consider that field emission has actually occurred from the all NMCNT pixels on the test sample in the same applied field. The actual emission area of NMCNT field emitters is 1600×0.000625cm^2 (1600 is the number of pixels on the test sample). Figure 10 shows a plot of the field-emission current density (ECD) versus the applied electric field, and the corresponding filed emission curve is shown in the inset. It is easy to find that all dots on the filed emission curve fit a single straight line well, which indicates the Fower-Nordheim-type field-emission behavior. The ECD of NMCNT emitters increases monotonically with the applied field, and the shape of the curve in Fig.5 is relatively smooth. When the applied electric field is enough high, the current gets saturated and remains constant. The highest current density was about 13mA·cm^{-2} at an applied electric field of 3.4Vμm^{-1} and the measured turn-on field to extract a current density of 10μA·cm^{-2} was 0.53 V μm^{-1}. Considering the fact that the emission area of NMCNT emitters was 1cm^2, this should be a rather strong filed emission performance in carbon-related cathodes. CNT films grown on various well treated or fabricated planar substrates with an ECD higher than 1mA·cm^{-2} were also reported, although it is of an actual emission area which is only several square millimeters. For example, ECD of~6A·cm^{-2} (at 7.4V·μm^{-1})and~25mA·cm^{-2} (at 4V·μm^{-1}) were demonstrated by CNT films grown on Fe/Al/TiN/Si and Ti substrates, with the emission areas being 10^{-4} and 0.04 cm^2, [50, 51].However, when the actual emission area reaches the magnitude of square centimeters, the ECD will drastically drop down to a scale from several tens to several hundreds of microamperes. For example, for CNT films grown on SiO_2/Si (0.8 cm^2), indium tin oxide coated soda lime glass (64cm^2) and glass modified by organic functional groups (1cm^2), the corresponding ECDs were only 42 μA·cm^{-2} (at 6 V·μm^{-1}), 62 μA·cm^{-2} (at 2 V·μm^{-1}) and 160μA·cm^{-2} (at 4.8 V·μm^{-1}) [52-54]. Based on these comparisons, it was concluded that the field electron emission from NMCNT field emitters was greatly enhanced, and the technical parameters, such as turn-on field and ECD, had almost satisfied the typical technical requirements for flat panel display operation [55]. The enhancement factor β was derived from the slope of the graph by assuming that the work function of CNTs was the same as that of graphite (5eV).The enhancement factor for the NMCNT-field-emitter was 27735. Such a β value is much higher than typical values reported for CNT cathodes, such as 400~1200 for CNTs on silicon and glass substrates, and 2600~3500 for highly ordered CNT arrays on porous aluminum oxide [56, 57]. Therefore, considering the actual emission area, the obtained turn-on field, current density, and the calculated field enhancement factor, the NMCNT emitters fabricated by CEMM is an ideal candidate cathode.

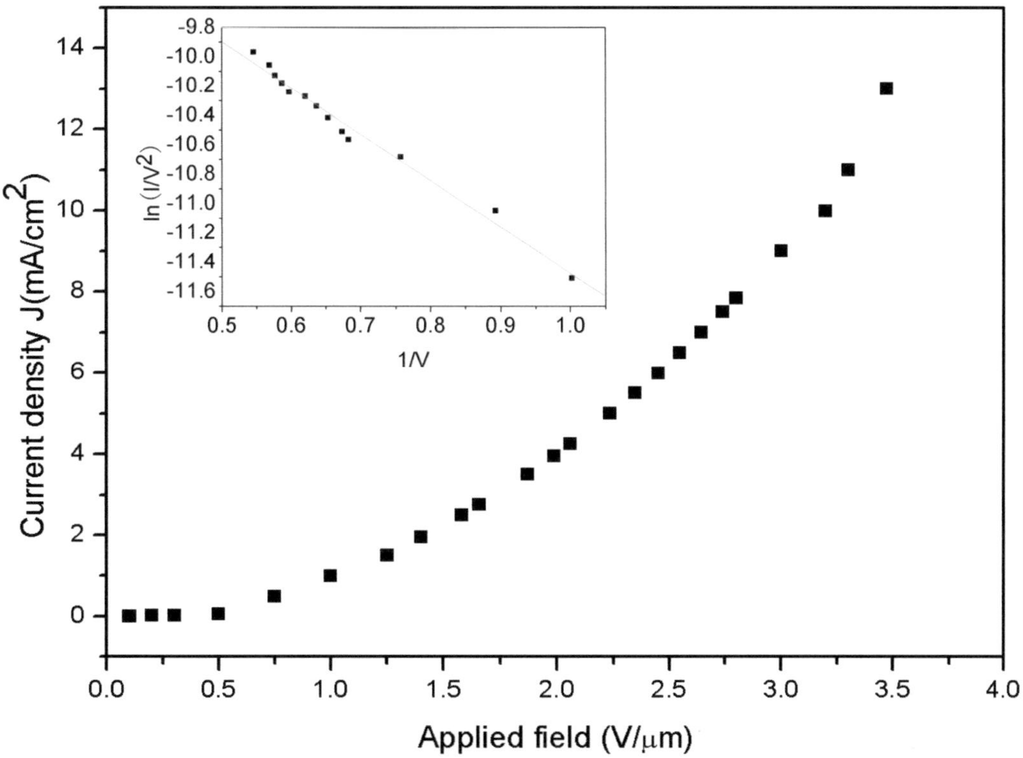

Fig. 10. ECD versus the applied electric field for the NMCNT field emitters fabricated by CEMM. Inset shows the Fowler–Nordheim plot.

3.3 Conclusions

In conclusion, we have fabricated large area NMCNT field emitters by CEMM at room temperature. Pretreated multi-walled CNT and Ni are deposited onto a Cr/Cu conducting layer by CCE; subsequently, protruding tips of CNTs are obtained by etching away a layer of Ni as emitters, followed by emitter pixels being formed by micromachining. Through the process of CEMM, CNTs with a clean surface are vertically embedded in the flat Ni substrate. Our field emitter shows relatively good field-emission properties such as high current density (13mA $\cdot$cm^{-2} at an applied electric field of 3.4 V μm^{-1}), low turn-on field (0.53 V μm^{-1}), and good stability (110 h for 10% degradation of current density from 400μA $\cdot$cm^{-2}). This excellent field emission performance is attributed to the uniform distribution of CNTs on the cathode, the strong adhesion of CNTs to Ni matrix and the flat surface of NMCNT field emitters. This method is not only simple in fabrication process, but also combine advantages of direct growth and screen printing. Above all, no further treatment is needed to initiate or augment field emission. The NWCNT field emitters can obtain practical applications such as backlight units of liquid crystal displays and cathode ray tubes.

4. Cu/CNT contact material prepared by CCE

The damage of electrical contact is a vital factor for the invalidation of integral electrical appliances [57–59], so materials used as electrical contacts in these applications must have a good combination of electrical conductivity, wearing qualities, and resistance to erosion and welding [60, 61]. Otherwise, the contacts will be eroded, which causes poor contact and arcing. Arcing takes place when contacts are in the process of establishing a current flow or interrupting the flow of current. Arc is characterized by high temperature and a high current density in the arc column. Because of the high temperature and mass flow, the contact material surface is severely corroded and eroded, which results in erratic contact resistance and material loss. Therefore, an electrical contact material should have high electrical and thermal conductivity, high melting point and high resistance to the environmental reaction, as well as high arc erosion to maintain contact integrity [62–65]. Cu is a good candidate material due to its high thermal and electrical conductivity, but it has a high coefficient of thermal expansion (CTE). Materials with low CTE and high thermal conductivity, such as Cu/CorCu/W composites, have improved the reliability of electronic devices [66, 67]. However, these composites are often too expensive for many applications. In addition, their machinability and the elaboration of thin sheets remain very difficult.
The CNTs reinforced Cu-matrix composites offer a good compromise between thermo-mechanical properties and thermal conductivity. Their advantages are :(i) lower density than copper, (ii) excellent thermal conductivity, (iii) low coefficient of thermal expansion and (iv) good machinability. Other advantages of Cu/CNT composites are adaptive thermal properties, which can be adjusted [68–70]. However, it is difficult to make CNTs disperse uniformly and combine well in the metal matrix [71], traditional techniques such as the powder metallurgic process would bring for midable technical hurdles to prepare high-quality metal- matrix CNT composites, which is ascribed to the low-strength interfacial adhesion [72]. As no melting and solidification of the matrix would be induced, CCE is one of the most important techniques for preparing high-quality-metal-matrix CNT composite. In addition, the CCE is compatible with the micro-electro-mechanical systems (MEMS) technology. The metal-matrix CNT composite films can be processed to various small volume and high-performance electric contacts by MEMS technology.
In this study, the CNT Cu-matrix composite films are prepared by CCE. SEM images show that the CNTs are dispersed uniformly and combined well in the Cu matrix. The Cu/CNT contact material displays good resistance to electric arc. We believe that the investigation should be helpful to the Cu/CNT contacts wide application for industry.

4.1 Experimental details

4.1.1 Preparation of Cu/CNT composite films

The pretreated CNTs ($1gL^{-1}$, $2gL^{-1}$, $3gL^{-1}$, $4gL^{-1}$, $5gL^{-1}$, $6gL^{-1}$) were added to the Cu electroplating baths, and six different plating baths were prepared for making a comparison. Then, these solutions having different concentrations were sonicated at 21 kHz for 3 h to produce homogeneous suspensions. The CNT Cu-matrix composite films were prepared by CCE using these CNT suspensions; the optimum plating baths are chosen by the quality comparison of these CNT composite films. All chemicals are chemically pure. A Cu plate with a purity of 99.98% is used as the anode.

4.2 Results and discussion

4.2.1 The choice of Cu electroplating baths and the combined state of CNT with the Cu matrix

Through the quality comparison of different CNT composite films fabricated by the six electroplating baths, we found that the CNT composite films have the best quality when the concentration of the CNTs in the Cu electroplating baths reaches 4 g L^{-1}. So this electrolyte concentration is chosen as the optimum plating bath for preparing the CNT Cu composite films. Figure11 shows the SEM images of the Cu-CNT composite films; it can be seen that the CNTs dispersed uniformly and combined well in the Cu matrix, the enhancements have not increased the porosity rate of composite films, and the surfaces of composite films are homogeneous.

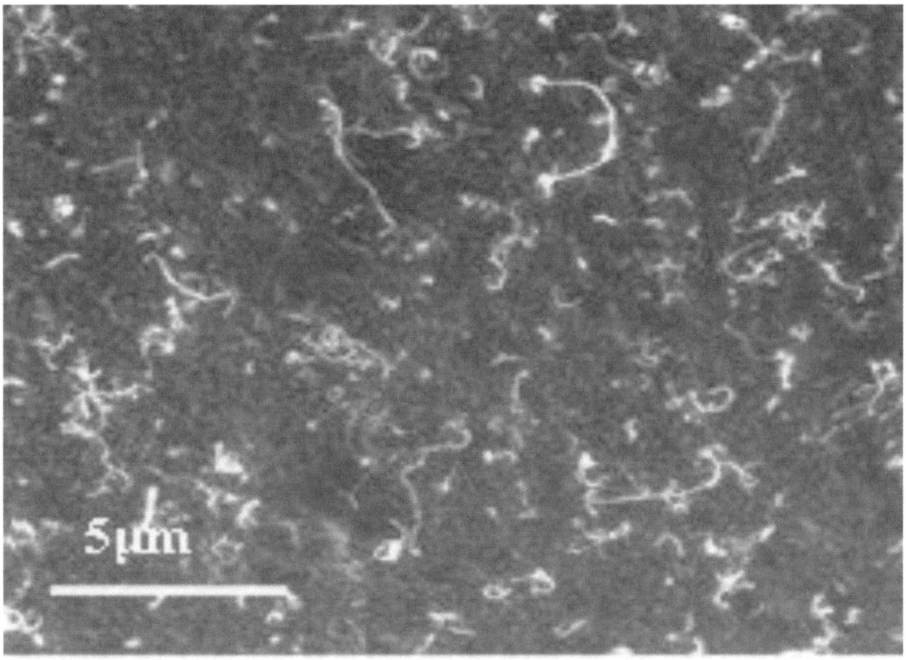

Fig. 11. The SEM images of CNT composite films

4.2.2 Mechanical properties of the Cu/CNT composite plating film

The hardness of the Cu-CNT composite films and a Cu film was tested by digital-micrographic hard-meter (model number: HXD-1000 TMB/LCD). Figure 12 shows the hardness change of the CNTs (a) (b) Cu composite films with the concentration variation of CNTs in the plating baths. From the graph, we found that the hardness of the CNF composite film was enhanced significantly with the increase of the CNTs in the plating baths, but for the CNT composite film it has not obviously been changed with the increase of the CNTs in the plating baths. The hardnesses of the CNT Cu composite films reach the maximum of 156 HV and 207 HV with the concentration of the CNTs of 4gL^{-1} and the concentration of the CNFs of 4 g L^{-1}in the plating baths, and about 13.9% and 51.1% higher than that of the pure copper plating film (137 HV). This result indicates that the CNTs/CNFs have significant effect on promoting the hardness of the Cu-matrix composite film, especially for the CNFs.

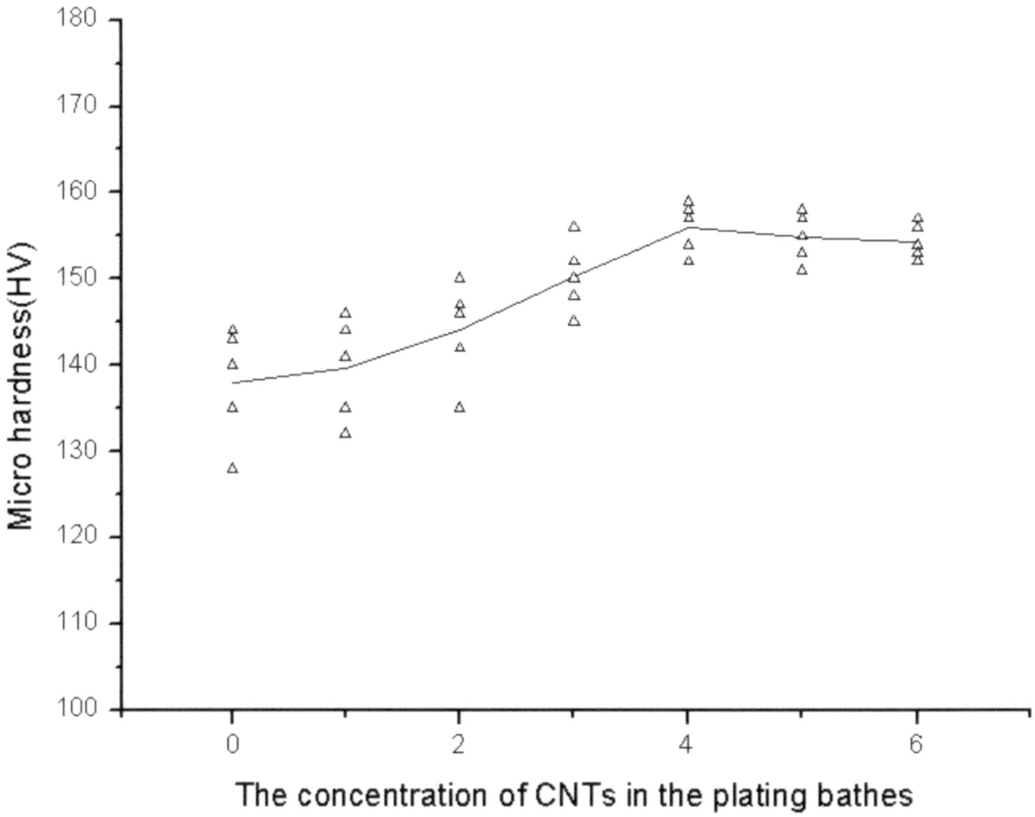

Fig. 12. The hardness change of CNT Cu composite films with the concentration variation of CNTs in the plating baths.

4.2.3 The electrical properties of the Cu/CNT composite films

The CNT composite films are tested by a contourgraph and semiconductor parameter analyser. The resistivity of the film can be expressed as $\rho= R.S/L$, where R is the electrical resistance, S and L are the cross–sectional area and the length of the composite plating films. The resistivities of the two different plating films are listed in table.2; it shows that the resistivities of the Cu/CNT composite plating films and Cu film are 2. 656 $\times 10^{-6}\Omega$ ·cm , 1.745 $\times 10^{-6}\Omega$ ·cm. The resistivity of the Cu/CNT composite plating film is higher than that of pure Cu film ,but lower than those of other Cu–matrix composites such as CuW 4. 35 $\times 10^{-6}\Omega$ ·cm 、CuMo 3. 571 $\times 10^{-6}\Omega$ ·cm.Theoretically, the Cu/CNT composite plating films should have better conductivity performance than that of pure Cu film. According to O. HJORTSTAM's report [73], it is possible that the room-temperature resistivity of CNT–metal composite is 50% lower than that of Cu when the CNT filling in the range of 50%–60% of the composite. The ultra-low resistivity is possible because the ballistic conducting CNTs have an electron mean free path several orders of magnitude longer than metals like Cu and Ag. The reason of the resistivity of CNT/Cu composite plating films fabricated by our method is higher than that of pure Cu film may be that the relatively low composite density of CNT in the Cu-matrix composite films limits the overall conductivity of CNT composite films [74].It needs our further research to resolve the problem and prepare ultra-low resistivity Cu/CNT composite.

species	resistance (Ω)	length of linetype film (cm)	cross-sectional area of linetype film (cm²)	resistivity (Ω·cm)
pure Cu films	10.50	53.25	8.850×10^{-6}	1.745×10^{-6}
Cu/CNTs composite plating films	15.80	53.25	8.950×10^{-6}	2.656×10^{-6}

Table 2. Resistivity of two different plating films

The thickness of the Cu/CNT composite films are controlled in the size rang of 1-10um by adjusting the process parameters of CCE and the same properties of composite films can be held; the CNT composite films are processed to various size films and employed for micro-relay by MEMS technology. Because of the stable processing technique and the compatibility of CCE with MEMS technology, the micro special contacts are directly integrated on the surface of the contact of MEMS micro relay.

In the process of arc erosion test, the closing force of contact is 1N, breaking force is 0.75N, circuit voltage is 1.5V, current is 1A and the frequency of breaking is 6000. There exists apparent electric spark when the contact was broken. Figure 13 shows the micrographs of electric contact samples before (a) and after the arc erosion test. The result shows that the arc erosion of the Cu/CNT contact is 2.7mg, and 22.9% lower than that of pure Cu (3.5mg).This implies that the CNTs can effectively improve the arc erosion of micro special contact.

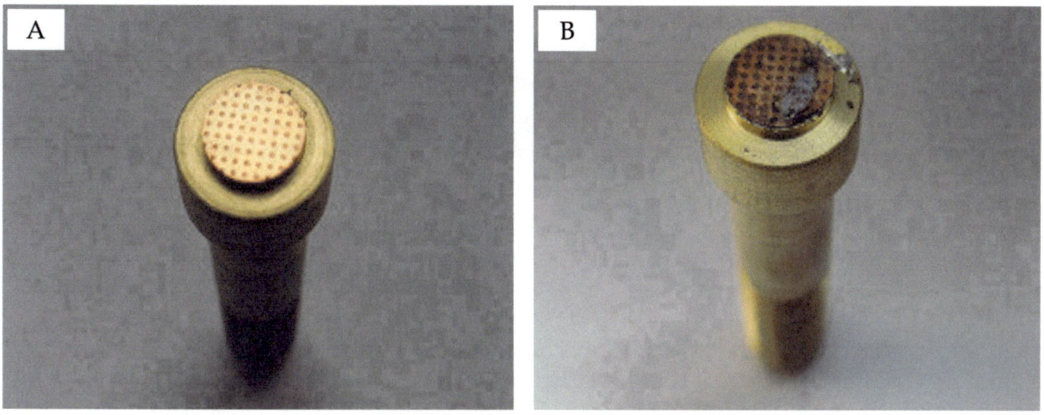

Fig. 13. The micrographs of electric contact samples before arc erosion being tested (A) and after arc erosion being tested (B)

4.3 Concluson

In the present research, the Cu/CNT matrix composite films are prepared by CCE. SEM images show that the CNTs dispersed uniformly in the Cu-matrix. Furthermore, the Cu-CNT composite films show relatively good physical properties such as the hardness of 156 HV, and about 13.9% higher than that of pure Cu plating film (137HV); the resistivity of the CNT composite plating films is 2.656 $\times 10^{-6}\Omega$ cm, and lower than those of other Cu-matrix composites such as CuW (4.35$\times 10^{-6}\Omega$ cm) and CuMo (3.571 $\times 10^{-6}\Omega$ cm). In addition, the arc erosion behavior of the Cu/CNT contact has been examined on an electric arc erosion apparatus. The result shows that the arc erosion loss of the Cu/CNT contacts is 2.7mg, and 22.9% lower than that of pure Cu (3.5mg) under the same conditions. We believe that the investigation should be helpful to the Cu/CNT wide application for industry.

5. CNT field emission micro-cathode arrays fabricated using CCE followed by micro-machining

On the basis of preparing the CNT emitters above, we propose a novel design of CNT field emission micro-cathode arrays. The 3D structure drawing is shown in Fig. 14, compared with previously reported CNT field emission micro-cathodes [33-36], this structure have relatively complete functional units, including bottom electrode down-lead current-limiting resistance CNT emitters supporting wall insulator layer suspension grid and focusing electrode. In this design, each structure layer is fabricated by using layer by layer lithography alignment process. All experiments are executed at room temperature.

5.1 Experimental details

The flow of the fabrication process for the triode-type CNT field emission micro cathode arrays is drawn in Fig.14 and the detail process is described as follows:

1. First, a patterned Au layer (1μm) is deposited on the Cr/Cu (10/40nm) seed layer as bottom electrode by photolithography and electroplating (Fig.14.a).
2. Then, a Al_2O_3 film (50nm) is sputtered on the Au electrodes as resistor layer, and then, the fabrication of the patterned CNT field emitters is preformed(Fig.14.b,c).
3. Afterward, a 5μm patterned Ni supporting wall (the reason for choosing Ni is its resistance to corrosion) is deposited by photolithography and electroplating (Fig.14.d).
4. Then a polysilicon insulator layer is deposited on the supporting wall to suppress the leakage current by sputtering (Fig.14.e).
5. Subsequently, patterned Ni gate electrode (1μm) electroplating is performed (Fig.14.f and g).
6. Finally, patterned focusing electrode is formed by electroplating with photolithography (Fig.14.h).

5.2 Results and discussion

The optical images of the integrated devices are shown in Fig.15 (A), (B) and their higher magnifications are shown in Fig.15(C), (D) and (E). It can be seen that the integrated CNT field emission micro-cathode arrays have uniform structure and good repeatability. The gate pores' diameter is ~1 μm and the height between the Ni metal gate and the CNT pattern edge is ~5 μm (Fig.15D).The pore density is about 1.2$\times 10^5$ /cm^2. The inset map in Fig.14E is a magnified view of a CNT emitter pixel; the CNTs are uniformly embedded in the pixel as shown in the inset map. The emitter pixel spacing shouldn't be too small,

small emitter pixel spacing would induce the field-screening effect between adjacent emitters. However, big emitter pixel spacing would decrease the resolution of field emitters. In the fabrication process of the integrated devices, it is convenient to adjust the size of the emitter pixels by masking and photolithography. The optimized distance between the every two adjacent CNT emitter pixels is 2.5um and the area of emitter pixel is 9 μm^2 in the present study.

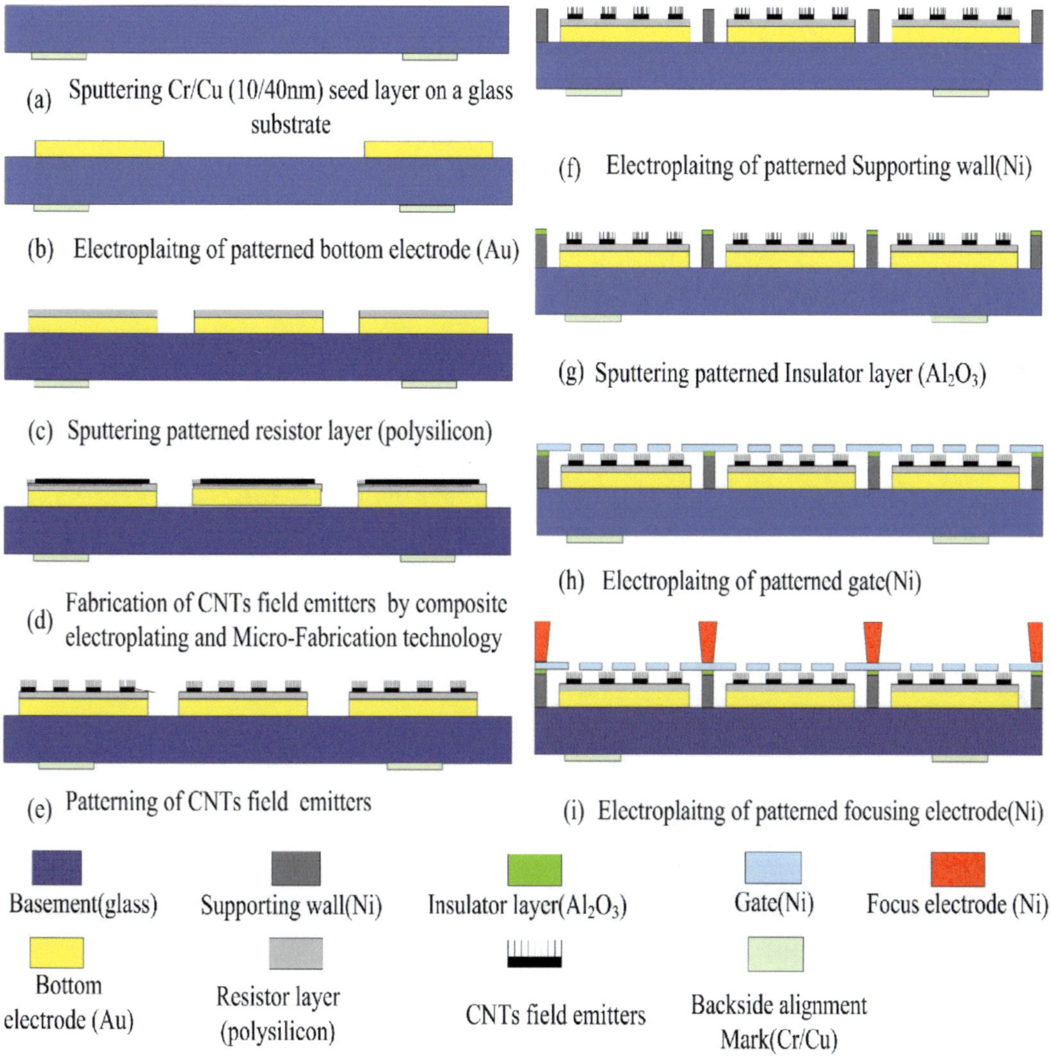

Fig. 14. Schematic diagrams of the fabrication process for the triode-type micro-gated CNT emitter arrays :(a)Patterned bottom electrode (Au) is deposited on basement by electroplating, (b) Resistor layer (Al_2O_3) is deposited on bottom electrode by sputtering, (c)Patterned CNT emitters are fabricated by electrochemical micromachining, (d) Electroplating of patterned Supporting wall (Ni),(e) Patterned insulator layer (polysilicon) is deposited on the supporting wall by sputtering, (f)(g) Gate(Ni) is fabricated by photolithography and electroplating, (h) The electroplating of Ni focusing electrode.

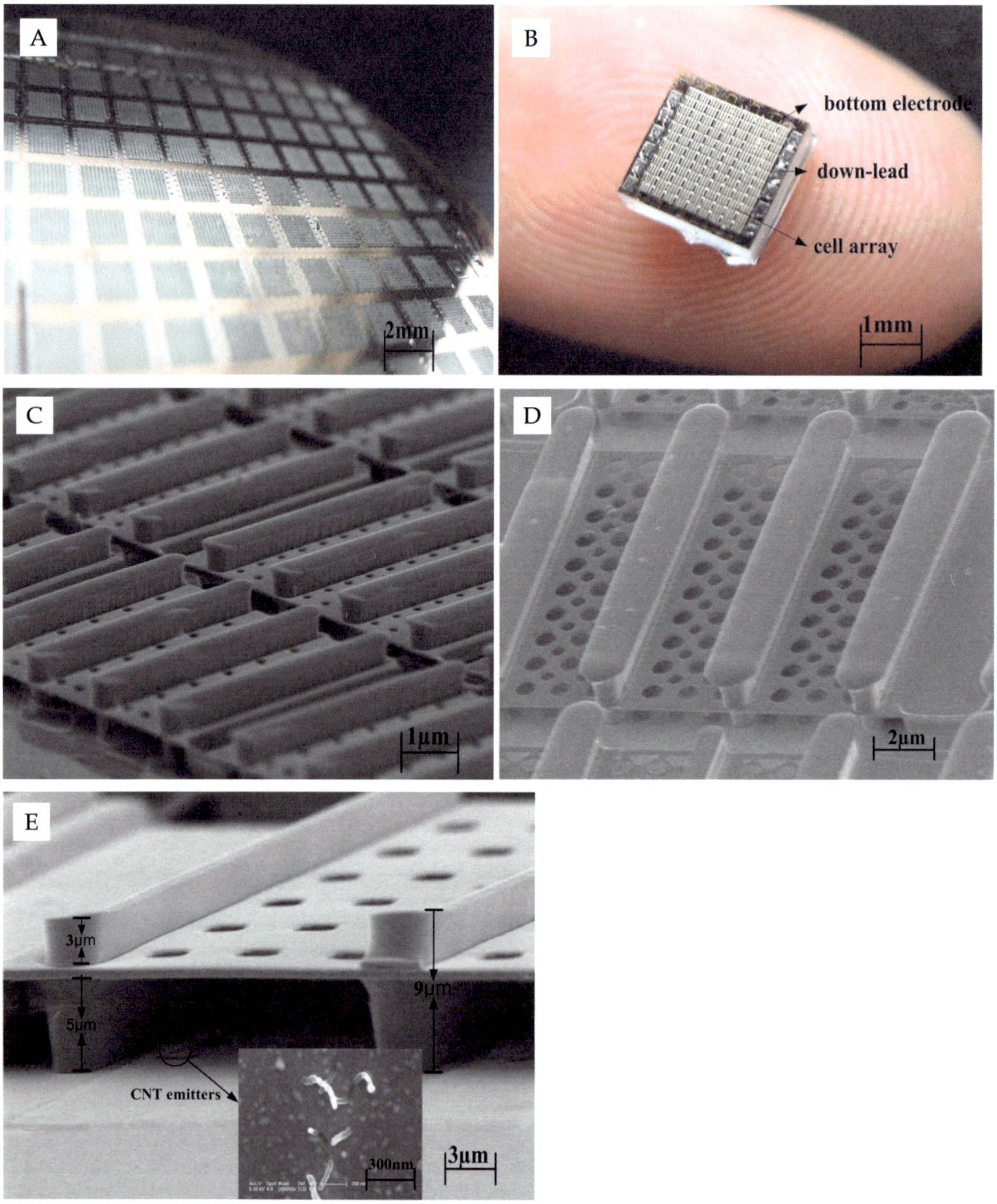

Fig. 15. The optical images of integrated devices (A, B) and their higher magnifications (C, D and E). The inset map in Fig.15E is a magnified view of a CNT emitter pixel.

Figure 16 shows a plot of the device field-emission current density (ECD) versus the applied electric field, and the corresponding field emission curve is shown in the inset. It is easy to find that all dots on the field emission curve fit a single straight line well, which indicates the Fower-Nordheim-type field-emission behavior. The ECD of the emitters increases monotonically with the applied field. When the applied electric field is high, the gate current gets saturated and remains constant. The highest gate current density is about 15.7mA $\cdot cm^{-2}$ at an applied electric field of 12V/μm and the measured turn-on field to extract a current density of 10μA $\cdot cm^{-2}$ is 2.4V $\cdot \mu m^{-1}$. The device characteristics are compared with those in other CNT emitter structures, although they are all fabricated by different structures and processes [79]. Pirio et al [80] observed turn-on voltages of 9–15 V, defined at 0.1 nA $\cdot cm^{-2}$. The low turn-on voltage was obtained with a sub-micrometer gate-to-tip distance realized using a self alignment process. However, the turn-on field remained at relatively high values of 18–30 V $\cdot cm^{-1}$. In Hu et al [81], a structure fabricated by screen-printed CNTs with a turn-on voltage of 40–45 V was obtained with ~ 5 μm gate-to-tip distance. Jang et al [82] obtained a turn-on voltage of 20 V (turn-on field ~ 1 V $\cdot \mu m^{-1}$) at ~10 nA $\cdot cm^{-2}$ with pasted CNTs. They all report, at most, mA $\cdot cm^{-2}$ range values for the maximum current density. On the other hand, Uh [83] obtained an exceptionally high current density of 275mA $\cdot cm^{-2}$ with a turn-on voltage of 38 V using catalytically grown CNTs. In summary, the triode emitter structure fabricated on a glass template using CCE and micromachining shows a very low turn-on voltage and a high current density, which are better than or, at least, compatible with those of other triode structures. The field enhancement factor (β) for the CNT emitters is derived from the slope of the graph by assuming that the work function of CNTs is found to be $2.4 \times 10^6 \cdot cm^{-1}$ under the assumption of the work function to be the same as that of graphite (4.5eV), which is calculated from the following equation: $\beta=2.84\times107\ \text{Ø}^{3/2}/S$, where Ø and S represent the work function of CNTs and the absolute value of the slope of the F–N plot. The field enhancement factor extracted from triode-type configuration is approximately two orders of magnitude higher than that of diode-type configuration fabricated by using the same method and also much higher than typical values reported for CNT cathodes, such as 400~1200 for CNTs on silicon and glass substrates, and 2600~3500 for highly ordered CNT arrays on porous aluminum oxide [84, 85]. The very high field enhancement factor confirms the high efficiency of the triode structure in electron extraction. The structure shows good field-emission properties, but the challenge of fabricating an applied device still remains including the further optimization of integral construction and preparation technology.

5.3 Conclusions

A new CNT field emission micro cathode array structure fabricated by CCE and micromachining is achieved. The relevant processing technology is also developed. Integrated CNT field emission micro-cathodes have intact structure and good repeatability. The structure revealed a very efficient performance as indicated by the high field enhancement factor and current density, low turn-on voltage and good emission stability. The micro-cathodes can obtain practical applications such as backlight units of liquid crystal displays and cathode ray tubes .This study laid a foundation on the device integration and cost-effective mass production, but it requires further optimization in the device configuration and processing.

6. The potential applications of CNT films prepared by CCE

Metal/CNT composite films produced by CCE are suitable for a wide range of applications; suggestions to date include field emission devices, biomedical sensors, contact materials and coatings as well as large surface area electrodes for fuel cells, capacitors and gas sensors. So far, the development of CNT-based devices from CNT films produced by the CCE method has been focused mainly on the field emission properties. It is well known that CNTs are promising candidates for field emission devices, due to their high aspect ratio, small size, structural and chemical stability and thermal conductivity; these features are responsible for a low emission threshold and high emission current densities compared to other alternatives.

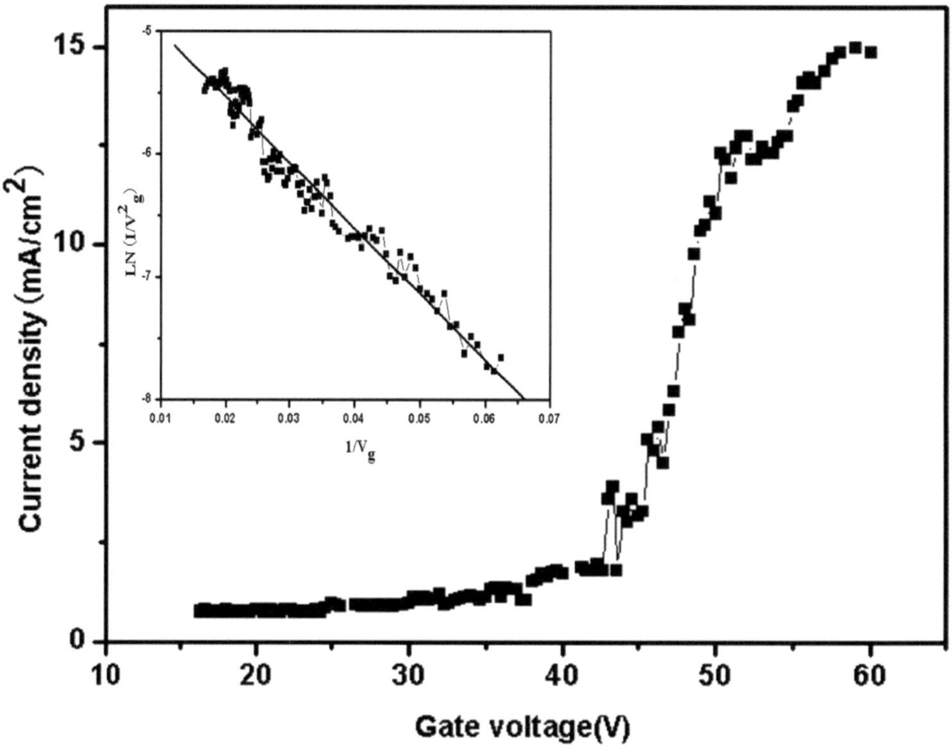

Fig. 16. The ECD versus the applied electric field for the CNT field emission micro-cathodes. Inset shows the Fowler–Nordheim plot.

6.1 The potential application in the displays

CNT field emission displays (CNT-FEDs) are promising for a range of situations including flat panel displays, cathode-ray tubes, and backlights for liquid crystal displays. CNT-FEDs have the potential to provide high quality moving images with low power consumption.

The technical parameters of field emission cathodes prepared by CCE in our present research, such as turn-on field and field emission current density, had almost satisfied the typical technical requirements for flat panel display operation. This suggests that CCE is a useful technique to obtain an electron source with minimal out-gassing and to make triode-type CNT field emission displays. Through the process of fabrication of Ni-matrix CNT field emitters using CCE in chapter 3, CNTs are vertically embedded in the flat Ni substrate. No further treatment is needed to initiate or augment field emission and the field emitters exhibit good field-emission properties such as high current density (13mA·cm^{-2} at an applied electric field of 3.4 V·μm−1), low turn-on field (0.53 V·μm−1), and good stability (110 h for 10% degradation of current density from 400μA·cm-2). In addition, we have successfully developed an effective fabrication method at room temperature for the mass production of triode-type CNT field emission micro cathode arrays. This technique combines CCE and micromachining, avoids the direct growth and screen-printing procedures conventionally used to fabricate such structures. Due to low cost and room temperature process, the technique is proven to be advantageous in mass production. Results of field emission testing show that the micro cathodes have perfect field-emission properties, such as high current density (15.7mA $\cdot cm^{-2}$), low turn-on field (2.5V $\cdot \mu m^{-1}$), and good stability (109 h for 10% degradation of current density from 400μA $\cdot cm^{-2}$). In addition, in the structure the leakage current of the gate has been reduced to near zero by coating the Al_2O_3 on the gate.

In addition, for CNT-FEDs and related devices, vertically aligned CNTs are preferred because they provide a low turn-on field as well as a uniform and stable electron emission. The CCE method has the potential to produce highly efficient CNT-based FED devices due to its ability to fabricate large, vertically aligned, and patterned nanotube arrays at low temperature.

CNT arrays by CVD; thermal CVD involves high processing temperatures so that glass or polymer substrates cannot be used. Although plasma-enhanced CVD can be used at lower temperatures, there are still difficulties in fabricating large area field emitter arrays.

6.2 The potential application in the modified electrode

In the area of electrochemistry, the combination of high aspect ratio, nanometer-sized dimensions, good electrical conductivity and low capacitance in the pristine state dictates that CNTs have the capability to make excellent electrodes. CNT film electrodes were first introduced for voltammetric analysis by Liu and his coworkers. Following this report, the electrocatalytic properties of CNT- modified electrodes became a hot topic. The basic electrode configuration of randomly dispersed, purified CNTs on the surface of a conducting support macroelectrode was used extensively for the electrochemical detection of a multitude of redox - active molecules in solution. The preparation of CNTs-modified electrodes has many problems. CNTs are insoluble in many solvents due to the large inter tube attraction energy. The CNTs dispersed by volatile solvents are often randomly dispersed on the surface. The conventional ways of CNT dispersion and the performances of interface are difficult to control. However, CNTs successfully dispersed in metallic matrixs are easily controlled by CCE. If CCE is utilized to prepare CNT modified electrode, it has several advantages which help improve the interface performance of the composite film. First, CNT can be homogeneously dispersed in the metallic matrixs; Secondly, metallic matrixs can be controllable etched by specific reagent, and make part of CNTs exposed on the surface of the metal/CNT composite without pollution; Thirdly, the density of CNTs in metal can be controlled by changing the density of CNTs in the plating bath and plating current density. Fourthly, the firm connection between CNT and metal matrix can be achieved and ensure a good electrical conductivity. In addition, the CCE is compatible with the MEMS technology. The micromachining technology helps us efficiently choose which area of the electrodes will be modified, thus making CNTs implantation technique feasible, controllable and reliable.

6.3 The potential application in the electrical contact.

As previously analysis in chapter 4, materials used as electrical contacts in electronic devices must have a good combination of electrical conductivity, wearing qualities, and resistance to erosion and welding. Otherwise, the contacts would be eroded to cause poor contact and arcing. Arcing takes place when contacts are in the process of establishing a current flow or interrupting the flow of current. Arc is characterized by high temperature and a high current density in the arc column. Because of the high temperature and mass flow, the contact material surface is severely corroded and eroded, which results in erratic contact resistance and material loss. Therefore, an electrical contact material should have high electrical and thermal conductivity, high melting point and high resistance to the environmental reaction, as well as high arc erosion to maintain contact integrity. Metals with high thermal and electrical conductivity are good candidate material, but they usually have high coefficient of thermal expansion (CTE). Materials with low CTE and high thermal conductivity, such as Cu/CorCu/W composites, have improved the reliability of electronic devices. However, these composites are often too expensive for many applications. In addition, their machinability and the elaboration of thin sheets still remain very difficult.

Furthermore, the metal/CNT composite films show relatively good physical properties. The hardness of the Cu/CNT composite film prepared by CCE is 156 HV and about 13.9% higher than that of pure copper plating film (137HV). The resistivity of the CNT/CNF composite plating film is 2. 656 $\times 10^{-6}\Omega\cdot cm$, and lower than those of other Cu-matrix composites such as CuW 4. 35 $\times 10^{-6}\Omega\cdot cm$ CuMo 3.571 $\times 10^{-6}\Omega\cdot cm$. In addition, Cu/CNT micro special electric contact has been designed and successfully fabricated by MEMS technology. The arc erosion behaviors of the Cu/CNT contacts have been examined on an electric arc erosion apparatus. The result shows that the arc erosion loss of the Cu/CNT contact is only 2.7mg, and 22.9% lower than that of pure Cu (3.5mg) under the same conditions. The Cu/CNT matrix specific contact can be fabricated with high yield and good reproducibility. High producing rate of finished products can effectively reduce the manufacturing cost of micro special electric contact and make them achieve practicable application. The thickness of the CNT/ Cu composite film are controlled in the rang of 1-10μm by adjusting the process parameters of CCE; The CNT composite films are processed to various films with different thickness and employed for micro-relay by MEMS technology. Because of the stable processing technique and the compatibility of composite electroplating with MEMS technology, the micro special contacts can be directly integrated on the surface of the contact of MEMS micro relay.

Exploitation of CCE in these and other applications will frequently rely on the attachment of functional groups or other nanostructures to the CNT surfaces. The combination of CNTs and nanocrystalline particles should have applications in field emission displays, nanoelectronic devices, drug delivery systems, antibacterial films, biosensors, photocatalytic nanostructures and other functional composites. We anticipate that CCE and combinations of CCE and other colloidal processing methods will play a significant role in the development of such metal/CNT composite nanostructures.

7. Conclusions

In this study, we considered the preparation of the metal/CNT composite using CCE method. The effects of pretreatments of the CNTs, such as purification activation and mechanical ball milling on the metal /CNT composite film were discussed. A detailed study of the Ni-matrix CNTs field emitter fabrication by CCE was presented. The CCE fabrication processes of the Cu/ CNT contact material and the CNT field emission micro-cathode arrays were followed. The potential applications of the resulting CNT structures and the scope for the future work are highlighted. Further developments of the CCE processes will allow the reliable fabrication of three dimensionally controlled nanostructures and nanocomposites either in the form of dense materials or with a required porosity; graded, aligned, and patterned features may also be incorporated as desired. It is believed that the investigation may be helpful to metal/CNT composite functional material and devices prepared by CCE wide application for industry.

8. References

[1] Iijima S. Helical microtubules of graphitic carbon. Nature 1991;354:56–8.
[2] Belin T, Epron F. Characterization methods of CNTs: a review. Mater Sci Eng B 2005;119:105–18.
[3] Gooding JJ. Nanostructuring electrodes with CNTs: a review on electrochemistry and applications for sensing. Electrochim Acta 2005;50(15):3049–60. Sp Iss.
[4] Iijima S. CNTs: past, present and future. Physica B 2002;323:1–5.
[5] Y. Qin, M. Hu, Field emission properties of electrophoretic deposition carbon nanotubes film, Appl. Surf. Sci. 255 (2009) 7618–7622.
[6] D.O. Kim,M.H. Lee, J.H. Lee, T.W. Lee, K.J. Kim, Y.K. Lee, Transparent flexible conductor of poly (methyl methacrylate) containing highly-dispersed multi-walled carbon nanotube, Org. Electron. 9 (2008)1–13.
[7] Berber S, Kwon YK, Tomanek D. Unusually high thermal conductivity of CNTs. Phys Rev Lett 2000;84(20):4613–6.
[8] Krishnan A, Dujardin E, Ebbesen TW, Yianilos PN, Treacy MMJ.Young's modulus of single-walled nanotubes. Phys Rev B1998;58(20):14013–9.
[9] Yan Chena, Fanli Menga, Minqiang Li， Novel capacitive sensor: Fabrication from carbon nanotube arrays and sensing property characterization.Sensors and Actuators B 140 (2009) 396–401
[10] M. Penza,R.Rossi,M.Alvisi, M.A. Signore, E.Serra, R. Paolesse, Metalloporphyrins-modified carbon nanotubes networked films-based chemical sensors for enhanced gas sensitivity. Sensors and Actuators B 144 (2010) 387–394.
[11] E. Dujardin, T. W. Ebbesen, H. Hiura, K. Tanigaki, "Wetting andnanocapillarity of CNTs," Science, 1994, vol. 265(5180), pp.1850-1852
[12] W. X. Chen, W. L. Chen, Z. D. Xu, Z. J. Liu, J. P. Tu, X. B. Zhang, et.al. "Characteristics of CNTs and High-quality Composite,"Acta Materiae Compositae Sinica, 2001, vol. 18(4), pp. 1-5
[13] W. F. Zhang, D. Zhu, "Electrodeposited nanocomposites: Research advances and applications," Materials Review, 2003, vol. 17, pp. 57-60
[14] Min Deng, Guifu Ding , Yan Wang, Huiqing Wu, Yuanjin Yao, Lida Zhu, Fabrication of Ni-matrix CNT field emitters using CCE and micromachining. CARBON 47 (2009) 3466 - 3471
[15] Min Deng, Guifu Ding, Yan Wang, YuchaoWang, HongWang and Shi Fu, MEMS-based CNT and carbon nanofiber Cu micro special electric contact. J. Micromech. Microeng. 19 (2009) 065001 (8pp)
[16] S. Khabazian, S. Sanjabi, The effect of multi-walled CNT pretreatments on the electrodeposition of Ni–MWCNTs coatings. Applied Surface Science xxx (2011) xxx–xxx)

[17] Singh C, Shaffer MS, Windle AH. Production of controlled architectures of aligned carbon nanotubes by an injection chemical vapourdeposition method. Carbon 2003;41(2):359–68.

[18] Andrews R, Jacques D, Qian DL, Rantell T. Multiwall carbon nanotubes: synthesis and applications. Accounts Chem Res2002;35(12):1008–17.

[19] Murakami Y, Chiashi S, Miyauchi Y, Hu MH, Ogura M, Okubo T,et al. Growth of vertically aligned single-walled carbon nanotubefilms on quartz substrates and their optical anisotropy.Chem PhysLett 2004;385(12):298–303.

[20] Shelimov KB, Esenaliev RO, Rinzler AG, Human CB, Smalley RE.Purification of single-wall carbon nanotubes by ultrasonically assistedfiltration. Chem Phys Lett 1998;282:429–34.

[21] Peng-Xiang Hou, Chang Liu, Hui-Ming Cheng, Purification of carbon nanotubes. CARBON 46 (2008) 2003 - 2025

[22] S.Costa,C.Tripisciano, E. Borowiak-Palen, R.J. Kalenczuk,Comparative study on purity evaluation of singlewall carbon nanotubes. Energy Conversion and Management 49 (2008) 2490–2493

[23].Orazio Vittorio, MSc, Vittoria Raff, Alfred Cuschieri, Influence of purity and surface oxidation on cytotoxicity of multiwalled carbon nanotubes with human neuroblastoma cells. Nanomedicine: Nanotechnology, Biology, and Medicine 5 (2009) 424–431

[24] Du CS, Heldebrant D, Pan N. Preparation of carbon nanotubes composite sheet using electrophoretic deposition process. J Mater SciLett 2002;21(7):565–8.

[25] Yu K, Zhu Z, Li Q, Lu W. Electronic properties and field emission ofcarbon nanotube films treated by hydrogen plasma. Appl Phys A2003;77:811–7

[26] Girishkumar G, Vinodgopal K, Kamat PV. Carbon nanostructures in portable fuel cells: single-walled carbon nanotube electrodes for methanol oxidation and oxygen reduction. J Phys Chem B 2004;108(52):19960–6.

[27] Girishkumar G, Rettker M, Underhile R, Binz D, Vinodgopal K,McGinn P, Kamat P. Single-wall carbon nanotube-based protonexchange membrane assembly for hydrogen fuel cells. Langmuir2005;21(18):8487–94.

[28] Barazzouk S, Hotchandani S, Vinodgopal K, Kamat P. Single-wall carbon nanotube films for photocurrent generation. A prompt response to visible light irradiation. J Phys Chem B 2004;108:17015–8.

[29] Kamat P, Thomas K, Barazzouk S, Girishkumar G, Vinodgopal K,Meisel D. Self-assembled linear bundles of single wall carbon nanotubes and their alignment and deposition as a film in a dc field.J Am Chem Soc 2004;126:10757–62.

[30] Lee CY, Chuang HM, Li SY, Lin P, Tseng TY. Characteristics and electrochemical performance of supercapacitors with manganese oxide-carbon nanotube nanocomposite electrode. J Electrochem Soc2005;152(4):A716–20.

[31] Jin YW, Jung JE, Park YJ, Choi JH, Jung DS, Lee HW, et al.Triode-type field emission array using carbon nanotubes and a conducting polymer composite prepared by electrochemical polymer-ization. J Appl Phys 2002;92(2):1065–8.

[32] Chen GZ, ShafferMSP, Coleby D, Dixon G, ZhouW, Fray DJ, et al.Carbon nanotube and polypyrrole composites: coating and doping.Adv Mater 2000;12(7):522–6.

[33] Jung MS, Jung HY, Suh JS. Horizontally aligned carbon nanotube field emitters having a long term stability. Carbon 2007; 45(15): 2917-2921

[34] Zhu L, Sun Y, Hess DW, Wong CP. Well-aligned open-ended carbon nanotube architectures: an approach for device assembly. Nano Lett 2006; 6(2):243–7

[35]Klinke C, Delvigne E, Barth JV, Kern K. Enhanced field emission from multiwall carbon nanotube films by secondary growth. J Phys Chem B 2005; 109(46):21677–80

[36] Wong YM, Kang WP, Davidson JL, Choi BK, Hofmeister W, Huang JH. Fabrication of aligned convex CNT field emission triode by MPCVD. Diam Relat Mater 2005; 14 (11): 2069 - 2073

[37] Nishuang Liu, Guojia Fang, Xiaoxia Yang, Wei Zeng, Chun Li, Mingjun Wang , Jun Lia, Xingzhong Zhao, Synthesis of patterned carbon nanotube arrays for field emission using a two layer Sn/Ni catalyst in an ethanol flame。Diamond & Related Materials 18 (2009) 1375–1380

[38] Nguyen Van Quy, Nguyen Duc Hoa,Wan Jun Yu, You Suk Cho,Gyu SeokChoi and Dojin Kim, The use of anodic aluminium oxide templates for triode-type carbon nanotube field emission structures toward mass-production technology. Nanotechnology 17 (2007) 2156–2160

[39] A.Mathur, S.S. Roy , Kiran S. Hazra , D.S. Misra , J.A. McLaughlin, Growth of carbon nanotube arrays using nanosphere lithography and their application in field emission devices. Diamond & Related Materials 19 (2010) 914–917.

[40] Do-Hyung Kim , Chang-Duk Kim, Hyeong Rag Leea , Effects of the ion irradiation of screen-printed carbon nanotubes for use in field emission display applications. Carbon 42 (2004) 1807–1812。

[41] Fan-Guang Zeng, Chang-Chun Zhu, Weihua Liu, Xinghui Liu, The fabrication and operation of fully printed Carbon nanotube field emission displays. Microelectronics Journal 37 (2006) 495–499.

[42] Milnoux E, Groening O, Teo KBK, Dalal S, Gangloff L, Schnell JP, et al. Achieving high-current carbon nanotube emitters. Nano Lett 2005; 5(11):2135–8

[43] Kang DW, Suh JS. Fabrication temperature effect of the field emission from closed and open tip carbon nanotube arrays fabricated on anodic aluminum oxide films. J Appl Phys 2004; 96(9):5234–8

[44] Min Qian, Tao Feng*, KaiWang, Hui Ding, Yiwei Chen, Zhuo Sun，A comparative study of field emission properties of carbon nanotube films prepared by vacuum filtration and screen-printing。Applied Surface Science 256 (2010) 4642–4646.

[45] Rujia Zou, Guannan Zou , Chunrui Wan, Shaolin Xue, Jian Liu , Guangping Ren , Improving the emission characteristics of a carbon nanotube film in NaCl electrolyte. Microelectronics Journal 40 (2009) 115- 119.

[46] Uh HS, Ko SW, Lee JD. Growth and field emission properties of carbon nanotubes on rapid thermal annealed Ni catalyst using PECVD. Diam Relat Mater2005; 14(7): 850-854

[47] Wang SC, Huang BC. Field emission properties of Ag/SiO2/carbon nanotube films by pulsed voltage co-electrophoretic deposition.Thin solid films2008; 26(11):436–9

[48] Fang-HsingWang, Tzu-Ching Lina,b, Shien-Der Tzeng, Field emission properties of carbon nanotube cathodes produced using composite plating Applied Surface Science 256 (2010) 7600–7605

[49] Kai Wang, Hui Ding, Yiwei Chen, Qiong Li, Zhuo Sun, Field emission of carbon nanotube films fabricated by vacuum filtration Min Qian, Tao Feng. Physica E 43 (2010) 462–465

[50] Chen Z, Engelsen D, Bachmann P K, Elsbergen V, Koehler I, Merikhi et al. High emission current density microwave-plasma-grown carbon nanotube arrays by postdepositional radio-frequency oxygen plasma treatment.Appl.Phys.Lett2005; 87(24): 3104-87
[51] Hahn J, Jung SM, Jung HY, Heo SB, Shin JH, Suh JS. Fabrication of clean carbon nanotube field emitters . Appl.Phys.Lett2006; 88(11): 3101-76
[52] Uh HS, Ko SW, Lee JD. Growth and field emission properties of carbon nanotubes on rapid thermal annealed Ni catalyst using PECVD. Diam Relat Mater2005; 14(7): 850-854
[53] Lee HJ, Lee YD, Cho WS, Ju BK, Lee YH, Han JH, et al. Field-emission enhancement from change of printed carbon nanotube morphology by an elastomer. Appl.Phys.Lett2005; 14(3): 850-854
[54] Jung MS, Ko YK, Jung DH, Choi DH, Jung HT, Heo JN et al. Electrical and field-emission properties of chemically anchored single-walled carbon nanotube patterns. Appl.Phys.Lett 2005; 87(1): 3114-80
[55] Amaratunga GAJ, Silva SRP. Nitrogen containing hydrogenated amorphous carbon for thin-film field emission cathodes.Appl.Phys.Lett2005; 68(18): 2529-7
[56] Xu X, Brandes GR. A method for fabricating large-area, patterned, carbon nanotube field emitters. Appl.Phys.Lett1999; 74(17): 2529-2
[57] ZHANGJ H, CHANYC, Research on the Contact Resistance, Reliability and Degradation Mechanisms of Anisotropically Conductive Film Interconnection for Flip-chip-on-flex Applications [J]. Journal of Electronic Materials, 2003, 32 (4) : 228～234.
[58] K.-C. Liao, C.-C. Chang, Applications of damage models to durability investigations for electronic connectors. Materials & Design, Volume 30, Issue 1, January 2009, Pages 194-199
[59] J. Das, S. M. Sivakumar, An evaluation of multiaxial fatigue life assessment methods for engineering components .International Journal of Pressure Vessels and Piping,Volume 76, Issue 10,August 1999, Pages 741-746
[60] Tadeusz Hejwowski, Erosive and abrasive wear resistance of overlay coatings. Vacuum,Volume 83, Issue 1, 4 September 2008, Pages 166-170 Chung, I.T. Han, A.Y. Cao, X.F. Zhang, X. Xiao, M.Q. Ding, D.M. Zhuang, C.L. Xu, B.Q. Wei, J. Liang, D.H. Wu, Materials Letters, 51, 2001,371–374
[61] E.Bousser, M.Benkahoul,L. Martinu, J.E. Klemberg-Sapieha, Effect of microstructure on the erosion resistance of Cr–Si–N coatings Surface and Coatings Technology.In Press, Corrected Proof,Available online 13 August 2008
[62] Phaedon Avouris, Carbon nanotube electronics. Chemical Physics, Volume 281, Issues 2-3, 1 August 2002, Pages 429-445
[63] R.C. Batra, W. Jiang, Analytical solution of the contact problem of a rigid indenter and an anisotropic linear elastic layer. International Journal of Solids and Structures, Volume 45, Issues 22-23,November 2008, Pages 5814-5830
[64] W. Wayne Chen, Q. Jane Wang, A numerical model for the point contact of dissimilar materials considering tangential tractions Mechanics of Materials,Volume 40, Issue 11,November 2008, Pages 936-948

[65] Riaz A. Mufti, Adrian Jefferies, Novel method of measuring tappet rotation and the effect of lubricant rheology. Tribology International,Volume 41, Issue 11, November 2008, Pages 1039-1048
[66] Xiaohong Yang, Shuhua Liang, Xianhui Wang, Peng Xiao, Zhikang Fan, Effect of WC and CeO2 on microstructure and properties of W–Cu electrical contact material. Journal of Refractory Metals & Hard Materials 28 (2010) 305–311
[67] Pierre-Marie Geffroy, Thierry Chartier, Jean-François Silvain, Preparation by tape casting and hot pressing of copper carbon composites films. Journal of the European Ceramic Society, Volume 27, Issue 1, 2007, Pages 291-299
[68] Korb, G., Buchgrader, W. and Schubert, T., Thermophysical properties and microstructure of short carbon fibre reinforced Cu-matrix composites made by electroless copper coating or powder metallurgical route respectively, IEEE/CPMT Berlin, Int'l Electronics Manufacturing Technology Symposium,27–28 April, 1998
[69] Praksan, K., Palaniappan, S. and Seshan, S, Thermal expansion characteristics of cast Cu based metal matrix composites. Composites Part A, 1997,28, 1019–1022.
[70] Kor´ab, J., Stef´anik, P., Kaveck´y, S., Sebo, P. and Korb, G., Thermal expansion of cross-ply and woven carbon fibre-copper matrix composites. Composites Part A, 2002, 33, 133–136.
[71] Huiqing Wu, Guifu Ding, Yuchao Wang, Ying Cao, Hong Wang, Chunsheng Yang, Composite Electrodeposition of Zinc and Carbon Nanotubes, IEEE Review On Advances In Micro/Nano And Molecular System (2006) Vol. 1, 211-212
[72] R.H.Baughman,A.A.Zakhidov,W.A.deHeer,Carbon nanotubes:the route toward application.Science,2002,vol.
297 (2),pp.787-792
[73] O. Hjortstam, P. Isberg ,S. S Oderholm ,H. Dai,Can we achieve ultra-low resistivityin carbon nanotube-based metal composites?Appl.Phys.A 78,1175-1179(2004)
[74] O.Chauvet,J.M.Benoit,B.Corraze,Electrical,magneto-transportandlocalization of charge carriers in nanocomposites based on carbon nanotubes.Carbon, Volume 42, Issues 5-6,2004, Pages 949-952
[75] Wu JF, Wyse M, McClain D, Thomas N, Jun J. Fabrication and Field Emission Properties of Triode-Type Carbon Nanotube Emitter Arrays. Nano Lett 2009; 9 (2): 595-600
[76] Nguyen VQ, Nguyen DH, Wan JY, Cho YS, Gyu SC, Kim DJ. The use of anodic aluminium oxide templates for triode-type carbon nanotube field emission structures toward mass-production technology. Nanotechnology2006; 17 (5):2156–2160
[77] Hahn J, Jung SM, Jung HY, Heo SB, Shin JH, Suh JS. Fabrication of clean carbon nanotube field emitters.Appl.Phys.Lett2006; 88(11): 3101-76
[78] Nguyen VQ, Nguyen DH, An MC, Cho YS, Kim DJ. A high-performance triode-type carbon nanotube field emitter for mass production. Nanotechnology2007; 18 (6): 345–361
[79] Nguyen VQ, Nguyen DH, An MC, Cho YS, Kim DJ. A high-performance triode-type carbon nanotube field emitter for mass production. Nanotechnology2007; 18 (6): 345-361
[80] Pirio G, Legagneux P, Pribat D, Teo BK, Chhowalla M, Amaratunga G A J, et al. Fabrication and electrical characteristics of carbon nanotube field emission

microcathodes with an integrated gate electrode. Nanotechnology2002; 13 (4): 143-147

[81] Hu WC, Yuan LM, Chen Z, Gong DW, Saito K. Fabrication and Characterization of Vertically Aligned Carbon Nanotubes on Silicon Substrates Using Porous Alumina Nanotemplates. J. Nanosci. and Nanotechnol 2002; 2(7): 203-207

[82] Jang Y T, Choi C H, Ju B K, Ahn J H , Lee Y H .Suppression of leakage current via formation of a sidewall protector in the microgated carbon nanotube emitter. Nanotechnology 2003; 14 (5):14-497

[83] Uh H S. Low Turn-On Voltage Field Emission Triodes with Selectively Grown Carbon Nanotube Emitters. J. Korean Phys Soc 2004; 45(5):1343 -1343

[84] Xu X, Brandes GR. A method for fabricating large-area, patterned, carbon nanotube field emitters. Appl.Phys.Lett1999; 74(17): 2529-2

[85] Wan JY, Cho YS, Choi GS, Do J. Patterned carbon nanotube field emitter using the regular array of an anodic aluminum oxide template. Nanotechnology 2005; 16 (5): 291–295

7

Assembly and Patterning of Single-Walled Carbon Nanotubes/Organic Semiconductors

Akira Baba[1], Kazunari Shinbo[1], Keizo Kato[1], Futao Kaneko[1],
Hirobumi Ushijima[2] and Kiyoshi Yase[3]
[1]*Center for Transdisciplinary Research, and Graduate School of Science and Technology, Niigata University*
[2]*Photonics Research Institute*
[3]*Nanosystem Research Institute, National Institute of Advanced Industrial Science and Technology (AIST)*
Japan

1. Introduction

Assembling composites of organic semiconductors and carbon nanotubes is a key approach for constructing a wide range of applications, such as the photoelectric conversion devices, biosensors, and electron storage devices (Liu et al., 2008; Chen et al., 2003). To assemble such composites, it is important to develop a method of solubilizing carbon nanotubes because they have low solubility in most solvents. Many groups have explored the solubilization properties of carbon nanotubes after chemical modifications involving their covalent bonding to organic materials (Chen et al., 1998; Baskaran et al., 2005). Carbon nanotubes with covalently linked porphyrin antennae have been developed as potential supramolecular donor-acceptor complexes for applications such as the photovoltaic devices and light-harvesting systems (Guldi et al., 2006). There have also been reports of noncovalent functionalizations of single-walled carbon nanotubes (SWNT) with aromatic compounds such as porphyrin or polyfluorene (Murakami et al., 2003; Tomonari et al., 2007; Nakashima, 2006; Nish et al., 2009). The solubilization mechanism involved π-π interactions between the side walls of the SWNT and aromatic compounds. Since charges can be injected from porphyrin into single-walled carbon nanotubes upon an irradiation of light, the composite has been used as photochemical solar cells (Chitta et al., 2007; Guldi et al., 2005; Hasobe et al., 2006). Sun and coworkers showed that porphyrine could adsorb on semiconducting single-walled nanotubes due to noncovalent interactions which might become an efficient method for the mass separation of semiconducting single-walled nanotubes from metallic single-walled nanotubes (Li et al., 2004). Bao and coworkers showed an enhancement of mobilities and high on/off ratio in organic semiconductor-CNT composites FET devices (Liu et al., 2008). Although fabricating ultrathin films from SWNT-chromophore composites is an important challenge in the development of optoelectronic device applications, there have been few reports on nanostructured assembled ultrathin films, particularly ultrathin films fabricated from noncovalently adsorbed carbon nanotube-chromophore composites.

In this chapter, we introduce the fabrication of nanostructured SWNT-organic semiconductor ultrathin films by a layer-by-layer (LbL) self-assembly approach and an enhancement of the photocurrent generation in the nanostructured corganic semiconductors-SWNT ultrathin films. The LbL self-assembly method, initially reported by Decher, is one of the most convenient techniques for fabricating molecularly controlled ultrathin multilayer films (Decher, 1991). The adsorption process involves the alternate deposition of cationic and anionic species from a solution (Baba et al., 2000, 2006, 2010; Advincula et al., 2003; Sriwichai et al. 2008). Both positively and negatively charged water-soluble organic semiconductor molecules were used for the solubilization of the SWNT. To investigate the composite ultrathin films properties, surface plasmon spectroscopy, UV-vis. spectroscopy, and fluorescence spectroscopy were employed as well as the cyclic voltammetric properties were studied. The photocurrent measurements were performed in photoelectrochemical cells in KCl aqueous solution using methyl viologen from chromophore-SWNT composite films on gold electrode as the electron acceptor molecule.

On the other hand, another important challenge for the nano-device applications is to pattern carbon nanotubes-organic semiconductor nanocomposites in large-scale. Patterned carbon nanotubes/thiol suspensions have been reported using both "top-down" and "bottom-up" techniques (Yan et al., 2007; Whang et al., 2004). Many groups have reported the microcontact-printed pattern with a variety of materials (Xia et al., 1998), since the first report by Kumar and Whitesides (Kumar & Whiteside, 2002). Recent reports have shown that selective and aligned carbon nanotubes were possible on functionalized surfaces by microcontact printing or PDMS transfer printing (Rao et al., 2003; Tsukruk et al., 2004; Ko et al., 2004; Ding et al., 2006; Meitl et al., 2004). While the microcontact printing technique is a versatile method to fabricate microstructured composite, "top-down" technique such as dip-pen nanolithography is also an attractive method as nanoscale manipulation of carbon nanotubes (Piner et al., 1999; Ginger et al., 2004). In the dip-pen nanolithography technique, AFM cantilever is inked with organic materials, which are then transferred to flat surfaces. Recently, Wang and co-workers have demonstrated that the microcontact printing and parallel dip-pen nanolithography technique allows one to pattern single-walled carbon nanotubes functionalized with COOH-terminated self-assembled monolayers (Wang et al., 2006). Another method, called fountain-pen nanolithography (FPN), has a potential to be effective large-area patterning technique, in which the modified AFM cantilever serves as a micro-pipet. In FPN method, the liquid ink is filled in reservoirs on AFM cantilevers, and then flows onto the surface when in contact with the surface (Kim et al., 2005). In this chapter, we also introduce the pattern formation and assembly of single-walled carbon nanotubes/ organic semiconductor composites. Microcontact printing (μCP) was done by a standard procedure. The PDMS stamp was inked with single-walled carbon nanotubes-organic semiconductor molecule in ethanol by drop-cast technique, dried under nitrogen, and applied to a cleaned glass slide surface. The dip-pen nanolithography and fountain-pen nanolithography techniques were also used for micro/nanopatterning of the composites and large-area pattern formation, respectively.

2.1 Solubilization of SWNT with water-soluble organic semiconductors

To examine the solubility of organic semiconductors-SWNT composites, UV-vis. absorption spectra were measured in aqueous solution. As organic semiconductor molecules, cationic

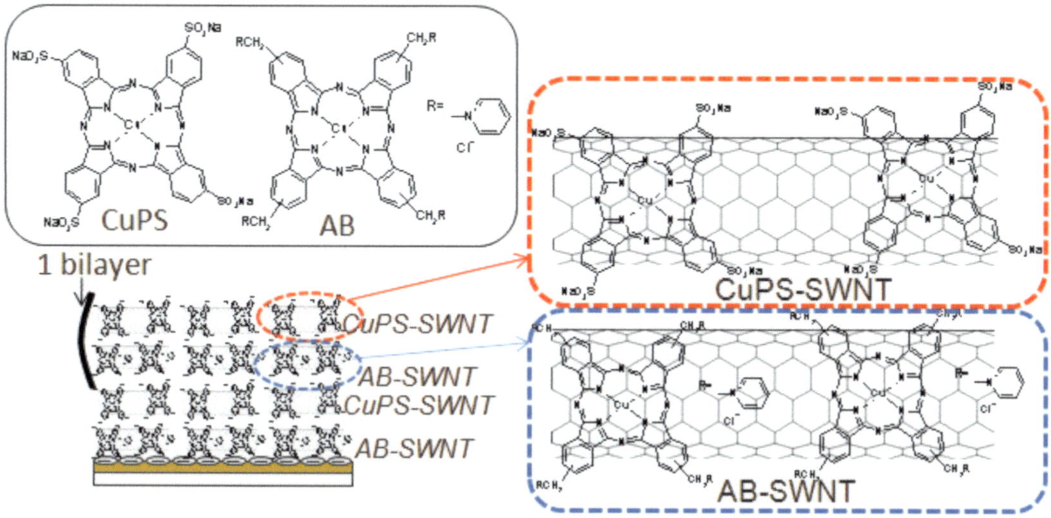

Fig. 1. Schematic drawing of the fabrication of nanostructured alcian blue, pyridine variant (AB)–SWNT/phthalocyanine-3,4′,4′′,4′′′-tetrasulfonic acid tetrasodium salt (CuPS)-SWNT LbL films. From Ref. [Baba et al. 2010] with permission.

alcian blue, pyridine variant (AB; Aldrich) and anionic Copper phthalocyanine-3,4′,4′′,4′′′-tetrasulfonic acid tetrasodium salt (CuPS; Aldrich) molecules were used (shown in the inset of Fig. 1). Figure 2(a) shows the UV-vis. absorption properties of AB and AB-SWNT composites after sonication for 3 h and after sonication (3 h)/centrifugation (1 h at rotation speeds of 7000 and 10000 rpm) in aqueous solution. Figure 2(b) shows the spectra for CuPS and CuPS-SWNT composites after sonication and after sonication/centrifugation. In the absorption spectra of both composites, peaks due to the Soret band (340 - 350 nm) and Q-band (610 – 620 nm) were clearly observed. After the sonication, an increase in the baseline of the broad absorption band was observed, indicating that both metallic SWNTs (400 - 600 nm region, corresponding to M_{11}) and semiconducting SWNTs (600 – 950 nm, corresponding to S_{22} transitions) were complexed with both AB and CuPS molecules. Since the bundles or aggregated complexes were mostly removed by centrifugation, the baseline gradually decreased as the rotation speed increased. On the other hand, the addition of SWNTs to AB or CuPS resulted in decreases in the peak intensity for both the Soret band and the Q-band, accompanied by a redshift of ca. 1 nm. These results suggest that an interaction between SWNTs and AB or CuPS is induced. The decrease in the peak intensities should be due to a decrease in the density of trapped electrons in AB or CuPS because they are transferred to SWNTs. Peak intensities then increased slightly as the rotation speed increased because of the removal of bundles or aggregates, in good agreement with the behavior of the baseline. Similar results were recently observed for SWNT-imidazole-porphyrin, SWNT-methyl viologen and SWNT-TiO2 (Kongkanand & Kamat, 2007) composites, which acted as donor-acceptor nanohybrid or electron storage systems in the SWNTs with Fermi level equilibration. Since a large number of SWNTs were removed at a higher rotation speed, a rotation speed of 6000 rpm was chosen for the fabrication of LbL films.

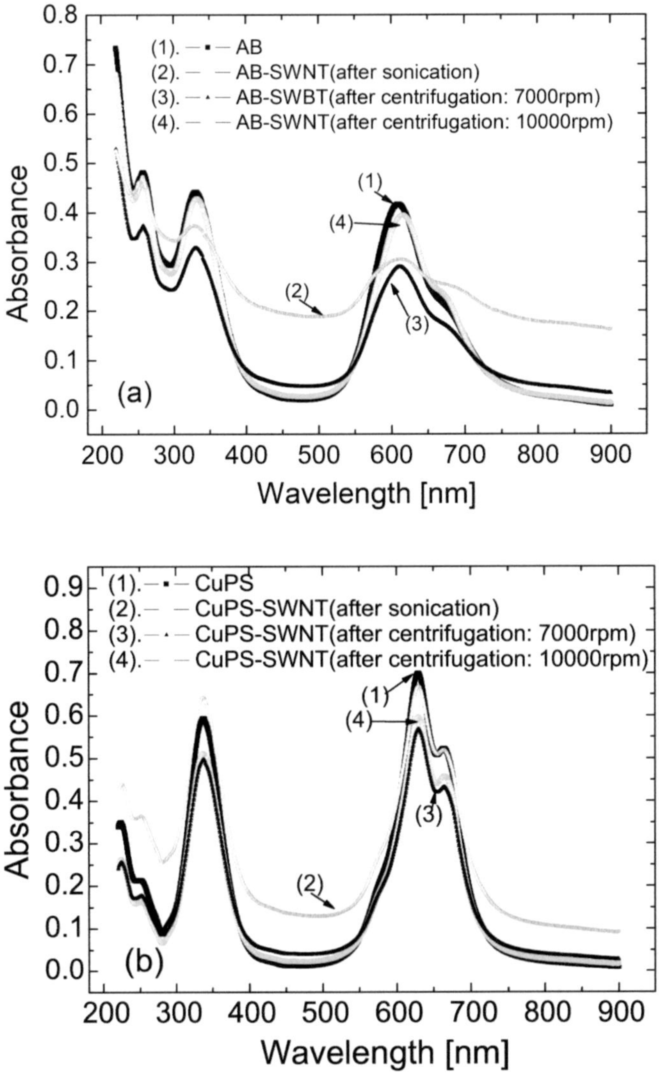

Fig. 2. UV-vis. absorption properties. (a) AB (0.25 mg/ml) and AB (0.1 mg/ml)-SWNT (0.25 mg/ml) composites after sonication for 3 h and after sonication (3 h)/centrifugation (1 h at rotation speeds of 7000 and 10000 rpm) in aqueous solution. (b) CuPS (0.25 mg/ml), CuPS (0.25 mg/ml)-SWNT (0.25 mg/ml) composites after sonication for 3 h, and after sonication (3 h)/centrifugation (1 h at rotation speeds of 7000 and 10000 rpm) in aqueous solution. From Ref. [Baba et al. 2010] with permission.

2.2 Fabrication of LbL ultrathin films

First, UV-vis. spectroscopy was performed to monitor the AB-SWNT/CuPS-SWNT multilayer formation. As schematically shown in Fig. 1, 0.5 bilayer indicates 1 layer of AB-SWNT or CuPS-SWNT composites. As can be seen in Figure 3, the change in absorbance for the first bilayer was larger than that for the subsequent bilayers. This result suggests that the AB-SWNT and CuPS-SWNT composites are poorly charged; thus, the amount of deposition on the APS-functionalized surface is larger than that on the composites. From second bilayer, the UV-vis. absorbance exhibited a monotonic increase with the number of bilayers,as shown in the inset, suggesting the successive deposition of the film during the assembly of the multilayers. As discussed in the previous section, the UV-vis. spectra of AB-SWNT and CuPS-SWNT in solution exhibited peaks at approximately 331 nm and 611 nm and at 338 nm and 630 nm, respectively. On the other hand, the Soret absorption band and Q-band in the multilayered film were observed at 334 nm and 619 nm, respectively, almost superpositions of the peaks due to the two phthalocyanine molecules (AB and CuPS). Furthermore, it should be noted that the baseline of the absorption at approximately 500 nm increased with the number of bilayers, while the baseline of the AB-CuPS LbL film (without SWNTs) remained almost constant (see supporting information), indicating that a well-ordered SWNT-phthalocyanine composite film was fabricated. SPR spectroscopy was also employed in order to study the *in situ* adsorption process of the composites. The stepwise increase in reflectivity with increasing number of bilayers indicated the deposition of a constant amount of AB-SWNT/CuPS-SWNT. The film thickness for the 5-bilayer film was estimated to be ca. 14.0 nm on the assumption of a dielectric constant of 2.1 + i0.2, giving an average thickness of 2.8 nm for each bilayer.

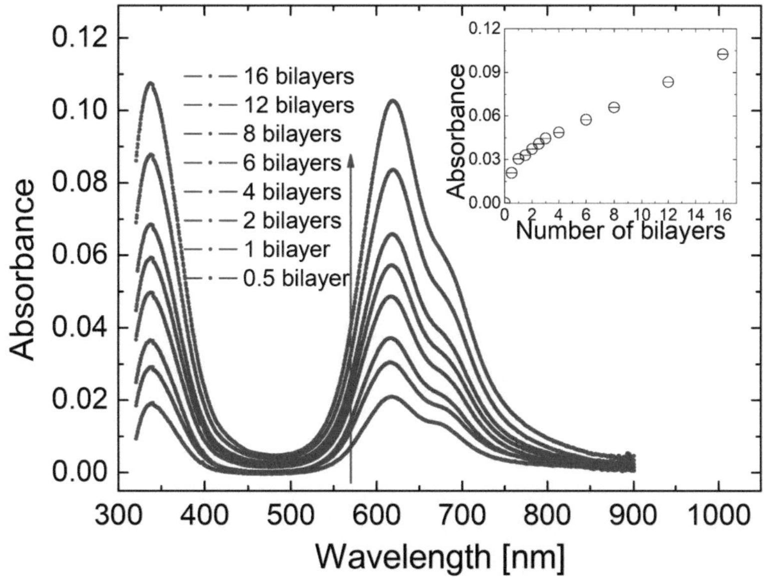

Fig. 3. UV-vis. spectrum of AB-SWNT/CuPS-SWNT LbL film on glass slide as a function of number of bilayers. (0.5 bilayer indicates 1 layer of AB-SWNT (or 1 layer of CuPS-SWNT).) The inset shows the absorbance of the UV-vis. peaks at 617.5 nm as a function of number of bilayers. From Ref. [Baba et al. 2010] with permission.

2.3 Photocurrent properties of LbL films

To study the photocurrent performance of the organic semiconductior molecules-SWNT LbL films, we prepared a photoelectrochemical cell. As the organic semiconductor molecules, cationic sodium copper chlorophyllin (SCC) and anionic 5,10,15,20-tetrakis (1-methyl-4-pyridinio) porphyrin tetra(p-toluenesulfonate) (TMPyP) were used. The photocurrent measurements were performed in KCl (5 mM) and methyl viologen (100 mM) aqueous solution. Methyl viologen was used as an electron acceptor molecule. Figure 4 shows the photocurrent response of 20 bilayers of TMPyP-SWNT/SCC-SWNT LbL film upon irradiation by visible light (490–740 nm), measured at −0.2 V. For comparison, the photocurrent response in 10 and 20 bilayers of TMPyP/SCC LbL film (no SWNT) was also measured under the same conditions (Figure 4(a)). As shown in this figure, upon irradiation by visible light in Q-band absorption region of chromophores, the cathodic current increased and when the irradiated light was turned off, the current shifted back to almost the initial level, although some fluctuation was observed. The amount of current increase for the TMPyP-SWNT/SCC-SWNT LbL film electrodes was 1.5–2 times more than the increase for the TMPyP/SCC LbL film, indicating that the SWNT is responsible for the enhancement of the photocurrent generation in the chromophore-layered films. These results suggest that the noncovalently adsorbed SWNT with chromophores provide effective charge separation in LbL films upon irradiation of visible light in Q-band absorption region. Once the excited electrons reach the SWNT from the choromophores, they are effectively transferred to methyl viologen in the solution. A schematic illustration of the photocurrent generation at the chromophores-SWNT electrodes is shown in Fig. 4(b) (Imahori et al., 2000; Guldi et al. 2005). It should be noted that the photocurrent generated in the multilayered system increases as the number of bilayers increases. Since the mechanism of the charge separation from the photoexcited state and the charge transfer to the methyl viologen is responsible for the presence of the SWNT, there should be SWNT pass-way to methyl violegen electrolyte solution in multilayered system. One possibility is that the TMPyP-SWNT/SCC-SWNT LbL films form a bulk hetero-like structure, so that the charge separation and the electron transfer can be generated in multilayered system. This explanation is reasonable because the chromophore-SWNT composite is not a well-ordered film structure. Furthermore, it is well known that LbL films often have interpenetrated structures (Decher, 1997; Yoo et al., 1998). Another possibility is that the electrons are transferred from SWNT to methyl viologen inside the film, since such small molecules can exist inside LbL films in electrolyte solution. This is reasonable because LbL films usually contain electrolytes in aqueous solution (Jiang et al., 2007; Baba et al. 2002). These results indicate that the TMPyP-SWNT/SCC-SWNT LbL multilayered film should have potential for effective photocurrent generation.

2.4 Patterning of organic semiconductor-SWNT composites

By using the solubilized SWNT-AB composites, a variety of patterning techniques were examined. First, microcontact printing technique was used to fabricate the patterned SWNT-AB composites. In this patterning, Si substrates were cleaned by piranha solution to create hydrophilic and negatively charged surface (Chrisey et al., 1996; Moriguchi et al., 1999; Im et al. 2000) as the AB molecule is positively charged. Figure 5 shows AFM topography images of SWNT-AB composites on Si substrates after the patterning by μCP method using 1μm line and 5 μm line PDMS stamps. In both cases, the patterned linewidth almost corresponded to the linewidth of PDMS stamps. In the case of patterning using 1 μm linewidth, the height was a few nm, which is slightly larger than the diameter of SWNT, indicating that the monolayer of SWNT-AB composites was transferred onto Si surface. As can be seen in this Figure, SWNT ranging from 500 nm to 3 μm in length were complexed with AB molecules. One can find the fact that the SWNT-AB composites with the length larger than 1 μm were confined to stay inside the 1 μm line space, resulting in the alignment

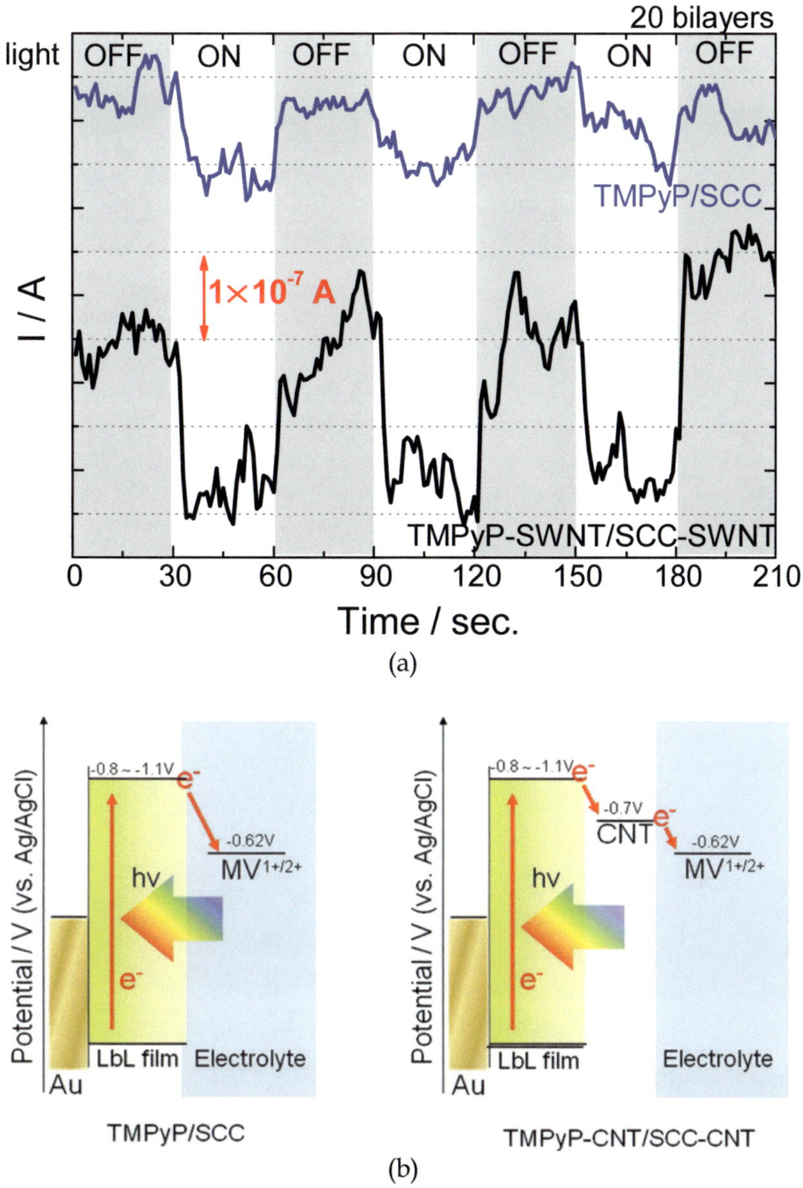

Fig. 4. (a) Photocurrent response upon irradiation by visible light (490–740 nm) to 20 bilayers of TMPyP-SWNT/SCC-SWNT and TMPyO/SCC (no SWNT) LbL film measured at −0.2 V(b) A schematic illustration of photocurrent generation at the chromophore-SWNT electrodes. From Ref. [Baba et al. 2010] with permission.

of SWNT along the direction. In the case of μCP using 5 μm linewidth, the SWNT-AB composites were more densely assembled on Si surface and contained a lot of longer SWNT as compared to 1 μm linewidth pattern. The schematic illustration is also shown in the figure. The average height of the pattern was ca. 15 nm which implies several SWNT-AB composites were overlaid on the surface. From these results, one can expect that the SWNT-

AB composites with length longer than the linewidth of PDMS hardly adsorb onto PDMS surface during the inking process, resulting in the random alignment after the patterning.

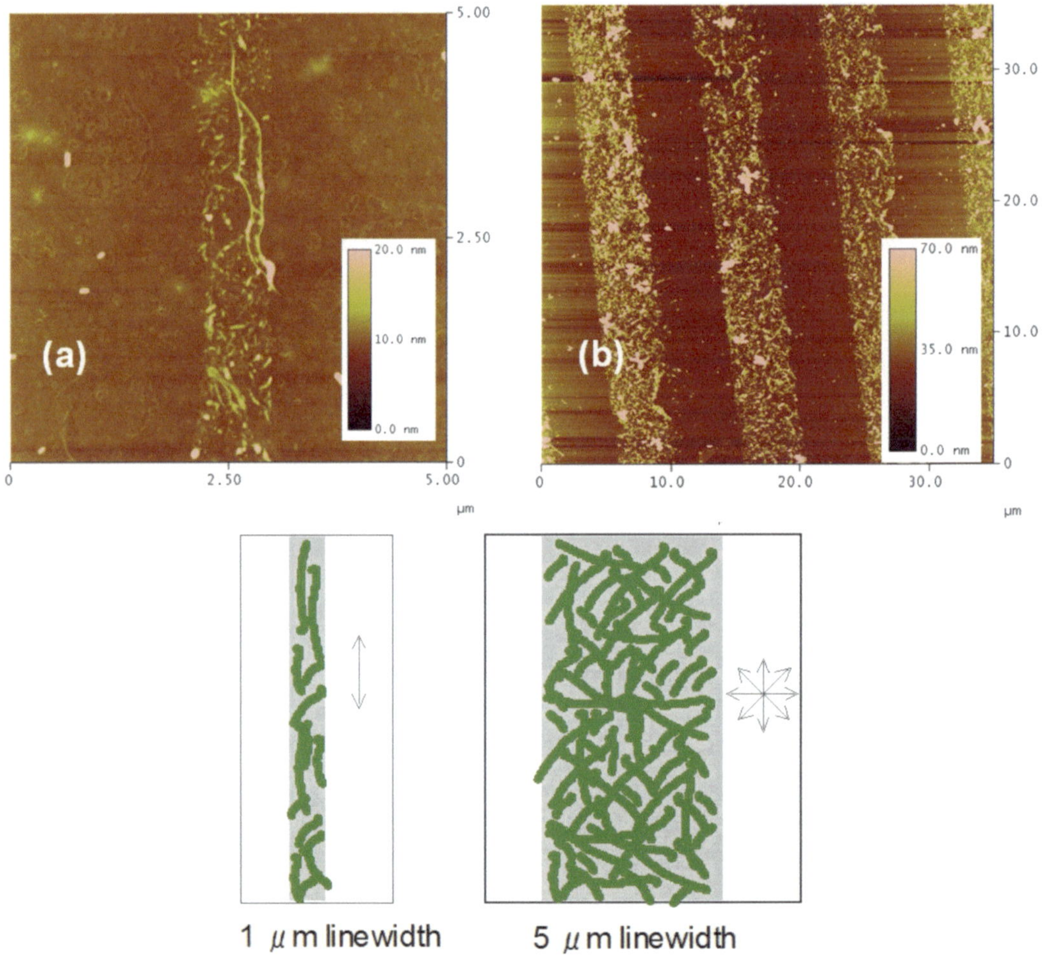

Fig. 5. AFM topographic images of μ-contact-printed SWNT-AB composites (1 μm linewidth (a) and 5 μm linewidth (b)), and schematic diagram of SWNT-AB alignment. From Ref. [Baba et al. 2009] with permission.

In order to fabricate a smaller size pattern of SWNT-AB composites, dip-pen nanolithography technique was applied. A silicon AFM cantilever was inked with SWNT-AB composites after the Si cantilever was cleaned by UV ozone cleaner. Figure 6(a) shows confocal laser microscopy image of the AFM cantilever after soaking with SWNT-AB composites ink solution in ethanol. A large black area at the left side is the cantilever tip, and the rest of the black area shows the adhesion of SWNT-AB composites on the AFM cantilever. As shown in this Figure, the composites were dispersed on AFM cantilever, though they were not homogenous dispersion. Furthermore, in order to confirm the presence of the SWNT-AB composites, AFM image of the inked AFM cantilever was taken by a tapping mode measurement (Fig. 6(b)). In order to prevent the interference between the inked AFM cantilever tip and the probing AFM cantilever tip during the tapping mode AFM measurement, the right side (squared area) was chosen to be observed. As shown in this figure, the SWNT-AB composites ranging from ca. 500 nm to 3 μm in length were indeed observed on the inked AFM cantilever, indicating that the inking with SWNT-AB composites in ethanol resulted in the adhesion onto Si AFM cantilever. Then, the nanoscale transfer of the composites onto a flat Si surface was attempted by using the inked AFM cantilever. Unlike a conventional DPN patterning technique using thiol molecules, SWNT-AB composites were hardly transferred onto Si surface at relative humidity of up to 70 %. Hence, DPN was driven at relative humidity of 70-80 %. A contact mode AFM was used for the writing, and the pattern was read out by a tapping-mode AFM measurement. Although it is hard to discern each nanotube, the patterned SWNT-AB composites were observed after DPN as shown in Fig. 7. One limitation in this measurement was the fact that the same cantilever was used for both the writing and reading, which should be improved by using a different cantilever (McKendry et al., 2002), hence the image in high resolution could not be obtained for this time. At the condition, it was possible to write 110 nm line in width. The height between A-B was about 11 nm which corresponds to several SWNT-AB composites, while the height of the circled area almost corresponds to a monolayer of SWNT-AB composites.

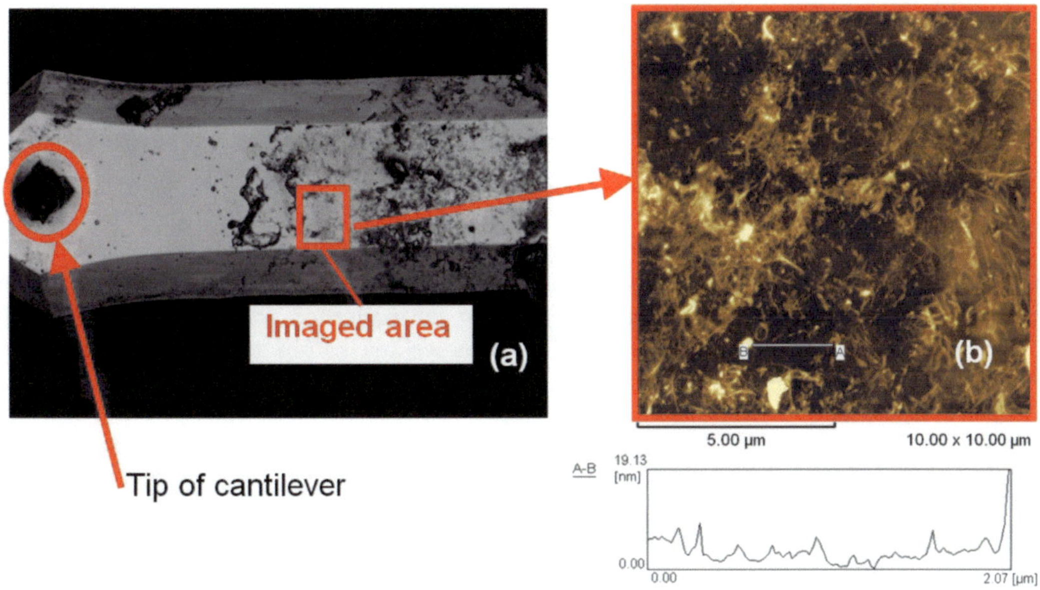

Fig. 6. Confocal laser microscopy image (a) and AFM topographic image (b) of SWNT-AB nanocomposites on AFM dip-pen cantilever.

As another strategy to fabricate micro/nanopatterning of SWNT-AB composites, a fountain-pen nanolithography technique was also attempted. The fountain-pen nanolithography has been done using microfluidic-based cantilevers. Aqueous SWNT-AB composites ink was fed in reservoir and was delivered to the top of the cantilever through the microchannel. As shown in Fig. 8, the composites solution was observed on the top of the cantilever which was delivered through the microchannel. In order to prevent the evaporation of aqueous

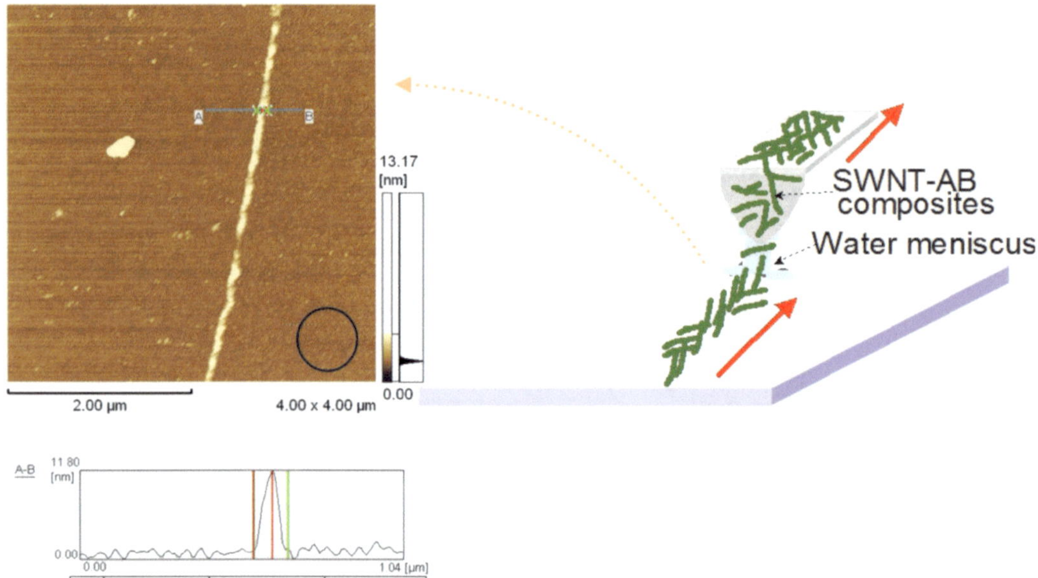

	Distance / µ m	Height / nm	Angle / deg.
— (red)	0.06	10.42	9.45
— (green)	0.11	0.18	0.09

Fig. 7. AFM topographic image on Si after dip-pen nanolithography of SWNT-AB nanocomposites and the schematic illustration. From Ref. [Baba et al. 2009] with permission.

SWNT-AB solution from the microchannel at room temperature, a small amount of glycerin was added in the composites solution. The patterning was performed to demonstrate the capability of the technique. As shown in Fig. 8, the dots of 8 µm diameter or the lines of 1.5 µm width and of 140 µm length was obtained in the CCD image, although the AFM image was not obtained due to the difficulty of finding the patterned area with another AFM instrument. In this experiment, 1 line or 1 dot was written within a few seconds after the cantilever was in contact with the surface, which showed the ability of large scale assembly of carbon nanotubes.

3. Conclusion

We have shown several assembly and patterning techniques of single-walled carbon nanotubes (SWNT)-organic semiconductor molecules composites. The organic semiconductor molecules-SWNT LbL films were fabricated with a convenient method that uses noncovalently adsorbed carbon nanotube-organic semiconductor composites. SWNT were solubilized with water-soluble charged organic semiconductor molecules, which were then used for electrostatic LbL multilayer fabrication. UV-vis. spectroscopy revealed the constant deposition of composite films. Cyclic voltammetry showed that the electroactivity of the hybrid film is enhanced by the incorporation of SWNT in the layered film. The slopes obtained from the peak currents for three film structures indicate that the SWNT facilitate electron transfer to the electrode/electrolyte solution. Organic semiconductor-SWNT LbL film electrodes exhibit an enhancement of photocurrent generation as compared with

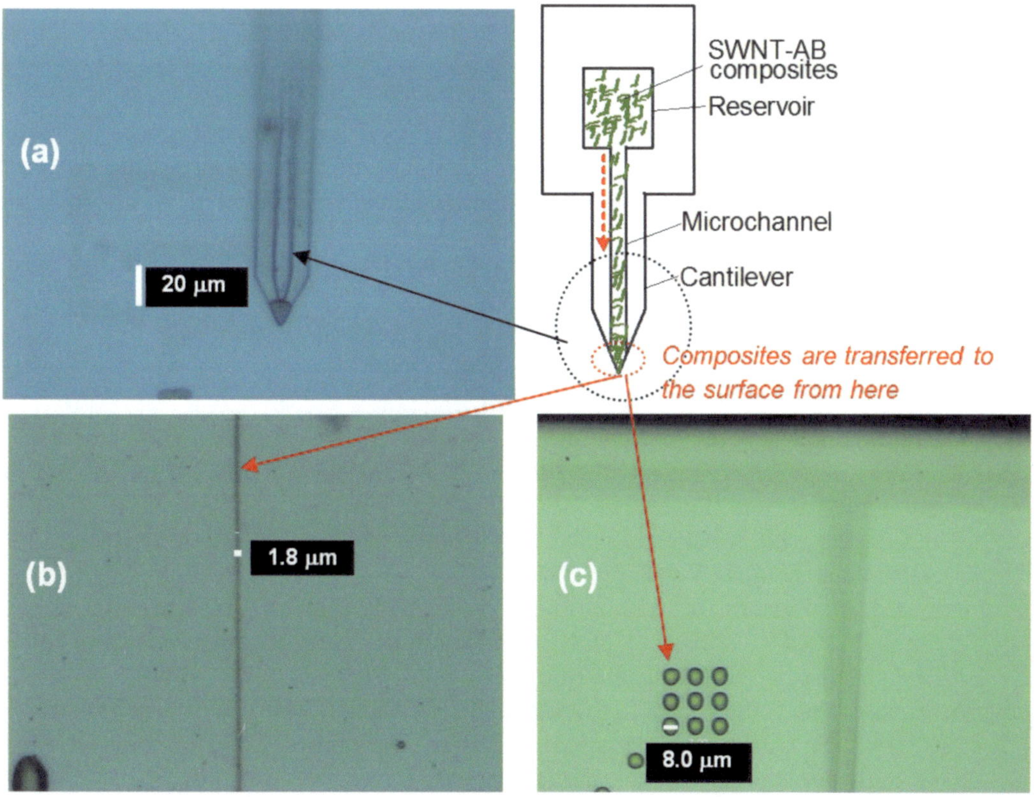

Fig. 8. CCD images of aqueous SWNT-AB composites in microchannel (top view) (a), patterned line (b), and patterned dots (c). From Ref. [Baba et al. 2009] with permission.

Organic semiconductor LbL (no SWNT) film electrodes, suggesting efficient charge separation and electron transfer in the system. These results suggest that the noncovalently adsorbed organic semiconductor-carbon nanotube composite is effectively assembled to form nanostructured films, which should provide new opportunities for photoelectric conversion devices. Furthermore, the pattern with 1 μm linewidth of the composites was obtained by microcontact printing method which tended to align the line direction. Nanometer scale patterns of SWNT- AB composites with 10 nm height was also obtained by dip-pen nanolithography on silicon surfaces. Fountain-pen nanolithography was also utilized for a possible application to fabricate large-scale patterning technique for organic semiconductor-carbon nanotubes composites.

4. Acknowledgments

We would like to thank all collaborators in this work at Niigata University and Photonics Research Institute, National Institute of Advanced Industrial Science and Technology (AIST). We gratefully acknowledge financial support from the Ricoh Company. A.B. would like to acknowledge funding from the Grant-in-Aid for Young Scientists (Start-up) from the Japan Society of Promotion Science.

5. References

Advincula, R.; Park, M. K.; Baba, A. & Kaneko, F. (2003). Photoalignment in Layer-by-Layer Ultrathin Films Molecularly Ordered Small Molecule Azobenzene Dyes. *Langmuir*, Vol. 19, pp. 654-665.

Baba, A.; Kaneko, F. & Advincula, R. C. (2000). Polyelectrolyte Adsorption Processes Characterized *In Situ* Using the Quartz Crystal Microbalance Technique : Alternate Adsorption Properties in Ultrathin Polymer Films. Colloids Surf. A, Vol. 173, pp. 39-49.

Baba, A.; Locklin, J.; Xu, R. S. & Advincula, R. (2006). Nanopatterning and Nano-Charge writing in Layer-by-Layer Quinquethiophene (5TN) /Phthalocyanine (CuPS) Ultrathin Films. *J. Phys. Chem. B*, Vol. 110, pp. 42-45.

Baba, A.; Park, M. K.; Advincula, R. C. & Knoll, W. (2002). Simultaneous Surface Plasmon Optical and Electrochemical Investigation of Layer-by-Layer Self-assembled Conducting Ultrathin Polymer Films. *Langmuir*, Vol. 18, pp. 4648-4652.

Baba, A.; Ponnapati, R.; Taranekar, P.; Knoll, W.; Advincula. R. (2010). Electrochemical Surface Plasmon Resonance (EC-SPR) and Waveguide Enhanced Glucose Biosensing with N-Alkylaminated Polypyrrole/Glucose Oxidase Multilayers. *ACS Appl. Mater. & Inter.*, Vol. 2, pp 2347–2354.

Baba, A.; Sato, F.; Fukuda, N.; Ushijima, H.; Yase, K. (2009). Micro/Nanopatterning of Single-Walled Carbon Nanotubes-Organic Semiconductor Composites. *Nanotechnology*, Vol. 20, pp85301(6pp).

Baba, A.; Kanetsuna, Y.; Sriwichai, S.; Ohdaira, Y.; Shinbo, K.; Kato, K.; Phanichphant, S.; Kaneko, F. (2010). Nanostructured Carbon Nanotubes/Copper Phthalocyanine Hybrid Multilayers Prepared Using Layer-by-Layer Self-Assembly Approach. *Thin Solid Films*, Vol. 518, pp2200–2205.

Baba, A.; Matsuzawa, T.; Sriwichai, S.; Ohdaira, Y.; Shinbo, K.; Kato, K.; Phanichphant, S.; Kaneko, F. (2010). Enhanced Photocurrent Generation in Nanostructured Chromophore/Carbon Nanotube Hybrid Layer-by-Layer Multilayers. *J. Phys. Chem. C*, Vol. 114, pp 14716–14721.

Baskaran, D.; Mays, J. W.; Zhang, X. P. & Bratcher, M. S. (2005). *J. Am. Chem. Soc.*, Vol. 127, 6916.

Chen, R. J.; Bangsaruntip, S.; Drouvalakis, K. A.; Kam, N. W. S.; Shim, M.; Li, Y. M.; Kim, W.; Utz, P. J. & Dai, H. J. (2003). *Proc. Natl. Acad. Sci. USA*, Vol. 100, 4984.

Chen, J.; Hamon, M. A.; Hu, H.; Chen, Y.; Rao, A. M.; Eklund & P C.; Haddon, R. C. (1998). *Science*, Vol. *282*, 95.

Chrisey, L. A.; Lee, G. U. & O'Ferrall, C. E. (1996). *Nucleic Acids Res.*, Vol. 24, 3031.

Chitta, R.; Sandanayaka, A. S. D.; Schumacher, A. L.; D'Souza, L. D.; Araki, Y.; Ito, O. & D'Souza, F. (2007). *J. Phys. Chem. C*, Vol. 111, 6947.

Decher, G. (1997). Science, Vol. 277, 1232.

Decher, G. & Hong, J. D. (1991). *Makromol. Chem. Makromol. Symp.*, Vol. 46, 321.

Ding, L.; Zhou, W.; Chu, H.; Jin, Z.; Zhang, Y. & Li, Y. (2006). *Chem. Mater.*, Vol. 18 4109.

Ginger, D. S.; Zhang, H. & Mirkin, C. A. (2004) *Angew. Chem. Int. Ed.*, Vol. 43, 30.

Guldi, D. M.; Rahman, G. M. A.,; Prato, M.; Jux, N.; Qin, S. & Ford, W. (2005). *Angew. Chem. Int. Ed.*, Vol. 44, 2015.

Guldi, D. M.; Rahman, G. M. A.; Jux, N.; Balbinot, D.; Tagmatarchis, N.; Prato, M. (2005). *Chem. Commun.*, Vol. 15, 2038.

Guldi, D. M.; Rahman, G. M. A.; Sgobba, V. & Ehl, C. (2006). Chem. Soc. Rev., Vol. 35, 471.

Hasobe, T.; Fukuzumi, S.; & Kamat, P. V. (2006). *J. Phys. Chem. B*, Vol. 110, 25477.

Im, H. J.; Kwon, P. O.; Kim, H. J. & Lee, S. H. (2000). *Macromolecules*, Vol. 33, 9606.

Imahori, H.; Norieda, H.; Nishimura, Y.; Yamazaki, I.; Higuchi, K.; Kato, N.; Motohiro, T.; Yamada, H.; Tamaki, K.; Arimura, M. & Sakata, Y. (2000). *J. Phys. Chem. B*, Vol. 104, 1253.

Jiang, G. Q.; Baba, A.; Ikarashi, H.; Xu, R. H.; Locklin, J.; Kashif, K. R.; Shinbo, K.; Kato, K.; Kaneko, F. & Advincula, R. (2007). *J. Phys. Chem. C.*, Vol. 111, 18687.

Liu, S.; Mannsfeld, S. C. B.; LeMieux, M. C.; Lee, H. W. & Bao, Z. (2008). Organic semiconductor-carbon nanotube bundle bilayer field effect transistors with enhanced mobilities and high on/off ratios. *Appl. Phys. Lett.*, Vol. 92, 053306.

Kim, K. H.; Moldovan, N. & Espinosa, H. D. (2005). *Small*, Vol. 1, 632.

Ko, H.; Peleshanko, S. & Tsukruk, V. V. (2004). *J. Phys. Chem. B*, Vol. 108, 4385.

Kongkanand, A. & Kamat, P. V. (2007). *ACS Nano*, Vol. 1, 13.

Kumar, A. & Whitesides, G. W. (2002). *Appl. Phys. Lett.*, Vol. 63, 1993.

Li, H. P.; Zhou, B.; Lin, Y.; Gu, L. R.; Wang, W.; Fernando, K. A. S.; Kumar, S.; Allard, L. F.; & Sun, Y. P. (2004). *J. Am. Chem. Soc.*, Vol. 126, 1014.

Liu, S.; Mannsfeld, S. C. B.; LeMieux, M. C.; Lee, H. W. & Bao, Z. (2008). *Appl. Phys. Lett.*, Vol. 92, 053306.

Liu, S.; Mannsfeld, S. C. B.; LeMieux, M. C.; Lee, H. W. & Bao, Z. (2008). Organic semiconductor-carbon nanotube bundle bilayer field effect transistors with enhanced mobilities and high on/off ratios. *Appl. Phys. Lett.*, Vol. 92, 053306.

McKendry, R.; Huck, W. T. S.; Weeks, B.; Fiorini, M. Abell, C. & Rayment, T. (2002). *Nano Lett.*, Vol. 2, 713.

Meitl, M. A.; Zhou, Y. X.; Gaur, A.; Jeon, S.; Usrey, M. L.; Strano, M. S. & Rogers, J. A. (2004). *Nano Let.*, Vol. 4, 1643.

Moriguchi, I.; Teraoka, Y.; Kagawa, S. & Fendler, S. (1999). *Chem. Mater.*, Vol. 11, 1603.

Murakami, H. ; Nomura, T. & Nakashima, N. (2003). *Chem. Phys. Lett.*, Vol. 378, 481.

Nakashima, N. (2006). *Sci. Tech. Adv. Mater.*, Vol. 7, 609.

Nish, A.; Hwang, J. Y.; Doig, J. & Nicholas, R. J. (2007). *Nat. Mater.*, Vol. 2, 640.

Piner, R. D.; Zhu, J.; Xu, F.; Hong, S. & Mirkin, C. A. (1999). *Science*, Vol. 283, 661.

Rao, S. G.; Huang, L.; Setyawan, W. & Hong, S. (2003). *Nature*, Vol. 425, 36.

Sriwichai, S.; Baba, A.; Deng, S. X.; Huang, C.; Phanichphant, S. & Advincula, R. C. (2008). Nanostructured Ultrathin Films of Alternating Sexithiophenes and Electropolymerizable Polycarbazole Precursor Layers Investigated by Electrochemical-Surface Plasmon Resonance (EC-SPR) Spectroscopy. *Langmuir*, Vol. 24, pp. 9017-9023.

Tomonari, Y.; Murakami, H. & Nakashima, N. (2006). *Chem. Eur. J.*, Vol. 12, 4027.

Tsukruk, V. V.; Ko, H. & Peleshanko, S. (2004). *Phys. Rev. Lett.*, Vol. 92, 065502.

Wang, Y.; Maspoch, D.; Zou, S.; Schatz, G. C.; Smalley, R, E, & Mirkin, C. A. (2006) *Proc. Natl. Acad. Sci. USA*, Vol. 103, 2026.

Whang, D.; Jin, S. & Lieber, C. M. (2004). *Jpn. J. Appl. Phys.*, Vol. 43, 4465.

Xia, Y. N. & Whitesides, G. M. (1998). *Annu. Rev. Mater. Sci., Vol.* 28, 153.
Yan, Y.; Chan-Park, M. B. & Zhang, Q. (2007). *Small*, Vol. 3, 24.
Yoo, D.; Shiratori, S. & Rubner, M. F. (1998). *Macromolecules*, Vol. 31, 4309.

8

Formation of a Silicon Carbide Layer on Vapor Grown Carbon Nanofiber® by Sol-Gel and Carbothermal Reduction Techniques

Kiyoshi Itatani[1], Jumpei Kita[1], Ian J. Davies[2], Hiroshi Suemasu[3], Hiroki Moriyasu[1] and Seiichiro Koda[1]
[1]Department of Materials and Life Sciences, Faculty of Science and Engineering, Sophia University, Tokyo
[2]Department of Mechanical Engineering, Curtin University, Perth
[3]Department of Engineering and Applied Sciences, Sophia University,Tokyo
[1,3]Japan
[2]Australia

1. Introduction

Recent years have noted a growing interest in carbon nanofibre (CNF), due to its excellent chemical, physical, and mechanical properties (Jacobsen et al., 1995). Therefore, CNF has been used as composite reinforcement fibre within polymer and ceramic matrices in order to take advantage of these characteristics (Maensiri et al., 2007). Similar to this, vapor grown carbon nanofibre (VGCF®) is a highly crystalline CNF synthesized by the gas-phase method and exhibits excellent thermal conductivity, electrical conductivity, strength, sliding properties (lubricity), and restoring force (resiliency). Since the handling of VGCF is known to be easier compared to that of carbon nanotubes (CNT) in processes such as dispersion, VGCF has found widespread application. Despite this, the utilization of VGCF has been hindered by several issues, namely the problems of poor dispersibility and fibre aggregation which are attributed to strong van der Waals interactions (Sun et al., 2003).
Previously, single-walled carbon nanotubes (SWNT) have been solubilized through the use of polyvinyl pyrrolidone (PVP) as a surfactant, due to a wrapping effect (O'Connell et al., 2001; Asai et al., 2003). Thus, a similar possibility may arise for the dispersion of VGCF in polar solvents through the use of PVP. Apart from this technique, the present authors (Kita et al., 2010) have treated VGCF with an acid treatment (*i.e.*, mixture of sulfuric and nitric acids) in order to enhance dispersibility within polar solvents, with this effect being attributed to the formation of COO^- groups on the VGCF surface. Thus, the VGCF surface properties were found to be modified from hydrophobic to hydrophilic as a result of acid treatment.
Once the surface properties of VGCF have been changed to hydrophilic in nature, further opportunities may exist for modifying the VGCF surface using an accelerated solid-liquid reaction. For example, the extremely poor oxidation resistance of VGCF (due to the intrinsic oxidizing nature of carbon) may be enhanced when the surface of VGCF is covered with an oxidation resistant material. Silicon carbide (SiC) is a promising candidate for such a material

due to its excellent oxidation, corrosion and creep resistance (Raj, 1993). On the basis of such a background, the present work describes the enhancement of oxidation resistance, due to the surface coating of VGCF with SiC *via* the carbothermal reduction of a silica precursor/silica layer formed from tetraethyl orthosilicate (TEOS) using the sol-gel method.

2. Experimental procedures

Commercially available VGCF® was used as the starting powder (Showa Denko, Tokyo, Japan). Firstly, the coating conditions for the formation of silica precursor on the surface of VGCF were examined using TEOS, *i.e.*, (i) coating the VGCF surface following a surfactant treatment with polyvinylpyrrolidone (PVP; surfactant) and (ii) coating the VGCF surface following acid treatment in order to form carboxyl groups (COO^-). The conditions for the coating of silica precursor on the surface of VGCF are shown in **Table 1**. Relating to the procedure for (i), VGCF (3 mg) was ultrasonically dispersed in 10 cm^3 water containing 2 ~ 10 mg PVP, whereas the hydrolysis of 0.002 ~ 0.05 cm^3 TEOS was conducted in the presence of 0.05 cm^3 concentrated HCl stirred at room temperature for one week. In contrast to this, for procedure (ii) the VGCF (2 mg) was placed into a mixture of concentrated sulfuric acid (2.5 cm^3) and 70% nitric acid (7.5 cm^3) and ultrasonicated for 1 h. The mixture was then refluxed at 40°C for 5 to 30 h and finally washed with distilled water and filtered off. The resulting VGCF coated with silica precursor was further heated at a temperature between 1300 and 1500°C for 1 h in order to form SiC-coated VGCF.
Crystalline phases were identified using an X-ray diffractometer (XRD; Model RINT2000PC, Rigaku, Tokyo) with monochromatic CuKα radiation generated at 40 kV and 40 mA. The substitution groups on the surface of VGCF were investigated using a Fourier transform infrared (FT-IR) spectrometer (Model IR Prestige-21, Shimadzu, Kyoto) by means of the KBr method. The heat resistance of chemically-modified VGCF in air was examined by thermogravimetry (TG; Model Thermo Plus TG8120, Rigaku, Tokyo). The morphologies of chemically-modified VGCF were observed using a field-emission scanning electron microscope (FE-SEM; S-4500, Hitachi, Tokyo) at an accelerating voltage of 5 kV, after the specimen surface had been coated with Pt-Pd using an ion coater (Model E-1030, Hitachi, Tokyo) in order to avoid charging effects. Furthermore, the microstructure of chemically-modified VGCF was observed using a transmission electron microscope (JOEL 2011; JEOL, Tokoy, Japan) at an acceleration voltage of 200 kV.

3. Results and discussion

3.1 Coating of silica precursor on the surface of VGCF

Prior to carbothermal reduction to form a SiC layer, conditions necessary for the coating of silica precursor on the surface of VGCF using TEOS were examined, *i.e.*, (i) coating of silica precursor following the modification of surface properties from hydrophobic to hydrophilic using PVP as a surfactant, and (ii) coating of silica precursor following an acid treatment to form carboxyl groups at 30°C for 48 h. Details for these procedures will be discussed in this section.

3.1.1 Utilization of PVP for the coating of silica precursor on the surface of VGCF

Firstly, the surfactant (PVP) was utilized not only for the dispersion of VGCF in the polar solvent but also for the interaction with TEOS. To this end, PVP was placed into the VGCF-dispersed and TEOS-hydrolyzed solutions in order to homogeneously coat the surface of VGCF. The experimental conditions were chosen on the basis of research previously reported (Asai et al., 2003). The expected mechanisms concerning the interaction of PVP with water and TEOS in these solutions are shown in **Fig. 1**. The site of $O^{\delta-}$ = in Equal layout interacts with $H^{\delta+}$ sites in water (H-O-H) (Fig. 1(b-1)) and $-OH^{\delta+}$ in the TEOS (Fig. 1(b-2)). Thus PVP is expected to aid dispersion of the VGCF in H_2O and TEOS in HCl solution due to the wrapping effect on the VGCF.

On the basis of experimental conditions listed in Table 1, VGCF (3 mg) was dispersed in a H_2O solution containing 2 to 10 mg PVP, whereas the amounts of TEOS, PVP and HCl were fixed at 0.05 cm^3, 50 mg and 0.05 cm^3, respectively (Sample Nos. 1 to 3). Typical FE-SEM micrographs of VGCF/silica precursor are shown in **Fig. 2**. Spherical particles with diameters of 5 ~ 10 μm, as well as acicular particles, were found to be present in the powder (Fig. 2(a)), with no significant coating of material noted on the surface of the acicular particles (Fig. 2(b)). The spherical and acicular particles were ascribed to be silica precursor and VGCF, respectively, and indicate the difficulty in homogeneous coating the surface of VGCF using PVP as a surfactant. It is suggested that the spherical particles of silica precursor were formed as a result of: (i) the formation of PVP micelle due to the excess amount over that for the surrounding of VGCF, and (ii) nuclei formation and subsequent growth of silica precursor entrapped by the PVP micelle (Wang et al., 2008; Woo et al., 2010; Duque et al., 2008).

	VGCF-dispersed solution			TEOS-hydrolyzed solution		
Sample No.	VGCF mg	PVP mg	H_2O cm^3	TEOS cm^3	PVP mg	HCl cm^3
1	3	2	10	0.05	50	0.05
2	3	5	10	0.05	50	0.05
3	3	10	10	0.05	50	0.05
4	3	2	10	0.05	1	0.05
5	3	5	10	0.05	1	0.05
6	3	10	10	0.05	1	0.05
7	3	2	10	0.02	1	0.05
8	3	5	10	0.02	1	0.05
9	3	10	10	0.02	1	0.05
10	3	10	10	0.015	1	0.05
11	3	10	10	0.010	1	0.05
12	3	10	10	0.005	1	0.05
13	3	10	10	0.002	1	0.05

Table 1. Experimental conditions for the dispersion of VGCF in the presence of PVP (VGCF-dispersed solution) and for the hydrolysis of TEOS (TEOS-hydrolyzed solution).

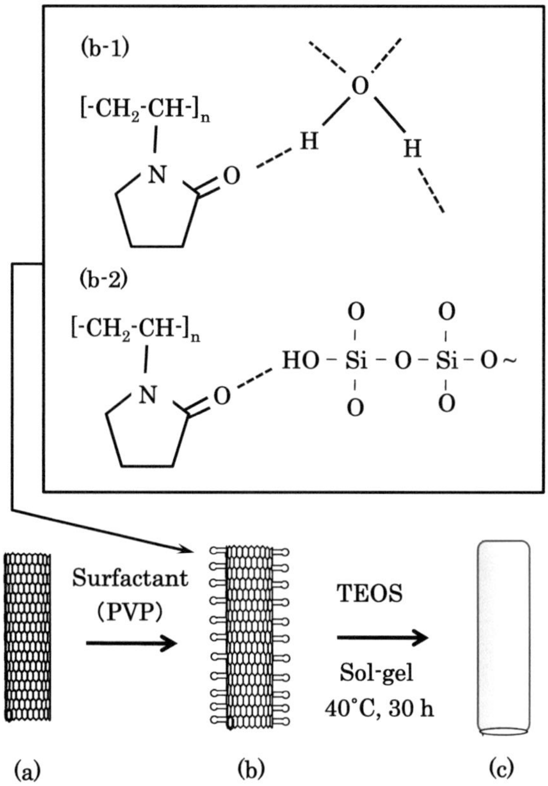

Fig. 1. Schematic diagram of the coating by silica precursor.

a. As-received VGCF
b. VGCF coated with PVP (b-1) Interaction with water (b-2) Interaction with TEOS
c. VGCF coated with silica precursor

Since the amount of PVP was considered to be slightly in excess of that required for the optimal coating of VGCF, the amount of PVP in the TEOS-hydrolyzed solution was reduced from 50 to 1 mg, and the conditions to homogeneously coat the surface of VGCF were examined (Sample Nos. 4 to 6). Typical FE-SEM micrographs of VGCF/silica precursor are shown in **Fig. 3**. It was noted that spherical particles with diameters of 5 ~ 10 μm, as well as acicular particles, were present in the powder (Fig. 3(a)), although no significant coating of material was observed on the surface of the acicular particles (Fig. 3(b)). As already mentioned, the spherical and acicular particles were believed to be silica precursor and VGCF, respectively, and further reinforce the difficulty in homogeneous coating the surface of VGCF using PVP as a surfactant.

Since little effect on the coating of silica precursor was apparent from the decrease in the amount of PVP, the amount of TEOS in the TEOS-hydrolyzed solution was reduced from 0.05 to 0.02 cm^3, with the conditions required to homogeneously coat the surface of VGCF being examined (Sample Nos. 7 to 9). Typical FE-SEM micrographs of VGCF/silica precursor have been shown in **Fig. 4**. Spherical particles with diameters of 5 ~ 10 μm, as well as acicular particles, were found to be present in the powder, and the number of spherical particles became small (Fig. 4(a)).

(a) Low magnification (b) High magnification

Fig. 2. Typical FE-SEM micrographs of VGVF and silica precursor formed by hydrolyzing TEOS at 40°C for 30 h (Sample No. 3).

(a) Low magnification (b) High magnification

Fig. 3. Typical FE-SEM micrographs of VGVF and silica precursor formed by hydrolyzing TEOS at 40°C for 30 h (Sample No. 6).

Furthermore, the surface of VGCF was partly coated by a layer of material (Fig. 4(b); see the arrow mark). As has already been mentioned, the spherical and acicular particles were believed to be silica precursor and VGCF, respectively, and thus the presence of a partly-coated silica precursor on the surface of VGCF indicates the possibility of homogeneous coating the surface of VGCF. Following this, the amount of TEOS was further reduced from

0.015 to 0.002 cm^3 in order to homogeneously coat the surface of VGCF with the results being shown in **Fig. 5**. According to the low magnification FE-SEM micrographs in this figure, spherical particles with diameters of approximately 1 ~ 3 μm, as well as acicular particles, were present in the powder, but the number of spherical particles was small (TEOS amount: 0.015 cm^3). The number of spherical particles was reduced as the amount of TEOS was reduced from 0.010 to 0.002 cm^3. According to the higher magnification FE-SEM micrographs in Fig. 5, no significant coating of TEOS was observed on the surface of the VGCF.

(a) Low magnification (b) High magnification

Fig. 4. Typical FE-SEM micrographs of VGVF and silica precursor formed by hydrolyzing TEOS at 40°C for 30 h (Sample No. 9).
Arrow mark in Fig. 4(b) indicates evidence of coating of silica precursor on the surface of VGCF.

In contrast to previous work (Asai et al., 2003) that suggested the silica precursor to have been successfully covered with SWNT, the application of a homogeneous coating on the surface of VGCF could not be achieved in the present research. The diameter of the VGCF (mean diameter: 150 nm) is approximately 15 times that of SWNT (mean diameter: 10 nm). Due to the significant difference in diameter between SWNT and VGCF, it seems to be possible for PVP to wrap the SWNT (*i.e.*, smaller diameter) but be difficult for PVP to wrap the VGCF (*i.e.*, larger diameter)(O'Connell et al., 2001). In the present case, therefore, it is concluded that the PVP may be significantly easier to form as spherical particles rather than as a layer on the surface of VGCF.

3.1.2 Acid treatment for the coating of silica precursor on the surface of VGCF

As shown in Section 3.1.1, the homogeneous coating of silica precursor could not be achieved using PVP. The authors therefore consider that the VGCF surface properties need to be modified from hydrophobic to hydrophilic in order to allow interaction between the VGCF and TEOS. In light of this, the authors planned to form carboxyl (*i.e.*, hydrophilic)

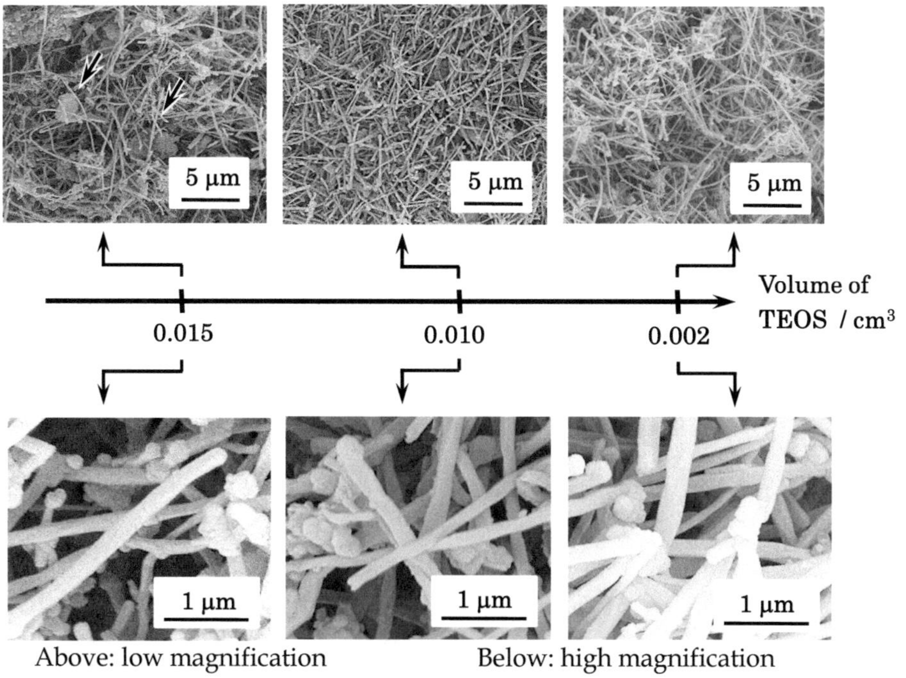

Fig. 5. Typical FE-SEM micrographs of VGCF and silica precursor formed by hydrolyzing TEOS at 40°C for 30 h (Sample Nos. 10 to 12).

groups and followed by the formation of silica precursor on the surface of VGCF, using TEOS. Work previously carried out on VGCF had included an acid treatment at 50°C for 24 h (Kita et al., 2010). In this work the temperature for the acid treatment of VGCF was reduced to 30°C in order to prevent damage to the VGCF surface, as this would negatively influence any additional modification to the VGCF surface that would be carried out using TEOS. Photographs indicating dispersion states of as-received and acid-treated VGCF in a polar solvent (acetone) are presented in **Fig. 6**, together with the FE-SEM micrographs. The as-received VGCF was seen to be sedimented to the bottom of the sample bottle whereas the acid-treated VGCF was still dispersed within the acetone after standing for 24 h. FE-SEM micrographs of VGCF showed that no marked difference in particle appearance was observed before and after the acid treatment. Since the VGCF could be dispersed in the polar solvent, the functional groups bonded to the surface of VGCF were examined using FT-IR spectroscopy. Typical FT-IR spectra of as-received and acid-treated VGCFs are shown in **Fig. 7**. In both cases, the absorption peaks appeared at 1550 and 1350 cm^{-1}, respectively. The absorption peaks at 1550 and 1350 cm^{-1} were both assigned to the COO- stretching vibration (Kubota et al., 2005), both of which are known to be typical modes of the carboxyl group. This fact demonstrates that carboxyl groups may be formed on the surface of VGCF as a result of the acid treatment. The imbalanced charge of $H^{\delta+}$ in the carboxyl group (-COOH) may thus interact with that of the carbonyl group (> C = O) in the acetone, thereby making possible the dispersion of acid-treated VGCF in acetone.

Previously, the oxidation of SWNT to form carboxyl groups has been examined by Zhang *et al.* (Zhang et al., 2003), who considered the reaction to comprise of three steps as follows: (i) an initial attack on the original existing active sites such as the $-CH_2$, -CH groups and

(a) (b)

Fig. 6. FT-IR spectra of (a) as-received VGCF and (b) VGCF acid-treated at 30°C for 48 h.

heptatomic rings, (ii) electrophilic addition at hexatomic-hexatomic boundaries, generating active sites or new defects (defect-generating step), and (iii) the breaking of the graphene structure around the generated active sites (defect-consuming step). The acid treatment, due to the presence of sulfuric acid, effectively produces electrophilic groups such as NO_2^+, which contribute to forming -COOH groups as follows:

$$H_2SO_4 + HNO_3 + C \rightarrow SO_2 + NO_2^+ + H_3O^+ + COO^- \quad (1)$$

Since the surface of VGCF could be modified to be hydrophilic through the formation of carboxyl groups, the conditions for the hydrolysis of TEOS were further conducted in order to coat the silica precursor on the surface of VGCF. In this case the heat-treatment temperature was fixed at 40°C whereas the time was changed from 5 to 30 h. Typical XRD patterns are shown in **Fig. 8**, together with FE-SEM micrographs. The crystalline phase detected by XRD was only carbon (JCPDS-ICDD Card, No. 41-1487) with the intensity of the carbon peaks increasing with hydrolysis time from 5 to 15 h but then being reduced upon further hydrolysis time to 30 h. Also of note was the broad XRD background in the 2θ range of 20 to 30°. According to the FE-SEM micrographs, the VGCF surface was increasingly covered with material as the TEOS hydrolysis time increased from 5 to 15 h at 40°C (Fig. 8(a) → (b); see the part enclosed with the square). On further increasing the hydrolysis time to 30 h, however, the VGCF became stuck together to form agglomerates.

The broad XRD background for VGCF in the 2θ range of 2 to 30° following hydrolysis of TEOS (see Fig. 8) indicates the presence of amorphous material with this being further investigated using FT-IR as shown in **Fig. 9**. The FT-IR spectra were found to contain

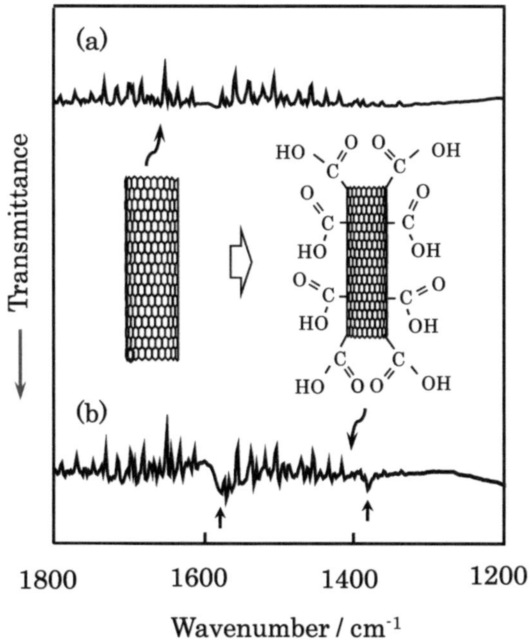

Fig. 7. Typical appearance and FT-IR micrographs of (a) as-received VGCF and (b) VGCF acid-treated at 30°C for 48 h.

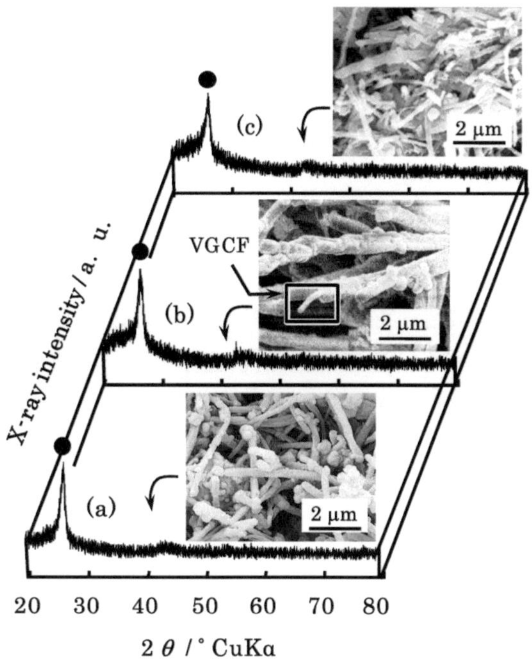

Fig. 8. Typical XRD patterns of VGCF coated with silica precursor at (a) 40°C for 5 h, (b) 40°C for 15 h, and (c) 40°C for 30 h after acid treatment at 30°C for 48 h. ●: C.

absorption peaks at 1200 (shoulder), 1047 and 947 cm^{-1} with the intensities of these absorption peaks increasing with hydrolysis time from 5 to 30 h. The absorption peaks at 1200 (shoulder) and 1047 cm^{-1} were assigned to Si-O-Si bonding, whereas the absorption peak at 947 cm^{-1} may be assigned to $Si-CH_3$ bonding (Navamathavan et al., 2008). Since the intensity of the absorption peak at 1047 cm^{-1} increased with hydrolysis time from 5 to 15 h, the bonding of Si-O-Si was concluded to have been enhanced with hydrolysis time. This fact indicates that the starting TEOS may be changed into silica *via* a silica precursor with increasing hydrolysis time.

Further to the results mentioned above, the formation of agglomerates following hydrolysis of TEOS at 40°C for 30 h indicates that an excess of silica precursor over the optimum amount required for the coating of VGCF contributes to unwanted adhesion between the VGCFs. An increased amount of silica precursor on the VGCF surface was also confirmed by the XRD patterns which indicated the X-ray intensity of carbon to have been reduced with increasing hydrolysis time of TEOS from 15 to 30 h. On the basis of these results, the optimum condition for hydrolysis of TEOS was determined to be 40°C for 15 h, with this being expected to result in homogeneous coating of silica precursor on the surface of VGCF.

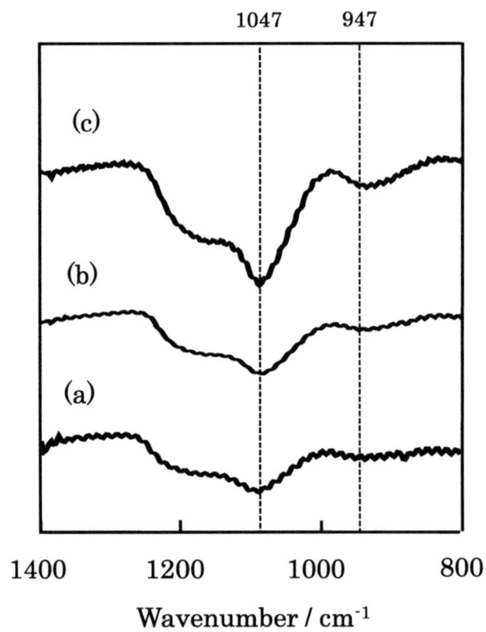

Fig. 9. Typical FT-IR spectra for VGCF coated with silica precursor at (a) 40°C for 5 h, (b) 40°C for 15 h, and (c) 40°C for 30 h after an acid treatment at 30°C for 48 h.

3.2 Carbothermal reduction for the coating of silicon carbide on the surface of VGCF

Since the homogeneous coating of silica precursor onto the surface of VGCF was successfully carried out by acid- and TEOS-hydrolysis treatments, the authors further examined the conditions for the carbothermal reduction of silica and/or silica precursor in order to form a SiC layer on the VGCF surface. Prior to coating with silicon carbide, the formation of a surface silica layer was confirmed by heating the VGCF with silica precursor layer at 850°C for 5 h in air. A typical XRD pattern is shown in **Fig. 10** together with a typical TEM micrograph. According to the XRD pattern, broad X-ray reflections appeared in the 2θ range of 20 to 40°. On the other hand, the TEM micrograph showed the presence of tube-like particles with thicknesses of approximately 20 nm.

Regardless of the broad XRD pattern, the precursor layer on the VGCF surface was throught to have been changed into silica (Tabatabaei et al., 2006; Wang et al., 2009) following heating at 850°C for 5 h in air. Moreover, the TEM micrograph in Fig. 10 indicated that the VGCF had been burned out to form tube-like particles. Due to the presence of the tube-like particles with homogeneous thicknesses of approximately 20 nm, the silica precursor was concluded to have been homogeneously covered onto the VGCF surface. Therefore, since the homogeneous coating of silica precursor and silica onto the surface of VGCF had been confirmed on the basis of XRD and TEM results, the conditions required for the carbothermal reduction of silica precursor/silica layer were also examined. The carbothermal reduction of TEOS-derived VGCF was conducted at a temperature between 1300 and 1600°C for 1 h with the results being shown in **Fig. 11**. Apart from the carbon reflection which appeared at 26.4°, no other reflections were noted for the carbothermal reduction of silica precursor/silica at 1300 and 1400°C for 1 h (Fig. 11(a) and (b)). On the other hand, β-SiC (JCPDS-ICDD Card. No. 29-1129), in addition to carbon, was detected when the carbothermal reduction had been conducted at 1500°C for 1 h (Fig. 11(c)). The FE-SEM micrograph in Fig. 11 indicated that the VGCF surface had been covered with

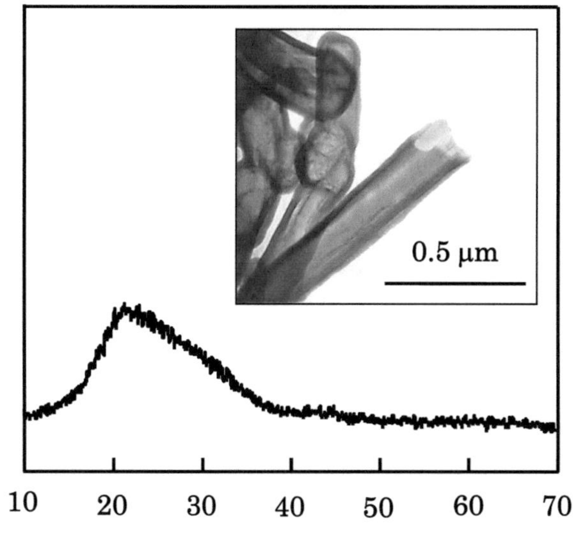

Fig. 10. Typical XRD pattern and TEM micrograph of VGCF coated with silica precursor at 40°C for 15 h and then heated at 800°C for 5 h in air.

material whilst the X-ray intensity due to β-SiC was increased as the carbothermal reduction increased from 1500 to 1600°C (Fig. 11(d)).

It should be noted that a broad X-ray background, attributed to the presence of SiO_2 in an amorphous state (Wang et al., 2009), was present in the range of 2θ = 20 to 30° for the cases of carbothermal reduction at 1300 and 1400°C (see Fig. 10). Furthermore, the reduction in intensity of the carbon reflection, simultaneously accompanied with an increased β-SiC reflection (Fig. 11(c) → (d)), suggests that the VGCF had been consumed to form β-SiC due to the carbothermal reduction from silica precursor/silica (Wang et al., 2009). This carbothermal reduction was presumed to have occurred as follows (Lin et al., 2003):

$$SiO_2 + 2C \rightarrow SiC + CO \quad (2)$$

The optimum temperature for the carbothermal reduction was considered to be 1500°C due to the presence of crystalline β-SiC at this temperature.

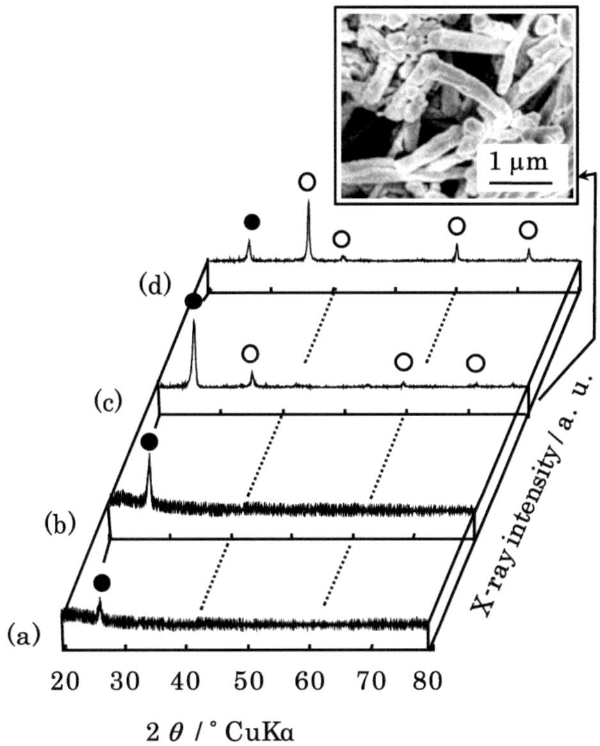

Fig. 11. Typical XRD patterns of VGCF coated with silica precursor carbothermally-reduced at (a) 1300°C, (b) 1400°C, (c) 1500°C, and (d) 1600°C for 1 h in Ar atmosphere. Note that the VGCF coated with silica precursor was prepared by heating the TEOS at 40°C for 15 h.
●: Carbon ○: β-SiC

FE-SEM, FE-STEM and TEM micrographs of SiC-coated VGCF are shown in **Fig. 12**. Whereas FE-SEM indicated the VGCF surface to have been homogeneously coated with material (Fig. 12(a)), the FE-STEM micrographs indicated the presence of spherical particles with diameters of 1 µm or less, in addition to the coating layer, as being present on the VGCF surface (Fig. 12(b)). Likewise, appearance of VGCF observed by TEM was similar to that observed by FE-STEM. Therefore, it was concluded that spherical particles appear to have stuck to the VGCF surface, in addition to the homogeneous layer. Such spherical particles, as well as the layer, were attributed to carbothermal reduction of an excess amount of silica precursor/silica over the amount required for the homogeneous coating of the VGCF surface.

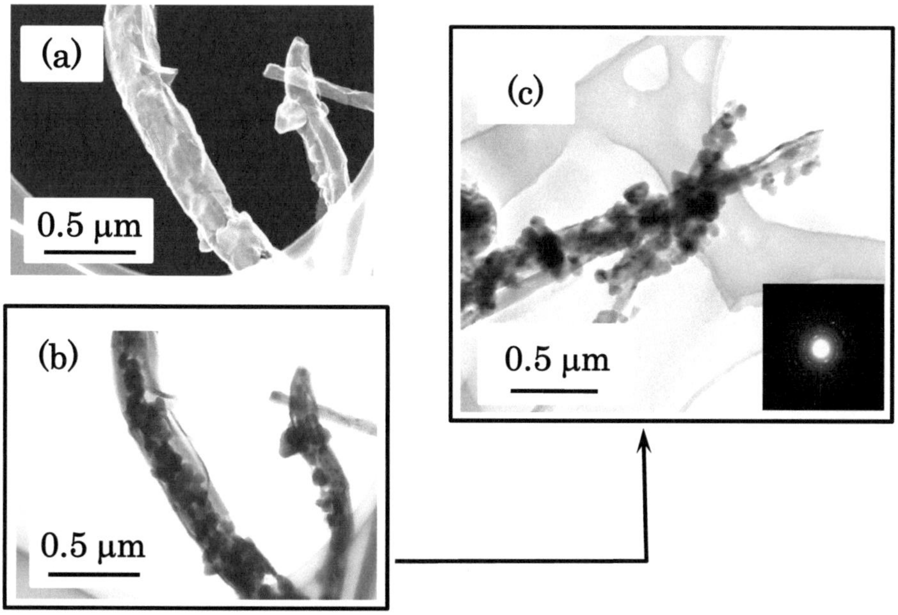

Fig. 12. Typical (a) FE-SEM, (b) FE-STEM, and (c) TEM micrographs of VGCF coated with silicon carbide obtained by carbothermally reduction of the silica precursor at 1500°C for 1 h in Ar atmosphere. Note that the VGCF coated with silica precursor was prepared by heating TEOS at 40°C for 15 h.

From the results presented above it was concluded that VGCF coated with silicon carbide could be prepared by making use of the acid-treated VGCF. A schematic diagram for the formation of the silicon carbide layer on the surface of VGCF *via* the formation of silica precursor/silica is shown in **Fig. 13**. The preparation process of the VGCF coated with silicon carbide could be divided into three parts, *i.e.*, (a) interaction of acid-treated VGCF with TEOS, (b) formation of silica precursor on the surface of acid-treated VGCF, and (c) formation of silica layer on the surface of acid-treated VGCF. In addition to this, excess spherical SiC particles were also included in the layer formed on the VGCF surface.

As a final step, VGCF with a SiC layer was heated in air in order to examine the oxidation resistance with the TG results being shown in **Fig. 14**. The mass loss of as-received VGCF started to occur at approximately 700°C and increased with a further increase in temperature (Fig. 14(a)). In contrast to this, the temperature at which the mass loss of acid-treated VGCF started was reduced to approximately 100°C (Fig. 14(b)), whereas mass loss of SiC-coated VGCF initiated at approximately 650°C and increased with a further increase in temperature (Fig. 14(c)). The mass losses of these VGCFs at 1000°C were arranged as follows: acid-treated VFCF (67.9%) > as-received VGCF (50.1%) > SiC-coated VGCF (45.7%). The significant mass loss in the case of acid-treated VGCF may be attributed to the formation of cracks or lattice defects on the surface of VGCF. The lower mass loss in the case of SiC-coated VGCF, compared to the case of as-received VGCF, demonstrates that the SiC had coated the VGCF surface.

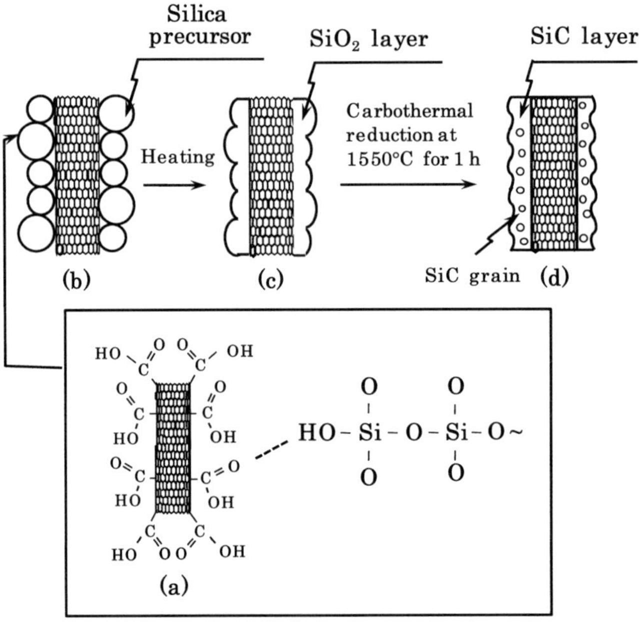

Fig. 13. Schematic diagram of the formation of a silicon carbide layer on the surface of VGCF *via* the formation of silica precursor/silica.

a. Interaction of acid-treated VGCF with TEOS
b. Formation of silica precursor on the surface of acid-treated VGCF
c. Formation of silica layer on the surface of acid-treated VGCF
d. Formation of silicon carbide layer on the surface of carbothermally-reduced VGCF, together with silicon carbide grains.

4. Conclusions

The conditions for the formation of a SiC layer on the surface of vapor grown carbon nanofiber (VGCF®) were examined by the sol-gel method and subsequent carbothermal reduction at a temperature between 1400 and 1600°C for 1 h in argon atmosphere. The results obtained were summarized as follows:

i. Two types of techniques to coat silica precursor onto the surface of VGCF by a sol-gel technique, using TEOS, were examined, *i.e.*, (i) the coating of silica precursor at 40°C for 30 h after the modification of surface properties from hydrophobic to hydrophilic, using PVP as a surfactant, and (ii) the coating of silica precursor at 40°C for 15 h after acid treatment (concentrated sulfuric and 70% nitric acids) to form hydrophilic carboxyl groups at 30°C for 48 h. Relating to the technique of (i), VGCF and spherical silica precursors were independently present in the powder with no significant coating of silica precursor noted on the VGCF surface. Relating to the technique of (ii), the coating of silica precursor was successfully conducted due to the formation of hydrophilic carboxyl groups.
ii. The carbothermal reduction of VGCF coated with silica precursor was further investigated at a temperature between 1300 to 1600°C for 1 h in Ar atmosphere. When the VGCF coated with silica precursor was carbothermally reduced at 1500°C for 1 h, a

notably crystalline SiC layer could be formed on the VGCF surface with this surface layer being found to enhance the heat resistance of VGCF in air.

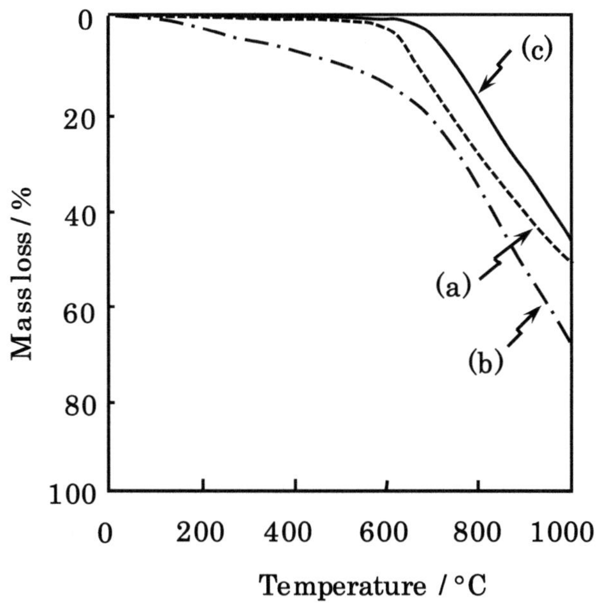

Fig. 14. TG curves of (a) as-received VGCF, (b) acid-treated VGCF, and (c) VGCF coated with silicon carbide obtained by carbothermally reducing the silica precursor at 1500°C for 1 h in Ar atmosphere. Note that the VGCF coated with silicon carbide was prepared by heating TEOS at 40°C for 15 h. Heating rate: 10°C·min^{-1}

5. References

Asai, M., Fujita, N., Sano, M. & Shinkai, S., (2003). Silica Gel Fabrication of [60]Fullerene Aggregates and Carbon Nanotubes Utilizing the Amphiphilic Nature of Poly(*N*-vinylpyrrolidone) As a 'Glue', *Journal of Materials Chemistry*, Vol. 13, pp. 2145-2149, ISSN 1364-5501

Jacobsen, R. L., Tritt, T. M., Ehrlich J. R. & Gillespie, D. J. (1995), Mechanical Properties of Vapor-Grown Carbon-Fiber, *Carbon*, Vol. 33, pp. 1217-1221, ISSN 0008-6223

Joint Committee on Powder Diffraction Standards-International Centre for Diffraction Data (JCPDS-ICDD), Card. No.29-1129, Swarthmore, PA, USA.

Joint Committee on Powder Diffraction Standards-International Centre for Diffraction Data (JCPDS-ICDD) Card. No.41-1487, Swarthmore, PA, USA.

Duque, J. G., Cognet, L., Parra-Vasquez, A. N. G., Nicholas, N., Schmidt, H. K. & Pasquali, M. (2008). Stable Luminescence from Individual Carbon Nanotubes in Acidic, Basic, and Biological Environments, *Journal of the American Chemical Society*, Vol. 130, pp. 2626-2633, ISSN 0002-7863

Kita, J., Suemasu, H., Davies, I. J., Koda S. & Itatani, K. (2010). Fabrication of Silicon Carbide Composites with Carbon Nanofiber Addition and Their Fracture Toughness, *Journal of Materials Science*, Vol. 45, Number 22, pp. 6052-6058, ISSN 0022-2461

Kubota, S., Nishikiori, H., Tanaka, N., Endo, M. & Fujii, T. (2005). Dispersion of Acid-Treated Carbon Nanofibers into Gel Matrices Prepared by the Sol-Gel Method. *Journal of Physical Chemistry B*, Vol. 109, pp. 23170-23174, ISSN 1520-6106

Lin Y.-J. & Tsang, C.-P., (2003), The Effects of Starting Precursors on the Carbothermal Synthesis of SiC Powders, *Ceramics International*, Vol. 29, Issue 1, pp. 69-75, ISSN 0272-8842

Maensiri, S., Laokul, P., Kinkaewnarong, J. & Amornkitbamrung, V. (2007). Carbon Nanofiber-Reinforced Alumina Nanocomposites: Fabrication and Mechanical Properties, *Materials Science and Engineering A*, Vol. 447, pp. 44-50, ISSN 0921-5093

Navamathavan, R., Kim C. Y. & Jung, A. S. (2008). Deposition and Characterization of Porous Low-Dielectric-Constant SiOC(-H) Thin Films Deposited from TES/O_2 precursors by Using Plasma-Enhanced Chemical Vapor Deposition. *Journal of the Korean Physical Society*, Vol. 53, pp. 351-356, ISSN 0374-4884

O'Connell, M. J., Boul, P., Ericson, L. M., Huffman, C., Wang, Y., Haroz, E., Kuper, C., Huffman, J., Ausman K. D. & and Smalley, R. E. (2001). Reversible Water-Solubilization of Single-Walled Carbon Nanotubes by Polymer Wrapping, *Chemical Physics Letters*, Vol. 342, pp. 265-271, ISSN 0009-2614

Raj, R. (1993). Fundamental Research in Structural Ceramics for Service Near 2000°C, *Journal of American Ceramic Society*, Vol. 76, Issue 9, pp. 2147-2174, ISSN 0002-7820

Sun J. & Gao L. (2003). Development of a Dispersion Process for Carbon Nanotubes in Ceramic Matrix by Heterocoagulation, *Carbon*, Vol. 41, Issue 5, pp. 1063-1068, ISSN 0008-6223

Tabatabaei, S., Shukohfar, A., Aghababazadeh, R. & Mirhabibi, A. (2006). Experimental Study of the Synthesis and Characterisation of Silica Nanoparticles via the Sol-Gel Method, *Journal of Physics: Conference Series*, Vol. 26, pp. 371-374, ISSN 1742-6588

Wang, K., Cheng Y.-B. & Wang, H. (2009). Growth of SiC Whiskers by Carbothermal Reduction of Mesoporous Silica-Carbon Composites Infiltrated With SiC Nuclei, *Journal of the Australian Ceramic Society*, Vol. 45, No. 1, pp. 10-12, ISSN 0004-881X

Wang, F., Liu, J., Yang, H., Luo, Z., Lv, W., Li, C. & Qing, S. (2008). Spherical Particles from Tetraorthosilicate (TEOS) Sol–gel Process with Dimethyldiethoxysilane (DDS) and Diphenyldiethoxysilane (DPDS) Addition, *Journal of Non-Crystalline Solids*, Vol. 354, Issues 45-46, 5047-5052, ISSN 0022-3093

Woo, H.-Y., Jung, W.-G., Ihm, D.-W. & *Kim,* J.-Y. (2010). Synthesis and Dispersion of Polypyrrole Nanoparticles in Polyvinylpyrrolidone Emulsion. *Synthetic Metals*, Vol. 160, pp. 588-591, ISSN 0379-6779

Zhang, J., Zou, H., Qing Q., Yang, Y., Li, Q., Liu, Z., Guo, X. & Du, Z. (2003). Effect of Chemical Oxidation on the Structure of Single-Walled Carbon Nanotubes. *Journal of Physical Chemistry B,* Vol. 107, pp. 3712-3718, ISSN 1089-5647

9

Microwave Dielectric Characterization of Carbon Nanotube Networks

Emmanuel Decrossas and Samir M. El-Ghazaly
University of Arkansas, 3217 Bell Engineering Center, Fayetteville, AR, USA

1. Introduction

Carbon nanotubes (CNTs) have been discovered in the early 1990s (Iijima, 1991). The earliest research has been focus on the synthesis of CNTs to produce sufficient amount of material necessary for experiments. The progress of common growth techniques including arc discharge and laser ablation of graphite pieces and later chemical vapor deposition (CVD) and plasma enhanced CVD has significantly improved the quantity as well as the quality of the CNTs (Meyyapan, 2005). In the last decade many companies have concentrated their efforts to develop and control those growth techniques to insure very high purity and uniform CNTs in length and diameter at a reasonable cost. The physical properties of carbon nanotubes (Saito, 1998) have demonstrated unique electronic structure depending on their dimensions and chiralities, which have been considered as a major discovery in semiconductor devices. In addition, due to strong covalent carbon-carbon bonding CNTs present very high tensile strength, thermal and electrical conductivity. Hence, those mechanical and electrical properties make them also very attractive to reinforce composites material for aerospace applications. Usually, carbon nanotubes are classified in two categories: Single-wall CNTs (SWCNT) composed of a single graphitic cylinder where the diameter varies from 0.7 to 2 nm and Multi-wall CNTs (MWCNTs) composed of several concentric graphitic layers where the diameter varies from 10 to 200 nm. Commercial carbon nanotubes are available in a powder form where they are entangled and randomly distributed.

Today, the growth of carbon nanotubes is well advanced and the main research interest has become CNT-based applications. The potential of carbon nanotube-based applications is currently investigated for developing novel transistors, scanning probes (M.J. O'Connell, 2006), mechanical sensors (Tung et al., 2007), field emission and lighting applications (Jorio et al., 2008). The knowledge of dielectric properties at low frequency (DC to 1MHz) as well as high frequency (above 1MHz) is the key for developing novel electronic devices such as radio frequency miniaturized resonators (El Sabbagh and El-Ghazaly, 2009a) as well as electronic model for future implementation (EL Sabbagh and El-Ghazaly, 2009c). Therefore, the electrical characterization at radio frequency/microwave (RF/M) bands becomes a challenge where common measurement techniques are inappropriate. For high frequency measurements, complex permittivity ($\varepsilon = \varepsilon' - j\varepsilon''$) is usually used to describe the dielectric properties of a material where the real part of the permittivity is related to the dielectric constant and the imaginary part can be associated to the conductivity.

Over 600 publications have reported the characterization of CNT-based composites in 2010 where carbon nanotubes are embedded in a host medium. However, those samples preparation considerably decreases the accuracy of the measurements and explains the difference between the reported results. In this chapter, a non destructive, low-cost, easily implemented and broad frequency band (from 10 MHz to 50 GHz) measurement setup is presented where only a small fraction of material in a powder form is necessary (Decrossas et al., 2010). Carbon nanotubes powder as furnished by the supplier is placed without host medium decreasing the unknown effects on the preparation of the samples. The extraction of the complex permittivity of CNT networks is investigated over a wide frequency band considering different powder densities. Then, the reported results are explained through the percolation theory (Stauffer and Aharony, 1992) as well as numerical and experimental studies where metallic particles are embedded in a dielectric medium (Decrossas et al., 2011). The conclusion summarizes the experimental reported results and is followed by a discussion on possible future research directions.

2. Description of the test structure

The measurement technique is realized by studying the propagation of electromagnetic waves in a circular waveguide shorted at the end filled with a dielectric medium. Therefore, incident electromagnetic waves are reflected after propagating in a medium and the study of these waves inform us about the electrical properties of the material under test (MUT).

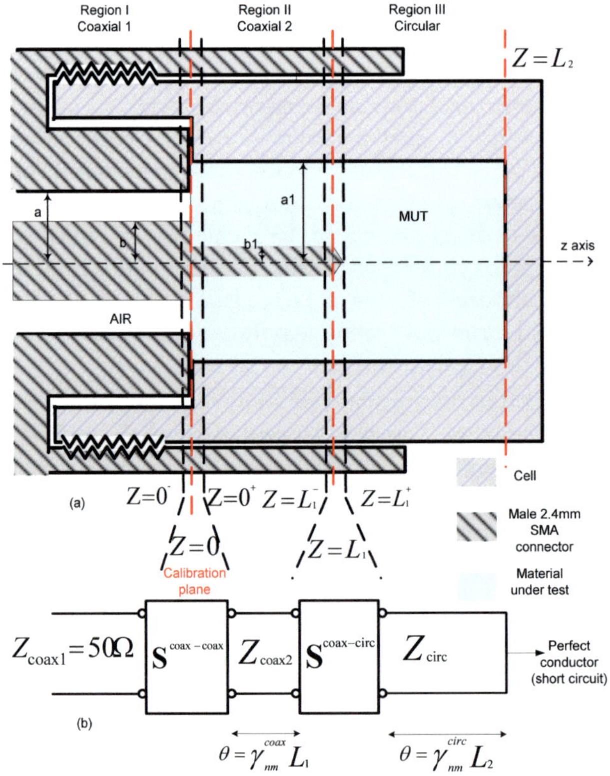

Fig. 1. (a) Schematic of the test structure. (b) Generalized scattering matrices building blocks.

The homemade test structure includes coaxial and circular discontinuities where the material under test filled the circular waveguide as shown in Fig. 1.The measurements are carried out using a performance network analyzer (PNA) capable of measurements from 10 MHz to 67 GHz. The hollow circular waveguide is connected to the PNA through a precision adapter via 1.85 mm coaxial cable. Copper well-known for its high conductivity has been chosen to build the cell holding the material under test as depicted in Fig. 2.

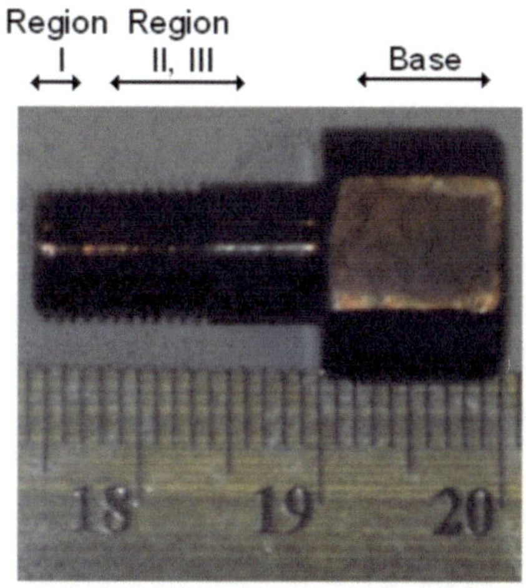

Fig. 2. Actual picture of the fabricated testing structure. Regions I, II, and III are the same as those shown in the schematic of Fig. 1. Dimensions of design parameters shown in Fig. 1 are: a = 1.2 mm, b = 0.52 mm, a_1 = 1.26 mm, b_1 = 0.254 mm, L_1 = 1.1 mm, L_2=5.3mm.

The adopted measurement technique presents several advantages: (a) only a small fraction of carbon nanotubes powder is required; (b) other parameters like density and temperature are easily controlled for repeatable measurement; (c) the powder is held in a closed structure to avoid fringing radiation. The extraction technique includes the computation of the reflection coefficient using the mode matching method where each discontinuity encounter by an incident wave is characterized by its generalized scattering parameters as shown in Fig. 1(b). The accuracy of the analysis comes from the consideration of all propagating and evanescent higher-order modes excited by the discontinuities. The model of the test structure using the mode matching technique (MMT) and the methodology to extract the complex permittivity have been previously described (Decrossas et al., 2011). If the problem to derive the reflection coefficient from the full-wave model is straightforward, the computation of the complex permittivity is obtained by an iterative method. An optimized gradient method solves this inverse problem based on comparison of the measured and simulated reflection coefficient using the mode matching technique.
In case of material in dry-powder form, density is an important parameter that affects the value of complex permittivity. The procedure to extract the permittivity of CNT networks involves also the study of the effect of the density of CNT networks inside the waveguide holder (Decrossas et al., 2011). This is carried out by increasing gradually the mass of added networks inside the holder. To calibrate the measurement setup and verify the method of extraction, experiments are first realized with silicon dioxide (SiO_2) as well as pure alumina powder (Al_2O_3) with particles size of approximately 50nm. Fig. 3 presents the real and

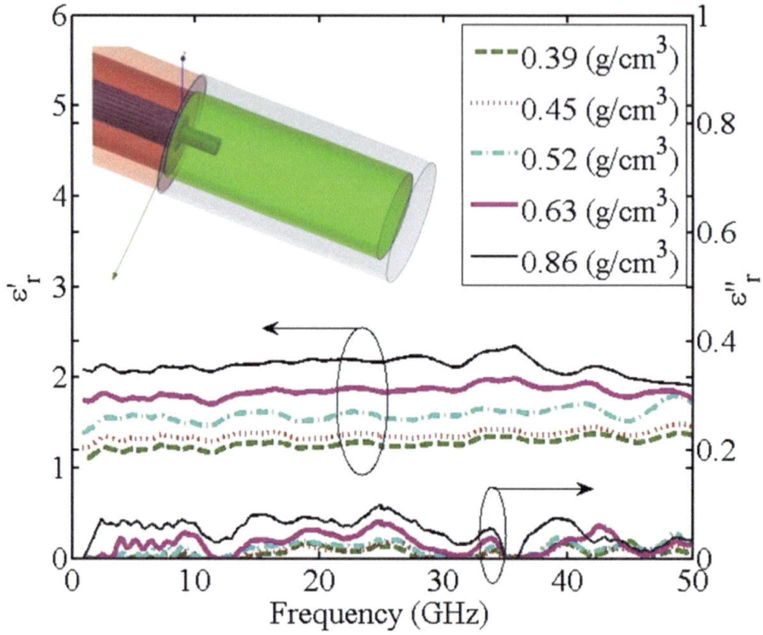

Fig. 3. Variation of extracted effective permittivity of pure pulverized SiO_2 versus frequency using the test setup as shown in Fig. 1.

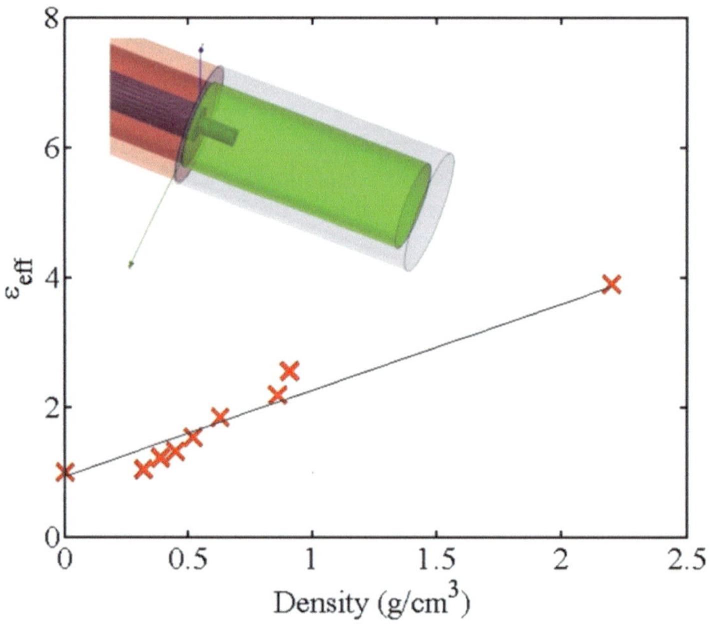

Fig. 4. Variation of extracted effective permittivity of pure SiO_2 versus density realized at frequency of 3GHz. The last point is the common accepted value for SiO_2, $\varepsilon_{eff} = 3.9$ for a density of 2.2 g/cm³.

imaginary part of the extracted complex relative permittivity for SiO_2 powder for various packing densities considering the test setup shown in Fig. 1. Fig. 4 shows the variation of the extracted real part of the effective permittivity of silicon dioxide function of the density at 3 GHz.

The density is computed by weighting the powder using an analytical balance divided by the known volume of the hollow circular waveguide $(\rho = \frac{w}{V} g / cm^3)$. Then the density is gradually increased by packing the powder inside the holder using a manual press. It exists different formulations to estimate the permittivity of bulk material from its pulverized form (Nelson, 1992). However, those formulations are more appropriate for particle size of 50-100μm. From experimental data the variation of the effective permittivity with the packing density for nano-particles suggests a linear dependence instead as shown in Fig. 4 and Fig. 6. Fig. 5 shows the variation of the real part of the effective permittivity of pure alumina powder (Al_2O_3) for various densities at low frequency up to 3 Ghz, it is interesting to observe the flat curves in those cases.

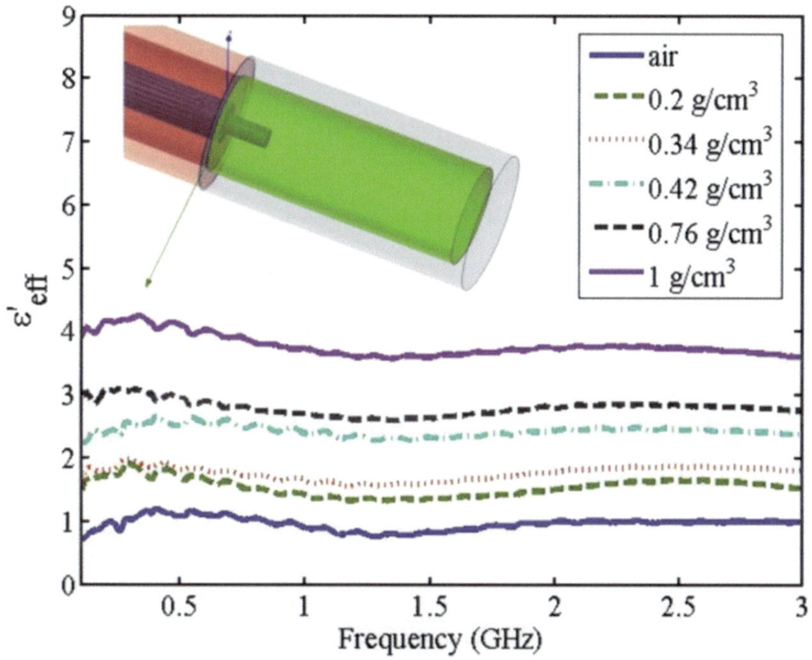

Fig. 5. Variation of extracted effective dielectric constant of pure Al_2O_3 powder versus frequency at different densities using the test setup shown in Fig. 1.

Similarly to the case of silicon dioxide in Fig. 4 the linear variation of the real part of the permittivity of pure alumina powder is presented in Fig. 6. The usual formulations to compute the bulk permittivity from pulverized material doesn't seem appropriate in case of nano-particles and overestimate the common reported value for bulk material. We have used dielectric material with low loss and well known properties to study the variation of the real part of the effective permittivity and provide a reference study to compare with the microwave characterization of carbon nanotube powder. The variation of the real part of the effective permittivity of composites material with densities is compared with the results shown in Fig. 6 to support our experimental studies.

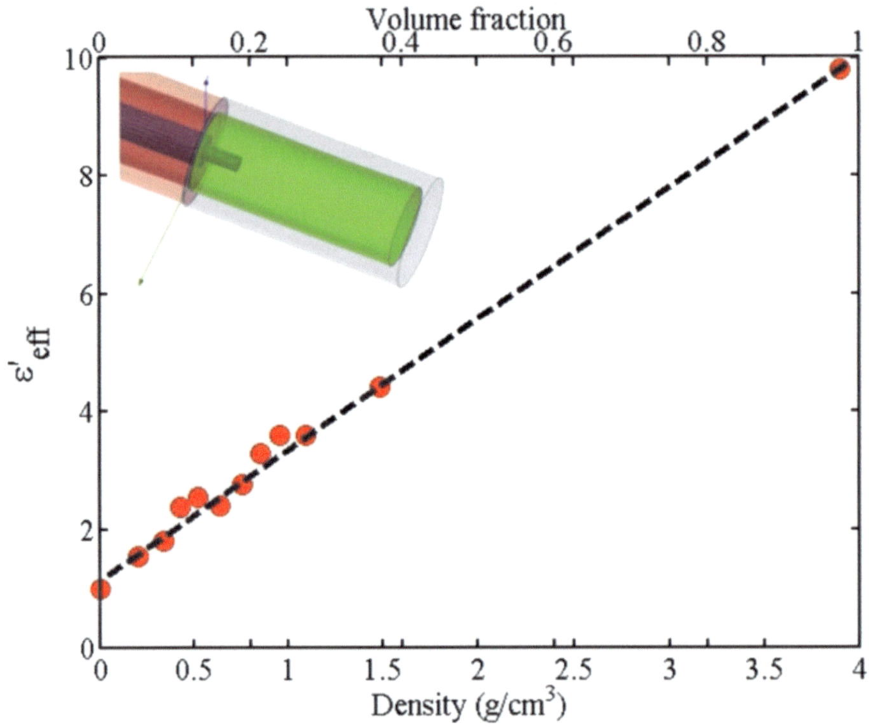

Fig. 6. Variation of extracted effective dielectric constant of pure Al_2O_3 versus density or the volume fraction at 3 GHz using the test setup shown in Fig. 1. The last point is the common accepted value ε'_{eff} = 9.8 at a bulk density of 3.9 g/cm³for pure alumina.

3. Complex permittivity of CNT network

Similarly it is possible to extract the complex effective permittivity of carbon nanotubes powder as furnished by the manufacturer (Sigma-Aldrich). The powder purity is 50 to 70 volume percentage as determined by Raman spectroscopy and scanning electron microscopy (SEM) and contains residual catalyst impurities nickel and yttrium. The lengths of CNTs are approximately 20 μm. In house SEM is carried out and it shows the presence multi-walled carbon nanotubes as well as single-walled CNTs within the sample. Fig. 7 to Fig. 10 present the variation of the real and imaginary part of the relative complex permittivity ($\varepsilon_r = \varepsilon_r^{'} - j\varepsilon_r^{''}$) over a broad frequency band (10 Mhz to 50 Ghz) for different densities (Decrossas et al., 2011).

The highest value of the real part of the relative complex permittivity obtained at low frequency is 1.6×10^5 and decreases continuously to finally reach 4.5 at 50 GHz. Similarly, the highest value of the imaginary part of the relative complex permittivity shown is 4.1×10^3 and decreases to 5 at 50 GHz. Fig. 7 is plotted on a log scale due the fast variation of the relative permittivity at low frequency. Then at high frequency the variation presented in Fig.

8 is reduced and suggests asymptotic values. The different scales highlight the frequency dependence of the relative complex permittivity of CNT networks. Similarly, the imaginary part of the relative permittivity is illustrated in Fig. 9 and Fig. 10.

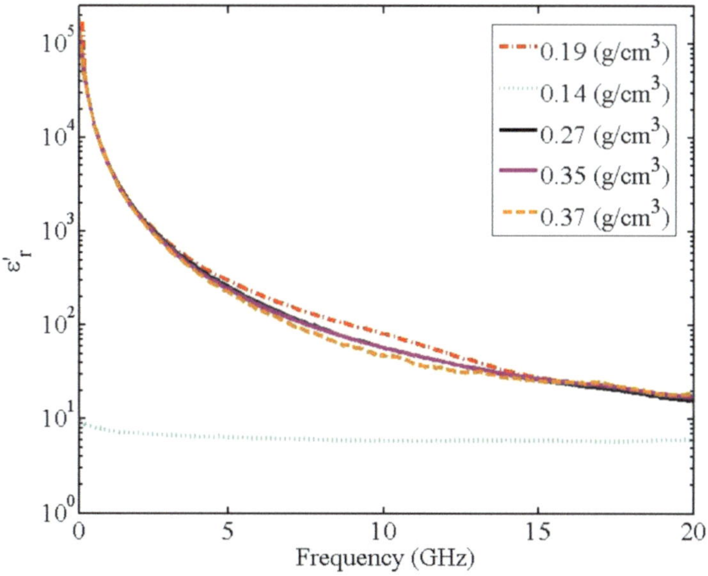

Fig. 7. Variation of the real part of the relative permittivity of carbon nanotubes networks for different densities, at high frequencies.

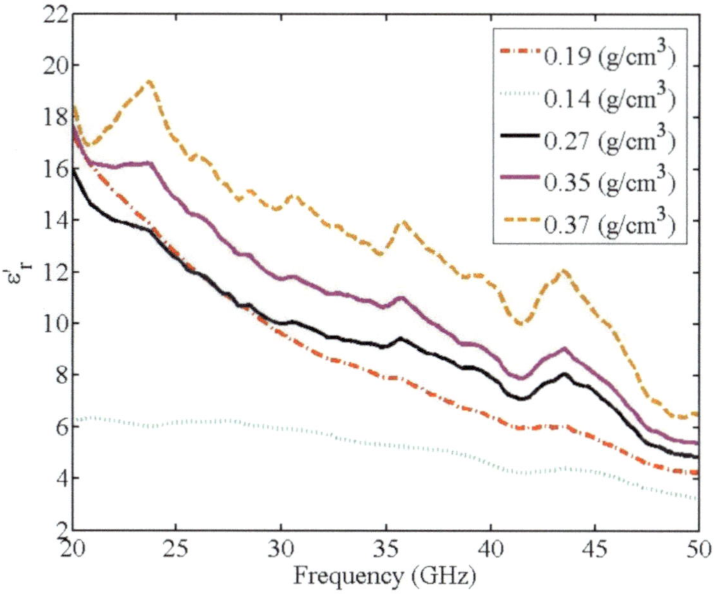

Fig. 8. Variation of the real part of the permittivity of carbon nanotubes networks at low frequencies considering various densities.

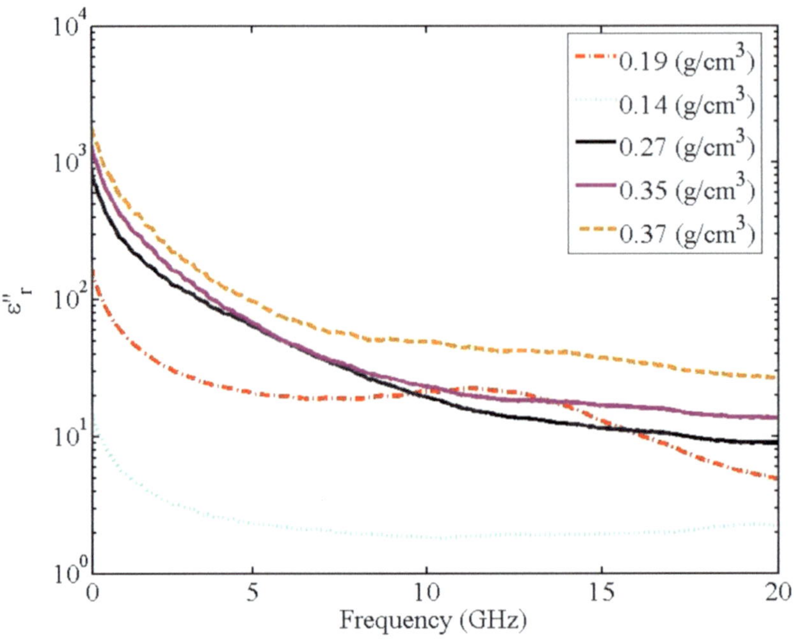

Fig. 9. Variation of the imaginary part of the relative permittivity of carbon nanotubes networks for different densities, at low frequencies.

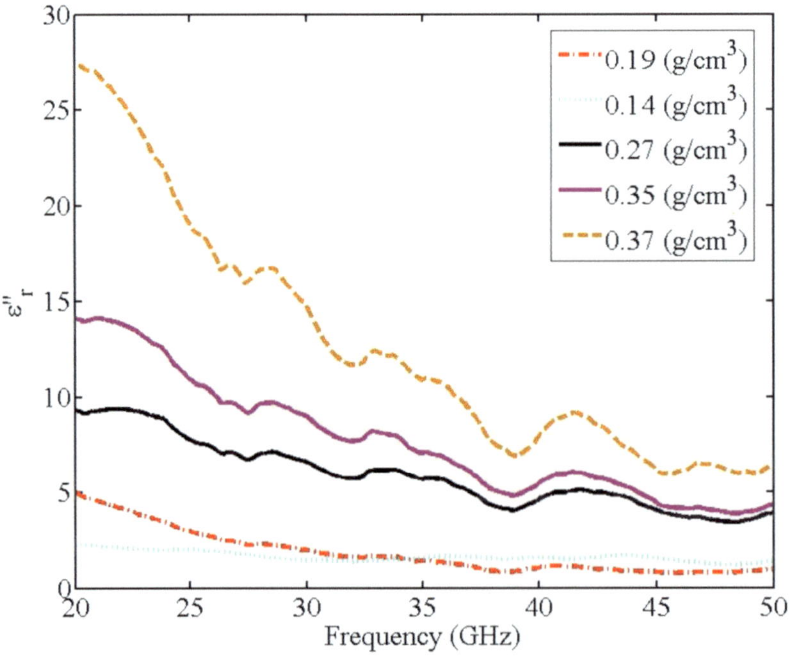

Fig. 10. Variation of the imaginary part of the permittivity of carbon nanotubes networks at high frequencies for various densities.

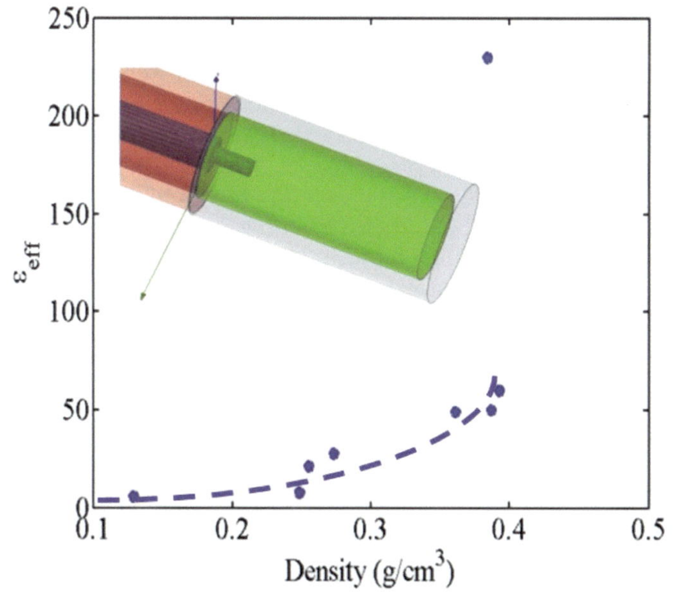

Fig. 11. Variation of the real part of the effective permittivity of carbon nanotube networks function of the density at 10 GHz.

The complex relative permittivity is continuously increasing versus the density of CNT networks as shown in Fig. 11 following the percolation theory (Stauffer and Aharony, 1992). The values obtained at high frequency are in agreement with the values reported (N.N. Al Moayed et al., 2007). Moreover, the large values at low frequency are also consistent with those large values reported (M.A. EL Sabbagh et al., 2009b, and H. Xu et al., 2007).
It should be noted the curve difference of the real part the permittivity function of the density between Fig. 11 and Fig. 4, Fig. 6. The percolation theory may provide a physical explanation about the unusual very high values of the dielectric properties. The high values obtained for the imaginary part of the complex relative permittivity can be attributed to high conductivity (Ebbensen et al., 1996) of carbon nanotubes using the following relation $\varepsilon_r'' = \frac{\sigma}{\omega\varepsilon_0}$ where σ is the conductivity of CNT and $\omega = 2\pi f$ the frequency relation.

The very large values of the real part of the permittivity are explained in the next section where metallic particles are embedded in a dielectric medium and the complex effective permittivity of the mixture is experimentally extracted (Decrossas et al., 2011b).

4. Complex permittivity of composites material by mixing metallic particles in dielectric medium

4.1 Finite element simulation

To study the variation of the dielectric constant, we have randomly embedded metallic spheres in a fictive dielectric (4.5-j*0.88). The metallic spheres considered as perfect electrical conductor have a radius of 0.5mm and are randomly positionned in the dielectric. Then, the effective permittivity is extracted using the mode matching method. The simulations are realized using the 3D full-wave modeling software HFSS™ by Ansoft.

It should be noted that MMT considers the material under test as an homogenous medium while the finite element method considers the interactions between the metallic spheres. For this reason we are only able to extract the effective complex permittivity of the mixture. In addition, by increasing the number of randomly positionned metallic spheres in the host medium considerably augment the computing simulation time and as long as 3 days are necessary to run the last simulation presented in Fig. 12. Those simulations even if the metallic balls dimensions cannot be compared to nano-particles are necessary to demonstrate the effect of the metallic particles in a dielectric medium.
By embedding metallic spheres in the medium, HFSS simulations present an enhancement of the effective dielectric constant as presented in Fig. 12. The extraction of the complex permittivity in addition to the comparison between the finite element method and the mode matching technique validate the process of electrical characterization of the material under test.

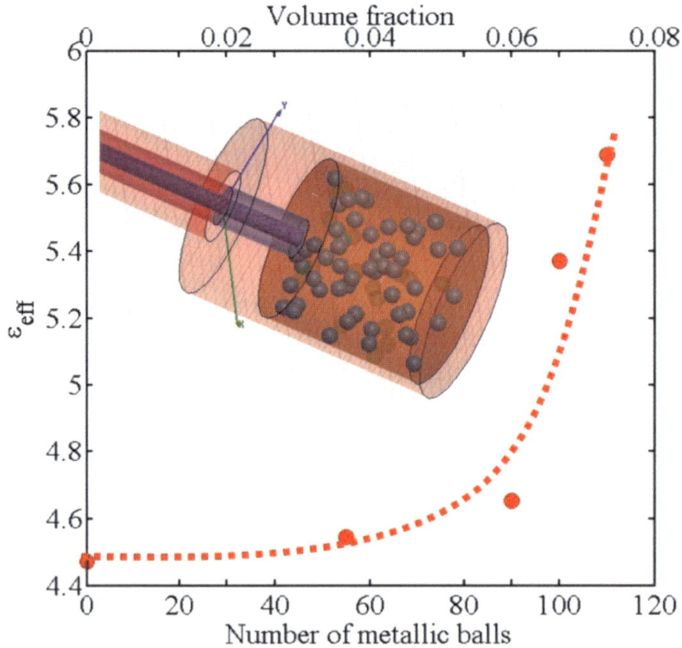

Fig. 12. Variation of the simulated effective permittivity using HFSS by increasing the spherical metallic particles in a fictive dielectric host medium.

4.2 Extracted complex permittivity of copper mixed with alumina particles composites material

The percolation behavior is verified experimentally by embedding metallic particles in a dielectric medium. This theory is usually applied to model the change of material where the increase of metallic particles creates some sort of conductive paths through the dielectric material. In other words, below the percolation threshold the material keeps behaving as a dielectric until the critical concentration point called percolation threshold is reached and the material becomes conductor. This phenomenon occurs at low frequency where the electromagnetic wavelength is larger than the interstice between metallic particles. So, we have concentrated our efforts to extract the effective complex permittivity up to 3GHz. A similar behavior exists in carbon nanotube networks where both conducting and

semiconducting CNTs as well as metal catalyst impurities are present. This work shows how embedded metallic particles enhance the effective relative permittivity of carbon nanotubes bundles due to metallic particle interactions in the dielectric medium below percolation threshold, providing a physical explanation on the usually reported very high value of the complex permittivity. The enhancement of the effective permittivity is realized by mixing copper particles with dry powder of alumina particles. The sketch in Fig. 13 illustrates the realized mixture which is dropped and packed in our setup to extract the effective permittivity of the mixture for different densities and concentrations.
The copper powder as furnished by the manufacturer (Alfa Aeser) has 99 % purity with particle sizes varying between 0.5-1.5μm. Two different weight percentages of copper mixed with alumina powder is studied to show the enhancement of the real part of the permittivity. (a) Copper has 75% weight obtained by mixing 0.5 g of Alumina with 1.5 g of Copper and (b) copper has 20% weight achieved by mixing 1 g of Alumina with 0.25 g of Copper.
It should be noted that copper particles tend to oxidize with air contact, consequently a thin layer of oxide copper (CuO) coats copper (Cu) particles. Fig. 14 illustrates the variation of the extracted complex effective function of the frequency for the 75% weight mixture.

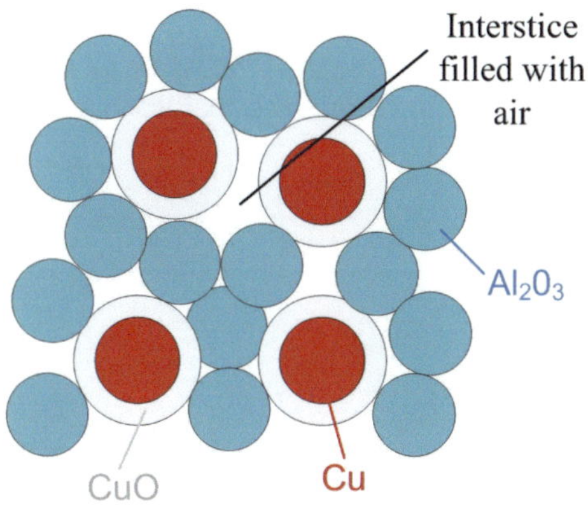

Fig. 13. Illustration of the realized composite material under test (not to scale). The size of the alumina (Al_2O_3) particles is 0.05μm while the size of the copper (Cu) particles is 0.5-1.5μm.

It should be noted that in Fig. 14, an enhancement of the real part is observed while the imaginary part related to the conductivity is still low which is characteristic of a dielectric composite material with a composition below percolation threshold.
Two mechanisms may explain this permittivity enhancement (Fiske et al., 1997, and Ravindran et al., 2006): a) the electrical field creates a surface charge polarization on metallic particles at the dielectric interface which yields an increase in capacitance. b) The dipole polarization contributes to the global permittivity when the electrical field frequency is slower than the relaxation frequency of the metallic particles. This phenomenon occurs at low frequency when the wavelength is larger than the metallic-semiconductor-metallic interactions. Fig. 15 presents a comparison of the extracted real part of the effective permittivity function of the density, highlighting the enhancement of the dielectric constant compared to pure alumina powder as reported in Fig. 6.
All those experiments have been realized considering spherical alumina and copper particles. However, some studies suggest that the shape of the metallic particles also influence the behavior of the effective permittivity (Link et al., 1999, and Guo et al., 2010). To illustrate this idea the next section discusses the mixture of alumina powder with carbon nanotubes.

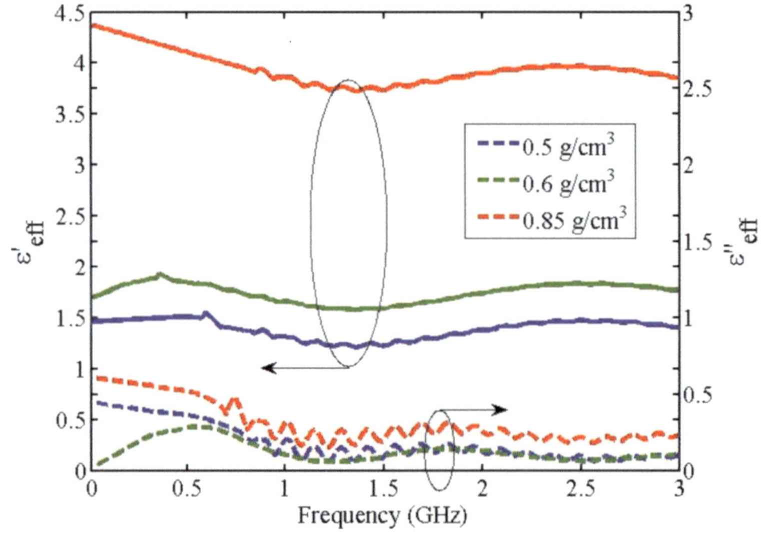

Fig. 14. Extraction of the complex effective permittivity ($\varepsilon_{eff} = \varepsilon'_{eff} - j\varepsilon''_{eff}$) considering the 75% weight mixture of Alumina-copper for different densities functions of the frequency using the test setup shown in Fig. 1.

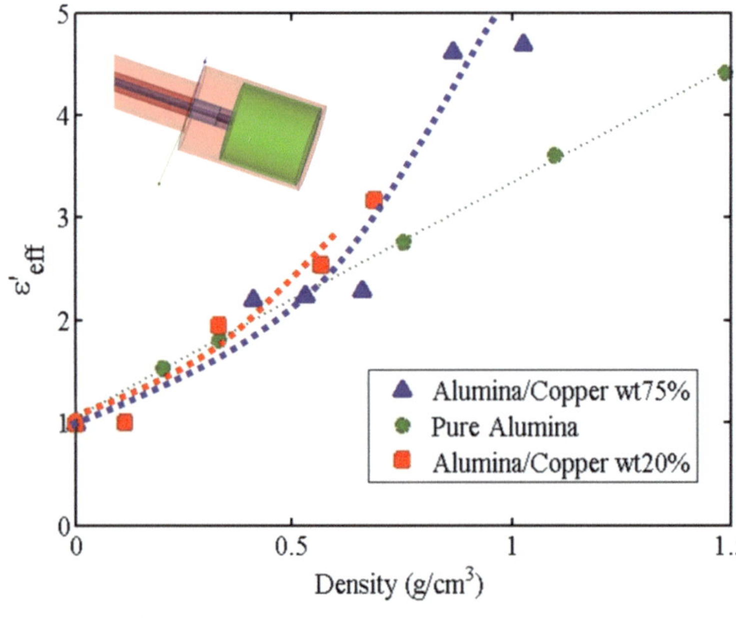

Fig. 15. Variation of the effective permittivity of the composite material function of the density realized at 60MHz and compared to pure alumina powder.

5. Complex permittivity of CNT-based composites material

The enhancement and the frequency dependence of the permittivity due to the presence of carbon nanotubes embedded in a dielectric medium are illustrated experimentally considering different concentrations and packing densities. Those experiments show the possibility to engineer novel composite materials by tuning the electrical response of this material under an electrical field (Liu et al., 2007). Controlling the dispersion, the size, the shape of those particles into the medium are the key of engineering those new composite materials. Fig. 16 is a picture of the prepared mixture of alumina and carbon nanotubes using a scanning electron microscope (SEM).

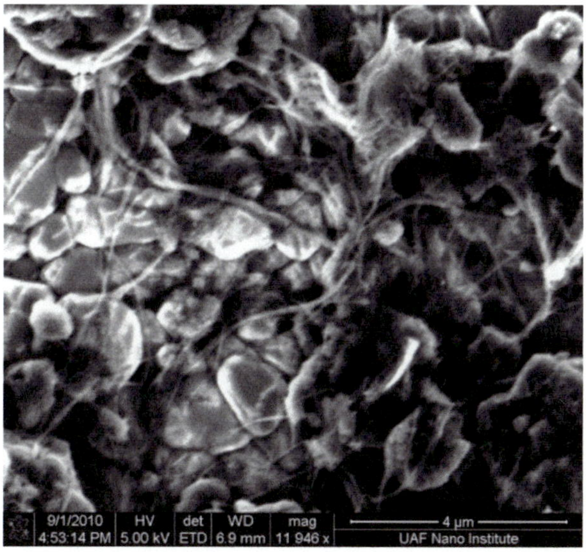

Fig. 16. SEM picture of the prepared mixture of alumina and carbon nanotubes.

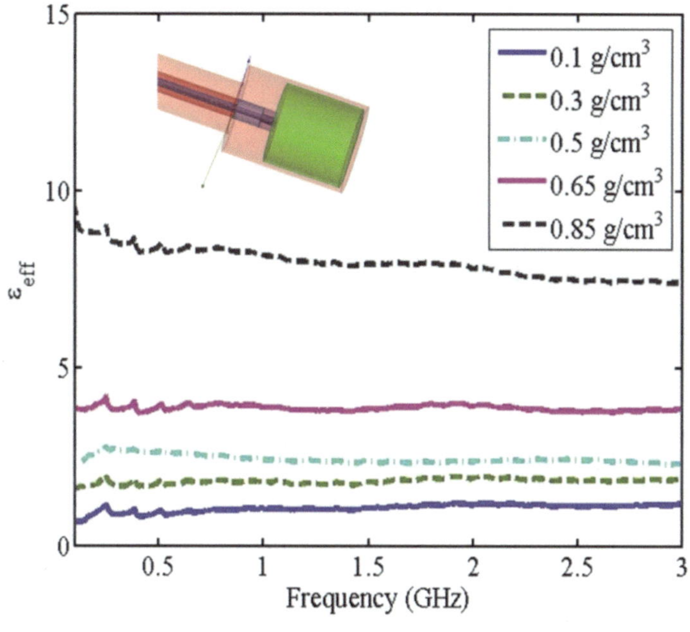

Fig. 17. Variation of extracted real part of the effective permittivity of carbon nanotubes mixed with alumina versus frequency using the test setup as shown in the inset.

The CNT-based composite material is realized by mixing 1g of alumina with 0.2g of carbon nanotubes powder. Fig. 17 illustrates the enhancement of the real part of the effective permittivity at various packing densities. It should be noted that at low density the variation of the effective permittivity shows a flat response indicating a low frequency dependence of the dielectric constant. However, when the density is 0.85g/cm³ in the hollow circular waveguide the dielectric constant decreases with the frequency. As explained in the previous section the permittivity enhancement at low frequency is due to the interfaces between the metallic and dielectric particles. When the frequency increases, i.e. the wavelength decreases, this phenomenon disappears and explains the slope of the dielectric plot at high density (black dashed curve in Fig. 17).

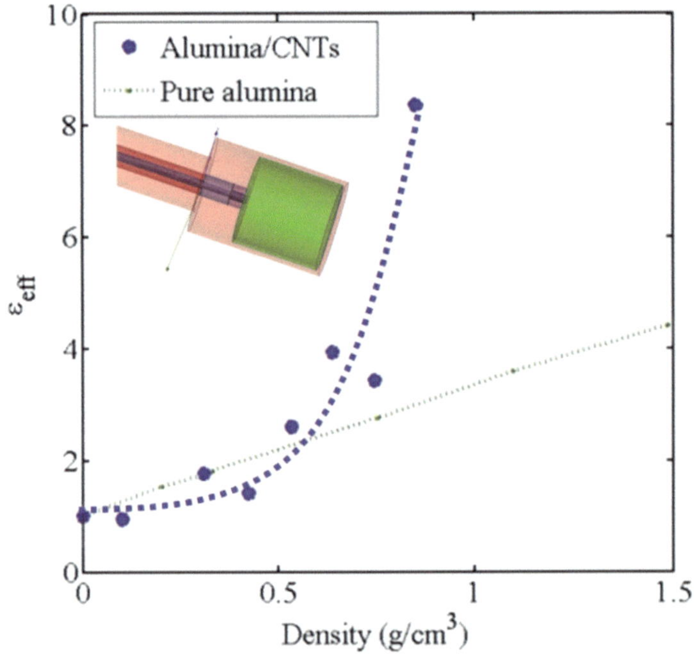

Fig. 18. Variation of the real part of the effective permittivity of CNTs mixed with alumina powder versus the density realized at 60MHz. The variation of the real part of the effective permittivity for pure alumina powder with density is provided as a comparison as reported in Fig. 6.

6. Conclusions

The electrical characterization at RF/microwave frequencies is the key to understand and model devices for future implementation. For this purpose, the frequency dependence of the effective permittivity for carbon nanotubes powder as well as CNT-based composites materials has been extracted. In this chapter, the frequency dependence of the complex permittivity has been demonstrated over a broad frequency band using only one testing setup. The carbon nanotubes powder is analyzed without samples preparation increasing the accuracy of the measurements. It appears that the microwave properties are sensitive to the samples preparation of composite materials due the dispersion of the metallic particles into the host medium. The experiments on composite material where metallic particles are mixed with a dielectric medium provide an explanation about the high value of the effective permittivity obtained for carbon nanotubes powder. The effective permittivity enhancement is explained based on the percolation theory caused by the multiple metallic-dielectric interfaces in the medium encountered by incident waves. The various results suggest that CNT-based composite material may reduce the size of radio frequency/microwave (RF/M) devices by engineering the dielectric constant of the substrate. The tunability of the real and imaginary part independently may have a great potential in novel RF/M devices such as smart material as well as dispersive media for electromagnetic shielding.

7. Acknowledgements

The authors thank M.A EL Sabbagh (Univ. of Syracuse, USA), V. Fouad Hanna (Univ. Pierre et Marie Curie-Paris 6, France), and H.A. Naseem (Univ. of Arkansas, Fayetteville) for fruitful conversations on this topic, D. Rogers for the fabrication of the test structure and M. Benamara for his experience about the preparation of SEM samples. This research was sponsored by the Army Research Laboratory and was accomplished under Cooperative Agreement Number W911NF-10-2-0072. The views and conclusions contained in this document are those of the authors and should not be interpreted as representing the official policies, either expressed or implied, of the Army Research Laboratory or the U.S. Government. The U.S. Government is authorized to reproduce and distribute reprints for Government purposes notwithstanding any copyright notation herein.

8. References

Al Moayed, N.N., Khan U.A., Obol, M., Gupta, S., & Afsar, M.N. (2007). Characterization of single- and multi-walled carbon nanotubes at microwave frequencies, *Proceedings of Instrumentation and Measurement Technology Conference*, pp. 1-4, ISBN 1-4244-0588-2, Warsaw, Poland, May 1-3, 2007.

Decrossas, E., EL Sabbagh, M.A., Fouad Hanna, V., & El-Ghazaly, S.M. (2010). Broadband characterization of carbon nanotube networks, *in IEEE International Symposium on Electromagnetic Compatibility*, pp. 208-211, ISBN 978-1-4244-6305-3, Fort Lauderdale, Florida, USA, July 25-30, 2010.

Decrossas, E., EL Sabbagh, M.A., Fouad Hanna, V., and El-Ghazaly, S.M. (2011). Rigorous Characterization of Carbon Nanotube Complex Permittivity over a Broadband of RF Frequencies, *in IEEE Electromagnetic Compatibility Trans.*, under review.

Decrossas, E., EL Sabbagh, M.A., Naseem, H.A., Fouad Hanna, V., & El-Ghazaly, S.M. (2011b) Effective Permittivity Extraction of Dielectric Nano-powder and Nano-composite Materials: Effects of Packing Densities and Mixture Compositions" in IEEE European Microwave Week, Manchester, UK, October 9-14, 2011.

Ebbensen, T.W., Lezec, H.J., Hiura, H., Bennett, J.W., Ghaemi, H.F., & Thio, T. (1996). Electrical conductivity of individual carbon nanotubes. *Nature*, Vol. 382, No. 6586, (July 1996), pp. 54 - 56.

EL Sabbagh, M.A, & El-Ghazaly, S.M. (2009a). Miniaturized carbon nanotube-based RF resonator, *in IEEE MTT-S Int. Microwave Symp. Dig.*, pp. 829-832, ISBN 978-1-4244-2803-8, Boston, MA, USA, June 7-12, 2009.

EL Sabbagh, M.A., El-Ghazaly, S.M., & Naseem, H.A. (2009b). Carbon nanotube-based planar transmission lines. *in IEEE MTT-S Int. Microwave Symp. Dig.*, pp. 353-356, ISBN 978-1-4244-2803-8, Boston, MA, USA, June 7-12, 2009.

EL Sabbagh, M.A, & El-Ghazaly, S.M. (2009c). Frequency-dependent circuit models of carbon nanotube networks, in *Electrical Performance of Electronic Packaging and Systems*, pp. 129-132, ISBN 978-1-4244-4447-2, Portland, OR, USA, October 19-21, 2009.

Fiske, T.J., Gokturk, H.S., & Kaylon, D.M. (1997). Percolation in magnetic composites. *Journal of Material Science*, Vol.32, No. 20, (October 1997), pp. 5551-5560, ISSN 0022-2461.

Guo, N., DiBenedetto, S.A., Tewari, P., Lanagan, M.T., Ratner, M.A., & Marks, T.J. (2010). Nanoparticle, size, shape, and interfacial effects on leakage current density, permittivity, and breakdown strength of metal oxide-polyolefin nanocomposites: experiment and theory. *In Chemistry of Material,* Vol. 22, No. 4, (February 2010), pp. 1567–1578.

Iijima, S. (1991). Helical microtubules of graphitic carbon. *Nature,* Vol. 354, (November 1991), pp. 56-58.

Jorio, A., Dresselhaus, M.S., & Dresselhaus, G. (2008). *Carbon Nanotubes: Advanced Topics in the Synthesis, Structure, Properties, and Applications,* Springer, ISBN 978-3540728643, New York, USA.

Link, S., Mohamed, M.B., & El-Sayed, M.A. (1999). Simulation of the optical Absorption spectra of gold nanorods as a function of their aspect ratio and the effect of the medium dielectric constant. *Journal of Physical Chemistry B,* Vol. 103, No. 16, (April 1999), pp. 3073-3077.

Liu, L., Matitsine, S., Gan, Y.B., Chen, L.F., & Kong, L.B. (2007). Frequency dependence of effective permittivity of carbon nanotube composites. *Journal of Applied Physics,* Vol. 101, No. 9, (May 2007), pp. 094106-1-7, ISSN 0021-8979.

Meyyapan, M. (2005). *Carbon Nanotubes: Science and Applications* (1st ed.), CRC Press LLC, ISBN 9780849321115, Florida, USA.

O'Connell, M.J., (2006). *Carbon Nanotubes Properties and Applications* (1st ed.), CRC Press LLC, ISBN 978-0849327483, Florida, USA.

Nelson, S.O. (1992). Estimation of permittivities of solids from measurements on pulverized or granular materials, In: *Dielectric Properties of Heterogeneous Materials*, A. Priou, Chap. 6, pp. 231-271, Elsevier, ISBN 9780444016461, New York, USA.

Ravindran, R., Gangopadhyay, K., Gangopadhyay, S., Mehta, N., & Biswas, N. (2006). Permittivity enhancement of aluminum oxide thin films with the addition of silver nanoparticles. *Applied Physics Letter,* Vol. 89, No. 263511, (December 2006).

Saito, R., Dresselhaus, G., & Dresselhaus, M.S. (1998). *Physical Properties of Carbon Nanotubes,* Imperial College Press, ISBN 978-1860942235, London.

Stauffer, D., & Aharony, A. (1994). *Introduction to Percolation Theory* (2nd ed.), Taylor and Francis, ISBN 978-0748402533, Washington DC.

Tung, S., Rokadia, H., & Li, W.J. (2007). A micro shear sensor based on laterally aligned carbon nanotubes. *In Sensors and Actuators A: Physical,* Vol. 133, No. 2, (February 2007), pp. 431–438.

Xu H., Anlage, S.M., Hu, L., & Gruner, G. (2007). Microwave shielding of transparent and conducting single-walled carbon nanotube films. *Applied Physics Letter,* Vol. 90, No. 183119, (May 2007).

10

Graphene Phytotoxicity in the Seedling Stage of Cabbage, Tomato, Red Spinach and Lettuce

Bunshi Fugetsu and Parvin Begum
Laboratory of Environmental Medical Chemistry
Graduate School of Environmental Science
Hokkaido University
Japan

1. Introduction

The use of nanostructures with unusual novel properties in agriculture (Joseph, 2006) and for technological applications has been an active and exciting area of research in recent years. Graphene, the most recently discovered carbon allotrope, is a two-dimensional building block of atomic thickness that can be stacked into three-dimensional graphite, rolled into one-dimensional nanotubes, or wrapped into zero-dimensional fullerenes (Geim & Novoselov, 2007). The unique electronic and transport properties of graphene (Ando, 2009), compatible with existing manufacturing processes (Xia et al., 2009), and the absence of the energy gap in the electronic spectra have opened up increasingly rich possibilities in the development of future electronic devices (Geim & Novoselov, 2007; Enderlein, 2010) and the graphene-based quantum electronics (Dragoman, 2009) that offer many benefits. If these trends in nanotechnology continue, graphene may ultimately be released into the aquatic, terrestrial, and atmospheric environments, where its fate and behavior are largely unknown. Exposure to nanoparticles in higher plants is expected to have an effect because these plants strongly interact with their atmospheric environments (Monica, 2009). The nanoparticles, with their ultra-small size, specific shape, geometric structure, and unique properties, may have the potential for increased toxicity (Bhabra, 2009; Kirchner et al., 2005; Nowack & Bucheli, 2007; Jia et al., 2005). Nanoparticles can drastically modify their physico-chemical properties compared to particles of bulk size (Nel et al., 2006). Carbon nanoparticles can penetrate plant cells (Lui et al., 2009; Khodakovskaya et al., 2011) and induce phytotoxicity at high doses (Stampoulis et al., 2009; Ghodake et al., 2010; D.H. Lin & Xing, 2007), leading some authors to conclude that certain carbon nanoparticles are not 100% safe. Therefore, great caution is suggested when considering the introduction of nanoparticle-based products to the market, and there is an urgent need for research related to the broad area of nanotoxicology. A recent study, for example, has pointed to the possible adverse effects of graphene on human health (Zhang et al., 2010) and in bacteria (Akhavan & Ghaderi, 2010).
In response to these concerns, we explored whether graphene can induce phytotoxicity at high doses in terrestrial plants grown hydroponically and exposed to varying concentrations (0 to 2000 mg/L) of graphene. Thus far, only one study has evaluated the

toxic effect of graphene on plant physiology and plant development, indicating possible adverse effects (Khodakovskaya et al., 2011). Khodakovskaya et al., (2011) reported insignificant toxicity of graphene in the growth of tomato, although they used only one low concentration (50 µg/mL) of chemically functionalized graphene with few layers and with a thickness of 2–5 nm. In a study by Says et al., (2006) carbon nanoparticle functionalization led to a remarkable decline in toxic effects. The effects of nanoparticles on different plant species can vary greatly with plant growth stages and method and duration of exposure and depend also on nanoparticle size, concentration, chemical composition, surface structure, solubility, shape, and aggregation (Nel et al., 2006).

Thus, to our knowledge, the possible adverse effects of graphene in terrestrial plants is almost totally unknown. Here, we use cabbage, tomato, red spinach, and lettuce as the selected crops to investigate the toxic effects of graphene and to identify appropriate target plant species for further studies associated with graphene. Potential targets include various terrestrial plants that are normally protected by specific barriers such as a cell wall. Targeting can be made more effective by prolonged exposure of different plant species to carbon nanoparticles and application of a high concentration, which can lead to aggregation on the root surface (C. Lin et al., 2009), penetration within the cells (Khodakovskaya et al., 2011), and a contribution to toxic effects (Stampoulis et al., 2009; Ghodake et al., 2010; S. Lin et al., 2009). This report is the first to describe the phytotoxic effect of graphene in terms of seedling growth, cell death, reactive oxygen species (ROS) generation, and morphology change. Our results showed the greatest toxic effect of graphene on cabbage and tomato, followed by red spinach, with no clear toxic effect for lettuce.

2. Results and discussion

A previous study uncovered evidence supporting that carbon-based nanoparticles have adverse effects on terrestrial plants (D.H. Lin & Xing, 2007) opening the way to further investigation. Consequently, we assessed the potential impact of graphene on the growth of tomato, cabbage, red spinach, and lettuce based on tests suggested and encoded by USEPA guidelines (USEPA, 1996), which include consideration of studies on seed germination and seedling growth (root and shoot growth, leaf number), often accompanied by other evaluations of cell death, ROS production, and morphological studies using scanning electron microscopy (SEM), useful for obtaining evidence of *in situ* symptoms of possible toxicity. Altogether, the current work investigating the potential effect of graphene demonstrates possible adverse effects on plants, underscoring the need for ecologically responsible disposal of graphene and for more research on the potential effects of graphene on agricultural and environmental systems.

2.1 Graphene analysis

Water-soluble graphene, commonly referred as graphene oxide with sodium ions as the counterions (solution pH 6.32), was used throughout this study. Atomic force microscopy (AFM) was used to evaluate the morphologies of the graphene; Figure 1 shows typical AFM images. The apparent heights of all the graphene observed were found to be around 1 nm, indicating that the graphene was fully exfoliated into individual sheets with the size of length × breadth ranged from 0.5 × 0.6 ~ 1.5 × 6.5 µm for 30 pieces of the graphene. Figure 2 A shows typical SEM image of the graphene, revealing the morphology of the graphene sheets. Figure 2 B shows a transmission electron microscope (TEM) image of the graphene, which exhibits a typical wrinkled structure (Stankovich et al., 2006) with the corrugation and scrolling that are fundamental to graphene (Meyer et al., 2007).

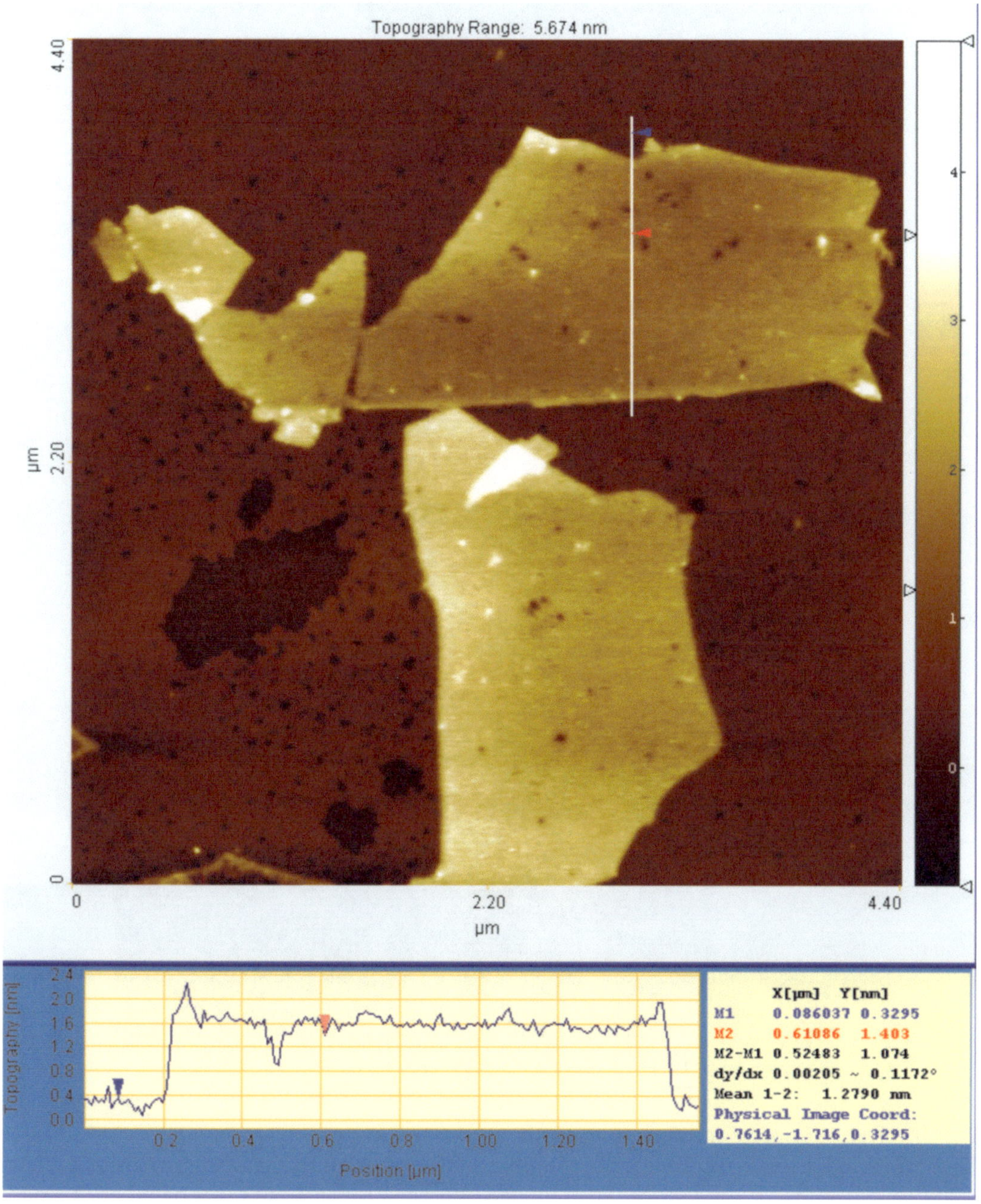

Fig. 1. AFM image of graphene.

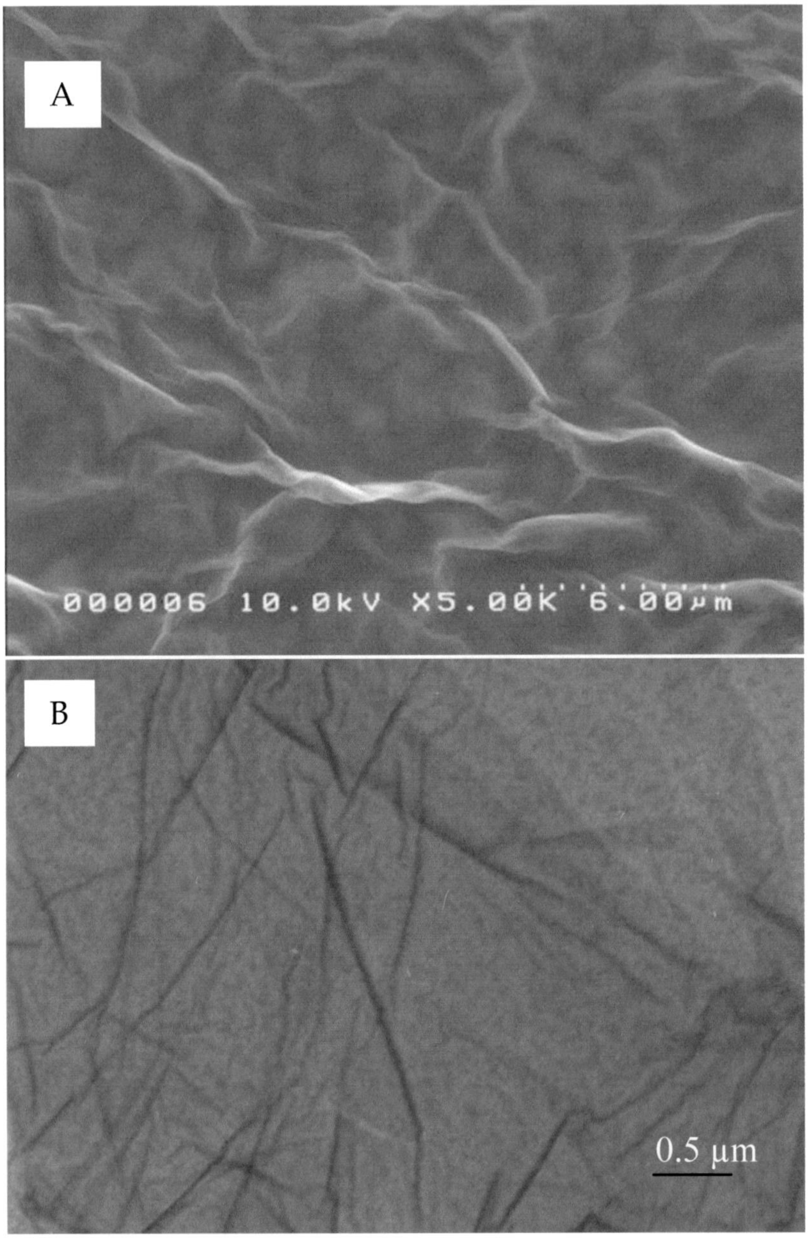

Fig. 2. (A) SEM image and (B) TEM image of graphene.

2.2 Graphene repression of plant growth

We conducted a series of tests of the potential effect of graphene on the growth of tomato, cabbage, red spinach, and lettuce. Treated and untreated seeds were germinated after 4 days of incubation at 25 °C in the dark. Fully germinated seedlings with developed cotyledons and root system are shown in Figure 3. Cotyledons and root system were retarded with increasing graphene concentration. Only in the case of lettuce was there no significant influence of graphene on the cotyledons and root system after 4 days (data not shown). On the other hand, graphene had a clear negative influence after 20 days on root and shoot length and biomass of tomato, cabbage and red spinach exposed to graphene. The observed influence depended on the concentration of graphene and the duration of the experiment. In hydroponics experiments, when plants were treated with different concentrations of graphene (0, 500, 1000, and 2000 mg/L), the 20-day-old plants were characterized by inhibition of plant growth and leaf number and leaf size decrease with increasing graphene concentration in comparison with control (Figure 4) and showed toxicity symptoms. No significant effect of graphene on the parameters tested (seed germination and growth) was noted in our experiment in the case of lettuce. Therefore, the following discussion is focused on the results obtained for the other three plants.

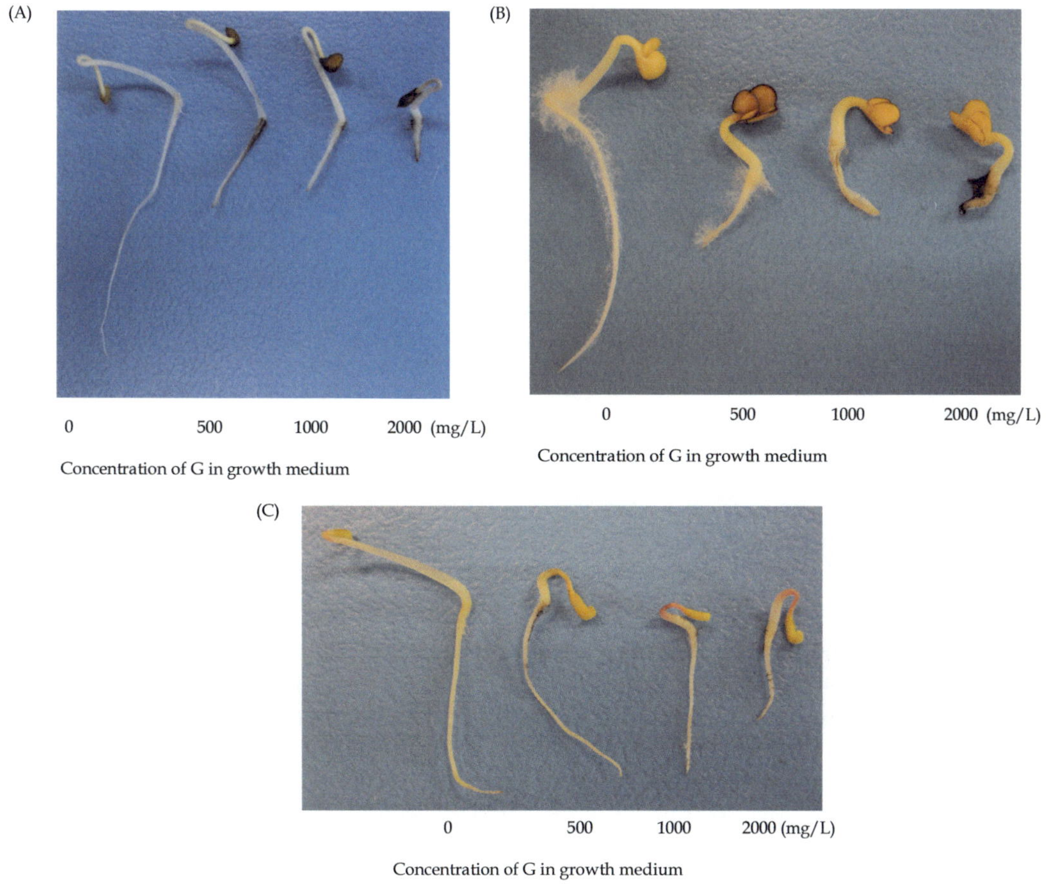

Fig. 3. Graphene (G) affected cotyledons and root systems. (A–C) Tomato, cabbage, and red spinach seeds after incubation with and without graphene solution on filter paper for 4 days, respectively.

Fig. 4. Effect of graphene (G) on growth and development of cabbage, tomato, and red spinach seedlings. (A–C) Tomato, cabbage, and red spinach seedlings were hydroponically grown in Hoagland media for 20 days with and without graphene, respectively.

The presence of graphene resulted in decreased root and shoot growth (Figure 5A, B). However, graphene at a lower concentration (500 mg/L) resulted in only a slight decrease in root and shoot length. A marked inhibition effect was observed with the highest concentration of graphene (2000 mg/L). In the case of cabbage, there was a significant increase in root and shoot growth inhibition greater than 78% and 61%, respectively, compared to control. The 2000 mg/L concentration graphene resulted in root and shoot growth inhibition of tomato of 46% and 53%, respectively, compared to control. Inhibition of shoot growth was noted in red spinach (13%) at 500 mg/L, and the highest concentration (2000 mg/L) resulted in further significant inhibition of shoot growth (39%). However, no significant effect of graphene on the root growth of red spinach at 500 mg/L was observed while root growth of red spinach was inhibited by 18% at the higher graphene concentrations (1000 and 2000 mg/L).

The presence of graphene resulted in decreased root and shoot weight (Figure 5C, D). With cabbage seedlings grown in hydroponic culture, the root and shoot weight was sensitive to

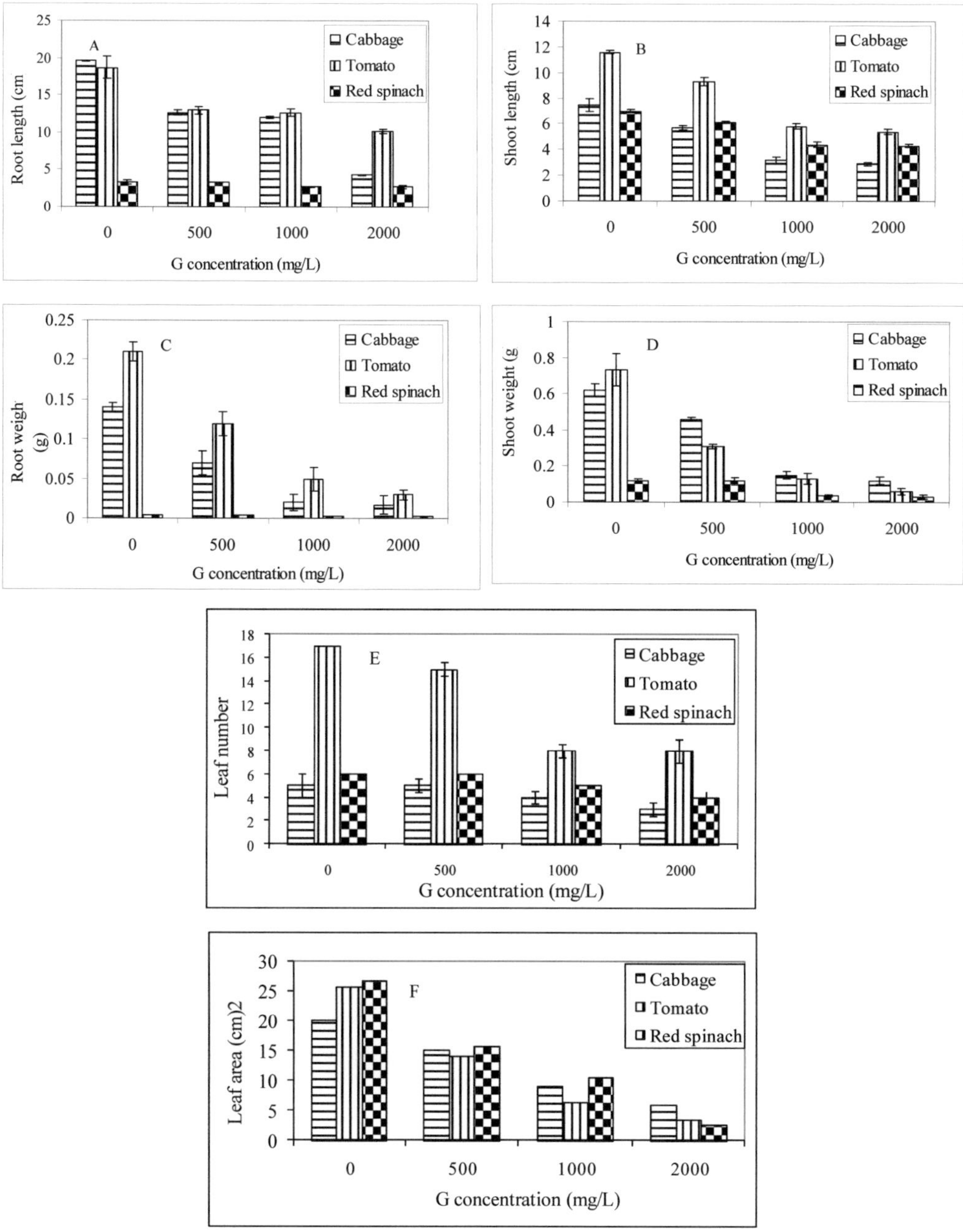

Fig. 5. Effect of graphene (G) on growth and development of cabbage, tomato, and red spinach seedlings. Twenty-day-old seedlings growing on Hoagland media with graphene (0, 500, 1000, and 2000 mg/L) were used for all measurements. Error bars represent standard deviation. (A) Root length, (B) shoot length, (C) root weight, (D) shoot weight, (E) leaf number, and (F) leaf area.

graphene, and weight decreased by 88% and 81%, respectively, at the highest concentration (2000 mg/L) compared to control. The root and shoot weight of tomato was also sensitive to graphene, and weight decreased by 86% and 92%, respectively, at the highest concentration (2000 mg/L) compared to control. With red spinach, the shoot weight appeared to be more sensitive to graphene compared to root; the root and shoot weight of red spinach decreased by 39% and 75% at the highest concentration (2000 mg/L), respectively, compared to control. Tomato seedlings were somewhat more sensitive to graphene than were cabbage and red spinach seedlings.

Graphene influenced the leaves of all tested plants in a dose-dependent manner (Figure 5E). The number of leaves decreased with increasing graphene concentration compared with control. For example, control tomato without graphene developed an average of 17 leaves each, whereas the treated tomato developed only 8 leaves each at 1000 and 2000 mg/L (Figure 5E). The leaf numbers of the treated plants were considerably decreased by 40%, 53%, and 33% at the highest graphene concentration (2000 mg/L) in cabbage, tomato, and red spinach, respectively. The leaves exhibited reduced size and wilting symptoms, as evidenced by visual observation (Figure 4). Of interest, the leaf area (visual observation, Figure 4, Figure 5F) of all treated plants was gradually reduced and continued to decline with increasing graphene concentration. A significantly reduced leaf area of the treated plants was clearly observed at the highest graphene concentration (2000 mg/L, Figure 5F) compared to control. Furthermore, the leaf area of cabbage was reduced by 25% at 500 mg/L and by 71% at 2000 mg/L compared to control, indicating a dose-dependent reduction. Tomato had an 88% leaf area reduction at 2000 mg/L (Figure 5F) and 45% at 500 mg/L. The leaf area of red spinach was reduced by 91% at 2000 mg/L compared to control (Figure 5F).

We found that cotyledons and the root system of cabbage, tomato, and red spinach were inhibited after four days and observed further inhibition in seedling growth at different graphene concentrations. Although 500 mg/L graphene had a slight effect, 2000 mg/L resulted in a notable effect on the seedling stage of cabbage, tomato, and red spinach. By comparison, graphene at 50 μg/mL increased growth rates with no sign of significant toxicity for tomato in the seedling stage (Khodakovskaya et al., 2011). Khodakovskaya et al., (2011) investigated the effect of four carbon materials (single-, multiwall CNTs, few-layer graphene materials, and activated carbon) on the seedling stage of tomato at 50 μg/mL. As noted, a high concentration of graphene was not employed in that study, and the authors used only one low concentration. Under such low-concentration conditions, metals play an essential role in the development of plants at the limit of plant tolerance. If an excess of metals is absorbed by plants, toxic effects can manifest, including growth reduction and abnormalities in cell division (Stiles, 1961), possibly because of the excess metals acting as cofactors for enzymes involved in the formation of intermediate metabolites.

2.3 ROS measurement and hydrogen peroxide (H_2O_2) detection

Plants continuously produce ROS as byproducts of various metabolic pathways, but the excess accumulation of ROS leads to oxidative stress and cell death (Apel & Hirt, 2004; Wen et al., 2008). ROS products, whether inside or outside the cell, can be key factors in the toxicological effects of nanostructure materials (Nel et al., 2006). CNTs are known to have phytotoxic effects in plant cells because of aggregation (C. Lin et al., 2009) causing cell death and accumulated ROS in a dose-dependent manner (Tan et al., 2009). Graphene may have the ability to generate ROS production, based on the similarities of some of the properties of graphene sheets and CNTs. Based on the assumption that graphene can be involved in ROS production in the leaf, we tested this possibility using 20-day-old leaves from test plants for the detection of ROS by means of the ROS-sensitive dye 2',7'-dichlorofluorescein diacetate (DCFH-DA).

We evaluated ROS production (oxidative stress) by means of H_2O_2 detection after infiltration with DCFH-DA of treated and untreated leaves. The accumulation of H_2O_2 was visualized under a microscope; it can be imaged under fluorescence microscopy after removal of DCFH-DA from the leaves by washing with PBS buffer (Figure 6A–F). Figure panels 6B, 6D, and 5F illustrate that graphene-treated leaves showed an increase in DCFH fluorescence compared to control leaves (Figure 6A, C, and E) of cabbage, tomato, and red spinach, respectively. Measurement of DCFH fluorescence by spectrofluorometer demonstrated the dose-dependent increase in ROS content in graphene-treated leaves (Figure 6G) compared to control. The accumulation of ROS in the leaf was measured with the excitation wavelength at 485 nm and emission wavelength at 522 nm. As Figure 6G shows, as the concentration of graphene in medium increased, a progressively enhanced DCFH response was observed, strongly suggesting that graphene can cause an oxidative stress reaction in plant cells.

As Figure 6G shows, in the case of red spinach, graphene and our control without graphene shared a similar fluorescence intensity, specificallyat the low graphene concentration (500 mg/L). This finding supports that a low concentration of graphene is not responsible for induction of ROS because metals play an essential role in the development of plants at the limit of plant tolerance. In contrast, graphene at a high concentration elicited a sharply increased ROS level compared to control.

We further tested the excess presence of ROS induced by graphene in 20-day-old leaves of cabbage, tomato, and red spinach by means of the ROS-sensitive dye 3,3'-diaminobenzidine (DAB). Infiltration of leaves with DAB allowed the detection of H_2O_2. The location of insoluble deep radish brown polymerization product produced when DAB reacts with H_2O_2 was visualized under a light microscope; it can be imaged after removal of chlorophyll from the leaves by boiling for 15 min in ethanol. Imaging of deep radish brown polymerization can be considered to indicate the accumulation of H_2O_2 (Figure 7B, D, and F) in cabbage, tomato, and red spinach, respectively, and formation of H_2O_2 would be expected as a result of the graphene induction. In control leaf (Figure 7A, C, and E) of cabbage, tomato, and red spinach, respectively, no distinctive deep radish brown polymerization of H_2O_2 was detectable. The plants exposed to concentrations of 1000 mg/L showed significant toxicity after 20 days, with the DAB assay indicating increasing ROS. The amount of formazan formation in leaves was determined at A_{700}. As shown in Figure 7G, higher levels of graphene (1000 mg/L) triggered production of more ROS. The excess ROS production at 1000 mg/L may be indicative of concentration-dependent ROS generation.

ROS production estimated by DCFH-DA was correlated to an increase in ROS predictable by DAB assays. However, the results of both assays demonstrated the direct presence of ROS that are produced inside the leaf in those plants grown with graphene. The ROS localization using DCFH-DA and DAB described above was also supported by the quantitative determinations of DCFH and deep brown polymerization at a higher concentration of graphene (1000 mg/L) (Figure 6G and Figure 7G). These findings are in agreement with those of Zhang et al., (2010) who exposed neural pheochromocytoma-

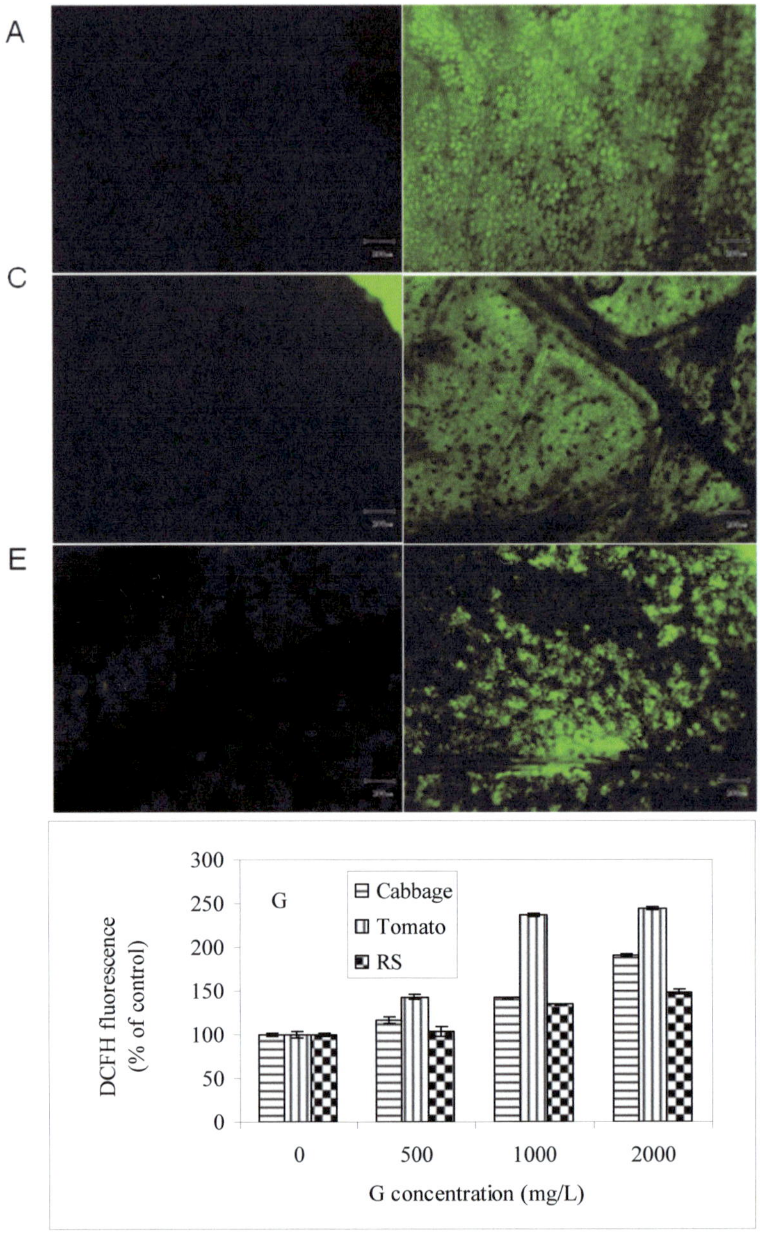

Fig. 6. Effect of graphene (G) on the generation of ROS in leaves tested by means of the ROS-sensitive dye DCFH-DA in cabbage, tomato, and red spinach seedlings. Twenty-day-old leaves treated with or without 1000 mg/L graphene were used for all measurements. Images were observed with fluorescence microscopy. Left lane (A), (C), and (E) cabbage, tomato, and red spinach leaves without graphene. Right lane (B), (D), and (F) cabbage, tomato, and red spinach leaves with graphene (1000 mg/L). The green signal indicates ROS generation (DCFH fluorescence) in the graphene-treated leaves. (G) Dose-related effects of graphene on ROS in treated leaves measured by DCFH fluorescence.

Fig. 7. Effect of graphene (G) on accumulation of H_2O_2 in leaves tested by means of the ROS-sensitive dye DAB of cabbage, tomato, and red spinach seedlings. Twenty-day-old leaves treated with or without 1000 mg/L graphene were used for all measurements. Left lane (A), (C), and (E) cabbage, tomato, and red spinach leaves without graphene, respectively. Right lane (B), (D), and (F) cabbage, tomato, and red spinach leaves with graphene (1000 mg/L), respectively. The brown staining indicates the formation of a brown polymerization product when H_2O_2 reacts with DAB. (G) Effect of graphene (1000 mg/L) on accumulation of H_2O_2 in treated leaves as measured using DAB.

derived PC12 cells to graphene and demonstrated that graphene-induced ROS were involved in the toxic mechanism. It has been reported that ROS are key signaling molecules and could be induced by many exogenous stimuli (Apel and Hirt, 2004).

Evidence for the accumulation of ROS induced by graphene includes the fact that the ability of graphene to reduce plant growth was positively correlated with both the generation of ROS and the graphene concentration (dose-dependent ROS production as estimated by DCFH-DA). Control plants that were devoid of graphene-produced ROS and consequently had not generated ROS did not show a reduction in growth. In the presence of a high graphene concentration in which the content of the nanoparticles exceeds the limit of plant tolerance, ROS may be over-generated, thereby decreasing plant growth. Thus, at least part of the toxic effects of graphene nanoparticles in plants likely results from an excess of induced ROS.

2.4 Evaluation of cell death

To evaluate the toxicity of graphene through cell membrane damage *in vivo* in hydroponic culture conditions, we used a cell death assay. The roots of treated and untreated seedlings were examined by measuring cell death using Evans blue staining and with O.D. values at 597 nm after a 20-day exposure to different concentrations (0, 500, 1000, and 2000 mg/L) of graphene. The measurement of the Evans blue extracted from roots showed a concentration-dependent induction (Figure 8A). With the concentration of graphene increasing to 1000 and 2000 mg/L in the case of cabbage, the amount of Evans blue uptake in root cells increased by about 5 and 12.3 fold, respectively, compared to that of control roots. A significant level of Evans blue uptake was observed after both 1000 and 2000 mg/L of graphene treatment of tomato roots, representing 11.5 and 14.4 fold, respectively, of that in control roots. The higher concentrations of graphene (1000 and 2000 mg/L) caused 8.5- and 11.8-fold higher uptake, respectively, of Evans blue in roots of red spinach compared to that in control roots. Thus, these results indicate that the loss of plasma membrane integrity occurred with exposure to graphene, as it was greater than in the control plants. Higher (1000 and 2000 mg/L) graphene concentrations caused severe stress on plant growth, inducing the excess generation of ROS. The effects of the ROS may be reflected in the rapid rise of Evans blue uptake. The toxic effect on plant roots appeared after exposure to the lower concentration of graphene, and differences between controls and graphene-treated plants became even more prominent at higher concentrations of graphene, as observed under a light microscope (data not shown).

We analyzed another cell death index to further demonstrate that graphene causes cell membrane damage (electrolyte leakage/plasma membrane sensitivity) in the treated leaf. Electrolyte leakage experiments indicated that the leaves of the graphene-treated plants had decreased membrane integrity (Figure 8B). For these experiments, we used 20-day-old seedlings with leaves, investigating the effect of increasing concentrations of graphene on membrane integrity. Cell membrane damage was noted at 500 mg/L in cabbage, tomato, and red spinach, and 1000 and 2000 mg/L resulted in further damage, indicating dose-dependent membrane integrity and disruption of the plasma membrane. In addition, as shown in Figure 8B, the results of the electrolyte leakage experiments indicated that the tomato had significantly greater leakage at the highest concentration (2000 mg/L) compared to cabbage and red spinach. A comparison of Figure 8A and 8B shows that enhancement of Evans blue uptake and loss of plasma membrane integrity gradually increased up to 2000 mg/L graphene under the same conditions. Evans blue staining showed that extensive cell death/membrane injury was induced in plant roots by different concentrations of graphene treatment. The electrolyte leakage assay confirmed the membrane injury resulting from graphene treatment. This result indicates that the loss of plasma membrane integrity was induced by graphene at the lower concentration and continued at higher concentrations. These results also suggest that intracellular ROS might have a crucial role in induction of cell death induced by graphene. It has been reported that the accumulation of ROS causes cell death, which is demonstrated by electrolyte leakage from cells (Kawai-Yamada et al., 2004).

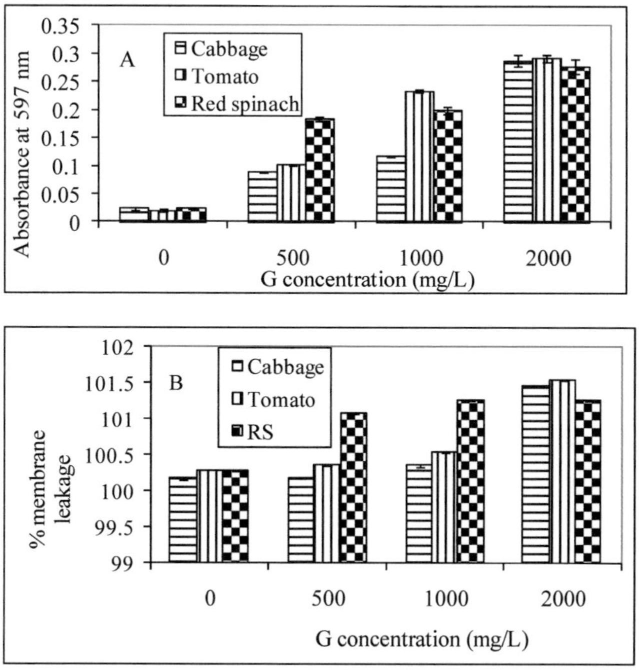

Fig. 8. Effect of graphene (G) on cell death of cabbage, tomato, and red spinach seedlings. Twenty-day-old seedlings growing on Hoagland media with graphene (0, 500, 1000, and 2000 mg/L) were used for all measurements. Error bars represent standard deviation. (A) Cell death determined by Evans blue and (B) cell death determined by electrolyte leakage.

The effect of graphene on cell death was evaluated by Evans blue staining and a conductivity method (% of membrane leakage). These two assays have been considered as indexes of cell death in plants (Del & Lam, 1998), although they give information only about plasma membrane integrity. The Evans blue assay is based on the accumulation of blue color inside dead cells (but not in living cells). The conductivity assay is based on electrolyte leakage from dying cells, with ion leakage used as a marker of cell death.

2.5 Morphological observation by SEM

To provide clues to graphene toxicity in the seedling stage of terrestrial plant roots, we next visualized the morphological alteration of cabbage, tomato, and red spinach using SEM of the treated roots and compared them to untreated controls (Figures 9, 10, and 11) respectively. SEM examination of the tested plants exposed to a higher concentration of graphene (1000 mg/L) revealed several morphological alterations compared to the control plants grown in Hoagland medium. At 20 days without exposure to graphene, the majority of the root surface appeared healthy; little to no alteration in morphology was observed. Regarding morphology at 1000 mg/L of graphene, root shape changed especially above the elongation area of cabbage and tomato (Figure 9 and 10) respectively. compared to control. Epidermis of the treated tomato (Figure 10B, C, E, and F) and red spinach (Figure 11B, D, and E) root was loosely detached or even completely detached compared to control, which was characteristic for membrane damage as determined by Evans blue. We note that specific differences could be observed after Evans blue staining at the light microscopic level (data

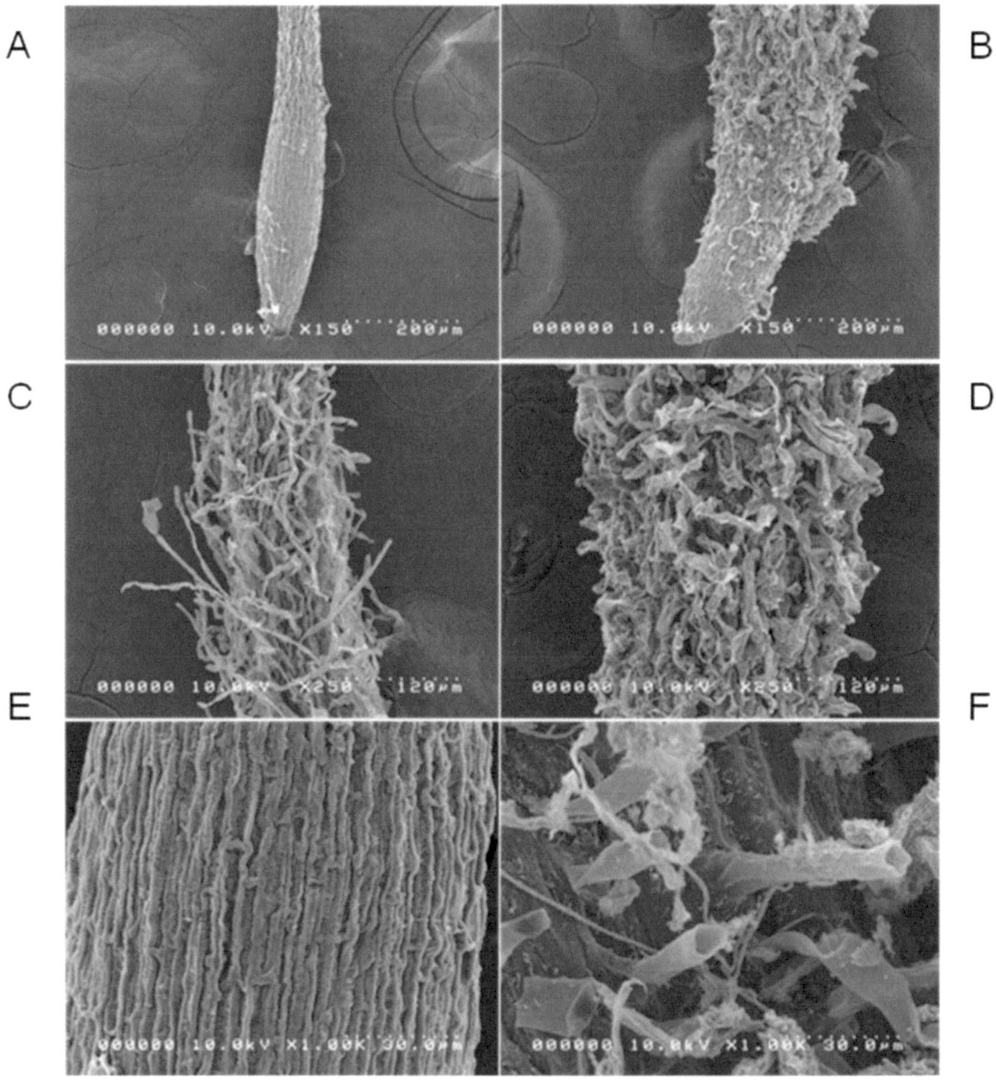

Fig. 9. Behavior of graphene (1000 mg/L) on the root surface of cabbage seedlings grown in Hoagland medium. (A, C, E) SEM image of the untreated control of cabbage root elongation, root hair zone and surface respectively. (B, D, F) SEM image of the G treated cabbage root elongation, root hair zone and surface respectively showing aggregates of G.

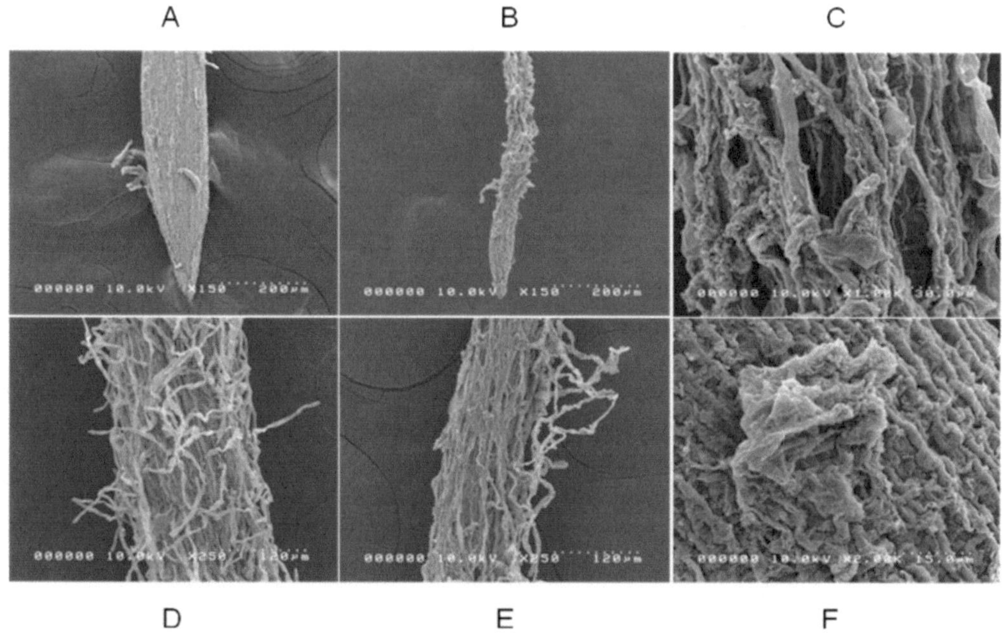

Fig. 10. Behavior of graphene (1000 mg/L) on the root surface of tomato seedlings grown in Hoagland medium. (A, D) SEM image of the untreated control of tomato root elongation and root hair zone respectively. (B) Root elongation zone of tomato root and (C, E, F) showing surface detachment and aggregates of G on the tomato roots surface treated with G.

not shown) with respect to the cell morphology between roots exposed to different concentrations of graphene. Thus, differences in cell death determined from Evans blue assays do reflect the sensitivity of cells to different concentrations of graphene, although the morphological alterations observed from light microscopy indicate a common cell death pathway. These results imply that the root membrane with a higher graphene concentration (1000 mg/L) had no tolerance to graphene stress. An aggregation of graphene was noted on the surface of the root of cabbage, tomato, and red spinach, indicating that aggregation of graphene contributes at least in part to its toxic properties, as is the case for CNTs (C. Lin et al., 2009).

The root surface area of the graphene-treated cabbage was obviously different from the untreated control (Figure 9). The root surface area of cabbage was significantly increased by graphene treatment, and it may be that an excess of graphene resulted in swelling. This observation is similar to results previously reported for *Origanum vulgare* by Panou-Filotheou & Bosabalidis (2004) and in cowpea by Kopittke et al., (2007). At a higher graphene concentration (1000 mg/L), the root hair growth of cabbage and red spinach was decreased compared to control.

The two forms of carbon nanoparticles, CNTs and graphene, have similar chemical composition and crystalline structure. Based on this fact, a reasonable inference is that the two would be similar in many ways, such as biological activity. But they are different in shape; CNTs are tubular while graphene is flat atomic sheets (Lui et al., 2009). As a result, atomically flat graphene nanostructures have been estimated to have an even stronger

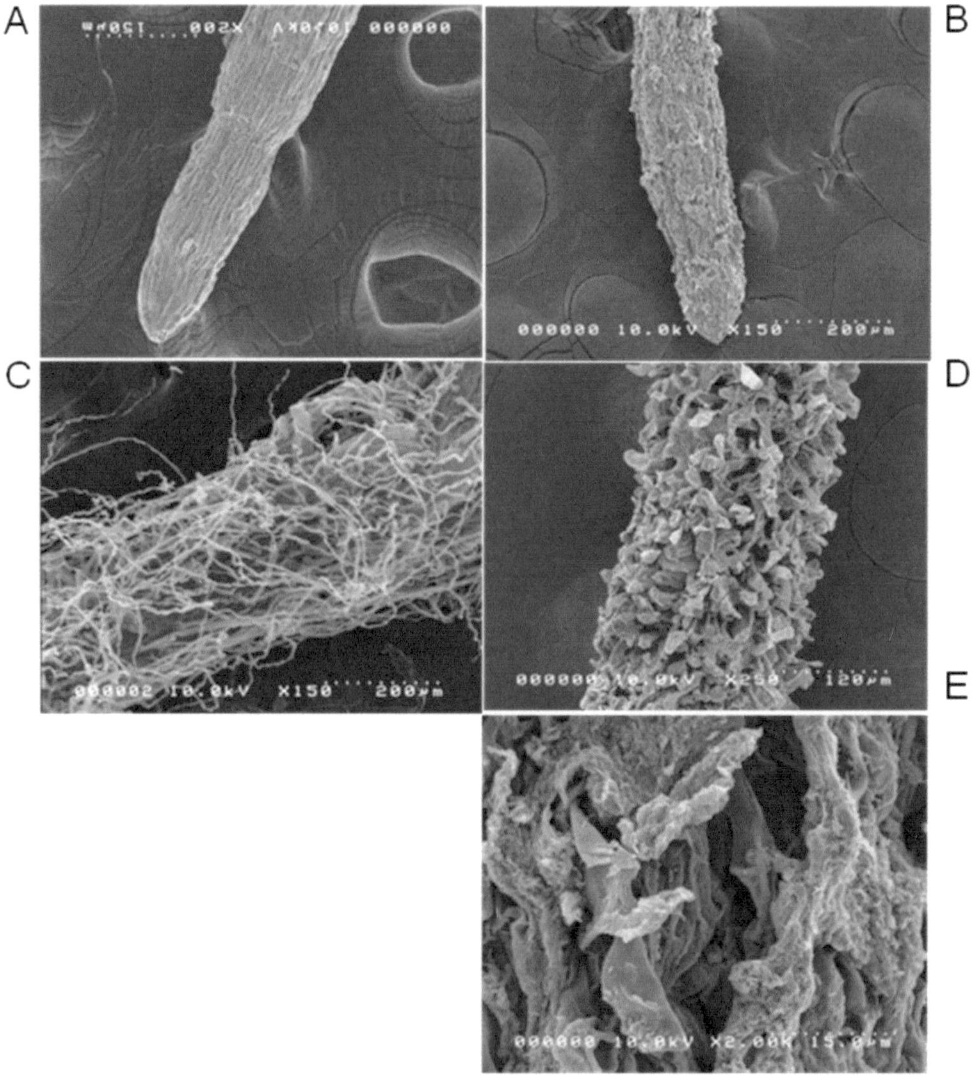

Fig. 11. Behavior of graphene (1000 mg/L) on the root surface of red spinach seedlings grown in Hoagland medium. (A, C) SEM image of the untreated control of red spinach root elongation, root hair zone and surface respectively. (B) Root elongation zone showing root damage (D) root hair zone and (E) surface showing aggregates of G on the root of red spinach.

interaction with biological systems than the tubular-shaped CNTs. There is evidence that CNTs could translocate to systemic sites (roots, leaves, and fruits) and engage in a strong interaction with the cells of tomato seedlings, resulting in significant changes in total gene expression in roots, leaves, and fruits (Khodakovskaya et al., 2011) and exerting toxic effects (Stampoulis et al., 2009; Ghodake et al., 2010; S. Lin et al., 2009). CNTs are also known to have phytotoxic effects in plant cells because of aggregation (C. Lin et al., 2009), causing cell death and accumulated ROS in a dose-dependent manner (Tan et al., 2009). Considering these aspects, it is not unexpected to find toxic effects of graphene on terrestrial plant species, as we did in tomato, cabbage, and red spinach in the current study. Oxidative stress has been regarded as a suitable mechanism for explaining the toxicity of graphene.

The phytotoxicity of graphene is likely the result of its ability to generate ROS as represented predominantly by hydrogen peroxide (H_2O_2) using the ROS-sensitive dyes DCFH-DA and DAB. Reports regarding graphene-induced oxidative stress are available (C. Lin et al., 2009). Concerning the relationship between graphene-mediated ROS generation and cell death in plants, we investigated physiological and morphological endpoints, including visual symptoms, ROS production, SEM observation, and cell death-related responses after exposure to graphene for 20 days. Graphene-treated leaves showed significant accumulation of H_2O_2 compared with control (Figure 6). This observation was closely associated with visible signs of necrotic damage lesions (Figure 4), cellular ROS accumulation monitored by DAB (Figure 7), and membrane damage (Figure 8B). The evidence of massive cell death (electrolyte leakage assay) as well as visual signs (Figure 4 and Figure 7) with the graphene-treated leaves demonstrated that at 20 days following exposure to graphene, necrosis was the predominant death response in treated leaf cells. In addition, the spatial pattern of H_2O_2 production was mainly located in the close proximity of visible symptomatic areas and the sites undergoing cell death.

Plant cell death occurs either by apoptosis or by necrosis. Necrosis is passive and characterized by a progressive loss of membrane integrity that results in cytoplasmic swelling and release of cellular constituents (Van Breusegem & Dat, 2006). Furthermore, we also detected the effect of graphene on the morphology of roots, finding that the epidermis of the treated tomato and red spinach roots was only loosely detached or even completely detached. Further, the aggregation of graphene was noted on the surface of the roots of all graphene (1000 mg/L)-treated plants, indicating that aggregation of graphene contributed at least in part to its toxic properties, as is the case for CNTs (C. Lin et al., 2009). On the basis of these results, we infer that graphene phytotoxicity involved the oxidative stress mechanism mediated through the necrotic pathway. In biological systems, ROS production and disturbances of cellular redox potentials have been found to be involved either directly or indirectly in the death of individual cells and/or the development of necrotic lesions (Pellinen et al., 2002; Yoda et al., 2006). Enhanced generation of ROS (hydrogen peroxide) and plasma membrane rupture have been traditionally associated with pathogenic events including necrotic cell death, which is recognized by morphological signs and is usually considered to be uncontrolled (Golstein & Kroemer, 2007).

3. Materials and methods

3.1 Chemicals and seeds

Chemicals used in this experiment were purchased from Kanto Chemical Co. Inc., Wako Pure Chemical Industries, Ltd., or Sigma Aldrich Inc., Japan. Seeds of three plant species (cabbage, *Brassica oleracea* var. capitata; tomato, *Lycopersicon esculentum*; and lettuce, *Lactuca sativa*) used in this study were purchased from Homac, Sapporo, Japan. Red spinach (*Amaranthus tricolor* L. and also *Amaranthus lividus* L.) seeds were purchased from Dhaka, Bangladesh.

3.2 Preparation of water-soluble graphene

Water-soluble graphene was obtained from natural graphite (SP-1, Bay Carbon) by using a modified Hummers and Offeman's method (Hummers & Offeman, 1958). In a typical treatment, 100 g of the graphite powders and 50 g of sodium nitrate and 2000 mL sulfuric acid were mixed in an ice bath. Next, 100 g of potassium permanganate was slowly added and well mixed. Once mixed, the suspension was placed in a water bath at 35 ± 3 °C and mixed for about 30 min. About 5 L of deionized water was added to the suspension, the temperature was increased to 90 ± 3 °C, and the suspension was mixed further for about 30 min. The suspension was finally diluted to about 10 L with warm deionized water and about 200 mL of 30 wt% H_2O_2. The warm suspension was filtered, and a yellow-brown filtered cake was obtained. The filtered cake was carefully washed with a large amount of warm deionized water and then dispersed in deionized water by mechanical mixing to prepare a stock graphene aqueous suspension containing approximately 2 wt% water-soluble graphene. An aqueous 0.1 mol/L sodium hydroxide solution was used to neutralize the graphene solution to a pH 6.32.

3.3 Seedling exposure

The sterile seeds of the selected plant species were soaked in solution with different concentrations (0, 500, 1000, and 2000 mg/L) of graphene overnight at 25 °C in the dark. Seeds were placed on wet filter paper and then exposed to 3 mL of the test solution with and without graphene and incubated at 25 °C until germination. After germination, the number of germinated seeds was counted. The seedlings were transferred to plastic pots containing 200 mL Modified Hoagland Medium (Hoagland & Arnon, 1950). Seedlings were treated with different concentrations of graphene solution (0, 500, 1000, or 2000 mg/L). At 2000 mg/L graphene was stable with very little settling.

3.4 Root, shoot growth, and leaf characters

After 20 days of exposure, roots and shoots were separated and washed with water to remove the growth medium and dried with wipes to remove the surface water. Their length and fresh weights were recorded. Leaf numbers were counted and leaf area measured using a RHIZO 2004b instrument.

3.5 ROS measurements and H_2O_2 detection

For visualization and measurement of ROS by spectroscopy, oxidatively sensitive probes DCFH-DA and DAB were used. Leaves of the treated and untreated plants were incubated in DCFH-DA for 2 h. After three PBS rinses, the images for visualization were captured using fluorescence microscopy (Olympus IX70). We performed a measurement of DCFH fluorescence intensity in the leaves using a spectrofluorometer (Hitachi F-4500) by suspending the leaves in PBS buffer with an excitation wavelength at 485 nm and emission wavelength at 522 nm. The values were expressed as % of fluorescence intensity relative to control. This experiment was performed without exposure to light.

H_2O_2 was detected using DAB by the method previously described by Thordal-Christensen et al., (1997). The fresh leaves from 20-day-old treated and untreated plants were placed in 1 mg/mL DAB-HCl, pH 3.2–3.8 and incubated under vacuum for ~8 h. DAB deposits were revealed after washing leaves in boiled 100% (v/v) ethanol for 15 min to decolorize the leaves except for the deep brown polymerization product from the reaction of DAB with H_2O_2. The images were observed and recorded using light microscopy (Transmit Light Microscope BX51, With Olympus DP72 Camera). The amount of formazan formation in leaves was measured spectrophotometrically at A_{700} (V-530 UV/UISNIR Spectrophotometer, Jasco, Japan) after leaves were ground in liquid nitrogen and solubilized in a mixture of 2 M KOH and DMSO at a ratio of 1:1.167 (v/v).

3.6 Evaluation of cell death

The cell death of the selected plant roots was evaluated by the method previously described by Baker & Mock (1994) using Evans blue (0.025% v/v) for 2 h after 20 days of exposure to different concentrations (0, 500, 1000, and 2000 mg/L) of graphene. For quantitative assessment, after several washes with water, the Evans blue was extracted using 1% (w/v) SDS in 50% (v/v) methanol at 50 °C for 15 min, and the absorbance (optical density) was measured spectrophotometrically at 597 nm (V-530 UV/UISNIR Spectrophotometer, Jasco, Japan). Cell death was expressed as absorbance of treated roots in relation to untreated roots (control).

Cell death was also evaluated by measurement of ion leakage from leaf. The percent of injury of the membrane was measured from the electrolyte leakage of treated and untreated plants by a conductivity method based on the procedure of Lutts et al., (1996) with some modifications. Leaf samples (100 mg) were placed in test tubes containing 10 mL of distilled water after three washes with distilled water to remove surface contamination. Test tubes were covered and incubated at room temperature (25 °C) on a shaker (100 rpm) for 24 h. Electrical conductivity of the solution (L_t) was determined. Samples were then autoclaved at 120 °C for 20 min, and a final conductivity reading (L_0) was obtained upon equilibration at 25 °C. Electrolyte leakage was defined as: Electrolyte leakage (%) = $(L_t/L_0) \times 100$. The extent of membrane injury (leakage of electrolytes) from cells was determined with a portable conductivity meter (pH/Cond Meter, Horiba D-54).

3.7 Morphological observation by AFM, SEM and TEM

For AFM, an Agilent Series 5500 AFM instrument was used. The samples were prepared by casting a diluted aqueous graphene suspension on the surfaces of mica. The images were obtained using the tapping mode at a scanning rate of 1 Hz. For SEM, a few drops of the graphene suspension were deposited on the aluminum stub, dried, sputter-coated, and observed using a Hitachi S-4000 SEM (Hitachi, Ibaraki, Japan) with an acceleration voltage of 10 kV. For SEM, the roots were fixed in 2.5% glutaraldehyde (GA) and 2% paraformaldehyde (PA) in 0.1 M phosphate buffer (PB), pH 7.4, postfixed in 1% osmium tetroxide, dehydrated, critical-point dried, sputter-coated, and observed. For TEM, the graphene nanosheets were homogeneously dispersed in 2-propanol under ultrasonication for 30 min. A few drops of the suspension were deposited on the TEM grid covered with a Formvar membrane, dried, and evacuated before analysis. The preparations were observed using a Hitachi H-800 TEM. The acceleration voltage was 75 kV.

3.8 Statistical analysis

Each treatment was conducted in triplicate. Phytotoxicity endpoints for all measurements were compared to those of the unexposed controls. Statistical analysis was performed using the Student' s t-test. Values of P • 0.01 were considered significant. Data are presented as mean ± SD (standard deviation).

4. Conclusion

In summary, we showed that graphene exerted toxic effects on terrestrial plant species such as cabbage, tomato, and red spinach. The overproduction of ROS induced by graphene could be responsible for significant plant growth inhibition and biomass reduction compared to control. Nel et al., (2006) reported that the production of ROS can be a key factor in the toxicological effects of nanostructure materials. Observation of accumulation ROS production by means of H_2O_2 visualization along with visible signs of necrotic damage lesions and evidence of a massive electrolyte leakage all indicated an oxidative stress mechanism mediated through the necrotic pathway, which requires further study. We suggest an evaluation of graphene toxicity to targets on terrestrial plant species and applying a prolonged exposure period with different concentrations to measure any potential risk.

5. Acknowledgment

We gratefully acknowledge Prof. Toru Mura and Prof. Shunitz Tanaka for providing necessary facilities during the study. We also gratefully acknowledge Yoshinobu Nodasaka and Natsumi Ushijima for their kind assistance in preparing the SEM and TEM samples. Many thanks go to Hongwen Yu for his assistance during the AFM study. The authors also wish to thank Mr. Egami Hiroshi for taking pictures of plant.

6. References

Akhavan, O. & Ghaderi, E. (2010). Toxicity of Graphene and Graphene Oxide Nanowalls Against Bacteria. *ACS Nano*, Vol.4, No.10, pp.5731–5736, DOI 10.1021/nn101390x

Ando, T. (2009). The Electronic Properties of Graphene and Carbon Nanotubes. *NPG Asia Materials*, Vol.1, No.1, pp.17–21, ISSN 18844049

Apel, K. & Hirt, H. (2004). Reactive Oxygen Species: Metabolism, Oxidative Stress, and Signal Transduction. *Annual Review of Plant Biology*, Vol.55, pp.373-399, DOI 10.1146/annurev.arplant.55.031903.141701

Baker, C.J. & Mock, N.M. (1994). An Improved Method for Monitoring Cell Death in Cell Suspension and Leaf Disc Assays Using Evans Blue. *Plant Cell, Tissue and Organ Culture*, Vol.39, No.1, pp.7–12, ISSN 01676857

Bhabra, G.; Sood, A.; Fisher, B.; Cartwright, L.; Saunders, M.; Evans, W.H.; Surprenant, A.; Lopez-Castejon, G.; Mann, S.; Davis, S.A.; Hails, L.A.; Ingham, E.; Verkade, P.; Lane, J.; Heesom, K.; Newson, R. & Case, C.P. (2009). Nanoparticles Can Cause DNA Damage Across a Cellular Barrier. *Nature Nanotech*nology, Vol.4, pp.876-883, DOI 10.1038/NNANO.2009.313

Breusegem, F.V. & Dat, J.F. (2006). Reactive Oxygen Species in Plant Cell Death. *Plant Physiology*, Vol.141, No.2, pp.384–390, DOI 10.1104/pp.106.078295

Del Pozo, O. & Lam, E. (1998). Caspases and Programmed Cell Death in The Hypersensitive Response of Plants to Pathogens. *Current Bio*logy, Vol.8, No.20, pp.1129-1132, DOI 10.1016/S0960-9822(98)70469-5

Dragoman, M. & Dragoman, D. (2009). Graphene-based Quantum Electronics. *Progress in Quantum Electronics*, Vol.33, No.6, pp.165-214, DOI 10.1016/j.pquantelec. 2009.08.001

Enderlein, C.; Kim, Y.S.; Bostwick, A.; Rotenberg, E. & Horn, K. (2010). The Formation of an Energy Gap in Graphene on Ruthenium by Controlling the Interface. *New Journal of Physics*, Vol.12, 033014 (9pp), DOI 10.1088/1367-2630/12/3/033014

Geim, A.K. & Novoselov, K.S. (2007). The Rise of Graphene. *Nature Materials*, Vol.6, No.3, pp.183–191, ISSN 1476-1122

Ghodake, G.; Seo, Y.D.; Park, D. & Lee, D.S. (August 2010). Phytotoxicity of Carbon Nanotubes Assessed by *Brassica Juncea* and *Phaseolus Mungo*. *Journal of Nanoelectronics and Optoelectronics*, Vol.5, No.2, pp.157-160 (4), DOI 10.1166/jno.2010.1084

Golstein, P. & Kroemer, G. (2007). Cell Death by Necrosis: Towards a Molecular Definition. *Trends in Biochemical Sciences*, Vol.32, No.1, pp.37-43, DOI 10.1016/j.tibs.2006.11.001

Hoagland, D.R. & Arnon, D.I. (1950). The Water Culture Method for Growing Plants Without Soil. The College of Agriculture, University of California, California Agricultural Experiment Station Circular 347, Berkeley, pp. 1-32

Hummers, W.S.Jr. & Offeman, R.E. (1958). Preparation of Graphitic Oxide. *Journal of American Chemical Socceity*, Vol.80, No.6, pp. 1339, DOI 10.1021/ja01539a017

Jia, G.; Wang, H.; Yan, L.; Wang, X.; Pei, R.; Yan, T.; Zhao, Y. & Guo, X. (2005). Cytotoxicity of Carbon Nanomaterials: Single-Wall Nanotube, Multi-Wall Nanotube, and Fullerene. *Environmental Science & Technology*, Vol.39, No.5, pp.1378–1383, DOI 10.1021/es048729l

Joseph, T., & Morrison, M. (2006). Nanotechnology in Agriculture and Food. Available from www. nanoforum.org

Kawai-Yamada, M.; Ohori, Y. & Uchimiya, H. (2004). Dissection of *Arabidopsis* Bax Inhibitor-1 Suppressing Bax-, Hydrogen Peroxide-, and Salicylic Acid-induced Cell Death. *The Plant Cell*, Vol.16, pp.21–32, DOI 10.1105/tpc.014613

Khodakovskaya, M.V.; Silvaa, K.de.; Nedosekinb, D.A.; Dervishic, E.; Birisa, A.S.; Shashkovb, E.V.; Galanzhab, E.I. & Zharovb, V.P. (2011). Complex Genetic, Photothermal, and Photoacoustic Analysis of Nanoparticle-plant Interactions. *Proceedings of the National Academy of Science (PNAS)* Vol.108, No.3, pp.1028–1033, DOI 10.1073/pnas.1008856108

Kirchner, C.; Liedl, T.; Kudera, S.; Pellegrino, T.; Javier, A.M.; Gaub, H.E.; Stolzle, S.; Fertig, N. & Parak, W.J. (2005). Cytotoxicity of Colloidal CdSe and CdSe/ZnS Nanoparticles. *Nano Letters*, Vol.5, No.2, pp.331–338, DOI 10.1021/nl047996m

Kopittke, P.M.; Asher, C.J.; Kopittke, R.A. & Menzies, N.W. (2007). Toxic Effects of Pb^{2+} on Growth of Cowpea (Vigna unguiculata). *Environmental Pollution*, Vol.150, No.2, pp.280–287, ISSN 0269-7491

Lin, C.; Fugetsu, B.; Su, Y. & Watari, F. (2009). Studies on Toxicity of Multi-Walled Carbon Nanotubes on *Arabidopsis* T87 Suspension Cells. *Journal of Hazardous Materials*, Vol.170, No.2-3, pp.578-583, DOI 10.1016/j.jhazmat.2009.05.025

Lin, D.H. & Xing, B.S. (2007). Phytotoxicity of Nanoparticles: Inhibition of Seed Germination and Root Growth. *Environmental Pollution*, Vol.150, No.2, pp.243-250, DOI 10.1016/j.envpol.2007.01.016

Lin, S.; Reppert, J.; Hu, Q.; Hudson, J.A.S.; Reid, M.L.; Ratnikova, T.A.; Rao, A.M.; Luo, H. & Ke, P.C. (2009). Uptake, Translocation, and Transmission of Carbon Nanomaterials in Rice Plants. *Small*, Vol.5, No.10, pp.1128-1132, DOI 10.1002/smll.200801556

Liu, Q.; Chen, B.; Wang, Q.; Shi, X.; Xiao, Z.; Lin, J. & Fang, X. (2009). Carbon Nanotubes as Molecular Transporters for Walled Plant Cells. *Nano Lett*er, Vol.9, No.3, pp.1007–1010, DOI 10.1021/nl803083u

Lui, C.H.; Liu, L.; Mak, K.F.; Flynn, G.W. & Heinz, T.F. (2009). Ultraflat Graphene. *Nature*, Vol.462, pp.339-341, DOI 10.1038/nature08569

Lutts, S.; Kinet, J.M. & Bouharmont, J. (1996). NaCl-Induced Senescence in Leaves of Rice (*Oryza sativa* L.) Cultivar Differing in Salinity Resistance. *Annals of Botany* (Lond), Vol.78, No.3, pp.389–398, DOI 10.1006/anbo.1996.0134

Meyer, J.C.; Geim, A.K.; Katsnelson, M.I.; Novoselov, K.S.; Booth, T.J. & Roth, S. (2007). The Structure of Suspended Graphene Sheets. *Nature,* Vol.446, No.7131, pp.60-63, DOI 10.1038/nature05545

Monica, R.C. & Cremonini, R. (2009). Nanoparticles and Higher Plants. *Caryologia,* Vol.62, No.2, pp.161-165, ISSN 0008-7114

Nel, A.; Xia, T.; Madler, L. & Li, N. (2006). Toxic Potential of Materials at the Nanolevel. *Science,* Vol.311, No.5761, pp.622-627, DOI 10.1126/science.1114397

Nowack, B. & Bucheli, T.D. (2007). Occurrence, Behavior and Effects of Nanoparticles in the Environment. *Environmental Pollution,* Vol.150, No.1, pp.5–22, DOI 10.1016/j.envpol.2007.06.006

Panou-Filotheou, H. & Bosabalidis, A.M. (2004). Root Structural Aspects Associated with Copper Toxicity in Oregano (*Origanum vulgare* subsp. *hirtum*). *Plant Science,* Vol.166, No.6, pp.1497–1504, DOI 10.1016/j.plantsci.2004.01.026

Pellinen, R.I.; Korhonen, M.S.; Tauriainen, A.A.; Palva, E.T. & Kangasjärvi, J. (2002). Hydrogen Peroxide Activates Cell Death and Defense Gene Expression in Birch, *Plant Physiology,* Vol.130, pp.549–560, DOI 10.1104/pp.003954

Sayes, C.M.; Liang, F.; Hudson, J.L.; Mendez, J.; Guo, W.; Beach, J.M.; Moore, V.C.; Doyle, C.D.; West, J.L.; Billups, W.E.; Ausman, K.D. & Colvin, V.L. (2006). Functionalization Density Dependence of Single-Walled Carbon Nanotubes Cytotoxicity in Vitro. *Toxicological Letter,* Vo.161, pp.135–142, DOI 10.1016/j.toxlet.2005.08.011

Stampoulis, D.; Sinha, S.K. & White, J.C. (2009). Assay-Dependent Phytotoxicity of Nanoparticles to Plants. *Environmental Science & Technology,* Vol.43, No.24, pp.9473-9479, ISSN 0013-936X

Stankovich, S.; Dikin, D.A.; Dommett, G.H.B.; Kohlhaas, K.M.; Zimney, E.J.; Stach, E.A.; Piner, R.D.; Nguyen, S.T. & Ruoff, R.S. (2006). Graphene-based Composite Materials. *Nature,* Vol.442, No.7100, pp.282-286, DOI 10.1038/nature04969

Stiles, W. (1961). Trace Elements in Plants, 3rd ed. Cambridge University Press, London. pp. 249

Tan, X.M.; Lin, C. & Fugetsu, B. (2009). Studies on Toxicity of Multi-Walled Carbon Nanotubes on Suspension Rice Cells. *Carbon,* Vol.47, No.15, pp.3479-3487, DOI 10.1016/j.carbon.2009.08.018

Thordal-Christensen, H.; Zhang, Z.; Wei, Y. & Collinge, D.B. (1997). Subcellular Localization of H_2O_2 in Plants. H_2O_2 Accumulation in Papillae and Hypersensitive Response During the Barley-Powdery Mildew Interaction. *The Plant Journal,* Vo.11, No.6, pp.1187–1194, ISSN 0960-7412

USEPA-Ecological Effects Test Guidelines (OPPTS 850.4200) (1996). Seed Germination Root Elongation Toxicity Test. Available from: http://www.epa.gov/publications

Wen, F.; Xing, D. & Zhang, L.R. (2008). Hydrogen Peroxide is Involved in High Blue Light-Induced Chloroplast Avoidance Movements in *Arabidopsis. Journal of Experimental Botany,* Vol.59, No.10, pp.2891-2901, DOI 10.1093/jxb/ern147

Xia, J.; Chen, F.; Li, J. & Tao, N. (2009). Measurement of the Quantum Capacitance of Graphene. *Nature Nanotechnology,* Vol.4, pp. *505–509,* DOI 10.1038/nnano.2009.177

Yoda, H.; Yamaguchi, Y. & Sano, H. (2006). Polyamine Oxidase is One of the Key Elements for Oxidative Burst to Induce Programmed Cell death in Tobacco Cultured Cells, *Plant Physiology,* Vol.142, pp.193–206, DOI 10.1104/pp.106.080515

Zhang, Y.; Ali, S.F.; Dervishi, E.; Xu, Yang.; Li, Z.; Casciano, D. & Biris, A.S. (2010). Cytotoxicity Effects of Graphene and Single-Wall Carbon Nanotubes in Neural Phaeochromocytoma-Derived PC12 Cells. ACS Nano, Vol.4, No.6, pp.3181-3186, DOI 10.1021/nn1007176

11

Carbon Nanotube Radio

Ujjal Kumar Sur
Department of Chemistry, Behala College,
India

1. Introduction

Carbon nanotubes (CNTs) are allotropes of carbon with a cylindrical nanostructure. Nanotubes have been synthesized with length-to-diameter ratio of up to 132,000,000 : 1, which is significantly greater than any other material. These cylindrical carbon molecules have novel properties which make them potentially useful in many applications in nanotechnology, electronics, optics, and other fields of materials science.

Nanotubes are members of the fullerene structural family (one dimensional form of fullerenes), which also includes the spherical buckyballs. Fullerenes were accidentally discovered by Kroto and Smalley in 1985, when they found strange results in mass spectra of evaporated carbon samples (Kroto et al. 1985). Since the discovery in 1991 by Iijima (Iijima, 1991), carbon nanotubes have been investigated by many researchers all over the world both from fundamental point of view and for future potential applications in the field of energy storage, molecular electronics, nanomechanical devices and composite materials. However, the real applications are still under development. Their name is derived from their size, since the diameter of a nanotube is on the order of a few nanometers (approximately 1/50,000th of the width of a human hair), while they can be up to 18 centimeters in length. Nanotubes are categorized as single-walled nanotubes (SWNTs) and multi-walled nanotubes (MWNTs). Figures 1 a) and b) show the scanning tunneling microscopic (STM) and transmission electron microscopic (TEM) images of a single-walled carbon nanotube. Figure 2 illustrates the scanning electron microscopic (SEM) image of carbon nanotube bundles. Single-walled nanotubes are an important variety of carbon nanotube because they exhibit electric properties that are not shared by the multi-walled carbon nanotube (MWNT). In particular, their band gap can vary from zero to about 2 eV and their electrical conductivity can show metallic or semiconducting behavior, whereas MWNTs are zero-gap metals. Single-walled nanotubes are the most likely candidate for miniaturizing electronics beyond the micro electromechanical scale currently used in electronics. The most basic building block of these systems is the electric wire, and SWNTs can be excellent conductors.

One useful application of SWNTs is in the development of the first intramolecular field effect transistors (FET). Multi-walled nanotubes (MWNTs) consist of multiple rolled layers (concentric tubes) of graphite. There are two models which can be used to describe the structures of multi-walled nanotubes. Single Walled Nanotubes (SWNTs) can be considered as long wrapped graphene sheets.

As stated before, nanotubes generally have a length to diameter ratio of about 1000 so they can be considered as nearly one-dimensional structures. Figure 3 shows some SWNTs with different chiralities. The difference in structure is shown at the open end of the tubes. More detailed, a SWNT consists of two separate regions with different physical and chemical properties. The first is the sidewall of the tube and the second is the end cap of the tube. The end cap structure is similar to or derived from a smaller fullerene, such as C60. Multi Walled Nanotubes (MWNTs) can be considered as a collection of concentric SWNTs with different diameters. The length and diameter of these structures differ a lot from those of SWNTs and, of course, their properties are also very different. Figure 4 illustrates the schematic diagrams of fullerenes, SWNTs and MWNTs. Electronic, molecular and structural properties of carbon nanotubes are determined to a large extent by their nearly one dimensional structure. The most important properties of CNTs and their molecular background are stated below.

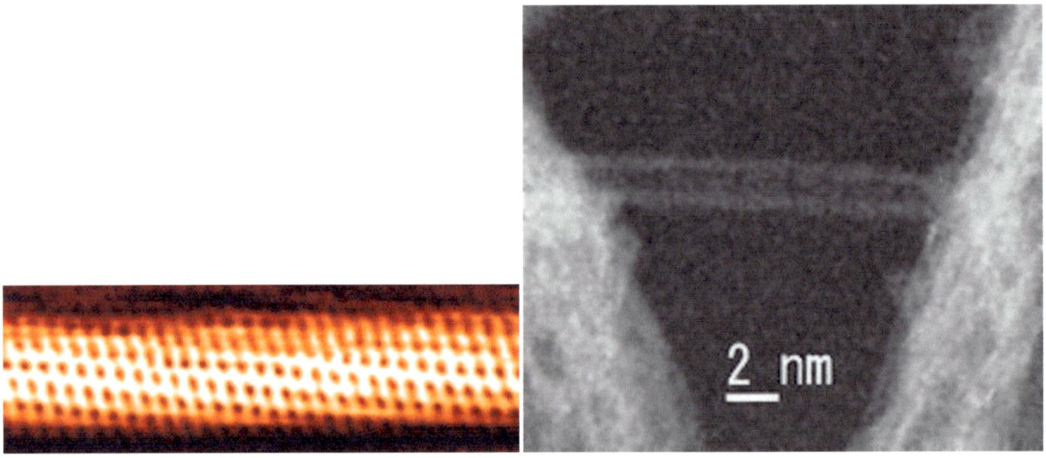

Fig. 1. a) An STM image of a single-walled carbon nanotube, b) Transmission electron microscopic image showing a single-walled carbon nanotube.

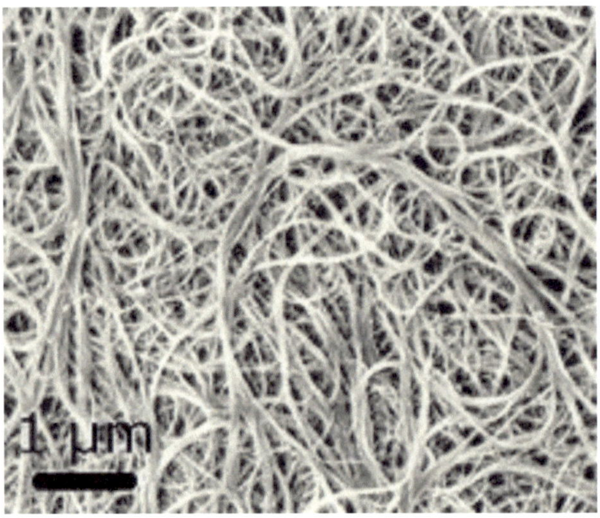

Fig. 2. Scanning electron microscopy image of carbon nanotube bundles.

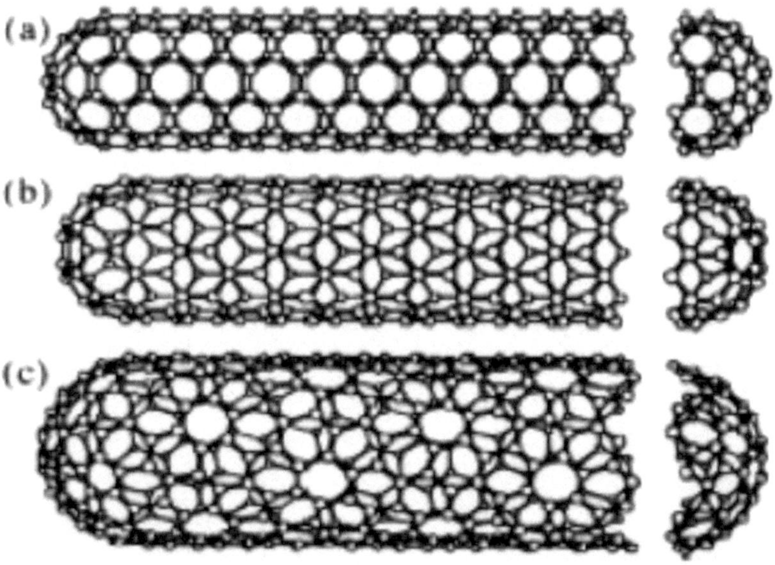

Fig. 3. Some SWNTs with different structures. (a) armchair structure (b) zigzag structure (c) chiral structure.

2. Properties of carbon nanotube

2.1 Chemical reactivity

The chemical reactivity of a CNT is, compared with a graphene sheet, enhanced as a direct result of the curvature of the CNT surface. Carbon nanotube reactivity is directly related to the pi-orbital mismatch caused by an increased curvature. Therefore, a distinction must be made between the sidewall and the end caps of a nanotube. For the same reason, a smaller nanotube diameter results in increased reactivity. Covalent chemical modification of either sidewalls or end caps has shown to be possible and the solubility of CNTs in different solvents can be controlled by this way. Though, direct investigation of chemical modifications on nanotube behaviour is difficult as the crude nanotube samples are still not pure enough.

2.2 Electrical conductivity

Depending on their chiral vector, carbon nanotubes with a small diameter are either semi-conducting or metallic. The differences in conducting properties are caused by the molecular structure that results in a different band structure and thus a different band gap. The differences in conductivity can easily be derived from the graphene sheet properties.

2.3 Optical activity

Theoretical studies have shown that the optical activity of chiral nanotubes disappears when the nanotubes become larger.Therefore, it is expected that other physical properties are influenced by these parameters too. Use of the optical activity might result in optical devices in which CNTs play an important role.

2.4 Mechanical strength

Carbon nanotubes have a very large Young modulus in their axial direction. The nanotube as a whole is very flexible because of the great length. Therefore, these compounds are potentially suitable for applications in composite materials that need anisotropic properties.

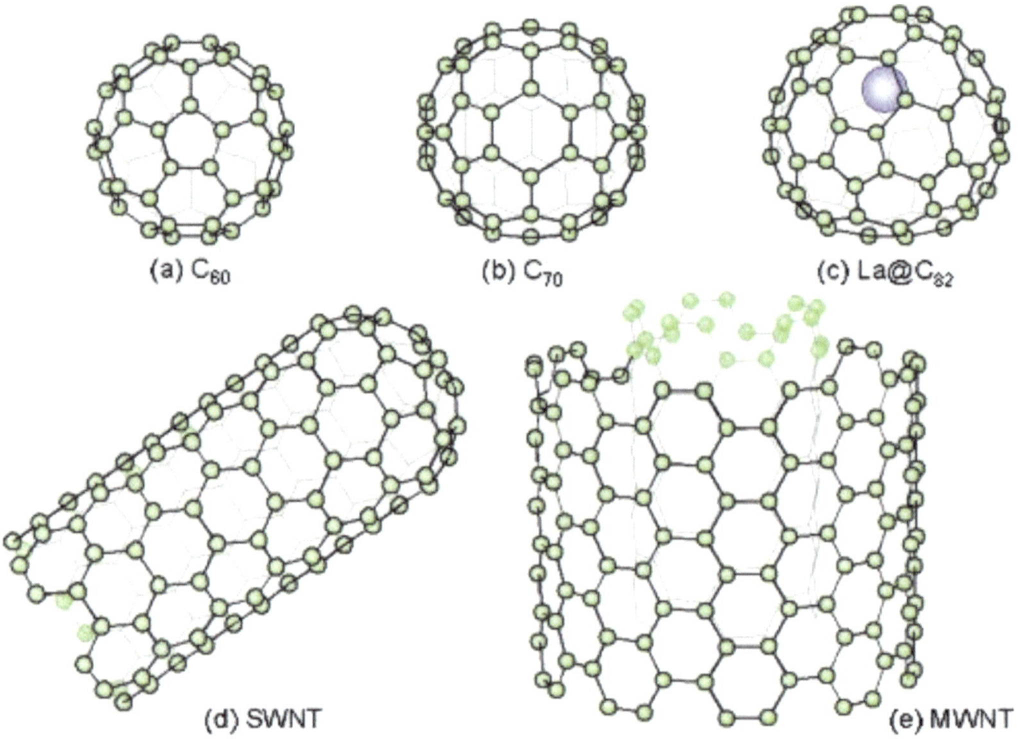

Fig. 4. Schematic diagrams of fullerenes and carbon nanotubes.

3. Synthesis and purification of carbon nanotube

3.1 Synthesis

Carbon nanotubes are generally synthesized by three main methods, arc discharge, laser ablation and chemical vapour deposition. In arc discharge technique, a vapour is created by an arc discharge between two carbon electrodes with or without catalyst. Nanotubes self-assemble from the resulting carbon vapour. In the laser ablation technique, a high-power laser beam impinges on a volume of carbon-containing feedstock gas (methane or carbon monoxide). At the moment, laser ablation produces a small amount of clean nanotubes, whereas arc discharge methods generally produce large quantities of impure material. In general, chemical vapour deposition (CVD) results in MWNTs or poor quality SWNTs. The SWNTs produced with CVD have a large diameter range, which can be poorly controlled. But on the other hand, this method is very easy to scale up, what favours commercial production. Table 1 shows a short summary of the three most common techniques for the synthesis of carbon nanotubes.

Figure 5 shows the photograph of the powder of carbon nanotube, while Figure 6 illustrates the growth of carbon nanotubes by plasma enhanced chemical vapour deposition method.

Techniques have been developed to produce nanotubes in sizeable quantities, including arc discharge, laser ablation, high pressure carbon monoxide, and chemical vapor deposition (CVD). Most of these processes take place in vacuum or with process gases. CVD growth of CNTs can occur in vacuum or at atmospheric pressure. Large quantities of nanotubes can be synthesized by these methods; advances in catalysis and continuous growth processes are making CNTs more commercially viable

Arc Discharge Method	Chemical Vapour Deposition	Laser Ablation
Introduced by Ajayan and Ebbesen in 1992	Developed by Endo et al. in 1993	Developed by Smalley in 1995
Typical yield 30 to 90 %	Typical yield 20 to 100 %	Typical yield upto 70 %
The method can easily produce both SWNTs and MWNTs	Easiest method to produce large-scale SWNTs	This method can produce SWNTs with good diameter control
This method is less expensive	This method is less expensive	This method is very expensive
The reaction product is not very pure and it requires further purification	The reaction product is quite pure	The reaction product is quite pure

Table 1. A summary of the three common techniques for the synthesis of carbon nanotubes

Fig. 5. Photograph of carbon nanotube powder

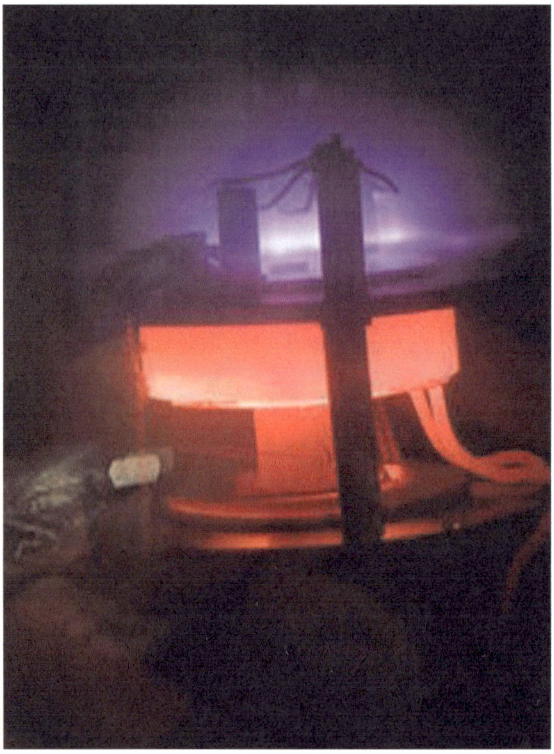

Fig. 6. Carbon nanotubes grown by plasma assisted CVD method

3.1.1 Arc discharge method

Nanotubes were observed in 1991 in the carbon soot of graphite electrodes during an arc discharge, by using a current of 100 amps, that was intended to produce fullerenes. However the first macroscopic production of carbon nanotubes was made in 1992 by Ajayan and Ebbesen (Ebbesen & Ajayan, 1992). Because nanotubes were initially discovered using this technique, it has been the most widely-used method of nanotube synthesis. The yield for this method is up to 30 percent by weight and it produces both single- and multi-walled nanotubes with lengths of up to 50 micrometers with few structural defects.

3.1.2 CVD method

During CVD, a substrate is prepared with a layer of metal catalyst particles, most commonly nickel, cobalt, iron, or a combination of all. The metal nanoparticles can also be produced by other ways, including reduction of oxides or oxides solid solutions. The diameters of the nanotubes that are to be grown are related to the size of the metal particles. This can be controlled by patterned (or masked) deposition of the metal, annealing, or by plasma etching of a metal layer. The substrate is heated to approximately 700°C. To initiate the growth of nanotubes, two gases are bled into the reactor: a process gas (such as ammonia, nitrogen or hydrogen) and a carbon-containing gas (such as acetylene, ethylene, ethanol or methane). Nanotubes grow at the sites of the metal catalyst; the carbon-containing gas is broken apart at the surface of the catalyst particle, and the carbon is transported to the edges of the particle, where it forms the nanotubes.

3.1.3 Laser ablation method
In the laser ablation method, a pulsed laser vaporizes a graphite target in a high-temperature reactor while an inert gas is bled into the chamber. Nanotubes develop on the cooler surfaces of the reactor as the vaporized carbon condenses. A water-cooled surface may be included in the system to collect the nanotubes. This process was developed by Dr. Richard Smalley and co-workers at Rice University (Smalley et al. 1995). They used a composite of graphite and metal catalyst particles (the best yield was from a cobalt and nickel mixture) to synthesize single-walled carbon nanotubes. The laser ablation method yields around 70% and produces primarily single-walled carbon nanotubes with a controllable diameter determined by the reaction temperature. However, it is more expensive than either arc discharge or chemical vapor deposition

3.2 Purification
A major problem with nanotube application is the purification of the reaction product along with the large-scale synthesis. The as-produced SWNT soot contains a lot of impurities. The main impurities in the soot are graphite (wrapped up) sheets, amorphous carbon, metal catalyst and the smaller fullerenes. These impurities will interfere with most of the desired properties of the SWNTs. Also in the fundamental research, it is preferred to obtain SWNTs or the impurities, as pure as possible without changing them. In order to understand the measurements better, the SWNT samples also have to be as homogeneous as possible. The common industrial techniques use strong oxidation and acid refluxing techniques, which have an effect on the structure of the tubes.

4. Potential applications of carbon nanotube

4.1 Energy storage
Graphite, carbonaceous materials and carbon fibre electrodes are commonly used in fuel cells, batteries and other electrochemical applications. Advantages of considering nanotubes for energy storage are their small dimensions, smooth surface topology and perfect surface specificity. The efficiency of fuel cells is determined by the electron transfer rate at the carbon electrodes, which is the fastest on nanotubes following ideal Nernstian behaviour.

4.1.1 Hydrogen storage
The advantage of hydrogen as energy source is that its combustion product is water. In addition, hydrogen can be easily regenerated. For this reason, a suitable hydrogen storage system is necessary, satisfying a combination of both volume and weight limitations. The two commonly used means to store hydrogen are gas phase and electrochemical adsorption. Because of their cylindrical and hollow geometry, and nanometre-scale diameters, it has been predicted that carbon nanotubes can store a liquid or a gas in the inner cores through a capillary effect. It is reported that SWNTs were able to meet and sometimes exceed this level by using gas phase adsorption (physisorption). Yet, most experimental reports of high storage capacities are rather controversial so that it is difficult to assess the applications potential. Another possibility for hydrogen storage is electrochemical storage. In this case, not a hydrogen molecule but an H atom is adsorbed. This is called chemisorption.

4.1.2 Lithium intercalation

The basic principle of rechargeable lithium batteries is electrochemical intercalation and deintercalation of lithium in both electrodes. An ideal battery has a high-energy capacity, fast charging time and a long cycle time. The capacity is determined by the lithium saturation concentration of the electrode materials. For Li, this is the highest in nanotubes if all the interstitial sites (inter-shell van der Waals spaces, inter-tube channels and inner cores) are accessible for Li intercalation. SWNTs have shown to possess both highly reversible and irreversible capacities. Because of the large observed voltage hysteresis, Li-intercalation in nanotubes is still unsuitable for battery application. This feature can potentially be reduced or eliminated by processing, i.e. cutting, the nanotubes to short segments. Carbon nanotubes are interesting intercalation hosts because of their structure and chemical bonding. Nanotubes might have a higher saturation composition than graphite as guest species can intercalate in the interstitial sites and between the nanotubes. Therefore, carbon nanotubes are expected to be suitable high energy density anode materials for rechargeable Li-ion batteries. Figure 7 shows the schematic diagram of graphite intercalated with lithium.

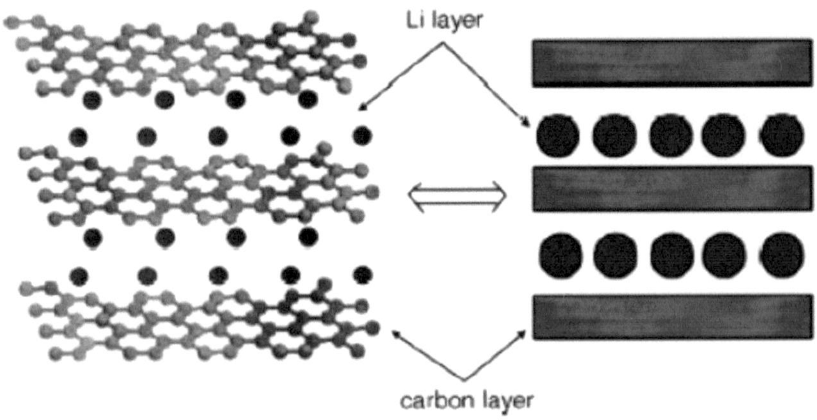

Fig. 7. Schematic diagram showing the intercalation of graphite with lithium

4.1.3 Electrochemical supercapacitors

Supercapacitors have a high capacitance and potentially applicable in electronic devices. Typically, they are comprised two electrodes separated by an insulating material that is ionically conducting in electrochemical devices. The capacity of an electrochemical supercapacitors inversely depends on the separation between the charge on the electrode and the counter charge in the electrolyte. Because this separation is about a nanometre for nanotubes in electrodes, very large capacities result from the high nanotube surface area accessible to the electrolyte. In this way, a large amount of charge injection occurs if only a small voltage is applied. This charge injection is used for energy storage in nanotube supercapacitors. Generally speaking, there is most interest in the double-layer supercapacitors and redox supercapacitors with different charge-storage modes. Carbon in general, and especially nanotubes, form an attractive material for electrochemical applications as they have a large active surface area. In addition, carbon is a relatively cheap, low density, environmentally friendly and highly polarisable material which makes application even more attractive.

Carbon nanotubes have been proposed as electrodes for supercapacitors. Different values of capacitance mainly depend on the kind and purity of the samples. For purified nanotubes specific capacitance varies from 5 to 80 F/g. Pure carbon nanotubes have a moderate surface area (120 to 400 m^2/g) because of their highly mesoporous character. The more graphitised nanotubes show smaller values of capacitance. However, presence of defects causes an increase of ability for accumulation of charges. There is a great difference in capacitance between SWNTs and MWNTs. However, the many different ways of producing the nanotubes also create differences in capacitance values. To increase the capacitance of nanotubes it is possible to increase the electrode surface area or to increase the pseudocapacitance effects obtained by addition of special oxides or electrically conducting polymers (ECP) like polypyrrole (PPy). The ECPs have the advantage of lower costs compared to oxides. Another advantage is that the pseudocapacitance effects of ECPs are quite stable.

4.2 Molecular electronics with CNTs

4.2.1 Field emitting devices

If a solid is subjected to a sufficiently high electric field, electrons near the Fermi level can be extracted from the solid by tunneling through the surface potential barrier. This emission current depends on the strength of the local electric field at the emission surface and its work function (which denotes the energy necessary to extract an electron from its highest bounded state into the vacuum level). The applied electric field must be very high in order to extract an electron. This condition is fulfilled for carbon nanotubes, because their elongated shape ensures a very large field amplification. For technological applications, the emissive material should have a low threshold emission field and large stability at high current density. Furthermore, an ideal emitter is required to have a nanometer size diameter, a structural integrity, a high electrical conductivity, a small energy spread and a large chemical stability. Carbon nanotubes possess all these properties. However, a bottleneck in the use of nanotubes for applications is the dependence of the conductivity and emission stability of the nanotubes on the fabrication process and synthesis conditions. Examples of potential applications for nanotubes as field emitting devices are flat panel displays, gas discharge tubes in telecom networks, electron guns for electron microscopes, AFM tips and microwave amplifiers.

4.2.2 Transistors

The field-effect transistor-a three-terminal switching device-can be fabricated using only one semiconducting SWNT. By applying a voltage to a gate electrode, the nanotube can be switched from conducting to an insulating state. Such carbon nanotube transistors can be coupled together, working as a logical switch, which is the basic component of computers.

4.2.3 Nanoprobes and sensors

Because of their flexibility, nanotubes can also be used in scanning probe instruments. Since MWNT tips are conducting, they can be used in STM and AFM instruments. Advantages are the improved resolution in comparison with conventional Si or metal tips and the tips do not suffer from crashes with the surfaces because of their high elasticity. However, nanotube vibration, due to their large length, will remain an important issue until shorter nanotubes can be grown controllably. Nanotube tips can be modified chemically by attachment of functional groups. Because of this, nanotubes can be used as molecular probes, with potential applications in chemistry and biology. Other applications are the following:

- A pair of nanotubes can be used as tweezers to move nanoscale structures on surfaces.
- Sheets of SWNTs can be used as electromechanical actuators, mimicking the actuator mechanism present in natural muscles.
- SWNTs may be used as miniaturised chemical sensors.

5. Carbon nanotube radio

5.1 Concept of radio device and historical background

The radio is a device based on wireless transmission of signals by modulating the frequencies of radio waves. Since its invention, the radio has been playing an important role in modern civilization due to its use in the mass media (audio/video), radar, cellular phones and wireless computer network. Historically, the Indian physicist Sir Jagadish Chandra Bose started his pioneering research work on the generation/detection of millimetre wave radiation more than 100 years ago (Bose, 1897). In 1897, he demonstrated his apparatus to detect radio frequency (RF) in the Royal Institution, London. Some concepts from his original 1897 paper have been incorporated into the 1.3 mm receiver of the NRAO 12 m radio telescope at Tuscon, Arizona, USA.

James Clerk Maxwell was the first to mathematically conceive the idea of electromagnetic waves (in 1873). After him, in the late 1880s, Heinrich Hertz tried to produce electromagnetic waves which had optical properties, based on Maxwell's hypothesis that visible light is just a fraction of the total electromagnetic spectrum. Hertz, and later Popoff and Oliver Lodge were all using 66 cm waves. Bose, in the 1890s, revolutionised this field of research by producing 5 mm waves, called microwaves or millimetre waves. These could be polarized using smaller crystals, than would be required if 66 cm waves were used, and thus Bose could show that electromagnetic waves do indeed have optical properties. In 1895, Marconi built a wireless system capable of transmitting signals at long distances (1.5 mi./2.4 km). In radio transmission technology, early public experimenters had made short distance broadcasts. Marconi achieved long range signalling due to a wireless transmitting apparatus and a radio receiver claimed by him. From Marconi's experiments, the phenomenon that transmission range is proportional to the square of antenna height is known as "Marconi's law". This formula represents a physical law that radio devices use. Marconi's experimental apparatus proved to be a complete, commercially successful radio transmission system. Shortly after the 1900s, Marconi held the patent rights for radio.

There was a gradual transformation in the dimension of radio from a bulky device with vacuum tube technology to a much smaller and sophisticated device based on solid state transistor technology. However, the transistor using silicon-based technology is reaching its physical limits. Carbon nanotube is an interesting nanomaterial with unique electrical and mechanical properties. Carbon nanotubes have been used as components in high-frequency electronics like high frequency field-effect transistors and RF detectors/mixers. Figure 8 shows the picture of the first wireless radio device fabricated by Professor Jagadish Chandra Bose (Figure 9 illustrates the image of Professor Bose).

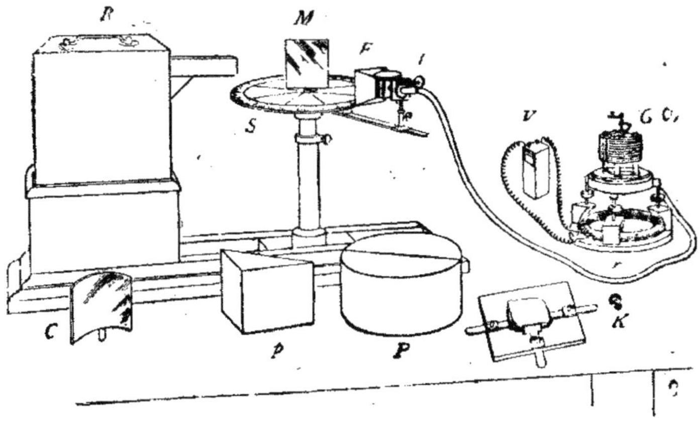

R, radiator ; S, spectrometer-circle ; M, plane mirror ; C, cylindrical mirror ; *p*, totally reflecting prism ; P, semi-cylinders ; K, crystal-holder ; F, collecting funnel attached to the spiral spring receiver ; *t*, tangent screw, by which the receiver is rotated ; V, voltaic cell ; *r*, circular rheostat ; G, galvanometer.

Fig. 8. Picture of the first radio device (measurement of millimeter waves) fabricated by Jagadish Chandra Bose

Fig. 9. Image of Professor Jagadish Chandra Bose

5.2 Development of carbon nanotube radio

Researchers in California have reported the development of the world's first working radio device based on nanosized radio-wave receivers and detectors fabricated from a single carbon nanotube. Although, other researchers have fabricated nano-sized radio detectors in the past, the above-mentioned study illustrates the demonstration of the RF components in an actual working radio system for the first time. Zettl and coworkers from Lawrence Berkeley National Laboratory, University of California, Berkeley, USA, have fabricated a fully integrated radio receiver from a single carbon nanotube (Zettl et al. 2007). This radio receiver has dimension orders of magnitude lower than any previous radio. A single carbon nanotube served as all the four essential components (antenna, tuner, amplifier and demodulator) of a radio at the same time. In the nanotube radio, antenna and tuner can operate in a completely different way compared to the traditional radio. The antenna receives signals via high-frequency mechanical vibrations of the nanotube rather than electrically. These vibrations become significant only when the frequency of the incoming wave matches the resonance frequency of the nanotube. The resonance frequency of the nanotube can be tuned to receive only a preselected band of the electromagnetic spectrum. Field emission of electrons from the tip of the charged nanotube has been utilized to detect vibrations and to amplify/demodulate the signal. Simulation studies have shown the distribution of electric field surrounding the nanotube radio during the operation of the radio. The field is strongest at the tip of the nanotube and varies as the nanotube vibrates. This effect allows the nanotube to demodulate the radio signal.

Zettl and co-workers have demonstrated the coarse tuning of the nanotube radio from a low-frequency, frequency-modulated (FM) radio band (~100 MHz), to much higher frequency bands (~350 MHz) along with the fully reversible fine tuning of resonance frequency of the nanotube. Carbon nanotubes with resonance frequencies from 10 to 400 MHz and quality factor (*Q*) ~500 were used in their experiments.

Fig. 10. TEM image of a single carbon nanotube protruding from an electrode. (Courtesy: Zettl and co-workers, Lawrence Berkeley National Laboratory and UC Berkeley, USA.)

A high-resolution transmission electron microscope (TEM) allows them to observe the nanotube radio in action. The TEM image (Figure 10) shows a single carbon nanotube of 1 μm length and 10 nm width protruding from an electrode. A direct current voltage source (battery) has been connected to the electrodes as a power source for the operation of the radio in vacuum of the order of 10^{-7} Torr. The nanotube starts to vibrate vigorously when a radio wave of specific frequency impinges on it. The nanotube radio has been used to receive and play music from FM radio transmissions and it faithfully reproduced the audio signal and the music was clearly audible to the human ear. Figure 11 illustrates the gradual transformation in size of the radio with time. It began with the Philco vacuum tube radio of the 1930s to the pocket-sized transistor radio of the 1950s and the recent single-chip radio found in cellphones and wireless sensors. This single nanotube radio is about 19 orders of magnitude smaller than the Philco vacuum tube radio of the 1930s.

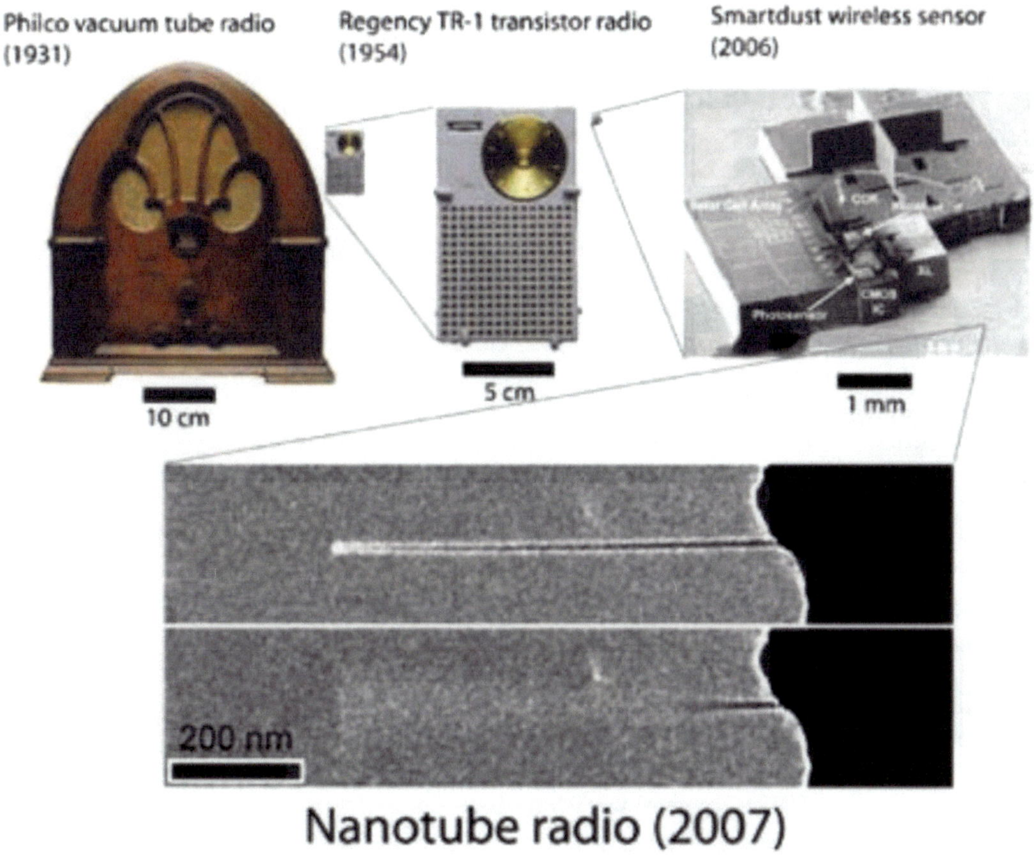

Fig. 11. Gradual miniaturization of radio with time. (Courtesy: Zettl and co-workers,Lawrence Berkeley National Laboratory and UC Berkeley, USA.)

Rutherglen and Burke from the University of California, Irvine, USA, have demonstrated a carbon nanotube to demodulate amplitude modulated signal with frequencies up to 100 kHz (Rutherglen & Burke, 2007). Their device represents a room-temperature, two-terminal nonlinear device, simpler than the previous three-terminal mixers/detectors. They incorporated the detector into a complete radio system for successful wireless transmission of music from an iPod to a speaker several feet away from the music player. The audio quality of the signal demodulated by the carbon nanotube was clear to the human ear. Their study demonstrates the possibility of fabricating other components of the radio at nanometer scale in the near future. This may finally lead to a fully integrated nanoscale wireless communication system.

Due to the small dimension, such a nanoscale device has potential medical applications. The nanotube radio can be inserted inside the living human cell to give a subcellular remote controlled interface. This special kind of nanotube radio can be utilized to image inside the cells in real time and under normal conditions. Zettl's team is currently working on the method to integrate this nanotube radio with biological systems, and this will help in understanding cell biology at the molecular level (Sur, 2008).

5.2 Development of carbon nanotube loudspeaker

In a recent development, researchers in China had claimed that they were able to develop a paper-thin loudspeaker that can either be attached to clothing, walls, windows or even can be inserted into the ear. These paper-thin loudspeakers were made from transparent carbon nanotube films and their work was published in the ACS journal Nano Letters (Fan et al. 2008). Professor Fan and his research group at the Tsinghua University, China developed the smallest loudspeaker of the world using CNT thin films by a simple fabrication process. At present, the loudspeakers available in the market are bulky and complex and by making these loudspeakers thinner and compact, these can be suitably used for new digital electronics devices and various other applications. Carbon Nanotubes (CNTs) are already being used for various applications including transmission of sound and music. Researchers mounted a carbon nanotube film on two electrodes and found that the system works as a loudspeaker. This CNT thin film can generate sounds by applying an audio frequency current through it. This phenomenon can be attributed to a thermoacoustic effect. The ultra small heat capacity per unit area of CNT thin films can lead to a wide frequency response range and a high sound pressure level. This CNT thin film can be a practical magnet-free loudspeaker, which has a very simple structure. The CNT thin film was pulled out from a superaligned CNT array grown on a 4 in. silicon wafer and put on two electrodes of a frame to make a loudspeaker (Figure 12a). The CNT thin film is composed of CNTs around 10 nm in diameter, which are sparsely parallel-aligned in the drawing direction as shown by the SEM image (Figure 12 b). It is tens of nanometers thick, extremely lightweight (typical mass per unit area is 1.5 μg/cm2), transparent (78% at 550 nm for a single-layer CNT thin film), and conductive (typical sheet resistance is around 1 kΩ per square for a single-layer CNT thin film). The ac impedance of a CNT thin film is pure resistance and shows no frequency dependence within 1 MHz. Several CNT thin films can be put together to make a large area loudspeaker. As CNT can be tailored into many shapes and sizes, one can fabricate loudspeakers with special functions. Figure 12d shows the cylindrical cage-like CNT thin film loudspeaker, which can emit sounds in all directions. Such a single-element loudspeaker can also be placed on a variety of rigid or flexible insulating surfaces. Two kinds of CNT thin films loudspeakers were fabricated for testing, i.e., one layer and four layer loudspeakers. In case of the CNT thin film loudspeaker, the sound generation can be understood with the aid of a thermoacoustic mechanism. The alternating current periodically heated the CNT thin films, resulting in a temperature oscillation. The temperature oscillation of the thin film excites the pressure oscillation in the surrounding air, resulting in the sound generation. In this process, it is the thermal expansion and contraction of the air in the vicinity of the thin film that produces sound, not the mechanical movement of the thin film itself. A direct consequence of this thermoacoustic mechanism is that the frequency of the output doubles that of the input. For an alternating current passing through, the CNT thin film will be heated during both positive and negative half-cycles, resulting in a double frequency temperature oscillation, as well as a double frequency sound pressure. Further these carbon nanotube based loudspeakers for small area devices are also suitable for buzzers or earphones. This research will lead many more applications of carbon nanotubes.

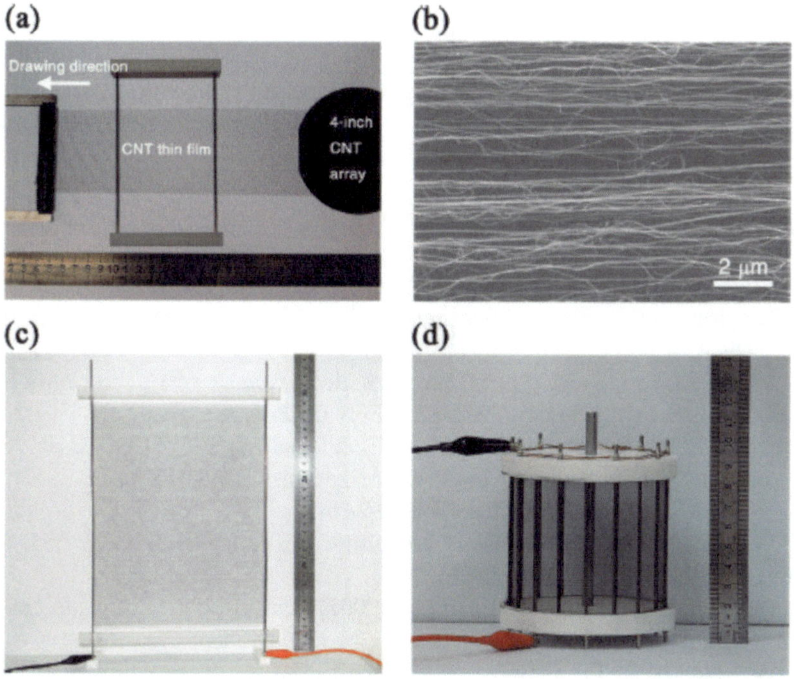

Fig. 12. Carbon nanotube thin film loudspeakers. (a) The CNT thin film was pulled out from a superaligned CNT array grown on a 4 in. silicon wafer and put on two electrodes of a frame to make a loudspeaker. (b) SEM image of the CNT thin film. (c) A4 paper size CNT thin film loudspeakers. (d) The cylindrical cage shape CNT thin film loudspeaker, which can emit sounds in all directions. (Reproduce with permission from American Chemical Society)

6. Conclusion

Researchers in California had developed a nanosized carbon nanotube based radio device. This single nanotube radio is about 19 orders of magnitude smaller than the Philco vacuum tube radio of the 1930s. It is expected that these carbon-nanotube based devices will replace the existing silicon (Si)-technology based devices in the near future providing an improvement in efficiency. On the other hand, researchers in China had developed a nanosized flexible, stretchable, transparent CNT thin film loudspeakers, which could open

up new applications of carbon nanotube based electronic devices such as loudspeakers. It is expected that scientists in the near future will able to integate the nanosized radio and loudspeaker with the possibility to fabricate a fully operating radio based on carbon nanotube. For further reading, see:
http://www.physics.berkeley.edu/research/zettl/projects/nanoradio/radio.html and the article published in Current Science (Sur, 2008).

7. Acknowledgement

We acknowledge financial support from the project funded by the UGC, New Delhi (grant no. PSW-038/10-11-ERO). I would like to thank Dr. Frank Marken, University of Bath and Professor V. Lakshminarayanan for their valuable comments. I would like to acknowledge Zettl and co-workers, Lawrence Berkeley National Laboratory and UC Berkeley, USA for giving permission to reproduce Fig. 10 & Fig. 11. I would like to thank the Director, Bose Institute, Kolkata for reproducing Fig. 8 & Fig. 9. I would like to thank American Chemical Society for giving permission to reproduce Fig. 12. Finally, this chapter is dedicated to the memory of my beloved brother Kallol Kumar Sur.

8. References

Bose, J.C., (1897) On the determination of the wavelength of electric radiation by a diffraction grating. *Proc. R. Soc. London*, Vol. 60, (December 1897), pp. 167–178, ISSN 1471-2946.

Ebbesen, T.W., & Ajayan, P. M., (1992) Large-Scale Synthesis of Carbon nanotubes. *Nature*, Vol. 358, (16 July 1992), pp. 220-222, ISSN 0028-0836.

Guo, T., Nikolaev, P.N., Thess A., Colbert, D.T., & Smalley, R.E., (1995) Catalytic Growth of Single-walled nanotube by laser vaporization. *Chemical Physics Letter,* Vol. 243, (8 September 1995), pp. 49-54, ISSN 0009-2614.

Iijima, S., (1991) Helical microtubules of graphitic Carbon. *Nature*, Vol. 354, (7 November 1991), pp. 56-58. ISSN 0028-0836.

Jensen, K., Weldon, J., Garcia, H. & Zettl, A., (2007) Nanotube Radio. *Nano Lett.*, Vol. 7, No. 12, (December 2007), pp. 3508–3511, ISSN 1530-6984.

Kroto, H. W., Heath, J. R., O'Brien, S. C., Curl, R. F., & Smalley, R. E. (1985). C60: Buckminsterfullerene. *Nature*, Vol. 318, (14 November 1985), pp.162–163, ISSN 0028-0836.

Rutherglen, C., & Burke, P., (2007) Carbon Nanotube Radio. *Nano Lett.*, Vol. 7, No. 11, (November 2007), pp. 3296–3299, ISSN 0028-0836.

Sur, U.K. (2008) World's first single carbon nanotube radio. *Current Science,* Vol. 94, No. 2, (25 January 2008), pp. 166-167, ISSN 0011-3891.

Xiao, L., Chen, Z., Feng, C., Liu, L., Bai, Z-Q., Wang, Y., Qian, L., Zhang, Y., Li, Q,, Jiang, K., & Fan, S. (2008) Flexible, Stretchable, Transparent Carbon Nanotube Thin Film Loudspeakers. *Nano Lett.*, Vol. 8, No. 12, (December 2008), pp. 4539–4545, ISSN 0028-0836.

Part 2

Carbon Nanotube-Based Composite Materials

12

Transparent Conductive Carbon Nanotube/ Binder Hybrid Thin Film Technology

Joong Tark Han, Hee Jin Jeong, Seung Yol Jeong
and Geon-Woong Lee
Korea Electrotechnology Research Institute
Republic of Korea

1. Introduction

Carbon nanotube (CNT)-based transparent conductive film (TCF) technologies have potential applications in electrostatic dissipation (ESD), electromagnetic interference (EMI) shielding, and transparent film heating, as well as in the development of alternative electrode materials for touch panels and e-papers in display technologies, solar cells, flexible electronic devices, automobiles, and optical devices. In particular, single-walled carbon nanotube (SWCNT) network films have been intensively studied for the development of alternative transparent conductive electrodes due to their excellent electrical properties, the flexibility of SWCNT networks, and their solution processability under ambient conditions (Wu et al., 2004; Kaempgen et al., 2005; Zhou et al., 2006). For such applications, the optoelectronic properties of SWCNT-based TCFs should optimally be controlled by the material properties of the nanotubes, including purity, diameter, chirality, defects, metallicity, and doping level (Geng et al., 2007). Organic materials, such as conjugated polymers, block copolymers, polyelectrolytes, pyrenes, DNA, and so on, may also be used in applications because CNTs display good dispersion and stabilization in a variety of solvent media and polymer matrices. To maintain good electrical and mechanical properties, as well as environmental stability (e.g., thermal and hydrothermal stability), SWCNTs must be hybridized or top-coated with binder materials, such as cross-linkable polymers, ceramic sols, or metal oxide sols. The electrical properties of SWCNT/binder hybrid thin films are sensitive to their surroundings and to the interfacial structure of the network film, and the interfacial interactions or interfacial tension among nanotubes, binder materials, and substrates can affect the optoelectronic and environmental properties of SWCNT-based TCFs.(Han et al. 2009)

Despite these attractive features, fundamental studies and several advances are needed for the practical application of high-performance CNT films. This chapter describes some of the research conducted over the past 3 years that addresses these and other challenges, with an emphasis on our own efforts. We begin with critical properties of binders in CNT/binder hybrid thin films and then describe the various binder materials that yield high-performance CNT-based films via molecular or interfacial engineering at the interface between CNTs and binder materials. We conclude with some discussion of future directions and the remaining challenges in CNT/binder hybrid thin film technologies.

2. Carbon nanotube/binder hybrid thin films

To fabricate CNT/binder hybrid thin films by spraying or spin-coating, CNTs must first be well-dispersed in an organic solvent, and the dispersion stability should be maintained after mixing with the binder materials or additives. The wettability of the components in the CNT/binder mixture solution with respect to the target substrate should be considered. Here, interfacial engineering concepts may be applied to balance the interactions at the interfaces between the CNTs, solvent, additives, binder materials, and substrates (Fig. 1).

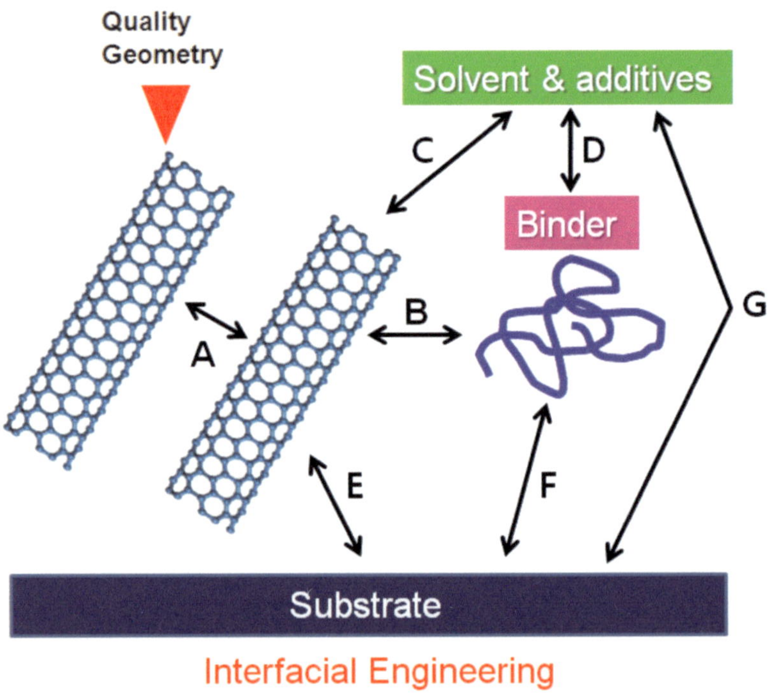

Fig. 1. Interfacial engineering in high-performance transparent conductive CNT/binder hybrid films.

The conductivity, σ_{DC}, of a disordered nanotube film depends on the number density of the network junctions, N_j, which in turn scales with the network morphology though the film fill-factor, V_f, the mean diameter of the bundles, $<D>$, and the mean junction resistance, $<R_J>$, (Hecht et al., 2006; Lyons et al., 2008; Nirmalraj et al.,2009),

$$\sigma_{DC} = \frac{K}{<R_J>} \frac{V_f^2}{<D>^3}$$

Here, K is the proportionality factor that scales with the bundle length. Note that changes in $<R_J>$ and V_f via hybridization with binder materials may be influenced by the wetting properties of the SWCNT films.

The changes in CNT film sheet resistance after hybridization with binder materials may be understood in terms of the quantity of binder material and the interfacial tension of the components (nanotubes and binder materials). The critical surface tension of a CNT sample falls within the interval 40–80 mN/m, and the cutoff value corresponding to cosθ = zero, γ_{max}, falls within the interval 130–170 mN/m (Dujardin et al. 1994, 1998). Liquids with $\gamma < \gamma_C$ yield complete wetting upon formation of a thin film. For $\gamma_C < \gamma < \gamma_{max}$, partial wetting of the liquid occurs. The liquid does not wet a surface for $\gamma > \gamma_{max}$. Most polymer materials with surface tensions of 30–50 mN/m wet CNT surfaces. Randomly oriented SWCNT network films include a large number of nanotube junctions. Such crossover sites attract polymeric materials via capillary effects (Dujardin et al. 1998). This means that the electrical properties of CNT/binder hybrid films can be controlled by modulating the interfacial tension between the CNT films and the binder materials or by modulating the quantity of binder material present. In addition, mixtures containing CNTs and a silane sol represent promising candidates for producing multifunctional coatings because the use of sol-gel chemistry to modify the properties of a gel with functionalized silane precursors has significant advantages. The sol-gel technique provides a method for fabricating ceramic materials and has been used to modify ceramic materials such as silica and TiO_2 with CNTs. This section presents four methods for modulating the optoelectronic and environmental properties of CNT/binder hybrid films based on interfacial and molecular engineering. The first method uses the concept of a critical binder content to optimize the amount of binder material present with respect to the mechanical and electrical properties of the films. The second method uses molecular engineering to minimize or decrease the sheet resistance of the films or to fabricate multi-functional films by adding insulating binder materials or metal oxides. The last method uses a strategy to control the optoelectronic properties of films by matching the wettability of the coating solution on the substrates.

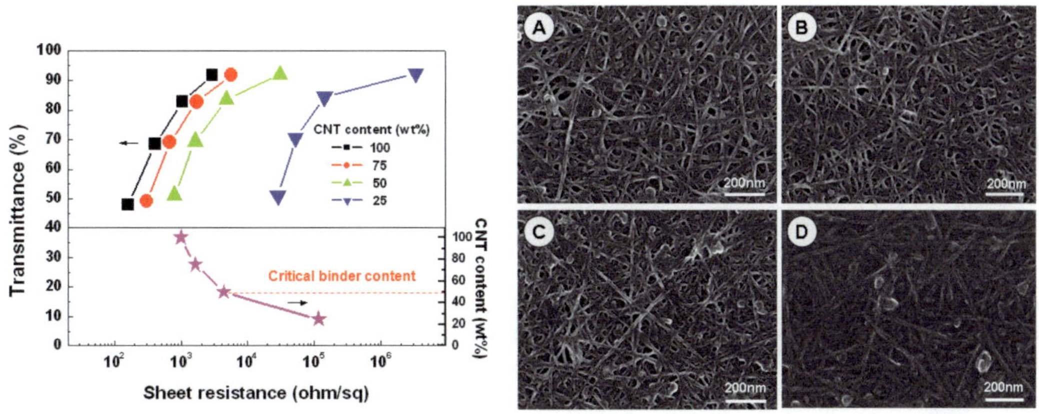

Fig. 2. Transmittance vs. sheet resistance for SWCNT/MTMS hybrid films containing various amounts of MTMS binder. FE-SEM images of SWCNT/MTMS hybrid films containing various amounts of CNTs: (a) 100 wt%, (b) 75 wt%, (c) 50 wt%, and (d) 25 wt%. (Han et al., 2009a)

2.1 Critical binder content

The transmittance and sheet resistance of spray-coated CNT/binder films depend on the quantity of deposited CNTs and binder material, and on the ratio between CNT and binder. A plot of the sheet resistance as a function of binder content shows that above a critical binder content (X_c), the sheet resistance increases dramatically (Han et al., 2009). The strength of the interactions between the nanotubes and binder materials is also an important parameter that determines X_c, thereby influencing the junction structure. Figure 2 shows a plot of the transmittance vs. sheet resistance of the SWCNT/binder hybrid films with various binder contents. In this experiment, a methyltrimethoxy silane (MTMS) sol with a moderate surface tension was used as a model binder material. Here, the sheet resistance increased dramatically at a critical binder content. In this system, the critical binder content, X_c, was approximately 50 wt%. Above X_c, the CNTs were fully covered with the binder material, as illustrated in the scanning electron microscopy image (Fig. 2), which increased the contact resistance between the CNT network and the probe and decreased tunneling between CNTs through the insulating binder layer between the CNT bundles.

2.2 Molecular engineering for CNT/binder hybrid thin films

Increasing the interaction strength between a binder material and CNT surfaces is expected to increase the distance between nanotubes in a network film due to penetration of the binder material into network junctions. To investigate this interfacial interaction effect, model binder materials are required. A silane sol was used in this study to take advantage of the significant benefits associated with using sol-gel chemistry to modify the properties of a gel using functionalized silane precursors (Brinker & Scherer, 1990). The intermolecular interactions between the nanotube surfaces were controlled using a series of model binder materials: tetraethoxysilane (TEOS), methyltrimethoxysilane (MTMS), vinyltrimethoxysilane (VTMS), and phenyltrimethoxysilane (PTMS), as shown in Fig. 3.

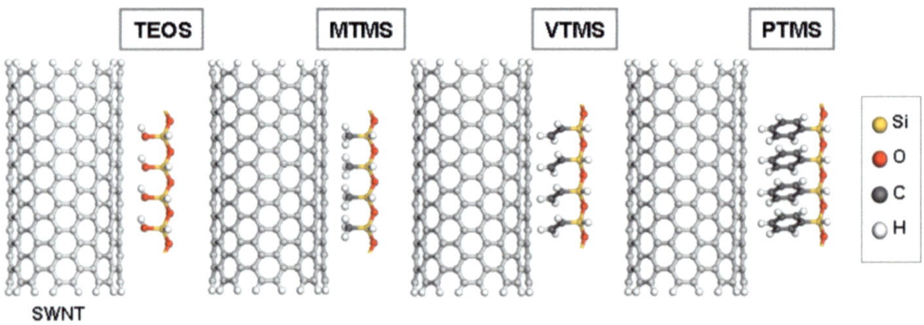

Fig. 3. A schematic diagram of the intermolecular interactions between SWCNTs and model binder materials: tetraorthosilicate (TEOS), methyltrimethoxysilane (MTMS), vinyltrimethoxysilane (VTMS), phenyltrimethoxysilane (PTMS). (Han et al., 2009a)

The unpaired electrons of the silanol groups of the TEOS sol did not significantly polarize the negative charges on the nanotube surface and did not form favorable interactions. Hydrophobic interactions can arise between the methyl groups in the MTMS sol and the nanotube surface (Gavalas et al. 2001). The vinyl groups in VTMS and the phenyl groups in PTMS can interact with SWCNT surfaces via π-π interactions (LeMieux et al., 2008). The phenyl rings of PTMS may provide the best interfacial surface for CNTs due to strong π-π interactions. Moreover, the surface tension of the MTMS/VTMS/PTMS sol was less than 30 mN/m (Tillman et al., 1998), and that of the TEOS sol was around 170 mN/m (Ulatowska-jara et al., 2009). Therefore, the intertube or interbundle distances in the SWCNT/binder hybrid films could be modulated using these binder materials. This property was directly correlated with the electrical properties of the SWCNT/binder film because the sheet resistance of the film resulted from the intrinsic resistance of the SWCNTs and the contact resistance at the junctions between nanotubes. The binder materials penetrated into the SWCNTs or the SWCNT bundles to increase the junction resistance. From this perspective, we expected the sheet resistance of the SWCNT/PTMS films to be the highest among all films tested because the PTMS increased the junction resistance in the network films.

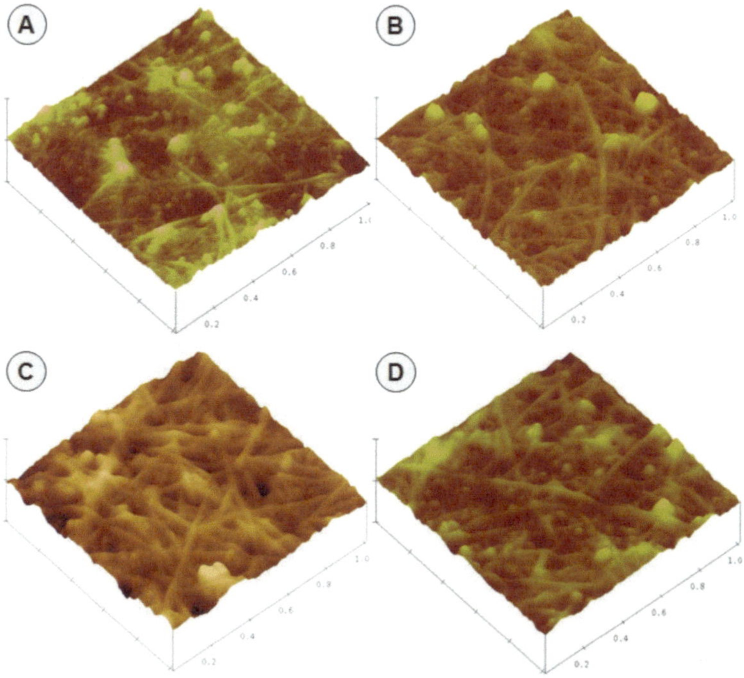

Fig. 4. AFM images of SWCNT/silane hybrid films: (a) SWCNT/TEOS, (b) SWCNT/MTMS, (c) SWCNT/VTMS, and (d) SWCNT/PTMS films. (Han et al., 2009a)

As expected, the sheet resistances of the films gradually increased in the order of SWCNT/TEOS < SWCNT/MTMS < SWCNT/VTMS films. However, the sheet resistance of the SWCNT/PTMS film was lower than that of the SWCNT/MTMS film, even though the CNTs appeared to be well-distributed and covered with the binder material (Fig. 4d). Aromatic molecules, such as the phenyl-terminated silane used here, have been reported to interact and bind selectively to metallic SWCNTs because the polarizability of this silane is larger than that of the semiconducting nanotubes (LeMieux et al., 2008). Therefore, R_s of the SWCNT/PTMS was lower than that of SWCNT/VTMS possibly due to interconnections between the nanotubes or nanotube bundles and the phenyl-functionalized silane sol via strong π-π interactions, which decreased the junction contact resistance. Raman spectral data provided evidence of bridging between the nanotubes and the PTMS sol. In a strongly aggregated state, for example a CNT network film without binder materials, van der Waals interactions between bundles dominated, whereas in a CNT/binder thin film, interactions between bundles and the functional groups of the binder materials influenced the Raman features. Binder materials with functional groups, such as nitro, amino, and chlorine groups, provided chemical doping effects via a charge transfer mechanism that influenced the conductivity of the nanotube films (Rao et al., 1997). However, in this system, doping effects were excluded, and the G^+ band was only slightly downshifted upon addition of the silane binder materials. This indicated that the functional groups acted as very weak electron-donating groups (CH_3, vinyl, phenyl) and the sheet resistances of the SWCNT/silane films were not significantly affected by charge transfer effects. Therefore, the dispersion state or the distance between nanotube bundles in the thin films appeared to dominate the conductivity in the CNT network films. The linewidth of the G+ band and the intensity ratio of the D and G bands were indicative of the degree of aggregation or bundling among the nanotubes. The enhanced resonance processes in the Raman scattering G band may have been due to exfoliation of the nanotubes, which decreased the D/G ratio of the G band. In addition, the relationship between the ratios I_D/I_{RBM} and I_D/I_G for laser excitation at 2.41 eV probed the aggregated state or the interbundle distances of bundles in the thin film network, assuming that the disorder defects were constant after hybridization (Liu et al., 2007). The high ratios of I_D/I_{RBM} and I_D/I_G indicated that the bundles were closely packed (Fig. 5d). The FWHM of the G^+ band of the films exhibited a similar trend in the D/G ratio. The sheet resistances of the various silane binders followed a trend opposite that of the D/G ratio and the G^+ band FWHM. These results, therefore, provide strong evidence that the average interbundle distance in the SWCNT/PTMS sol hybrid films did not differ from that in the pristine and SWCNT/TEOS sol hybrid films. The SWCNT bundles were presumably bridged by the strong interactions between the CNTs and the phenyl groups of PTMS, which contributed to the enhanced conductivity of the SWCNT networks, even though the CNTs were fully covered with insulating material, as determined by the top-view image. Such precise control over the optoelectronic properties of the SWCNT/binder films may be useful for fabricating high-performance conductive thin films, with ramifications for understanding the fundamental intermolecular interactions in carbon material science.

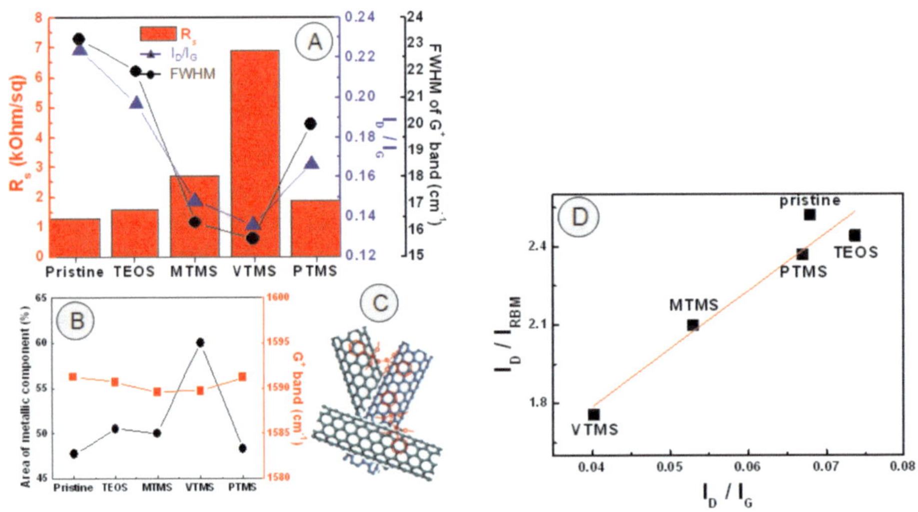

Fig. 5. (a) The correlations between the Raman spectral band at 1.96 eV (D/G ratio, FWHM of the G+ band) and the Rs (with an optical transmittance of 85%) for pristine SWCNTs and SWCNT/silane films. (b) Metallic components extracted from the G-band and G-band shift at 1.96 eV. (c) An illustration of the possible interactions between the SWCNTs and PTMS. (d) Correlation between the ratios I_D/I_{RBM} vs. I_D/I_G at 2.41 eV. (Han et al., 2009a)

2.3 Transparent, conductive, superhydrophobic CNT/binder hybrid films

If the wettability of conductive CNT films with high transmittance could be controlled via a superhydrophobicity (with a contact angle (CA)> 150°)-to-superhydrophilicity (CA< 5°) transition, this technology could potentially meet the needs of a wide range of applications that require multifunctional coatings (e.g., in optoelectronic devices, structural coatings, etc.). Many authors have focused on the fabrication and understanding of superhydrophobic surfaces, particularly those based on CNTs. However, most studies have not considered the optical properties of such CNT-based superhydrophobic surfaces. For applications in optical devices, transparency is one of the most important characteristics.

In nature, the leaves of many plants exhibit super water repellency (super-hydrophobicity) and are cleaned completely during a rain shower via the rolling of surface water droplets, which remove dirt and debris (self-cleaning) (Barthlott & Neinhuis, 1997). The unusual wetting characteristics of superhydrophobic surfaces are governed by both the chemical composition and the geometric microstructure of that surface. Wettability can be decreased or increased by creating a local structure that has a large geometric surface area in three dimensions relative to the projected two-dimensional area (Wenzel, 1936; Cassie & Baxter, 1944). Control over the wettability and optical properties may be achieved using mixed solutions containing CNTs and silane sols to produce multifunctional coatings. CNT networks control the nanostructure of the films, and silane compounds introduce a variety of chemical moieties on the top surface to provide particular mechanical properties. Recently, we presented, for the first time, a facile method for creating transparent, conductive, superhydrophobic (or superhydrophilic) films from a one-component CNT/silane sol solution (Fig. 6). The stable CNT/silane sol solution relied on the intermolecular interactions between the hydroxyl groups of the H_2O_2-treated CNTs and the silanol groups of the silane sol. Moreover, the superhydrophobicity of the transparent (T > 90%) conductive films was enhanced by introducing nanoparticles into the coating solution

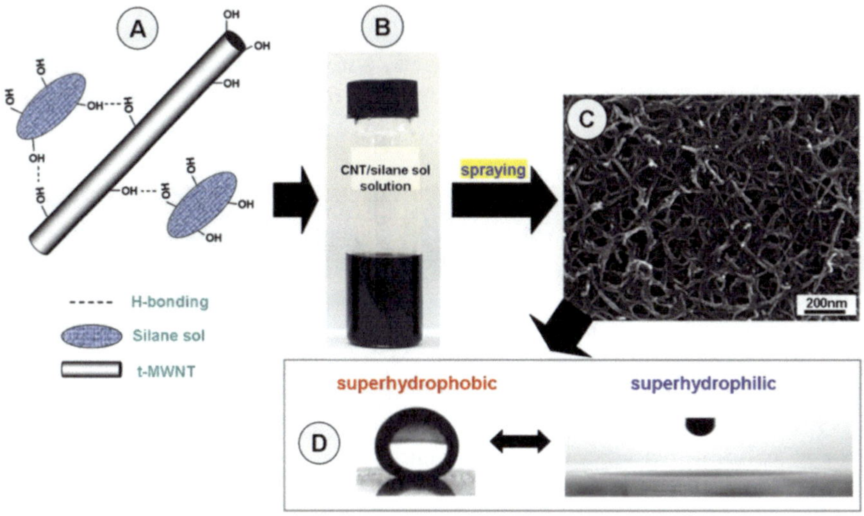

Fig. 6. (A) Schematic diagram of the hydrogen bond-driven stabilization of a CNT solution. (B) Image of a stabilized CNT/silane sol solution. (C) FE-SEM image of a spray-coated CNT/silane hybrid film. (D) Water droplets on this film. (Han et al., 2008)

(Fig. 7). The combination of the transparency and conductivity of CNTs with the chemical functionality of the silane binder would be beneficial to a wide range of CNT-based film applications, for example, development of self-cleaning optoelectronic coatings, transparent film heaters, electrostatic discharge coatings, and EMI shielding.

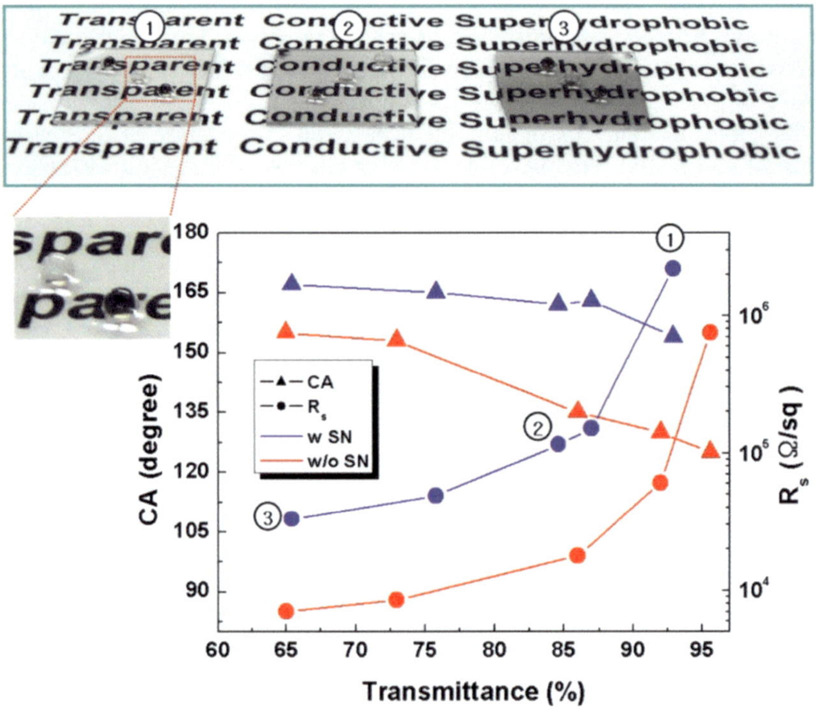

Fig. 7. Water CA (triangles) and sheet resistance (R_s) (circles) versus transmittance of CNT/silane hybrid films (silane content = 70 wt%) without (red) or with (blue) silica nanoparticles. The upper image shows water droplets on transparent conductive films (the numbers shown in this image correspond to those in the plot). (Han et al., 2008)

2.4 Hybridization with metal oxide

CNTs have been used to prepare a variety of hybrid materials that enhance the stability and functionality of CNT-based films by incorporating organic materials or inorganic oxides, such as SiO_2, TiO_2, SnO_2, and ZnO. A successful strategy for fabricating the SWCNT/metal oxide films should employ a reliable means for forming stable solutions of SWCNTs and the metal oxide sol. The dispersion stability of SWCNTs functionalized with carboxylate groups (SWCNT-COOH) strongly depends on the ionic strength and pH of the solution.(Zhao et al. 2002) At pH < 3.0, SWCNTs are protonated, and they aggregate due to van der Waals forces and hydrogen bonding between protonated carboxylic acid groups. At pH > 3.0, mutual repulsion between tubes with charged carboxylic groups stabilizes the SWCNT dispersion. Organic or inorganic materials that contain amine groups can promote aggregation of the SWCNTs-COOH through hydrogen-bonded network formation.
In particular, titania layers provide efficient shielding to prevent penetration of oxygen or moisture into the electronically active layer.(Lee et al. 2007) Uniform titania coatings on

CNT films constitute a potentially useful approach to enhancing the thermal and thermo-hydrostatic stabilities of CNT network films. Titania also acts as an electron transport material due to its high n-type carrier density and high electron mobility, which minimize junction resistance within the film network after hybridization to a binder material. Until now, titania coatings on CNT surfaces have been applied using highly functionalized multiwalled carbon nanotubes (Slazmann et al. 2007 & Gomathi et al. 2005) or benzyl alcohol (BA)-assisted noncovalent methods (Eder et al. 2008). However, functionalization by acid treatment decreases the conductivity of films. The BA method is not suitable for preparing TCF coating solutions because BA does not disperse SWCNTs in organic solvents and cannot stabilize titania sols during the coating process.
We recently reported that a complex formed between acetylacetone (acac, stabilizer of titania sol) and titanium in a titania precursor sol could be used to form a uniform titania layer on nanotube surfaces via hydrophobic interactions (Fig. 8). The thickness of the titania sol layer was controlled by varying the quantity of titania sol used in the solution. TEM images demonstrated formation of a uniform titania layer coating several nanometers thick on the surfaces of the SWCNTs. However, in the absence of acac, irregular titania formed because titanium atoms interacted selectively with carboxyl groups on the nanotube surfaces and amorphous carbon. The titania layer dramatically enhanced the thermal stability of the SWCNT films. The SWCNTs were easily oxidized at temperatures above 350°C, and the network in the SWCNT films was found to be disconnected (Fig. 9d). In contrast, the SWCNTs wrapped with a titania layer were stable under heating, as shown in Fig. 9c. Moreover, the titania layer provided positive effects on the electrical properties of the films via doping effects that operated under a charge transfer mechanism. Titania withdrew electrons from the nanotube surfaces, resulting in enhanced conductivity of the nanotubes. The D-band in the Raman spectra of functionalized SWCNT samples usually contains a broad peak upon which

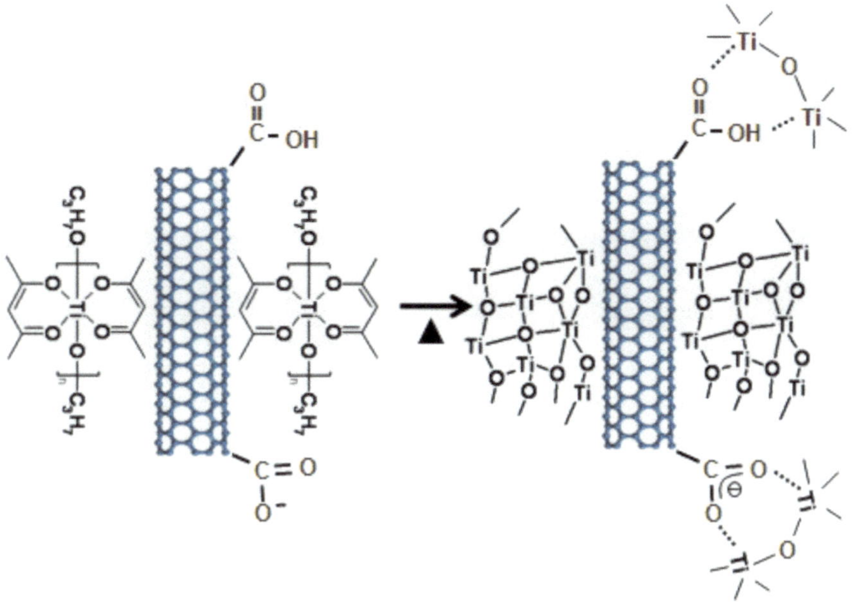

Fig. 8. Mechanism for the noncovalent coating of SWCNTs with a titania layer, followed by removal of acetylacetone molecules by thermal treatment. (Han et al., 2010)

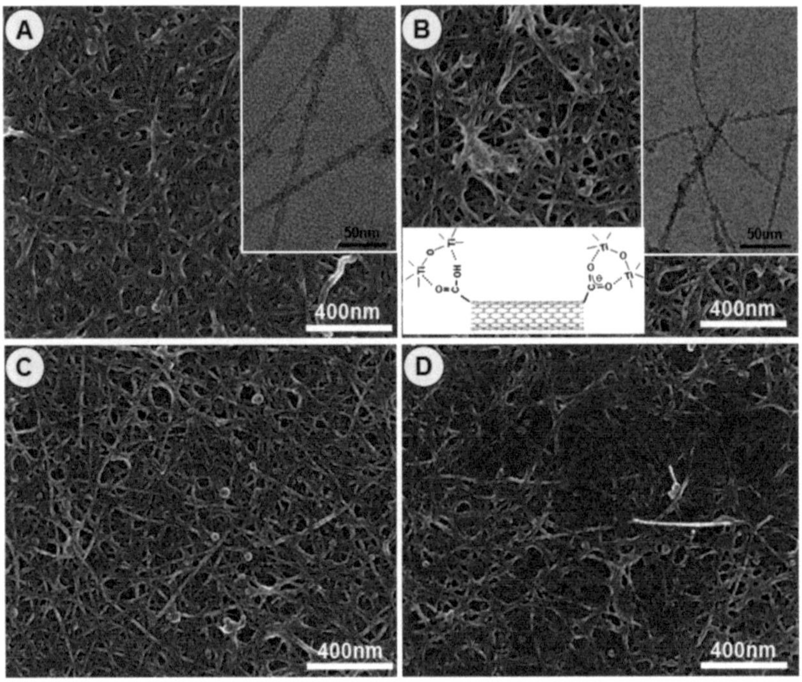

Fig. 9. SEM images of (a), (c) SWCNT/titania sol with acac and (b), (d) SWCNT/titania sol without acac containing 50 wt% titania sol; (a), (b) cured at 150°C, and (c), (d) baked at 350°C for 1 h. Right inset images in (a) and (b) show TEM images. Left bottom images in (b) show the chemical environment of the nanotube surface without acetylacetone. (Han et al., 2010)

is superimposed a sharper peak. The broad feature arises from amorphous carbon, and the sharper feature arises from carbon nanotubes. The narrowing of the D-band of titania-wrapped SWCNT (SWCNT@titania) films and the decrease of the carboxyl C1s peak in XPS after heating at 300°C indicated the removal of amorphous carbon without oxidation of the functionalized SWCNTs. The removal of amorphous carbon also decreased the sheet resistance of the SWCNT@titania films. Moreover, the ultrathin titania layer on the SWCNTs protected against water molecule absorption.

2.5 Wettability-controlled conductive films

Transparent conductive coatings based on CNTs are currently made using membrane filtration or spraying techniques. Spray application over a large irregular area is advantageous for high-throughput fabrication. Here, the wettability of the CNT/binder coating solutions on the substrates should be controlled during fabrication of highly transparent conductive thin films, because the film thickness is optimally smaller than several hundred nanometers. In this respect, the surface free energy of the substrate affects the surface properties and interfacial interactions, such as adsorption, wetting, and adhesion. Control over the wettability and optical properties may be achieved using a mixture of CNTs and silane sol, which is a promising candidate for producing multifunctional coatings. Sol-gel chemistry offers several advantages when used to modify the properties of a gel with functionalized silane precursors. Recently, we studied the

surface energy effects on the optoelectronic properties of CNT/binder hybrid films on glass substrates modified with silane layers containing various end functionalities. The CAs of silane-modified glasses were 67° for an NH_2-functionalized surface, 96.5° for a CH_3-functionalized surface, and 112° for a CF_3-functionalized surface (Fig. 10).
The sheet resistances gradually decreased with increasing wettability of the coating solution on the substrates. Although the transmittance of the films changed very weakly (T changed from 92.3% to 91.2% in moving from a CF_3-functionalized to an OH-functionalized surface), the sheet resistance of the film on the OH surface was an order of magnitude smaller than the counterparts prepared on a CF_3-functionalized surface, giving a very low surface energy. This result is significant because the sheet resistance can change dramatically for high transmittance films. SEM images of the CNT/MTMS sol hybrid films clearly showed that the hydrophilic surfaces were more homogeneous than the hydrophobic surfaces. A decreased surface energy increased the heterogeneity of the surface morphology. In particular, the most hydrophobic surfaces (containing CF_3 groups) clearly showed a dewetted pattern after spray-coating, which may explain the slightly higher transmittance of the film. Nevertheless, the sheet resistance of this film was sufficient for transparent ESD films. The CNTs were macroscopically connected with a fractal dimension of 1.77 for the film surface. The dark regions in the SEM images indicate the low-CNT-density areas (mostly binder materials), as shown in Fig. 11. The low sheet resistance and high transmittance of the film prepared on a CF_3-functionalized surface was explained in terms of the submicrometer-scale disconnect between CNTs, as shown in Fig. 15d. These results indicated that the sheet resistances of highly transparent CNT/binder hybrid films were easily modulated by controlling the wettability of the CNT/binder mixture solutions on the substrate. Previous studies by Kim et al. also attempted to improve the transparency of CNT films by adjusting the CNT network density using a two-dimensional colloidal crystal template. (Kim et al. 2008)

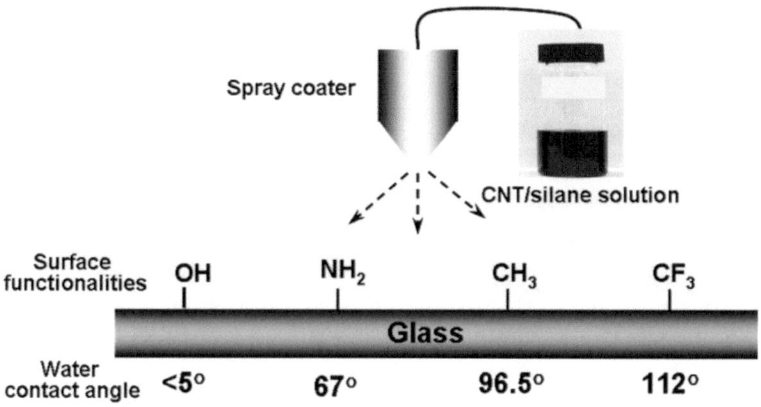

Fig. 10. Schematic representation of the spray coating of FWCNT/silane solutions on surface-modified model substrates. (Han et al., 2009b)

These results have important implications for the fabrication of highly transparent conductive films from CNTs and binder solutions. Although we used a polar solvent and a hydrophilic binder material in this study, our method is applicable to a variety of coating solutions prepared using other solvents and binder materials on various substrates, such as poly(ethylene terephthalate), polyether sulfone, and polycarbonate. Moreover, we suggest that the transparency of CNT/binder films can be improved by manipulating the CNT density in the film, which can be achieved by adjusting the wettability of the coating solution or by forming dewetted areas with different surface energies, because the conductivity and transparency of a film depend primarily on the CNT density.

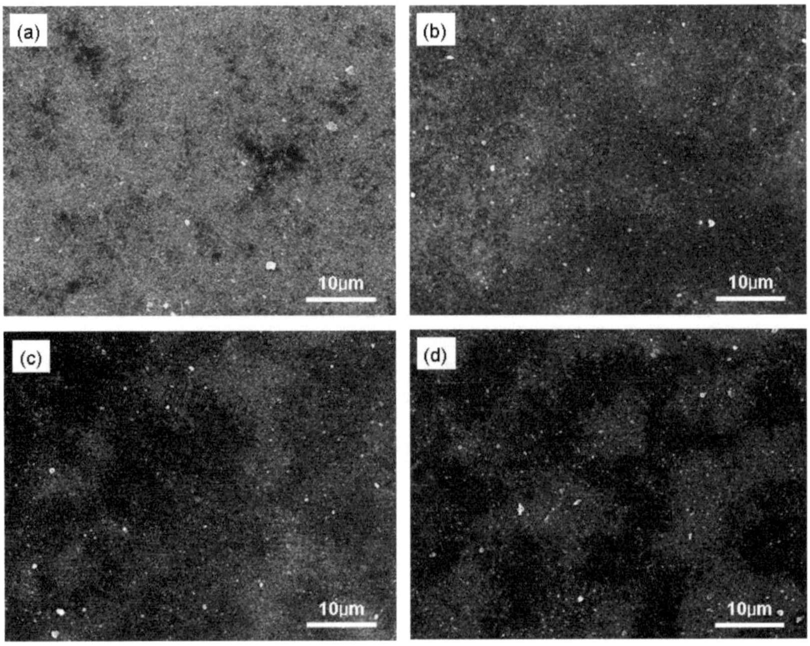

Fig. 11. Scanning electron microscopy images of CNT/MTMS thin films on various substrates; the surface functionalities are: (a) OH, (b) NH_2, (c) CH_3, and (d) CF_3. (Han et al., 2009b)

3. Summary

Research into CNT/binder hybrid thin films over the last few years has yielded significant progress in controlling the optoelectronic properties of the films by modulating the balance of interactions at the interfaces among the components: CNTs, solvent, additives, binder materials, and substrates. A critical binder content was identified, above which the sheet resistance increased dramatically, and this value was found to depend on the interfacial tension between the CNTs and binder materials. At the same time, effective methods were developed for minimizing or decreasing the sheet resistance by adding insulating binder materials. The transparent, conductive, superhydrophobic coating technology relies upon controlling the surface nanostructure and chemical state of the surface. The thermal and environmental stability of the SWCNT films were enhanced by noncovalent wrapping by a titania layer. The optoelectronic properties of the CNT/binder hybrid films were modulated by controlling the wettability of the coating solutions on the substrate.

Significant challenges to this technology remain. First, strategies for minimizing the junction resistance in a random network structure must be developed for applications such as high-performance CNT-based TCFs. Second, improved hybridization methods using various ceramic oxides or metal oxides are needed to use these films in multifunctional electronic devices, such as sensors, actuators, and thin film heaters.

4. References

Brinker, C. J. & G. W. Scherer, (1990) Sol-Gel Science; the physics and chemistry of sol-gel processing, Academic Press, ISBN 0-12-134970-5, San Diego, USA

Barthlott, W.; Neinhuis, C. (1997) Purity of the sacred lotus, or escape from contamination in biological surfaces, *Planta* Vol. *202*, pp. 1-8, ISSN 0032-0935

Dujardin, E.; Ebbesen, T. W.; Hiura, H.; Tanigaki, K. (1994) Capillarity and Wetting of carbon nanotubes. *Science* Vol. *265*, pp. 1850-1852, ISSN 0036-8075

Dujardin, E.; Ebbesen, T. W.; Krishnan, A.; Treacy, M. M. J. (1998) Wetting of single shell carbon nanotubes. *Adv. Mater.* Vol. 10, pp. 1472-1475, ISSN 1521-4095

Eder, D.; Windle, A. H. (2008) Carbon-inorganic hybrid materials: The carbon-nanotube/TiO_2 interface, *Adv. Mater.* Vol. 20, pp.1787-1793, ISSN 1521-4095

Gavalas, V. G.; Andrews, R.; Bhattachrayya, D.; Bachas, L. G. (2001) Carbon nanotube sol-gel composite materials, *Nano Lett.* 2001, Vol. 1, pp.719-721, ISSN1530-6984

Geng, H. Z.; Kim, K. K.; Lee, K.; Kim, G. Y.; Choi, H. K.; Lee, S. D.; An, K. Y.; Lee, Y. H. (2007) Dependence of material quality on performance of flexible transparent conducting films with single-walled carbon nanotubes. *NANO* Vol. 2, pp. 157-167, ISSN 1793-2920

Gomathi, A.; Vivekchand, S. R. C.; Govindaraj, A.; Rao, C. N. R. (2005) Chemically bonded ceramic oxide coatings on carbon nanotubes and inorganic nanowires, *Adv. Mater.* Vol. 17, pp. 2757-2761, ISSN 1521-4095

Han, J. T.; Kim, S. Y.; Woo, J. S.; Lee, G. -W. (2008) Transparent, conductive and superhydrophobic films from stabilized carbon nanotube/silane sol mixture solution *Adv. Mater.* Vol. 20, pp. 3724-3727 ISSN 1521-4095

Han, J. T.; Kim, S. Y.; Jeong, H. J.; Jeong, S. Y.; Lee, G. -W. (2009a) Molecular engineering to minimize the sheet resistance increase of single-walled carbon nanotube/binder hybrid conductive thin films. *J. Phys. Chem. C* Vol. 113, pp. 16915-16920, ISSN 1932-7447

Han, J. T.; Kim, S. Y.; Jeong, H. J.; Lee, G. -W (2009b) Wettability controlled fabrication of highly transparent and conductive carbon nanotube/silane sol hybrid thin films, Ind. Eng. Chem. Res. Vol. 48, pp. 6303-6307, ISSN 0888-5885

Han, J. T.; Kim, J. S.; Jeong, H. D.; Jeong, H. J.; Jeong, S. Y.; Lee, G. -W. (2010) Non-covalent titania wrapping of single-walled carbon nanotubes for environmentally stable conductive films, *J. Mater. Chem.* Vol. 20, pp. 8557-8562, ISSN 0959-9428

Hecht, D.; Hu, L. B.; Grüner, G. (2006) Conductivity scaling with bundle length and diameter in single walled carbon nanotube networks. *Appl. Phys. Lett.* Vol. 89, pp. 13112/1-13112/3, ISSN 003-6951

Kaempgen, M.; Duesberg, G. S.; Roth, S. (2005) Transparent carbon nanotube coatings. *Appl. Surf. Sci. Vol.* 252, pp. 425-429, ISSN 0169-4332

Kim, M. H.; Choi, J. Y.; Choi, H. K. ; Yoon, S. M. ; Park, O. O. ; Yi, D. K. ; Choi, S. J. ; Shin, H. -J. (2008) Carbon nanotube network structuring using two-dimensional colloidal crystal templates, *Adv. Mater.* Vol. 20, pp. 457-461, ISSN 1521-4095

Lee, K.; Kim, J. Y.; Park, S. H.; Kim, S. H.; Cho, S.; Heeger, A. J. (2007) Air-stable polymer electronic devices, *Adv. Mater.* Vol. 19, pp. 2445-2449, ISSN 1521-4095

LeMieux, M. C.; Roberts, M.; Barman, S.; Jin, Y. W.; Kim, J. M.; Bao, Z. (2008) Self-sorted, aligned nanotube networks for thin-film transistors, *Science* Vol. 321, pp. 101-104, ISSN 0036-8075

Liu, Y.; Gao, L.; Sun, J. (2007) Noncovalent functionalization of carbon nanotubes with sodium lignosulfonate and subsequent quantum dot decoration, *J. Phys. Chem. C* Vol. 111, pp. 1223-1229, ISSN 1932-7447

Lyons, P. E.; De, S. ; Blighe, F.; Nicolosi, V.; Pereira, L. F. C.; Ferreira, M. S.; Coleman, J. N. (2008) The relationship between network morphology and conductivity in nanotube films. *J. Appl. Phys.* Vol. 104, pp. 044302/1-044302/8, ISSN 0021-8979

Nirmalraj, P. N.; Lyons, P. E.; De, S.; Coleman, J. N.; Boland, J. J. (2009) Electrical connectivity in single-walled carbon nanotube networks. *Nano Lett.* Vol. 9, pp. 3890-3895, ISSN ISSN1530-6984

Rao, A. M.; Eklund, P. C.; Bandow, S.; Thess, A.; Smalley; R. E. (1997) Evidence for charge transfer in doped carbon nanotube bundles from Raman scattering *Nature* Vol. *388*, pp. 257-259, ISSN 0028-0836

Salzmann, C. G.; Llewellyn, S. A.; Tobias, G.; Ward, M. A. H.; Huh, Y.; Green, M. L. H. (2007) The role of carboxylated carbonaceous fragments in the functionalization and spectroscopy of a single-walled carbon nanotube material, *Adv. Mater.* Vol. 19, pp. 883-887, ISSN 1521-4095

Tillman, N.; Ulman, A.; Schildkraut, J.S.; Penner, T.L. (1998) Incorporation of phenoxy groups in self-assembled monolayers of trichlorosilane derivatives: Effects on film thickness, wettability, and molecular orientation. *J. Am. Chem. Soc.* Vol. 111, pp. 6136-6144, ISSN 0002-7863

Ulatowska-jara, A.; Hołowacz, I.; Wysocka, K.; Podbielska, H. (2009) Silica-based versus silica-titania sol-gel materials-comparison of the physical properties: surface tension, gelation time, refractive index and optical transmittance, *Optica Appl.* Vol. *XXXIX*, pp. 211-220, ISSN 0078-5466

Wu, Z. C.; Chen, Z.; Du, X.; Logan, J. M.; Sippel, J.; Nikolou, M.; Kamaras, K.; Reynolds, J. R.; Tanner, D. B.; Hebard, A. F.; Rinzler, A. G. (2004) Transparent, conductive carbon nanotube films, *Science* Vol. *305*, pp. 1273-1276, ISSN 0036-8075

Zhao, W.; Song, C.; Pehrsson, P. E. (2002) Water-soluble and optically pH-sensitive single-walled carbon nanotubes from surface modification, *J. Am. Chem. Soc.* Vol. 124, pp. 12418-12419, ISSN 0002-7863

Zhou, Y.; Hu, L.; Grüner, G. (2006) A method of printing carbon nanotube thin films. *Appl. Phys. Lett.* Vol. 88, pp. 123109/1-123109/3, ISSN 003-6951

13

Fabrication and Applications of Carbon Nanotube-Based Hybrid Nanomaterials by Means of Non-Covalently Functionalized Carbon Nanotubes

Haiqing Li and Il Kim
The WCU Center for Synthetic Polymer Bioconjugate Hybrid Materials
Department of Polymer Science and Engineering, Pusan National University
Korea

1. Introduction

Carbon nanotubes (CNTs) including single-walled CNTs (SWCNTs) and multi-walled CNTs (MWCNTs) are allotropes of carbon with cylindrical nanostructures. These cylindrical carbon molecules exhibit many facinating properties including high aspect ratio and tubular geometry, which provides ready gas access to a large specific surface area and percolation at very low volume fractions. They also possess extraordinary mechanical, thermal, electrical and optical properties, which support CNTs as ideal building blocks in hybrid materials with potentially useful in many applications in nanotechnology, electronics and optics [Capek, 2009]. By templating against CNTs, a variety of functional components, such as metal nanoparticles (NPs), quatum dots, inorganic oxides and organic species, can be used to decorate CNTs sidewalls or fill CNTs matrix, forming varied CNT-based hybrid nanomaterials [Eder, 2010]. These yielded hybrids generally exhibit synergistic properties, which greatly optimize the technological potentials of CNTs and enable them to be applied in more versatile areas. However, CNTs generally exist in the form of solid bundles, which are entangled together giving rise to a highly complex network. Together with the chemically inert surfaces, pristine CNTs tend to lack of solubility and be difficult manipulated in any solvents, which have imposed great limitations to the use of CNTs as templates to assemble diverse functional components. Therefore, to efficiently fabricate CNT-based nanohybrids, it is necessary to activate the graphitic surfaces of CNTs. In this direction, two types of CNT-surface-functionalization strategies, covalent and non-covalent methodologies, have been extensively explored in the recent decades.

The end caps of CNTs (when not closed by the catalyst particles) tend to be composed of highly curved fullerene-like hemispheres, which are therefore highly reactive, as compared with the sidewalls [Niyogi et al., 2002]. The sidewalls themselves contain defective sites such as pentagon-heptagon pairs called Stone-Wales defects, sp^3-hybridized defects and vacancies in the nanotube lattice [Hirsch, 2002]. These intrinsic defects provide versatile alternatives to covalently modify the CNTs by means of varied organic chemistry. For instance, Tessonnier et al. [Tessonnier et al., 2009] recently explored to functionalize

MWCNTs with amino groups by deprotonation-carbometalation and subsequent electrophilic attack of bromotriethylamine. Sidewall functionalization also can be achieved by ozonolysis of CNTs followed by treatment with varied reagents [Banerjee & Wong, 2002]. Dissolved lithium metal in liquid ammonia was also used to hydrogenate SWCNTs [Pekker et al., 2001]. In addition, free radicals generated by decomposition of organic peroxide in the presence of alkyl iodides have been used to modify small-diameter SWCNTs [Peng et al., 2003]. More recently, we have developed a rapid, facile and green strategy to modify the pristine CNTs with hydroxyl groups by means of plasma treatment technique [Li et al., 2009]. Note that this surface-modification method effectively avoids the use of any toxic organic solvents or additional surfactants, which not only lowers the production cost but also simplifies the preparation procudures. Although these pioneering methodologies have been extensively explored, the traditional oxidation strategy is still the most common and efficient route to functionalize CNTs so far. In such sidewall modification process, the intrinsic defects of CNTs are supplemented by oxidative damage to the nanotube framework by strong acids which leave holes functionalized with oxygenate functional groups such as carboxylic acid, ketone, alcohol, and ester groups [Chen et al., 1998]. In particular, the treatment of CNTs with strong acids such as nitric acid or with other strong oxidizing agents including $KMnO_4/H_2SO_4$, oxygen gas, $K_2Cr_2O_7/H_2SO_4$ and OsO_4 [Banerjee, et al., 2005], tends to open these tubes and to subsequently generate oxygenated functional moieties that serve to tether many different types of chemical functionalities, such as polymers, inorganic oxides, and metal nanoparticles, onto the ends and defect sites of CNTs, yielding a wide range of CNT-based nanohybrids with extensive applications.

For example, Salavagione et al. [Salavagione et al., 2010] directly grafted poly(vinyl chloride) onto the carboxylic groups modified MWCNT surfaces through esterification reactions in an efficient "grafting to" method. Pei et al. [Pei et al., 2007] successfully grafted poly(2-hydroxyethyl methacrylate) (PHEMA) brushes to the MWCNTs surfaces by means of a surface-initiated reversible addition and fragmentation chain transfer (RAFT) polymerizations, yielding well dispersed CNT/polymer hybrid nanostructures. After hydrolysis of PHMA in the presence of HCl, poly(methacrylic acid) grafted MWCNTs were achieved and showed higher loading capacities for metal ions such as Ag^+. Beside these, a variety of polymerization techniques, such as in-situ radical, anionic, emulstion, Ziegler-Natta and electrochemical polymerizations, have been extensively explored to surface-graft diverse polymer chains from covalently surface-modified CNTs [Tasis et al., 2006]. For the fabrication of CNT/inorganic oxide hybrid nanostructures, numerous studies have been involved. Bottini et al. [Bottini et al., 2005] explored to graft tetraethyl or tetramethyl-orthosilicate (TEOS or TMOS) onto carboxylic acid groups contained CNTs obtained under concentrated HNO_3 oxidizing conditions, forming coupling aninopropyltriethyoxysiane functionalized CNTs through a carboxamide bond. On the basis of these surface-modified CNTs, silica beads were generated and decorated along the CNTs by a sol-gel process in the presence of ammonia water. More recently, Zhang et al. [Zhang et al., 2009] explored a facile route to assemble 3-(trimethoxsilyl)-1-propanethiol modified silica nanoparticles onto the sidewalls of oxygenated moieties contained MWCNTs in the presence of poly(ethylene oxide)-*block*-poly(propylene oxide)-*block*-poly(ethylene oxide), resulting in the formation of nonionic nanofluid hybrid materials. In addition, the suitable surface modification of CNTs also provide promising substrates for the deposition of varied noble metal NPs. As a typical example, Gu et al. [Gu et al., 2009] further modified oxygenated MWCNTs with imidazole salts motifs whose counterions allow to be exchanged with metallic ions. Upon reducion reactions, those metal ions are in-situ transformed to metal NPs, yielding CNT/metal nanostructures with good electrochemical properties. Han et al. [Han et al., 2004] demonstrated a simple and effective alternative to assemble monolay-capped metal NPs onto the CNT surfaces via a combination of hydrophobic and hydrogen-bonding interactions between the capping/mediating shell of metal NPs and CNT sidewalls. The loading and distribution of NPs on CNT sidewalls can be well-controlled depending on the relative concentration of metal NPs, CNTs and mediating or linking agents. As another representative example, we recently explored an effective protocol to fabricate CNT-based nanohybrids, in which hydroxyl groups were introduced onto the sidewalls of pristine SWCNTs by means of plasma treatment technique.[Li et al., 2009] Followed by a co-condensation process between those hydroxyl groups bearing on the SWCNTs and TEOS (or together with MPTO), a uniform SiO_2 and thiol groups-functionalized SiO_2 coating on the CNTs can be fabricated effectively. By means of SWCNT@SiO_2-SH, a stable SWCNT@SiO_2/Ag heterogeneous hybrid has been generated via in-situ growth process in the absence of any additional reducing agents

Although the conventional covalent CNT-surface-modification methodologies such as strong oxidizing acids treatments can introduce a variety of organic groups on the CNTs surfaces which can serve as effective media to tether or immobilize varied functional components to produce versatile hybrid nanostructures, those introduced functional groups tend to be with limited control over their number, type and location. Moreover, such treatment processes generally cause the surface etching and shortening of CNTs, resulting in the compromise of the electronic and mechanical properties thus suppress their extensive applications. In addition, the deposition of functional components on such covalently surface-modified CNTs often leads to the non-uniform coatings owing to the non-uniform functionalities on the such modified CNT surfaces. Therefore, to achieve uniform coatings on CNTs sidewalls, recently developed non-covalent (non-destructive) methods have provided more facile and efficient alternatives to homogeneously functionalize CNT sidewalls by means of van der Waals interactions, hydrogen bonding, π-π stacking, or electrostatic interactions in the presence of CNT-surface-modifiers such as small molecular surfactants and polymers. In those non-covalently functionalization processes, CNT-surface-modifiers play key roles which not only endow the CNTs with certain dispersity in solvents, but also act as "bridges" to integrate various of functional components onto the CNT surfaces to generate varied CNT-based nanohybrids. Moreover, such resultant hybrids generally exhibit synergistic properties while still reserving nearly all the intrinsic properties of CNTs.

In the recent years, a variety of CNT-surface-modifiers have been developed and utilized to non-covalently functionalize CNTs to create versatile CNT-based hybrid nanomaterials targeted to specific applications (Scheme 1). In this chapter, the recent advances in the use of those non-covalent surface-modifiers for the fabrication of CNT-based hybrid nanomaterials are overviewed.

2. Small molecular CNT-surface-modifiers

To date, many small molecular CNT-surface-modifiers such as some amphiphilic molecules (surfactants) including ionic surfactants and aromatic compounds have been widely utilized to non-covalently functionalize CNTs surfaces (see Scheme 1). In the case of

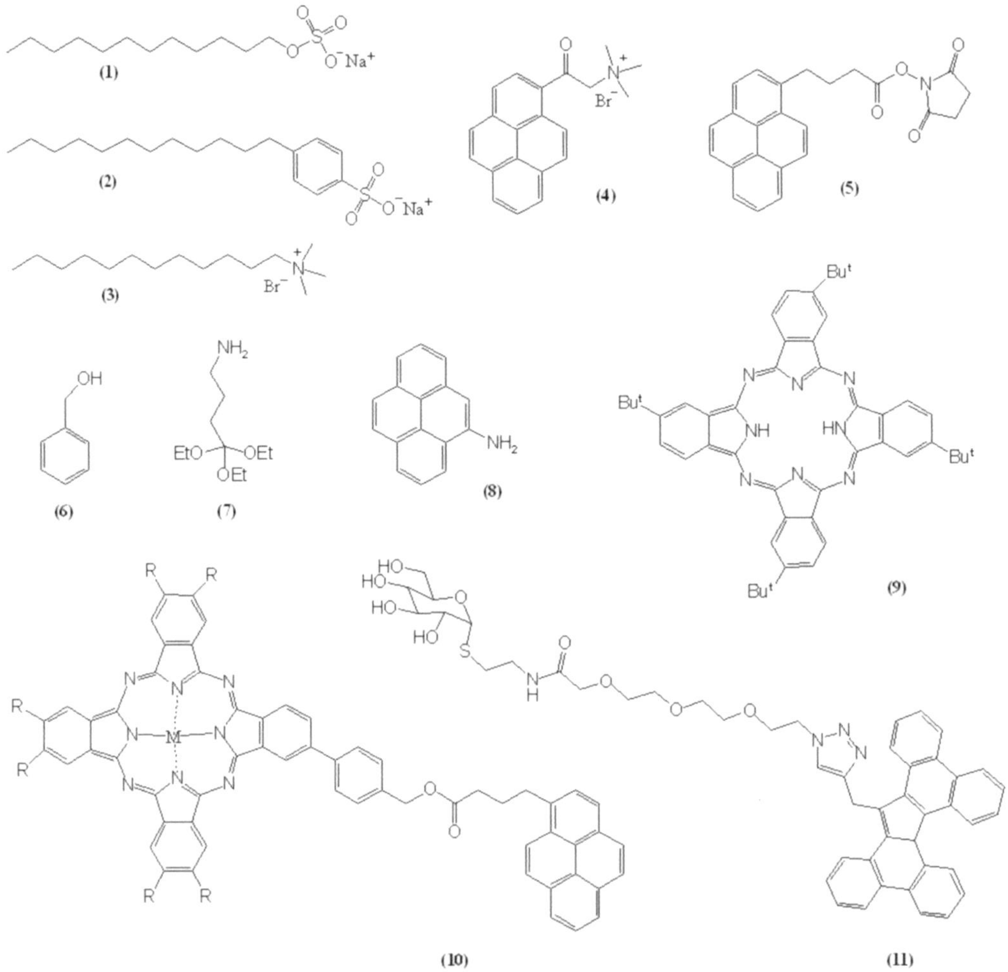

Scheme 1. Different types of small molecular CNT-surface-modifiers.

small molecular surfactants, their hydrophobic parts tend to be adsorbed onto the CNT surfaces by means of diverse hydrophobic interactions, while the hydrophilic parts point towards and interact with the surrounding media. Those non-covalent interactions can effectively solubilise CNTs in certain solvents and prevent them from the aggregation into bundles and ropes. Moreover, those hydrophilic parts provide platforms for the integration of functional components onto the CNT sidewalls to achieve diverse hybrid nanostructures.

2.1 Ionic surfactants

For the ionic small molecular CNT-surface-modifiers, anionic sodium dodecylsulfate (SDS) surfactant has received the most enormous studies. It has found SDS arranged into rolled-up half-cylinders with the alkyl-groups of each molecule pointed towards the MWCNTs [Richard et al., 2003]. Such striation patterns on the sidewalls of MWCNTs were related to the presence of the long alkyl chains and are unaffected by the nature of hydrophilic groups.

It also believed that the simple alkyl chains of surfactants such as SDS, sodium dodecyl sulfonate (SDSA), dodecyltrimethylammonium bromide (DTAB) formed non-specific hydrophobic interactions with CNTs, which result in the loose packing of surfactant molecules around CNTs [O'Connell et al., 2002; Moore et al., 2003]. In addition, the length and shape of the alkyl chains of surfactants also play key roles for the efficiency of the interaction of such surfactants with CNTs: longer and more branched alkyl groups are better than linear and straight ones, respectively [Wenseleers et al., 2004; Islam et al., 2003].

On the basis of those surface-modified CNTs, varied CNT-based hybrid nanostructures have been fabricated. For instance, using SDS as non-covalent CNT-surface-modifiers not only greatly enhance the dispersion of CNTs in water but also provide negative charges to the CNT surfaces, which make SDS-modified CNTs very useful for mediating the attachment of metal NPs on their surfaces. Following this direction, gold NPs were successfully in-situ generated and attached onto the SDS-modified MWCNTs, forming heterogeneous nanostructures [Zhang et al., 2006]. In addition, those surface-charged SDS-modified MWCNTs can be easily layer-by-layer assembled onto the indium tin oxide-coated glass plates mediated by the oppositely charged polyelectrolyte. Similarly, Lee et al. [Lee et al., 2005] decorated in-situ synthesized Pt NPs onto the sidewalls of SDS-functionalized CNTs. The resulting CNT/Pt hybrids exhibited high activity towards the oxidation of methanol.

Whisitt et al. [Whitsitt & Barron, 2003] evaluated different surfactants for their ability to facilitate the deposition of silica NPs onto SWCNT surfaces in the acid conditions. By using anionic SDS, silica NPs were deposited around the bundles of SWCNTs to form coated ropes, while the use of cationic DTAB enabled a significantly better deposition and de-bundling of SWCNTs so that individual nanotubes were coated. They proposed that this effect is the consequence of the pH stability of the SWCNT/surfactant interaction. Acidification of a SWCNT/SDS solution results in the immediate formation of SWCNT ropes, while the SWCNT/DTAB interaction is far less susceptible to the changes of pH. Based on the SDS-modified SWCNTs, an optically homogeneous SWCNT/silica gel has also been fabricated via a sol-gel process [Zamora-Ledezma et al., 2008]. The resultant gel displays a strong fluorescence signal in the NIR, thus it is good candidate for the development of new opto-electronic devices with extended possibilities of processing, especially into thin films.

Besides the CNT/metal NPs and CNT/oxides hybrids, CNT/polymer nanostructures also can be achieved by means of ionic-surfactant modified CNTs. For example, Yang et al. [Yang et al., 2006] used sodium dodecylbenzene sulfonate (SDBS) to exfoliate SWCNT bundles into individual nanotubes with good dispersity in aqueous media. It was found that SDBS-functionalized SWCNTs can adsorb acrylonitrile monomers on their surfaces. After a conventional in-situ radical polymerization and a subsequent hydrolysis reaction, poly(acrylic acid) (PAA) chains were grafted onto the SWCNT sidewalls, producing pH-responsive SWCNT/PAA hybrid with controlled solubility in water depending on pH.

2.2 Aromatic-group-contained molecules

In contrast to the alkyl-chain-contained surfactants, aromatic-group-contained molecules are capable of forming more specific and directional π–π stacking interactions with graphitic surfaces of CNTs. This fact has been evidenced by the comparing results between the use of

SDS and SDBS [Zhang et al., 2006]. It was demonstrated that the presence of phenyl ring made SDBS more effective for the solubilisation of CNTs than SDS although they possess the same length of alkyl chains. Therefore, aromatic-group-contained molecules have been widely utilized to surface-modify CNTs. A typical example involves the use of benzene alcohol to non-covalently functionalized CNTs have been well demonstrated by Eder et al. [Eder & Windle, 2008a, 2008b]. They have found that the π-π interactions of benzene ring enable this surfactant to be adsorbed onto the CNTs' sidewalls. Simultaneously, the hydrophilic hydroxyl groups bearing on the benzyl alcohol-modifed CNTs provid effective platforms for the hydrolysis of the titanium precursor to yield CNT/titania hybrid nanostructures with quite uniform titania coatings. After removal of CNT cores from CNT/titania nanohybrids via calcination treatment, anatase and rutile titania nanotubes can be achieved. This work also showed that benzyl alcohol strongly affected the phase transition from anatase to rutile, providing very high specific surface areas.

Recent studies have shown that the surfactants containing polyaromatic components such as pyrene generally demonstrate more affinity for the CNT surfaces compared with the simple aromatic compounds, resulting in the formation of more stable CNT sols. So far the related researches have been under intense investigations. For example, Bogani et al. [Bogani et al., 2009] synthesized pyrene-functionalized single-molecule magnets (SMMs) and non-covalently bridged them onto the CNT sidewalls, generating the first CNT/SMMs hybrids in conditions compatible to the creation of electronic devices. This wok paves a way to the construction of "double-dot" molecular spintronic devices, where a controlled number of nanomagnets are coupled to an electronic nanodevice, and to the observation of the magneto-Coulomb effect. As another typical example, Li et al. [Li et al., 2006] explored to use 1-aminopyrene to non-covalently modify MWCNT sidewalls. Those amino moieties-contained CNTs exhibited specific adsorption capacities towards different NP precursors via electrostatic interactions and/or preferential affinity under appropriate conditions (Fig. 1). Followed by in-situ reduction or sol-gel processes, a wide range of NPs such as Pt, CdS, and silica were in-situ formed and decorated onto the sidewalls of CNTs with high specificity and efficiency. In addition, inspired by the immobilization of biomolecules onto CNT surfaces in a reliable manner, a bifunctional molecule, 1-pyrenebutanoic acid succinimidyl ester was synthesized and applied to non-covalently functionalize SWCNT surfaces (compound 5 in Scheme 1) [Chen et al., 2001]. By means of nucleophilic attack reactions, various protein and biological molecules such as enzymes can be subsequently covalently attached onto the surface-modified CNTs with a high degree of control and specificity. These surface-modified SWCNTs also can immobilize varied NPs such as ferritin, streptevidin and Au NPs.

Heterocyclic porphyrins and their derivatives are another class of polyaromatic molecules with specific π-π interactions with CNTs. Tetrabutyl-substituted phthalocyanine can non-covalently adsorbed on CNT surfaces, forming nano-sized clusters which presumably consist of aggregated phthalocyanine molecules [Wang et al., 2002]. It was also found the CNTs can fade the colour of phthalocyanin solution in chloroform depending on the relative weight of CNTs in the composites. More recently, a new type of pyrene (Py)-substituted phthalocyanines (Pcs) including ZnPc-Py and H_2Pc-Py (compound 10 in Scheme 1) were synthesized and utilized to non-covalently functionalize SWCNTs via π-π interactions between the pyrene groups and CNTs, forming stable electron donor-acceptor SWCNT/ZnPc-Py and SWCNT/H_2Pc-Py hybrids [Bartelmess et al., 2010]. Encouraged by

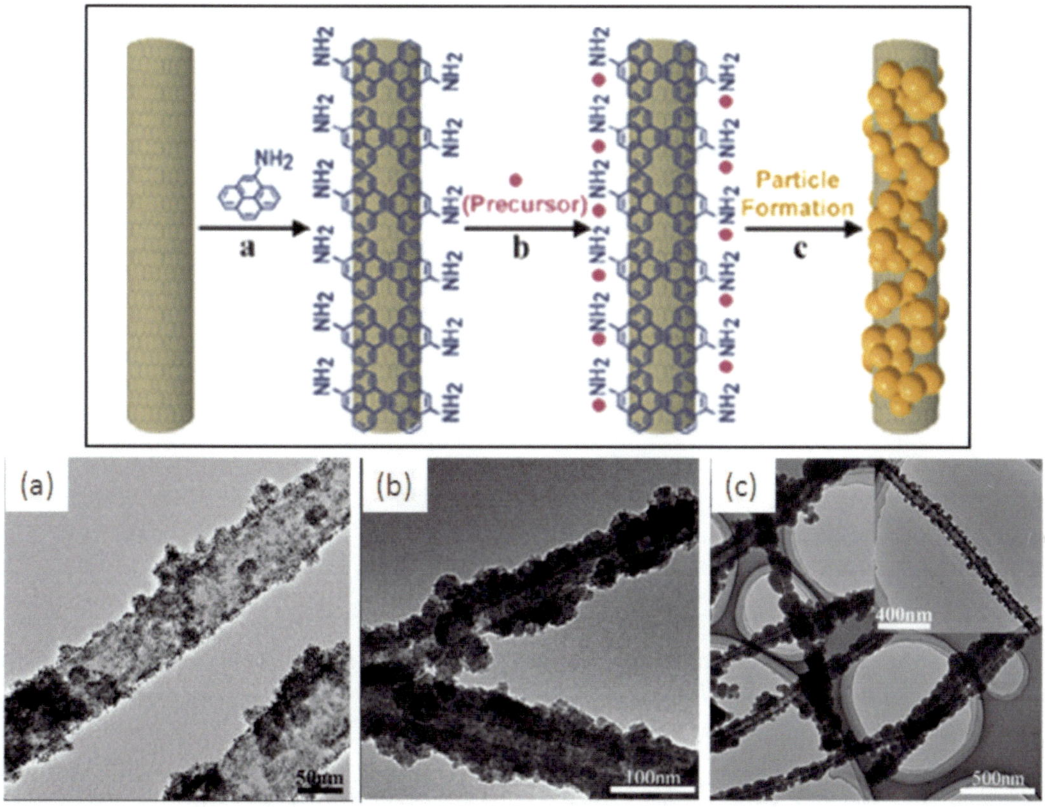

Fig. 1. (Top) scheme for the preparation of CNT/NPs hybrids on the basis of 1-aminopyrene-modified CNTs; (bottom) TEM images of (a) CNT/Pt NPs, (b) CNT/CdS NPs, and (c) CNT/silica NPs. Reprinted with permission from Ref [Li et al., 2006]. Copyright 2006 Wiley-VCH.

the phoptoinduced electron-transfer features, SWCNT/ZnPc-Py and SWCNT/H_2Pc-Py have been integrated into photoactive electrodes within the photoelectrochemical cells, revealding stable and reproducable photocurrents with monochromatic internal photoconversion efficiency values for SWCNT/ZnPc-Py as large as 15 and 23% without and with an applied bias of +0.1 V. In addition, Assali et al. [Assali et al., 2010] synthesized a new SWCNT-surface-modifier amphiphilile consists of a polyaromatic component resembling a butterfly topology with open wings, and a carbohydrate-tethered tetrabenzo(*a,c,g,i*)fluorene (Tbf) segment (compound 11 in Scheme 1). The resulting compounds exhibited more effective capacity to exfoliate MWCNTs in water than the pyrene-based amphiphilic carbohydrates, since the much stronger π-π interactions between the SWCNTs and Tbf groups. This enhanced interaction can be most likely ascribed to the ability of butterfly-like polyaromatic structure of Tbf to fit more effectively on the CNT surfaces. It is also found that the resulting surface-modified SWCNTs with a multivalent sugar exposition on their surface display selective binding with appropriate biological receptors.

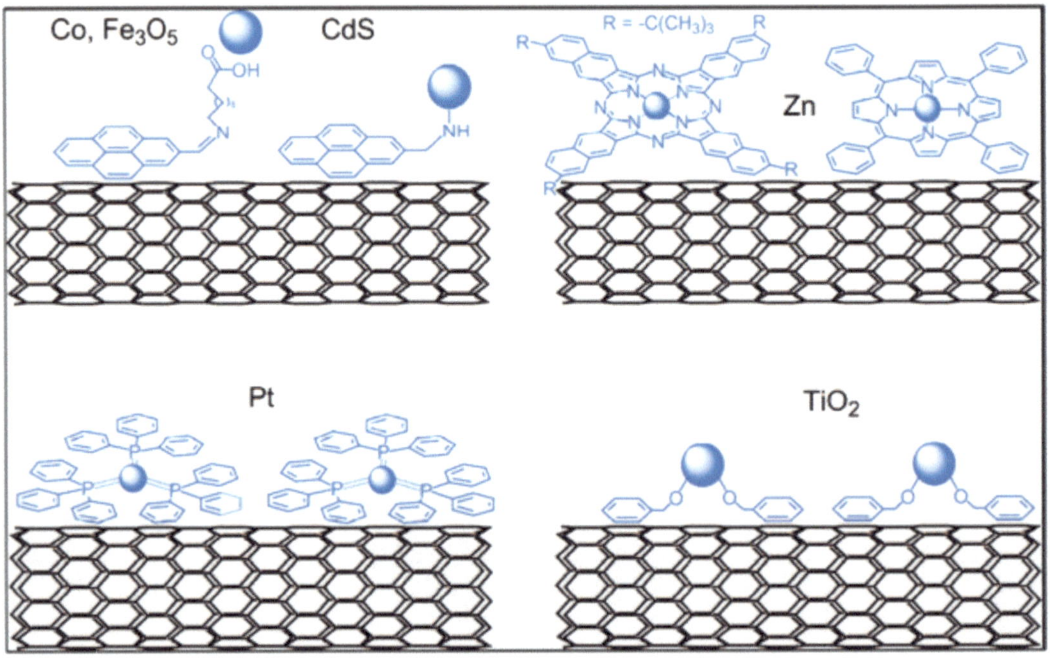

Fig. 2. Examples of linking agents and ligands used to attach inorganic NPs to pristine CNTs via π–π stacking interactions. Reprinted with permission from Ref. [Eder, 2010]. Copyright 2010 American Chemical Society.

In addition, direct assembles of aromatic-compound-stabilized-NPs onto CNT surfaces through π–π stacking interactions provides a more facile route to fabricate CNT-based nanohybrids (Fig. 2) [Eder, 2010]. Such employed aromatic compounds tend to be terminated with functional moieties such as amine, thiol and carboxylic acid groups, which can interact with surfaces of specific NPs and thus stabilize them. Simultaneously, the remained aromatic ends enable those NPs to be anchored onto the CNT sidewalls by π–π stacking interactions, resulting in the formation of a variety of hybrid nanostructures. Following this strategy, Ou et al. [Ou & Huang, 2006] described the fabrication of CNT/Au NPs composites in aqueous solution using 1-pyrenemethylamine as the interlinker. The alkylamine substituent of 1-pyrenemethylamine binds to a Au NP, while the pyrene chromophore is noncovalently attached to the sidewall of a CNT via π–π stacking interaction. Such Au NPs with diameters of 2-4 nm can be successfully assembled on the MWCNT surfaces in a quite high density. It was also found that the attachment of Au NPs onto the CNT surfaces can largely quench the photoluminescence of 1-pyrenemethylamine and lower its emission intensity. Similarly, CdS, Co, Fe_3O_4, Pt and TiO_2 NPs have also been directly assembled onto the CNT surfaces, yielding versatile hybrid nanostructures [Eder, 2010].

2.3 Other small molecular non-covalent CNT-surface-modifiers

Besides the ionic and aromatic-groups contained molecules, some other small molecular surfactants also have been utilized to non-covalently functionalize CNTs aimed to create varied CNT-based nanohybrids. For instance, Bourlinos et al. [Bourlinos et al., 2007] wetted pristine CNTs with vinyl silane molecules via non-covalent interactions between the vinyl groups and CNT surface. After condensation to an oligomeric siloxane network and subsequent calcinations, silica nanoparticles with diameter ranging from 5 to 12 nm were generated and well-dispersed onto the CNT surfaces. Another approach to noncovalently modify MWCNTs was performed by embedding the CNTs within the polysiloxane micelles [Wang et al., 2006]. After a condensation process, a unifrom polysiloxane shell formed around the CNT sidewalls. It was also found that the Au NPs can be in-situ generated and attached on the polysiloxane shells upon heating $HAuCl_4$ aqueous solution at 100 °C. Prolonging the heating process, the growing Au NPs can be further jointed and form continuous Au nanowires along the CNTs.

3. Polymeric CNT-surface-modifiers

Although a large number of hybrid nanostructures have been built on the basis of non-covalently surface-modify CNTs with small molecules, such resultant nanohybrids tend to lack of stability owing to the limit interaction sites between the small molecules and CNT sidewalls. As a promising alternative choice, amphiphilic linear polymers are often used to non-covalently functionalize CNT sidewalls, since they not only reduce the entropic penalty of micelle formation, but also have a significantly higher energy of interaction than small molecules with CNTs. So far, several types of such polymeric CNT-surface-modifiers have been developed. They can be categorized into polyelectrolytes and non-ionic polymers.

3.1 Polyelectrolytes

The choice of polyelectrolytes for non-covalent functionalization of CNTs endows CNT surfaces with positively or negatively charged properties, which provide a variety of opportunities to generate varied CNT-based hybrid nanostructures. This type of polymer generally contains multiple aromatic motifs which allow them to be directly attached onto CNT sidewalls via π–π stacking interactions and polymer-wrapping techniques. For example, the hydrolyzed poly(styrene-*alt*-maleic anhydride) (hPSMA) can be non-covalently adsorbed onto CNT surfaces from aqueous solutions via hydrophobic interactions [Carrillo et al., 2003]. Such attached hPSMA layer contained carboxylic groups, which were used as handles to further covalently attach poly(ethyleneimine) (PEI) and a cross-linked polymer bilayer was formed. These cross-linked polymer layers greatly enhanced the stability of the resultant CNT/polymer hybrids. By simply repeating these steps, a multilayered polymeric film consisting of alternate polyanionic and polycationic layers can be built up. On the basis of the terminated PEI layers, negatively charged Au NPs can be immobilized on the surfaces of CNT/polymer hybrids by means of electrostatic interactions. Another typical polyelectrolyte for CNT-surface-modification has been explored by Mountrichas et al. [Correa-Duarte et al., 2004]. They have synthesized an amphiphilic polystyrene-*b*-poly(sodium (2-sulfamate-3-carboxylate)isoprene) (PSHI) copolymer and utilized them to non-covalently functionalize MWCNTs. The hydrophobic polystyrene block of the polymer can interact with CNT sidewalls via π–π stacking and wrapping. While the hydrophilic polyelectrolyte block stands on the CNT surface towards the surrounding media, which not only enables the PSHI-modified CNTs to be well dispersed in water, but also provides anionic environment to cap cationic ions such as Cd^{2+}. Followed by the addition of thioacetamide, CdS NPs were in-situ generated and attached onto the CNTs surfaces, leading to the formation of a CNTs/PSHI-CdS ensemble as a stable aqueous solution.

LBL techniques provide effective routes to assemble varied polyelectrolytes and charged functional components onto the CNT sidewalls to create versatile hybrid nanostructures. For instance, poly(styrene sulfonate) (PSS) containing both benzene groups and negatively charged sulfonate groups enable them to effectively functionalize CNTs surfaces, yielding a stable dispersion of individual CNT in water. By means of negatively charged sulfonate groups, the cationic poly(diallyldimethylammonium chloride) (PDDA) can be homogenously adsorbed onto the surfaces of PSS-modified CNTs through the electrostatic interactions [Mountrichas et al., 2007]. Followed by LBL processes, negatively charged NPs such as Au@silica can be closely packed onto such surface-modified CNT sidewalls in a controllable manner depending on the number of layers deposited.

3.2 Non-ionic polymers

Although polyelectrolyte-modified CNTs provide versatile approaches to fabricate CNT-based hybrid nanostructures, they tend to be sensitive to the surrounding conditions such as pH value and ionic strength. These greatly limit the fabrication of CNT-based hybrids in varied reaction conditions. To avoid these intrinsic limitations, non-ionic linear and dendritic polymers have also been explored to non-covalently functionalize CNTs.

3.2.1 Linear polymers

An important type of non-ionic linear polymers for non-covalently functionalization of CNTs is block copolymer. Selective adsorption of block copolymer triggers a repulsion among the polymer-decorated CNTs and stabilizes the exfoliated CNTs in the dispersion. It has been found that the solubility of CNTs can be effectively manipulated the composition of the utilized copolymer. For example, Smally group used polyethylene oxide-polypropylene oxide-polyethylene oxide (Pluronic PEO-PPO-PEO) triblock copolymers, selecting molecules with large PEO molecular weight to provide steric stabilization [Moore et al., 2003]. Similarly, Shvartzman-Cohen demonstrated that a large variety of di- and triblock copolymers in selective solvent (aqueous and organic) conditions also excellent stabilizing ability to SWCNTs [Shvartzman-Cohen et al., 2004a]. They also suggusted that a proper choice of the polymer molecular weight may result in dimensional selectivity enabling purification of SWCNTs from mixtures of non-nanometric objects [Shvartzman-Cohen et al., 2004b].

More recently, Zou et al. [Zou et al., 2008a, 2008b] have explored to disperse CNTs in varied solvents and PS matix using conjugated block copolymer of poly(3-hexylthiophene)-*b*-polystyrene (P3HT-*b*-PS). In such dispersion processes, P3HT blocks attached to CNT surfaces through π–π stacking interactions, while PS blocks located at the outermost surface of CNTs. The good solubility of PS blocks in various organic solvents (chloroform, tetrahydrofuran and toluene) and the compatibility with the PS matrix enhanced the solubility of CNTs in organic solvents and the dispersibility in PS matrix. Similarly, they also utilized a series of block copolymers such as P3HT-*b*-poly(methyl methacrylate), P3HT-*b*-poly(acrylic acid) and P3HT-*b*-poly(poly(ethylene glycol) methyl ether acrylate) (P3HT-b-PPEGA, Fig. 3) to non-covalently modified CNTs, and dispersed such surface-modified CNTs into various solvents and polymer matrices [Zou et al., 2009]. In addition, oligothiophene terminated poly(ethylene glycol) was also demonstrated to enable non-covalently functionalized CNTs and well disperse them in aqueous media [Lee et al., 2007]. As a more complexed system, Kim et al. [Kim & Jo, 2010] applied poly(vinyl benzyloxy ethyl naphthalene)-*graft*-poly(methyl mathacrylate) as compatibilizer, in which naphthalene units interact with MWCNTs via π-π interactions. While the poly(methyl mathacrylate) units of the compatibilizer are miscible with poly(styrene-*co*-acryonitrile) (SAN) matrix, which enable MWCNTs to be homogeneously dispersed in SAN matrix even in the presence of small amount of compatibilizer. Those resultant composites exhibited greatly improved mechanical properties and electrical conductivity as compared with those of composites without compatibilizer, since both the homogeneous dispersion of CNTs in SAN matrix and good interfacial adhesion between SAN and non-covalently compatibilizer-modified CNTs.

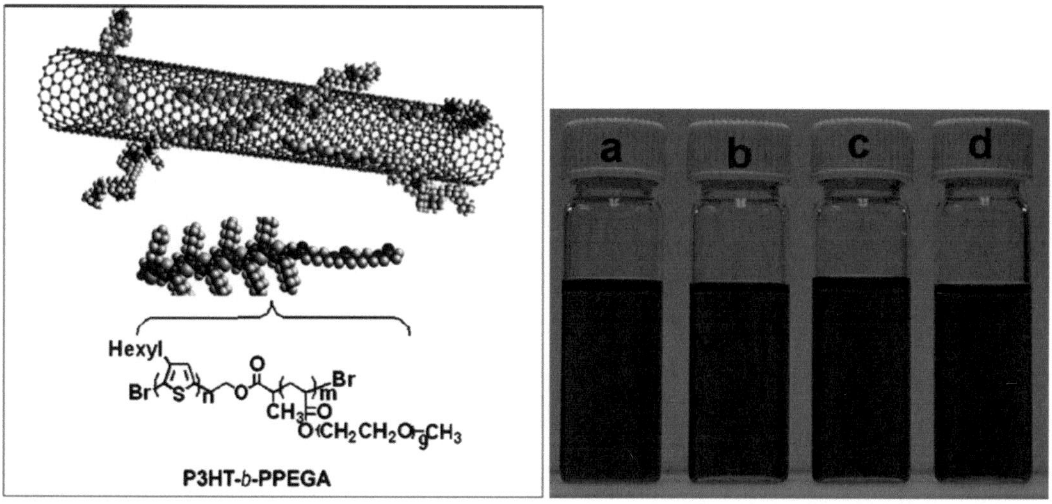

Fig. 3. (Left) Schematic illustration of dispersing and functionalizing CNTs by conjugated P3HT-b-PPEGA block copolymer; (Right) photographs of P3HT-b-PPEGA dispersed MWCNTs in a) chloroform, b) toluene, c) methanol, d) DMF. Reprinted with permission from Ref [Zou et al., 2009]. Copyright 2009 Wiley-VCH.

Besides the amphiphilic block copolymers, linear homopolymers with or without functional moieties, such as poly(vinyl pyrridone) and phospholipid-polyethylene glycol (PL-PEG) also can be used to non-covalently functionalize CNTs via relatively weak wrapping interactions [O'Connell et al., 2001; Welsher et al., 2009]. Using PL-PEG, Welsher et al. [Welsher et al., 2009] successfully debundled SWCNTs in aqueous media by means of a stalibilizer-exchange process. The resulting SWCNTs suspension demonstrated an increase in quantum yield of more than one order of magnitude, while still maintaning the high biocompatibility. More importantly, the near-infrared photoluminescence emission of such modified-SWCNTs allow them to be used to perform cell imaging at a quite low dose.

Beyond the conventional synthetic linear polymers, biomacromolecules such as DNA can non-covalently modify CNTs sidewalls as well. So far, several groups have reported that DNA strands strongly interact with CNTs to form stable hybrids that can be effectively dispersed in aqueous solutions. In the case of wrapping CNTs with single-stranded DNA, DNA tends to self-assemble into a helical structure around individual nanotubes in such a way that the electrostatics of DNA-CNT hybrid depends on tube diameter, electronic properties, paricular DNA sequence and length of sequence, enabling not only to seperate metallic fractions from semiconducting tubes but also to perform a diameter-dependent seperation via ion exchange chromatography [Zheng et al., 2003; Tu et al., 2009]. Mediated by chemical linkers, DNA can be physically attached onto CNT surfaces, which provides an indirect alternative to fabricate CNT/DNA hybrids. For example, using pyrene methylammonium compound as a chemical linker, 2/3 of CNTs were anchored with DNA strands by means of electrostatic interactions between the ammonium moieties of linkers and the phosphate groups of DNA backbones [Xin et al., 2003]. Similarly, Taft et al. [Taft et al., 2004] attached pyrene-modified oligonucleotides onto the CNT sidewalls though hydrophobic interactions. To visualize the immobilized DNA strands, complementary sequences were thiolated and attached to Au NPs, which offers a direct visualization strategy to analyze CNT/DNA conjugates by scanning electron microscopy.

For CNT/DNA hybrid nanostructures, the combination of DNA-based biomolecular recognition principles and outstanding electronic properties of CNTs make them very ideal for the construction of electrochemical sensors, biosensors and electronic devices. Also, the DNA functionalization of CNTs holds interesting prospects in various fields including solubilization in aqueous media, nucleic acid sensing, gene-therapy and controlled deposition on conducting or semiconducting substrates. The advances in the relavant field have been well reviewed in the recent publications [Daniel et al., 2007; Jacobs et al., 2010; Zhang et al., 2010].

3.2.3 Dendritic polymers

Although the conventional linear polymers including polyelectrolytes and block copolymers have been extensively used to non-covalently functionalize CNTs, there exist two main shortcomings derived from the intrinsic properties of such employed polymers: 1) the amphiphilic block copolymers tend to form free micelles which are hardly removed from the CNT/polymer suspensions; 2) the polyelectrolytes are very sensitive to the surrounding media, which hinders the extensive manipulation of CNT/polymer hybrids in varied conditions. Moreover, targeted to assembly of varied functional components onto CNT surfaces, most of the reported CNT-surface-modifiers generally lack of multifunctionalities and thus those non-covalently modified CNTs were only efficient to anchor either certain metal NPs or inorganic oxides. Therefore, more general type of CNT modifiers is highly desired to fabricate various types of stable and versatile CNT-based nanohybrids.

In our group, we recently synthesized pyrenyl moieties decorated hyperbranched polyglycidol (pHBP) and employed them as a novel CNT-surface-modifier *via* non-covalent processes [Li et al., 2010]. The pHBP macromolecule consists of dendritic units, linear polyether segments and numerous terminal hydroxyl and pyrene groups, which provides a three-dimensional dendritic globular architecture (Fig. 4). In comparison with the reported modifiers, the unique molecular structure of pHBP molecules makes them particularly suitable for fabricating versatile CNT-based nanohybrids by means of non-covalent techniques owing to the following reasons: 1) multiple pyrene moieties bearing on the periphery of pHBP can be tightly attached onto the CNT surfaces through π-π stacking interactions; 2) the uniform molecular composition of pHBP allows free pHBP to be easily removed from the CNT/pHBP sols using pure solvent; 3) the dendritic polyether structure of pHBP provides void-containing electron-negative environment, which is well-suitable to attract metal ions to be *in-situ* reduced to form and accommodate metal NPs; 4) the hydroxyl

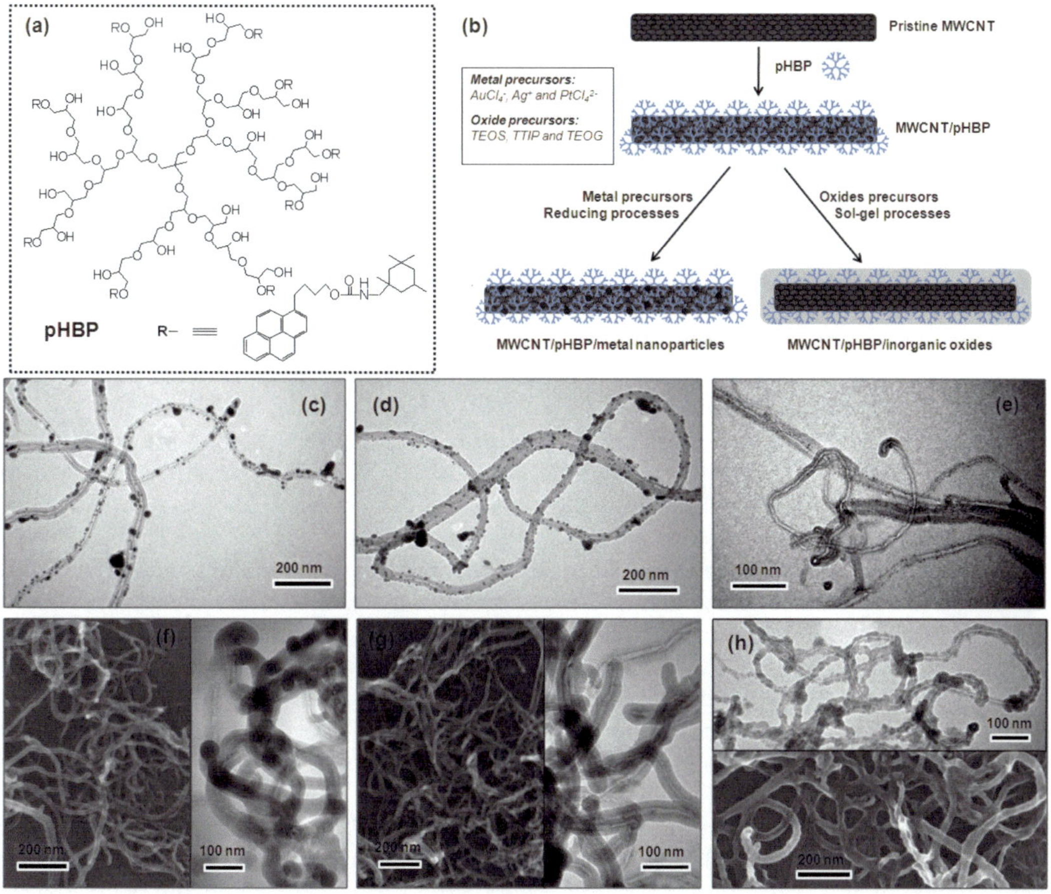

Fig. 4. Schematic illustration of molecular structure of pHBP (a) and the preparation of varied CNT/metal NPs and CNT/oxide hybrid nanostructures based on the non-covalently modified CNTs with pHBP (b); TEM images of CNT/pHBP/Au NPs (c), CNT/pHBP/Ag NPs (d), and CNT/pHBP/Pt NPs (e) hybrid nanostructures; SEM and TEM images of CNT/pHBP/SiO_2 (f), CNT/pHBP/GeO_2 (g), and CNT/pHBP/TiO_2 (h) nanofibers. Reprinted with permission from Ref [Li et al., 2010]. Copyright 2010 Wiley-VCH.

groups bearing on the periphery of pHBP facilitate the *in-situ* nucleation and growth of inorganic oxides via a sol-gel process, resulting in the formation of CNT/pHBP/inorganic oxides nanohybrids. This is the first report on the non-covalent functionalization of CNTs using three-dimensionally spherical macromolecules towards a general synthetic strategy for the efficient fabrication of CNT/pHBP/metal NPs and CNT/pHBP/inorganic oxide nanofibers. We also found that both the CNT/pHBP/Pt NPs and the CNT/pHBP/Au NPs hybrids showed excellent catalytic activity towards the reduction of 4-nitrophenol. In addition, rhodamine 6G was successfully incorporated into the CNT/pHBP/SiO_2 matrix, resulting in fluorescent nanohybrids. It is believed that these as-synthesized CNTs-based heterogeneous nanohybrids are promising for a wide range of applications in catalytic, energetic and bioengineering filed.

4. Conclusions

This chapter provides a comprehensive description for the fabrication of various CNT-based hybrid nanostructures mediated by a wide range of non-covalent CNT-surface-modifiers including small molecular surfactants, functional polymers and biomacromolecules. In such fabrication processes, the choices of non-covalent CNT-surface-modifiers not only enable the CNTs to be homogeneously suspended in various media, but also provide a wealth of opportunities to assembly additional functional components on CNT sidewalls to generate versatile hybrid nanostructures.

Although the great advances in the non-covalent assembly of functional components onto the CNT sidewalls have been achieved, several critical issues still remain to be suitably addressed. One of the most challenging topics is how to guarantee a mechanical stability of CNT-based hybrids obtained via non-covalent routes. In comparison with the covalent bonding, non-covalent attraction is less sufficient to tightly tether functional components onto CNT sidewalls. In this direction, it is a promising pathway to design novel CNT-surface-modifiers which either can interact with CNTs via relatively strong attractions and/or provide multiple interaction sites with CNT sidewalls. The second challenge involves the effectively non-covalent de-bundling and dispersion of SWCNTs in varied media. Although SWCNTs possess more excellent properties compared with MWCNTs, they tend to aggregate into bundles and ropes due to the strong van der Waals interactions between each other derived from their high polarizability and smooth surface. So far, substantial efforts have been done to overcome that shortcoming. Unfortunately, relatively less satisfied outcomes involving the obtainment of individual SWCNTs have been achieved, which sets a main barrier to produce more SWCNT-based hybrids targeted extensive applications. In this case, the design and synthesis of new type of CNT-surface-modifiers are highly desired.

In addition, a successful application of CNT-based hybrid nanomaterials and their implementation into the market requires a strong improvement in methodology to ensure reproducibility and better understanding the structure-property relationship. Considering the possible health and safety issues aroused by the use of nanomaterials in biological and medical field, it offers a promising resolution to develop complete green protocols to fabricate such hybrid nanomaterials. In the meantime, it is equally important to address the biocompatibility of the hybrid materials by the further detailed studies on their toxicology and exposure.

5. Acknowledgment

This work was supported by grants-in-aid for the *World Class University Program* (No. R32-2008-000-10174-0) and the *National Core Research Center Program* from MEST (No. R15-2006-022-01001-0), and the *Brain Korea 21 program* (BK-21).

6. References

Capek, I. (2009). Dispersion, Novel Nanomaterial Sensors and Nanoconjugates Based on Carbon Nanotubes. *Advances in Colloidal and Interface Science*, 150, 63-89.

Eder, D. (2010). Carbon Nanotube-Inorganic Hybrids. *Chemical Reviews*, 110, 1348-1385.

Niyogi, S.; Hamon, M. A.; Hu, H.; Zhao, B.; Bhowmik, P.; Sen, R.; Itkis, M. E. & Haddon R. C. (2002). Chemistry of Single-Walled Carbon Nanotubes. *Accounts of Chemical Research*, 35, 1105-1113.

Hirsch, A. (2002). Functionalization of Single-Wall Carbon Nanotubes. *Angewandte Chemie International Edition*, 14, 1853-1859.

Tessonnier, J.-P.; Villa, A.; Majoulet, O.; Su, D. S. & Schlögl R. (2009). Defect-Mediated Functionalization of Carbon Nanotubes as a Route to Design Single-Site Basic Heterogeneous Catalysts for Biomass Conversion. *Angewandte Chemie International Edition*, 48, 6543-6546.

Banerjee, S. & Wong, S. S. (2002). Rational Sidewall Functionalization and Purification of Single-Walled Carbon Nanotubes by Solution-Phase Ozonolysis. *The Journal of Physical Chemistry B*, 106, 12144-12151.

Pekker, S.; Salvetat, J. P.; Jakab, E.; Bpnard, J. M. & Forro, L. (2001). Hydrogenation of Carbon Nanotubes and Graphite in Liquid Ammonia. *The Journal of Physical Chemistry B*, 105, 7938-7943.

Peng, H.; Reverdy, P.; Khabashesku, V. N. & Margrave, J. L. (2003). Sidewall Functionalization of Single-Walled Carbon Nanotubes with Organic Peroxides. *Chemical Communications*, 362-363.

Li, H.; Ha, C.-S. & Kim I. (2009). Fabrication of Carbon Nanotube/SiO_2 and Carbon Nanotube/SiO_2/Ag Nanoparticles Hybrids by Using Plasma Treatment. *Nanoscale Research Letters*, 4, 1384-1388.

Chen, J.; Hamon, M. A.; Hu, H.; Chen, Y.; Rao, A. M.; Eklund, P. C. & Huddon, R. C. (1998). Solution Properties of Single-Walled Carbon Nanotubes. *Science*, 282, 95-98.

Banerjee, S.; Hemraj-Benny, T. & Wong, S. S. (2005). Covalent Surface Chemistry of Single-Walled Carbon Nanotubes. *Advanced Materials*, 17, 17-29.

Salavagione, H. J.; Martínez, G. & Ballesteros, C. (2010). Functionalization of Multi-Walled Carbon Nanotubes by Steroselective Nucleophilic Substitution on PVC. *Macromolecules*, 43, 9754-9760.

Pei, X.; Hao, J. & Liu, W. (2007). Preparation and Characterization of Carbon Nanotubes-Polymer/Ag Hybrid Nanocomposites via Surface RAFT Polymerization. *The Journal of Physical Chemistry C*, 111, 2947-2952.

Tasis, D.; Tagmatarchis, N.; Bianco, A. & Prato, M. (2006). Chemistry of Carbon Nanotubes. Chemical Reviews, 106, 1105-1136.

Bottini, M.; Tautz, L.; Huynh, H.; Monosov, E.; Bottini, N.; Dawson, M. I.; Bellucci, S. & Mustelin, T. (2005). Covalent Decoration of Multi-Walled Carbon Nanotubes with Silica Nanoparticles. *Chemical Communications*, 758-760.

Zhang, J.-X.; Zheng, Y.-P.; Lan, L.; Mo, S.; Yu, P.-Y.; Shi, W. & Wang, R.-M. (2009). Direct Synthesis of Solvent-free Multiwall Carbon Nanotubes/Silica Nonionic Nanofluid Hybrid Material. *ACSNANO*, 3, 2185-2190.

Gu, Y.; Hou, X.; Hu, H.; Yu(F), H.; Wang, L. & Zhou, F. (2009). Controlled Loading of Gold Nanoparticles on Carbon Nanotubes by Regenerative Exchange. *Materials Chemistry and Physics*, 116, 284-288.

Han, L.; Wu, W.; Kirk, F. L.; Luo, J.; Maye, M. M.; Kariuki, N. N.; Lin, Y.; Wang, C. & Zhong, C.-J. (2004). A Direct Route toward Assembly of Nanoparticle-Carbon Nanotube Composite Materials. *Langmuir*, 20, 6019-6025.

Richard, C.; Balavoine, F.; Schultz, P.; Ebbesen, T. W. & Mioskowski, C. (2003). Supramolecular Self-Assembly of Lipid Derivatives on Carbon Nanotubes. *Science*, 300, 775-778.

O'Connell, M. J.; Bachilo, S. M.; Huffman, C. B.; Moore, V. C.; Strano, M. S.; Haroz, E. H.; Rialon, K. L.; Boul, P. J.; Noon, W. H.; Kittrell, C.; Ma, J.; Hauge, R. H.; Weisman, R. B. & Smalley, R. E. (2002). Band Gap Fluorescence from Individual Single-Walled Carbon Nanotubes. *Science*, 297, 593-596.

Moore, V. C.; Strano, M. S.; Haroz, E. H.; Hauge, R. H. & Smalley, R. E. (2003). Individually Suspended Single-Walled Carbon Nanotubes in Various Surfactants. *Nano Letters, 3*, 1379–1382.

Wenseleers, W.; Vlasov, I. I.; Goovaerts, E.; Obraztsova, E. D.; Lobach, A. S. & Bouwen, A. (2004). Efficient Isolation and Solubilization of Pristine Single-Walled Nanotubes in Bile Salt Micelles. *Advanced Functional Materials*, 14, 1105-1112.

Islam, M. F.; Rojas, E.; Bergey, D. M.; Johnson, A. T. & Yodh, A. G. (2003). High Weight Fraction Surfactant Solubilization of Single-Wall Carbon Nanotubes in Water. *Nano Letters, 3*, 269–273.

Zhang, M.; Su, L. & Mao, L. (2006). Surfactant Functionalization of Carbon Nanotubes (CNTs) for Layer-by-Layer Assembling of CNT Multi-Layer Films and Fabrication of Gold Nanoparticle/CNT Nanohybrid. *Carbon*, 44, 276-283.

Lee, C.-L.; Ju, Y.-C.; Chou, P.-T.; Huang, Y.-C.; Kuo, L.-C. & Oung, J.-C. (2005). Preparation of Pt Nanoparticles on Carbon Nanotubes and Graphite Nanofibers via Self-Regulated Rreduction of Surfactants and Their Application as Electrochemical Catalyst. *Electrochemistry Communications*, 7, 453-458.

Whitsitt, E. A. & Barron, A. R. (2003). Silica Coated Single Walled Carbon Nanotubes. *Nano Letters, 3*, 775-778.

Zamora-Ledezma, C.; Añez, L.; Primera, J.; Silva, P.; Etienne-Calas, S. & Anglaret, E. (2008). Photoluminescent Single Wall Carbon Nanotube-Silica Composite Gels. *Carbon*, 46, 1253-1255.

Yang, D.; Hu, J. & Wang, C. (2006). Synthesis and Characterization of pH-Responsive Single-Walled Carbon Nanotubes with A Large Number of Carboxy Groups. *Carbon*, 44, 3161-3167.

Eder, D. & Windle, A. H. (2008). Carbon-Inorganic Hybrid Materials: The Carbon-Nanotube/TiO_2 Interface. Advanced Materials, 20, 1787-1793.

Eder, D. & Windle, A. H. (2008). Morphology Control of CNT-TiO_2 Hybrid Materials and Rutile Nanotubes. Journal of Materials Chemistry, 18, 2036-2043.

Bogani, L.; Danieli, C.; Biavardi, E.; Bendiab, N.; Barra, A.-L.; Dalcanale, E.; Wernsdorfer, W. & Cornia, A. (2009). Single-Molecule-Magnet Carbon-Nanotube Hybrids. *Angewandte Chemie International Edition*, 48, 746-750.

Li, X.; Liu, Y.; Fu, L.; Cao, L. Wei, D. & Wang Y. (2006). Efficient Synthesis of Carbon Nanotube-Nanoparticle Hybrids. *Advanced Functional Materials*, 16, 2431-2437.

Chen, R. J.; Zhang, Y.; Wang, D. & Dai, H. (2001). Noncovalent Sidewall Functionalization of Single-Walled Carbon Nanotubes for Protein Immobilization. *Journal of the American Chemical Society*, 123, 3838-3839.

Wang, X.; Liu, Y.; Qiu, W. & Zhu, D. (2002). Immobilization of Tetra-*tert*-Butylphthalocyanines on Carbon Nanotubes: A First Step towards The Development of New Nanomaterials. *Journal of Materials Chemistry*, 12, 1636-1639.

Bartelmess, J.; Ballesteros, B.; Torre, G.; Kiessling, D.; Campidelli, S.; Prato, M.; Torres, T. & Guldi, D. M. (2010). Phthalocyanine-Pyrene Conjugates: A Powerful Approach toward Carbon Nanotube Solar Cells. *Journal of the American Chemical Society*, 132, 16202-16211.

Assali, M.; Leal, M. P.; Fernández, I.; Romero-Gomez, P.; Baati, R. & Khiar, N. (2010). Improved Non-Covalent Biofunctionalization of Multi-Walled Carbon Nanotubes Using Carbohydrate Amphiphiles with A Butterfly-Like Polyaromatic Tail. *Nano Research*, 3, 764-778.

Eder, D. (2010). Carbon Nanotube-Inorganic Hybrids. *Chemical Reviews*, 110, 1348-1385.

Ou, Y.-Y. & Huang, M. H. (2006). High-Density Assembly of Gold Nanoparticles on Multiwalled Carbon Nanotubes Using 1-Pyrenemethylamine as Interlinker. *The Journal of Physical Chemistry B*, 110, 2031-2036.

Bourlinos, A. B.; Georgakilas, V.; Zboril, R. & Dallas, P. (2007). Preparation of A Water-Dispersible Carbon Nanotube-Silica Hybrid. *Carbon*, 45, 2136-2139.

Wang, T.; Hu, X.; Qu, X. & Dong, S. (2006). Noncovalent Functionalization of Multiwalled Carbon Nanotubes: Application in Hybrid Nanostructures. *The Journal of Physical Chemistry B*, 110, 6631-6636.

Carrillo, A.; Swartz, J. A.; Gamba, J. M.; Kane, R. S.; Chakrapani, N.; Wei, B. & Ajayan, P. M. (2003). Noncovalent Functionalization of Graphite and Carbon Nanotubes with Polymer Multilayers and Gold Nanoparticles. *Nano Letters*, 3, 1437-1440.

Correa-Duarte, M. A.; Sobal, N.; Liz-Marzán, L. M. & Giersig, M. (2004). Linear Assemblies of Silica-Coated Gold Nanoparticles Using Carbon Nanotubes as Templates. *Advanced Materials*, 16, 2179-2184.

Mountrichas, G.; Pispas, S. & Tagmatarchis, N. (2007). Aqueous Carbon-Nanotube-Amphiphilic-Block-Copolymer Nanoensembles: Towards Realization of Charge-Transfer Processes with Semiconductor Quantum Dots. *Small*, 3, 404-407.

Moore, V. C.; Strano, M. S.; Haroz, E. H.; Hauge, R. H. & Smalley, R. E. (2003). Individually Suspended Single-Walled Carbon Nanotubes in Various Surfactants. *Nano Letters*, 3, 1379-1382.

Shvartzman-Cohen, R.; Nativ-Roth, E.; Baskaran, E.; Levi-Kalisman, Y.; Szleifer, I. & Yerushalmi-Rozen, R. (2004). Selective Dispersion of Single-Walled Carbon Nanotubes in the Presence of Polymers: the Role of Molecular and Colloidal Length Scales. *Journal of the American Chemical Society*, 126, 14850-14857.

Shvartzman-Cohen, R.; Levi-Kalisman, Y.; Nativ-Roth, E. & Yerushalmi-Rozen, R. (2004). Generic Approach for Dispersing Single-Walled Carbon Nanotubes: the Strength of A Weak Interaction. *Langmuir*, 20, 6085-6088.

Zou J.; Chen, H.; Chunder, A.; Yu, Y.; Huo, Q. & Zhai, L. (2008). Preparation of A Superhydrophobic and Conductive Nanocomposite Coating from A Carbon-Nanotube-Conjugated Block Copolymer Dispersion. *Advanced Materials*, 20, 3337-3341.

Zou J.; Liu, L.; Chen, H.; Khondaker, S. I.; McCullough, R. D.; Huo, Q. & Zhai, L. (2008). Dispersion of Pristine Carbon Nanotubes Using Conjugated Block Copolymers. *Advanced Materials*, 20, 2055-2060.

Zou, J.; Khondaker, S. I.; Huo, Q. & Zhai, L. (2009). A General Strategy to Disperse and Functionalize Carbon Nanotubes Using Conjugated Block Copolymers. *Advanced Functional Materials*, 19, 479-483.

Lee, J. U.; Huh, J.; Kim, K. H.; Park, C. & Jo, W. H. (2007). Aqueous Suspension of Carbon Nanotubes via Non-Covalent Funcitonalization with Oligothiophene-Terminated Poly(Ethylene Glycol). *Carbon*, 45, 1051-1057.

Kim, K. T. & Jo, W. H. (2010). Noncovalent Functionalization of Multiwalled Carbon Nanotubes Using Graft Copolymer with Naphthalene and Its Application as A Reinforcing Filler for Poly(Styrene-*co*-Acrylonitrile). *Journal of Polymer Science Part A: Polymer Chemistry*, 48, 4184-4191.

O'Connell, M. J.; Boul, P.; Ericson, L. M.; Huffman, C.; Wang, Y.; Haroz, E.; Kuper, C.; Tour, J.; Ausman, K. D. & Smalley, R. E. (2001). Reversible Water-Solubilization of Single-Walled Carbon Nanotubes by Polymer Wrapping. *Chemical Physics Letters*, 342, 265-271.

Welsher, K.; Liu, Z.; Sherlock, S. P.; Robinson, J. T.; Chen, Z.; Daranciang, D. & Dai, H. (2009). A Route to Brightly Fluorescent Carbon Nanotubes for Near-Infrared Imaging in Mice. *Nature Nanotechnology*, 4, 773-780.

Zheng, M.; Jagota, A.; Strano, M. S.; Santos, A. P.; Barone, P.; Chou, G. S.; Diner, B. A.; Dresselhaus, M. S.; Mclean, R. S.; Onoa, G. B.; Samsonidze, G. G.; Semke, E. D.; Usrey, M. & Walls, D. J. (2003). Structure-Based Carbon Nanotube Sorting by Sequence-Dependent DNA Assembly. Science, 302, 1545-1548.

Tu, X.; Manohar, S.; Jagota, A. & Zheng, M. (2009). DNA Sequence Motifs for Structure-Specific Recognition and Separation of Carbon Nanotubes. *Nature*, 460, 250-253.

Xin, H. & Woolley, A. T. (2003). DNA-Templated Nanotube Localization. *Journal of the American Chemical Society*, 125, 8710-8711.

Taft, B. J.; Lazareck, A. D.; Withey, G. D.; Yin, A.; Xu, J. M. & Kelley, S. O. (2004). Site-Specific Assembly of DNA and Appended Cargo on Arrayed Carbon Nanotubes. *Journal of the American Chemical Society*, 126, 12750-12751.

Daniel, S.; Rao, T. P.; Rao, K. S.; Rani, S. U.; Naidu, G. R. K.; Lee, H.-Y. & Kawai, T. (2007). A Review of DNA Functionalized/Grafted Carbon Nanotubes and Their Characterization. *Sensors and Actuators B*, 122, 672-682.

Jacobs, C. B.; Peairs, M. J. & Venton, B. J. (2010). Review: Carbon Nanotubes Based Electrochemical Sensors for Biomolecules. *Analytica Chimica Acta*, 662, 105-127.

Zhang, Q. D.; Piro, B.; Noël, V.; Reisberg, S. & Pham, M.-C. (2010). Applications of Carbon Nanotubes to Electrochemical DNA Sensors: A New Strategy to Make Direct and Selective Hybridizaion Detection from SWNTs. *Advances in Natural Sciences: Nanoscience and Nanotechnology*, 1, 045011 (8pp).

Li, H.; Jo, J. K.; Zhang L.; Ha, C.-S.; Suh, H. & Kim, I. (2010). A General and Efficient Route to Fabricate Carbon Nanotube-Metal Nanoparticles and Carbon Nanotube-Inorganic Oxides Hybrids. *Advanced Functional Materials*, 20, 3864-3873.

14

Novel Carbon Nanotubes-Based Hybrid Composites for Sensing Applications

Nicola Donato, Mariangela Latino and Giovanni Neri
Dept. of Matter Physics and Electronic Engineering
University of Messina
Dept. of Chemical Science and Technologies
University of Rome Tor Vergata
Dept. of Industrial Chemistry and Materials Engineering
University of Messina
Italy

1. Introduction

In this chapter book is reported about the development and applications of carbon nanotube (CNT)-based hybrid composites material for gas sensing devices. Gas sensors are employed in many applications spanning from security, environmental and pollution monitoring, healthcare, indoor and outdoor fields. The great number of practical applications is due to their low cost, small dimensions, easiness of use and attitude for being organized in arrays. Metal oxides and conductive polymers are the conventional sensing elements in use today. Hybrid composites are materials offering new advantages and are promising candidates for the development of high performance sensor devices. Further, with the advent of new synthesis methods at the nanoscale, this let chemical gas sensors to improve their properties and figures of merit. Indeed, owing the small size and large surface to bulk area ratio of the grain nanoparticles, this results in an increased sensitivity. However, dealing with the fabrication of sensors with nanostructured materials some problems must be taken into account, such as the synthesis and deposition methods of the sensing material (finalized to optimize grain size, porosity, film thickness, etc.) and the optimization of the sensor parameters (operating temperature, sensitivity, response and recovery time).
In the chapter book there will be reported the main transduction phenomena involved on the working conditions of resistive sensor devices based on hybrid composites. Then the authors will focus the attention on two composite typologies: inorganic/CNT and organic/CNT composite materials.
The first topic, about inorganic-carbon nanotube composites, will deal about the development of sensing materials based on metal oxide/CNTs composites. It will be shown as it is possible to enhance gas sensing properties towards specific gas targets using CNTs as conductive media to help to transduce any adsorption/chemical reaction on the semiconducting layer into an electrical response, i.e. by means of resistive sensors. The case that will be reported, is about the development of resistive devices obtained by employing a Pt/TiO_2/CNT composite as sensing layer for monitoring high hydrogen concentration in inert atmosphere at near room temperature.

The second topic is about the development of novel organic/CNTs semiconductor composites for resistive gas sensors. Organic materials based on π-conjugated molecules are intensively studied nowadays in the field of organic semiconductors as a complement to the shortcomings of inorganic semiconductors. Organic semiconductors offer, with respect to current inorganic-based technology, greater substrate compatibility, device processability, flexibility, large area coverage, and reduced costs. However, in many cases, the reduced conductivity of these materials hindered their use in resistive sensors. Carbon nanotubes provide in this case an easy method to modulate their electrical transport properties. A diaza-perylene/CNT composite will be reported as an example in the monitoring of vapour of protic solvent (e.g. water, acetic acid).

2. Main transduction and sensing mechanisms for resistive gas sensors

In the last years, the ask for real-time, compact and low cost chemical gas sensors has been increased, due to their employment in many fields of science and technology. The applications span from homeland and work-place security, antiterrorism and defence scenarios, to automotive and biomedical employments. For instance, homeland security and defence applications need effective portal monitoring, chemical weapons sensing, and water quality testing. Devices that can serve as personal exposure monitors, provide advance warning of food spoilage, and enable breath analysers to uncover pre-symptomatic disease are also under development and diffusion in the market. The increasing demand for small scale solid-state sensors for automotive exhaust gas mixtures are also of great interest due to the improvements of the emission control legislations. The turning point for these devices is to satisfy the requirements of high levels of sensitivity and specificity in small, economical packages. These requirements are mainly fulfilled by resistive solid state sensors, making these devices the most promising in the market for gas and volatile organic compounds (VOCs) detection.

In the planar configuration, they are composed of a porous thick/thin film as a gas sensing element, deposited onto a ceramic substrate with interdigitated electrodes (Fig. 1).Resistive solid state sensors performances have been improved by optimizing the sensing layers properties, for example by using nanostructured materials (G. Neri, 2010). It is well known that the sensing characteristics of a chemoresistive sensor depend primarily on the nature of the sensing layer. In this respect, most of conventional sensors used films based on metal oxides such as SnO_2, ZnO, or conducting polymers. In the first type of sensors, i.e. metal oxide semiconductor (MOX) sensors, power supply is generally required to heat the sensitive layer deposited on the interdigitated contacts to an operating temperature of few hundred degree Celsius e.g. from 200° C to 1000° C supplied by means of a heater. The latter operate at lower temperatures, are highly sensitive and selective, but their life time is limited. Additionally, some organic materials are not compatible with micro-electronic fabrication technologies and, therefore, not suitable for large scale production. The most important aspect to focus is that gas sensors need to operate at elevated temperatures to avoid selectivity problems. The most suitable sensing material is that providing the right balance between sensor performance and power consumption (lower operating temperatures).

The employment of carbon nanotubes, as intrinsic sensing materials or as composite materials, where a metal oxide or polymer is deposited on carbon nanotubes (Y. P. Sun et al., 2002), provide an easy diffusion for chemical gas accessing through over the bulk material. Indeed, the use of CNTs can bring some advantages such as introducing identical open gas nano-channels through bulk material, achievement of a great surface to volume ratio, and providing good gas-adsorption sites due to inside and outside of metal oxide/CNTs composites.

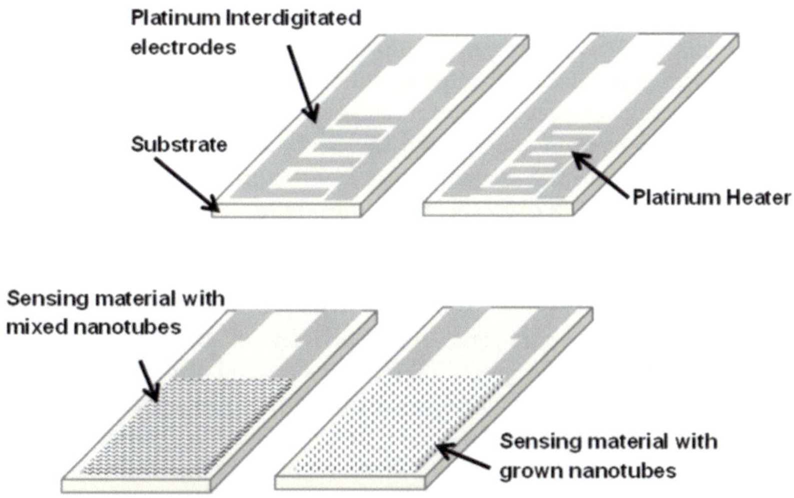

Fig. 1. Planar structure of chemoresistive sensor with top side coated thick/thin film and heater on the backside.

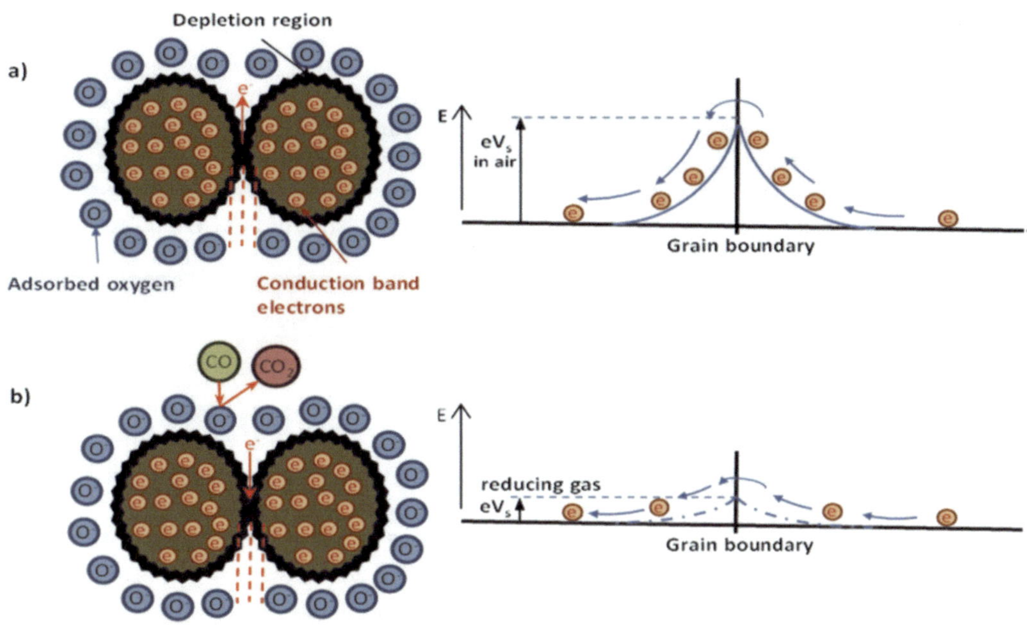

Fig. 2. Mechanism of gas sensing operating on n-type metal oxide semiconductors with reducing gases.

The transduction mechanism for gas detection on resistive sensors is based on the variation of the electrical resistance (conductance) due to target gas adsorption (Y. Shimizu and M. Egashira, 1999). By measuring the resistance variation it is possible to determine the presence and the concentration of the gas in the ambient surrounding the sensor. To better understand the operating sensing mechanism, the responses of resistive gas sensors have been widely explored with respect to surface adsorption, chemical reaction, and resulting conductivity changes. Several sensing mechanisms are possible, on the basis of the different sensing material. With conventional n-type metal oxide semiconductors-based sensors (e.g. ZnO, SnO_2) the response to the presence of a target gas relies on the surface reactions which occur between adsorbed oxygen species and the probed gas (A. Gurlo et al, 2006). In air, point a), oxygen adsorbed on the surface traps free electrons because of its high electron affinity, forming a potential barrier at the grain boundaries, determine the electrical resistance value. When the sensor is exposed to an atmosphere containing reducing gases, e.g. hydrocarbons, CO, ethanol, etc., point b), the gas molecules adsorbs on the surface and reacts with active oxygen species. Reactions with surface oxygen species will vary depending upon the temperature and the reactivity of the sensing material. These reactions/interactions decrease the potential barrier allowing electrons to flow more easily, thereby reducing the electrical resistance, as a function of the concentration of the target gas.By employing polymer films as sensing layer, weak intermolecular interactions, rather than redox chemistry, can be envisaged as the main pathway of the sensing mechanism. In some cases these interactions require specialized functions of the molecules, making these chemoresistive sensors more selective than those based on metal oxide semiconductors.
The development of nanostructured sensing materials can be considered the right trade-off between the traditional approach and the new one, devoted to the development of new sensing materials. In such scenario carbon nanotubes can be employed by mixing them with polymer host, or by functionalizing them to specific gas targets by means of coating them with nanostructured metal oxides or directly grown on the transduction substrates.
This variety of sensing materials and of transduction mechanisms, contributes to develop chemoresistive sensors with enhanced performance tuned towards specific applications. Some examples from our laboratory are here reported, highlighting how carbon nanotubes can be used to modulate the conduction of the sensing layer and/or to enhance sensing properties toward specific gas target or volatile organic compounds of chemoresistive gas sensors.

3. Role of carbon nanotubes in hybrid sensing materials

In general, nanostructured materials offer a huge number of possibilities to enhance the property of sensing of the developed devices. Carbon nanotubes, for example, are suitable candidates to serve as a constituent in functional materials as they provide properties such as a well defined structure, high chemical stability, high surface area and a good thermal conductivity.
Since the discovery of CNTs in 1991 by Iijima (S. Iijima, 1991), a great deal of effort has been devoted not only to study the chemistry and fundamental electronic and physical properties but also the application and integration of this new class of materials into electronic devices. They are viewed as single or multiple sheets (from 2 to 50) of graphene to build single-walled carbon nanotubes (SWCNTs),with a diameter in the range of 0.4–3 nm, or coaxial multiwalled carbon nanotube (MWCNTs) with an interlayer spacing of 0.34 nm and

diameter in the range from 4 to 30 nm. The length of nanotubes is in the range of several hundred micrometers to millimeters.

Here we focused our attention only on the MWCNTs and, otherwise not specified, the discussion is referred to this nanotube typology. From an electrical point of view, MWCNTs can be divided in two broad classes - metals and semiconductors. In metallic conductors, electric current generally flows freely and there is no energy gap between the valence and the conducting states. In a semiconductor, such an energy gap exists and therefore a higher voltage is needed to make electric current flow. Whether a carbon nanotube is a metal or semiconductor depends upon their diameter and chirality.

Semiconducting carbon nanotubes have been proposed for applications as chemical and biochemical sensors (Kong et al, 2000; Star et al, 2003; Kauffman et al, 2008). CNTs-based gas sensors have received considerable attention because of their outstanding properties towards a wide variety of gases that may be detected, such as high sensitivity and especially lower operating temperature, compared with the other types of gas sensors (Pengfei et al., 2003; Valentini et al., 2003).

To date CNTs have shown sensitivity towards such gases as NH_3, NO_2, H_2, CH_4, CO, SO_2, H_2S, and O_2. CNTs-based gas sensing utilizes a change in an electrical property due to adsorption of gas molecules as the output signal. The response is attributed to the p-type conductivity in semiconducting MWCNTs and the electrical charge transfer is found to be the major sensing mechanism at low temperature. Target gas molecules (e.g., NH_3, NO_2, etc.) directly adsorb onto the CNT surface inducing electron transfer and changing the electrical conductivity of the nanostructure. Figure 3 depict in a schematic way as the conductance of a carbon nanotube is modified when electron donor (NH_3) or acceptor (NO_2) gaseous molecules are in the atmosphere surrounding the sensor.

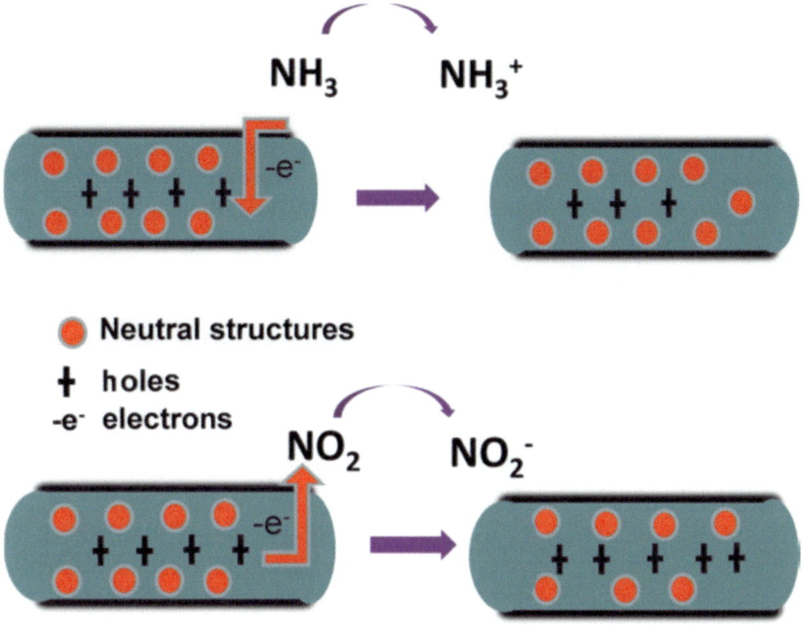

Fig. 3. Mechanism of gas sensing operating on p-type CNT semiconductors with reducing/oxidizing type gases.

NH_3 transfers electrons to the underlying nanotube structure. As MWCNTs behave as a p-type semiconductor, the transferred electrons will recombine with 'hole' carries, thereby decreasing the charge carrier concentration in the support and a consequent increasing the electrical resistance. In the presence of NO_2, the reverse mechanism takes place and as a result of the presence of this gas the carbon nanotube electrical resistance decreases.

In order to enhance the sensitivity of resistive gas sensors, the addition of semiconducting metal oxides (SnO_2, TiO_2, V_2O_5, etc.) or polymers to carbon nanotube network (and vice versa, coating carbon nanotubes by metal oxides or polymers) has been widely exploited today (W.-D. Zhang and W.-H. Zhang, 2009).

The examples reported in the above references demonstrated that the presence of carbon nanotubes as a component in nanocomposites can help to improve the sensing properties. From a practical point of view, MWCNTs act as conductive media to transduce adsorption/chemical reaction processes into an electrical response, able to be measured and recorded with conventional or custom electronics.

The interaction between the carbon nanotube and the matrix is then essential in order to have a synergic action. This is favoured, making the carbon nanotube surface more reactive by creating functional groups on its. A functionalization process, for example by a strong oxidation in HNO_3 of the nanotubes, is then often required to produce COOH and OH groups along the sidewall and the caps of the CNTs. These functional groups can physically/chemically interact with functional groups in the matrix, resulting in a strong interfacial adhesion and a better dispersion of surface modified CNTs in the polymer or metal oxide matrix. Obviously, these groups can act as anchoring sites for deposition of film of polymer or metal oxides on the surface of the carbon nanotube.

4. Cases studies in the development of selected hybrid composite sensing materials

This section is mainly focused on the development of two sensing material composite typologies: inorganic/MWCNT and organic/MWCNT ones.

The first case, will deals about the development of sensing materials based on metal oxide/MWCNTs composites. Metal oxides are well known for their electrical semiconducting properties that make them suitable for sensing applications. However, the full potentiality of these materials is only partially exploited, owing by the difficulty inherent to high resistance values they present at low temperatures. The possibility of an efficient electrical conductance promotion when adding a conductive second phase, such as MWCNTs, opened many opportunities. It is very important further to recognize that the amount of such second phases required to get a reasonable reduction in resistivity is very smaller. Due to their high aspect ratio, quite low percolation thresholds have been reported for metal oxide/CNTs composites (E. Flahaut, et al., 2000). Conductivity values after percolation seem to depend more on the CNT type and purity as well as on the composite processing procedure.

In the case study here reported, we focused first our attention about the development of resistive devices obtained by employing a Pt-doped TiO_2/MWCNT composite as sensing layer for monitoring high hydrogen concentration in inert atmosphere at near room temperature. Hydrogen-based systems truly promise a futuristic energy scenario, where hydrogen fuel will be used in fuel cells for civil transportation and in rockets for space vehicles. All these applications necessitate then the development of hydrogen sensor devices, which allow its safe and controlled use (V. M. Aroutiounian, 2005). Numerous metal oxide thin films work at high temperature in air as effective resistive hydrogen sensors, allowing detection of this gas down to very low concentration. However, many of these devices need ambient oxygen to operate, and moreover they fail to detect hydrogen at high concentrations due to saturation effects. Then, there is also great interest to detect high concentrations of hydrogen in reducing or inert atmosphere, by means of devices operating at temperature as low as possible, preferably room temperature.

The second topic is about the development of novel organic/MWCNTs semiconductor composites for resistive gas sensors. The preparation of organic materials stable in air, solution-processable and with suitable electrical conductivity, is a fundamental step in organic electronics. To obtain this, MWCNTs can be dispersed in organic material hosts to enhance the conduction properties, offering an attractive route to introduce new electronic properties based on electronic interactions between the two composite constituents (M.S.P. Shaffer and A.H. Windle, 1999).

4.1 Hydrogen sensor based on Pt/TiO_2/MWCNT composites

TiO_2/carbon nanotube composites attracted more attention because of their potential in gas sensing. Moreover, such composites, exhibiting enhanced photocatalytic activity, because CNT could acts as an electron sensitizer and donator to accept the photo-induced electron (e^-) into the conduction band of TiO_2 particles under UV light irradiation, are promising candidate to fabricate UV-photoactivated sensors.

We prepared CNT/TiO_2 composites with different MWCNTs loading ranging from 0 to 70 wt%,by a sol-gel method. MWCNTs were first functionalised by a nitric acid treatment at 110°C for 18 h, in order to create on the surface of the nanotubes, suitable functionalized groups, facilitating the interaction with titania. The typical morphology of the resulting composite materials is reported in the SEM and TEM micrographs of Fig. 3, showing the titania particles dispersed in the mats of carbon nanotubes.

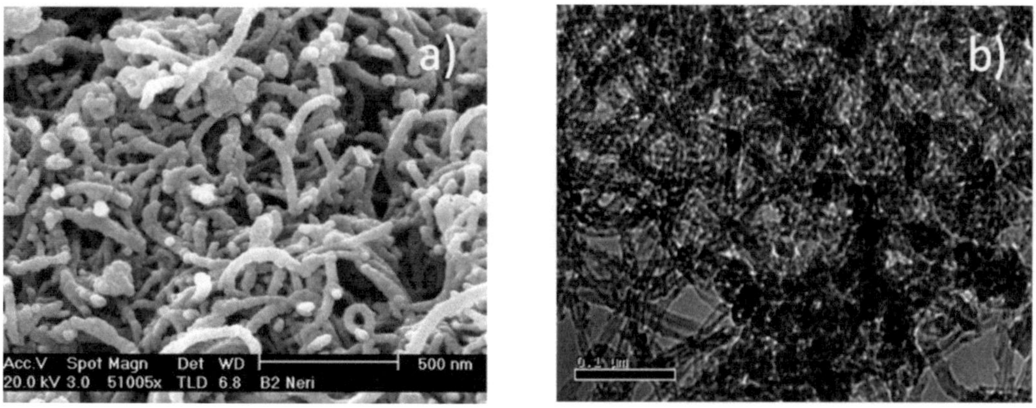

Fig. 4. a) SEM and b) TEM micrographs showing the morphology of the TiO_2/MWCNT composite.

The comparison of XRD spectra registered on all samples with titania reference-spectra showed only the presence of the diffraction peaks of crystalline anatase. Micro-Raman analysis also evidenced a marked similarity with the spectrum of anatase.

Subsequently, 2 wt% of Pt was introduced by the wetness impregnation method, by using a solution of platinum acetylacetonate in acetone. The resulting Pt/TiO_2/MWCNT composites were treated at 200°C for 2 h in a mixture of 5% H_2 in argon. XRD and micro-Raman spectra of the ternary composites do not differ from those reported for the parent TiO_2/MWCNT. Pt diffraction peaks are absent, suggesting the noble metal is in an amorphous state and/or having very small size. In agreement, small platinum nanoparticles wrapped on the composite surface were observed by transmission electron microscopy.
Sensing tests were carried out in an apparatus, interfaced with a PC, composed by a stainless steel box, where is allocated the sensor, and connected to gas supply (He and H_2) and a power supply able to set and control the operating temperature (in the range 25-100 °C), and measuring the resistance values of the sensor when it is maintained under H_2/He mixtures flow (100mL/min) by means of a resistance meter. The sensor response is given as $[(R-R_0)/R_0]\cdot 100$, where R_0 is the resistance recorded under He and R the resistance recorded under H_2/He mixture. Before carrying out the sensing measurements, sensors were pre-treated "in situ" upon different experimental conditions. This treatment allowed to obtain a better baseline stability (Fig. 5). In this regard, it can be hypothesized that during storage in air, Pt particles oxidize at least in the surface and a reductive treatment at higher temperature "in situ" is necessary to establish the optimal working steady-state conditions.

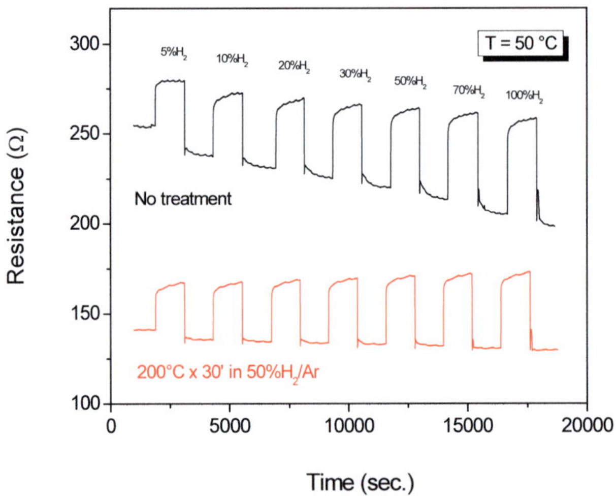

Fig. 5. Resistance variations registered for the Pt/TiO_2/MWCNT composite sensor, after different "in situ" reducing treatments.

The behaviour of the composite sensors was, as expected, strongly related to Pt presence, MWCNTs loading and working temperature. Pt was found to be essential to have a measurable response to H_2. As platinum is known to be an efficient catalyst for hydrogen dissociation, it is plausible that its role is related to the promotion of this reaction on the metal surface at low temperature. The hydrogen atoms so formed are much reactive, favouring the hydrogen sensing mechanism at near room temperature.
The best sensing performance were obtained for a MWCNTs loading of about 35 wt% (Fig. 6a). Furthermore, the comparison of the response of the Pt/TiO_2/MWCNT sensor, with that of TiO_2/MWCNT and Pt/MWCNT sensors, clearly indicated that hydrogen monitoring take place only with the three-component sensor, suggesting a synergic action between the metal

oxide, noble metal and carbon nanotubes. A linear relationships between the response and hydrogen concentration was well established for the sensor having a MWCNTs loading of about 35 wt% (Fig. 6b).

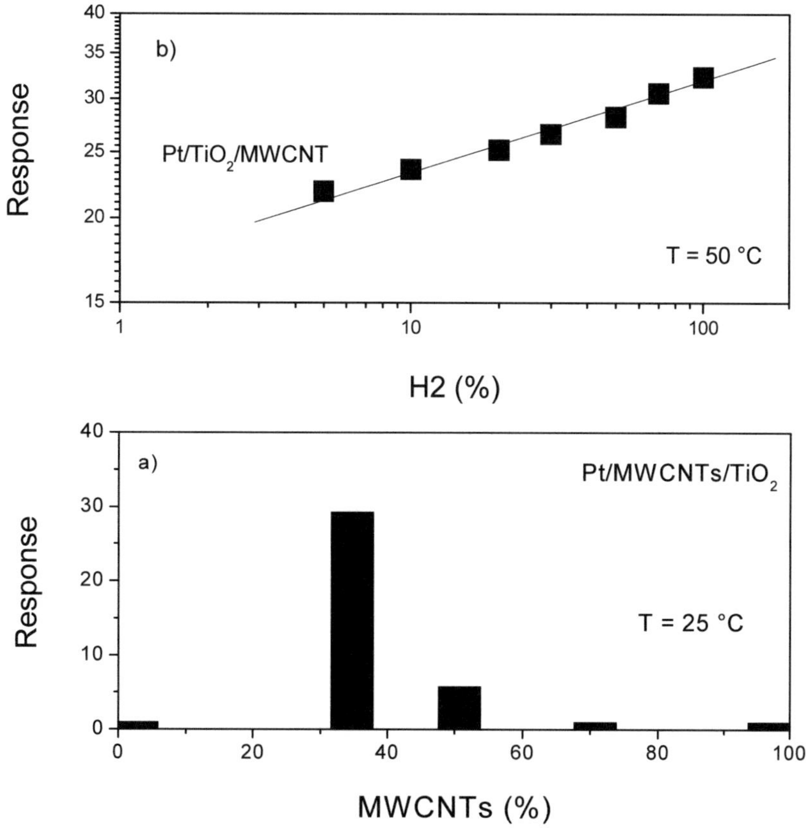

Fig. 6. a) Response to hydrogen of sensors having different carbon nanotube loading. b) Calibration curve of the $Pt/TiO_2/MWCNT$ sensor with a carbon nanotube loading of about 35 wt%.

In order to formulate an hypothesis about the sensing mechanism, it is should be remarked that the sensor measurements were conducted in absence of oxygen. Then, the usual mechanism involving the removal of chemisorbed oxygen by hydrogen cannot be invoked in this case. We think the sensor response can be attributed to a "spill-over" mechanism, in which hydrogen molecules are chemisorbed and dissociated by platinum, and finally spill out of the Pt, diffusing into the TiO_2 surface layer. This agrees with the fact that both the platinum and TiO_2 are necessary for obtain the sensor response. MWCNTs act providing a preferential pathway to the current flow and a larger specific surface area for the adsorption of hydrogen atoms.

According to current view, once the active hydrogen atom is chemically adsorbed at the interstitial positions in the oxide lattice structure, partial electron charge transfer occurs to the n-type TiO_2 and the conductance should increases (O. K. Varghese et al., 2003). Vice versa, we observed that the sensor resistance increases in hydrogen. In literature this has been explained suggesting a p-type behaviour of the metal oxide (F. Hossein-Babaei et al.,2005), or an H→TiO_2/MWCNT charge transfer which reduces the hole concentration in the carbon nanotube structure, increasing the resistivity in the system (R. A. Guirado-Lopez et al., 2007).

In order to get a deeper insight into the sensing mechanism more systematic investigations are however necessaries, aimed at clarifying the role of MWCNTs-TiO_2 interaction. The response can arises from the modulation of nanotube themselves, the junctions between platinum electrodes and the nanotubes, or between two adjacent nanotubes. For example, it has suggested that NH_3 mainly interacts with carbon nanotubes themselves In contrast, other authors suggested that the modulation of nanotube metal electrode junctions influence the response to NH_3, or that both nanotube channels and nanotube electrode junctions play a role in the detection process of NH_3 (A. Salehi-Khojin et al., 2011).

Due to the extremely practical importance, further investigations are also planned aiming to optimise the formulation of the sensing layer devoted to the development of a prototypal hydrogen sensing device.

4.2 Organic/MWCNTs composites: semiconductor composite for resistive gas sensors

In the framework of a research oriented to study new π-electron organic systems for sensing applications, we have focused the attention on the aza-helicenes (with 5 or 6 or even more ortho-condensed benzenic or pyridinic rings) and aza-perylenes (with 5 benzenic or pyridinic peripheric rings) compounds. These organic molecules appear very promising for gas sensing, due to the presence of a reactive nitrogen functionality on the pyridinic ring, which could be able to interact reversibly with specific gases. Most important, the possible existence of a charge transfer might considerably alter the electronic structure of the conduction channels in the highly conjugated molecule system, modifying as a consequence the transport properties making them very attractive for fabricating resistive chemosensors.

Recently, several authors have explained the electrical properties of composites in which CNTs are dispersed in an insulating polymer matrix based on percolation theory (Allaouia et. al, 2002). The use of CNTs in a polymer composite is found to reduce the percolation threshold to a very low filler volume than in the case of carbon black/polymer matrix.

Here, we report data obtained in the investigation of the electrical and sensing properties of 7,8-diazabenzo[ghi]perylene (DABP)/CNTs composites. The molecular structure of DABP is shown in Figure 7. DABP was synthesized through photochemical reactions carried out as reported elsewhere (Bazzini, C., et al, 2005).

DABP is a dark orange solid material with high melting point (226-234°C), and results stable in air. It shows moreover a good solubility in many organic solvents, allowing preparing processable solutions for thin films deposition. However, notwithstanding the large conjugation, the conductivity of DABP is low, and not suitable for the application in chemoresistive sensors. Therefore the organic material was mixed with a suitable amount of multi-walled carbon nanotubes, in order to enhance the conductivity of the sensing film.

The composite sensor was fabricated depositing by drop-coating from dichloroethane solution thin films of DABP/MWCNTs on an alumina substrate provided with interdigitated electrodes. The sensor was preliminary conditioned in air for 2 h at room temperature. Electrical tests were carried out at room temperature (25 °C). I/V curves were acquired by means of a Keithley 2400 source meter. The relative humidity (RH) and organic vapours were generated by using a bubbler connected to a mass flow control system, in order to vary their concentration.

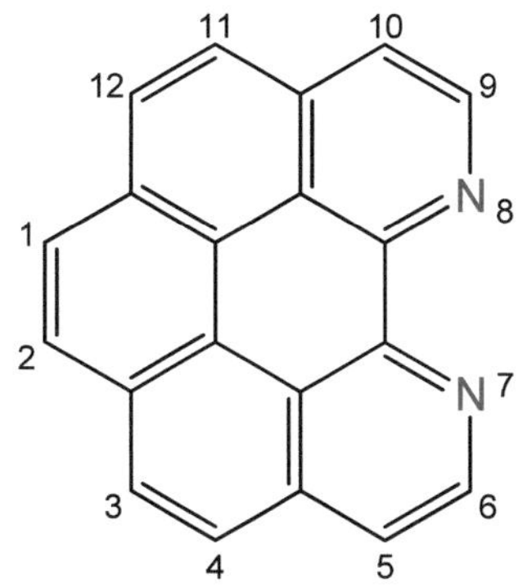

Fig. 7. Molecular structure of 7,8-diazabenzo[ghi]perylene (DABP).

Electrical properties of the pure DABP film and DABP/MWCNTs composite, were studied in order to evaluate the effect of the CNTs on the films conductivity. Figure 8 shows the output currents of a DABP/MWCNTs composite film at room temperature as a function of input bias.

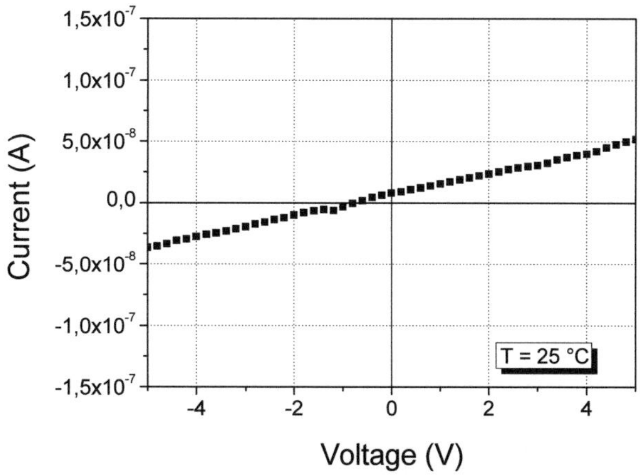

Fig. 8. I/V characteristics of a DABP/MWCNTs composite film.

While the output signal of the pure DABP film is not measurable due to its very low conductivity, the output signal is remarkably enhanced with the addition of carbon nanotubes, thus indicating the improvement in conductivity of the active layer. The dependence of conductivity on the CNT content in the composites is in agreement with previously reported results of polymer/CNT composites in which the conductivity exhibits percolation behaviour (J. N. Coleman et al., 1998).

The effect of humidity on the electrical characteristics of the DABP/MWCNTs composite has evidenced a large variation of the sensor current with humidity change, suggesting that it can be used as resistive sensor for monitoring relative humidity (Figure 9).

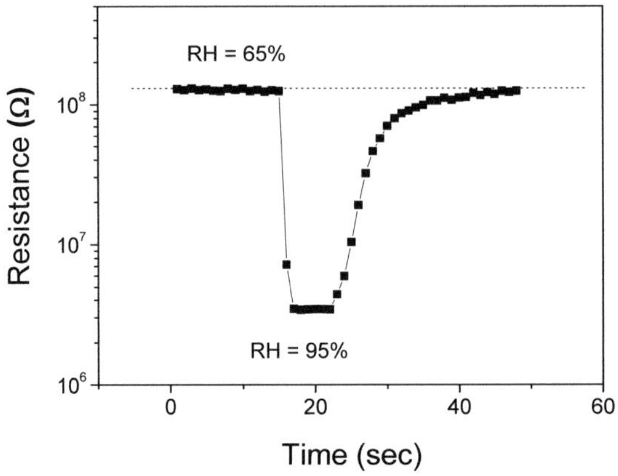

Fig. 9. Transient response of the composite sensor to different relative humidity variations.

It can be noted a strong decreases of resistance as the relative humidity value increase. Moreover, the response is well reversible, exhibiting a fast response/recovery time.

Experiments have also shown that no significant variations of current occurs on devices based only on DABP or MWCNTs, confirming that changes observed on the composite sensor are due to the synergic action between carbon nanotubes and the organic material.

According to other authors (J. Roithová et al. 2007), we explained this behaviour assuming that DABP manifests a strong affinity for protons. The electron transferring (hopping) assisted by the presence of a proton trap such as the heterocyclic nitrogen may be considered responsible of the resistance variations observed in humid ambient.

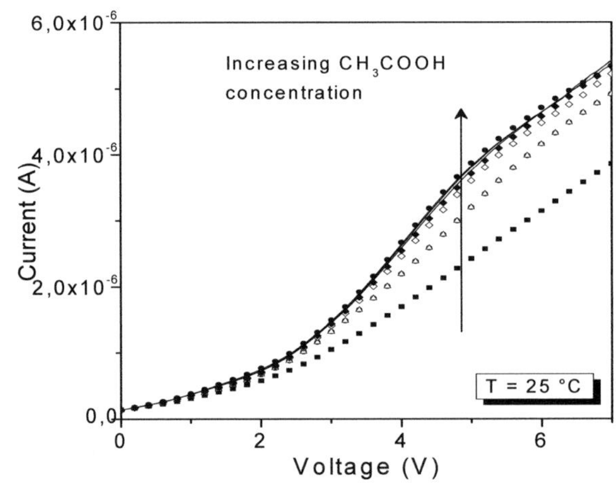

Fig. 10. I/V characteristics of DABP/MWCNTs composite film-based sensor at different acetic acid concentrations.

To support the above hypothesis, experiments aimed to investigate the effect of gases of different chemical nature (e.g. cyclohexane and acetic acid) have been carried out. The results of the tests have shown that the DABP/MWCNTs composite film-based sensor displays large variations of the electrical resistance with varying the concentration of acetic acid (Figure 10).

No significant variations were instead observed when cyclohexane vapours contacted the sensor, confirming that the interaction between the nitrogen(s) of the DABP and proton of water and acetic acid is the mechanism leading to the current variation registered. This opens the possibility to design and develop selective sensors for protic solvents.

5. Applications of MWCNT's based sensors

MWCNT's based gas sensors are one of the turning point for the development of high performance gas sensors. While there are many research activities focused on this direction, few are the commercial applications based on low cost reliable devices. These devices should have, among the others, low power consumption. The main contribution to power consumption of such sensor devices is the part employed in setting and controlling the optimal operating temperature value and, by a consequence of this, to reduce the power consumption is need to work at operating temperature values close to room temperature ones.

Nanomix employs carbon nanotubes based network to develop sensing systems for biomedical applications. This company developed a breath analysis system for the monitoring of nitric oxide (NO) as biomarker that has been shown to be correlated with asthma, an inflammatory airway disease. The breath analysis is one of the emerging biomedical applications involving gas sensors. By monitoring specific targets (biomarkers) in the alveolar air it is possible to monitor several diseases in the metabolic functions. This technique could be a cheaper and low intrusive test for early diagnosis of many human diseases allowing a large scale prevention and improving the quality of life (Nanomix website, 2011).

Applied Nanotech Holdings Inc. developed Enzyme coated Carbon Nanotubes (ECNT) as a small dimensions enzymatic biosensor suited for medical, environmental and chemical analysis. The carbon nanotubes are employed in the development of a three dimensional network promoting enhanced enzyme loading and increasing the analyte exposure. By choosing the proper enzyme for the specific analyte it is possible to enhance the performance of the device. These kinds of devices can be employed in arrays as electronic tongues, and they can analyse a large number of analytes from a single drop of blood or urine for metabolic analysis or as electronic nose for toxic gas detection and chemical warfare sensing (Applied Nanotech Holdings Inc. website, 2011).

6. Acknowledgments

The work was funded by the University of Messina under the framework of the project PRA Interdisciplinare 2007, n. PRME07AWAK.

7. References

Allaouia, A., Baiab, S., Cheng, H. M., & Baia, J. B.(2002). Mechanical and electrical properties of a MWNT/epoxy composite, *Compos. Sci. Technol.* 62 1993–8

Aroutiounian, V. M., Hydrogen detectors, *ISJAEE*, 23, 2005, pp. 21-31.

Bazzini, C., Brovelli, S., Caronna, T., Gambarotti, C., Giannone, M., Macchi, P., Meinardi, F., Mele, A., Panzeri, W., Recupero, F., Sironi, A., & Tubino. (2005) Synthesis and characterization of some Aza[5]helicenes, *R. Eur. J. Org. Chem.*, 1247.

Coleman, J. N., Curran, S., Dalton, A. B., Davey, A. P., McCarthy, B., Blau, W., & Barklie, R. C. (1998). Percolation-dominated conductivity in a conjugated-polymer-carbon-nanotube composite, *Phy. Rev.* B 58, R7492.

Flahaut, E., Peigney, A., Laurent, Ch., Marliére, Ch., Chatel, F., & Rousset, A. (2000). Carbon nanotube-metal-oxide nanocomposites: microstructure, electrical conductivity and mechanical properties, *Acta Mater.* 48, pp. 3803–3812.

Guirado-Lopez, R. A., Sanchez, M., & Rincon, M. E., Interaction of acetone molecules with carbon-nanotube-supported TiO_2 nanoparticles: possible applications as room temperature molecular sensitive coatings. *J. Phys. Chem.* C 2007, 111, 57-65.

Gurlo, A., Barsan, N., & Weimar, U. (2006). *Gas sensors based on semiconducting metal oxides, in Metal Oxides: Chemistry and Applications.* CRC Press. 683-738.

Hossein-Babaei, F., Keshmiri, M., Kakavand, M., Troczynski, T.(2005) A resistive gas sensor based on undoped p-type anatase, *Sens. Actuators B*, 110, 28–35.

Iijima, S. (1991). Helical microtubules of graphitic carbon , *Nature*, 354, 56-58

Kauffman, D. R., & Star, A. (2008). Carbon Nanotube Gas and Vapor Sensors, *Angew. Chem. Int. Ed.*, 47, 6550 - 6570

Kong, J.; Franklin, N. R.; Zhou, C.; Chaplin, M. G.; Peng, S.; Cho, K.; & Dai, H. (2000), Nanotubes Molecular Wires as Chemical Sensors, *Science*, 287, 622-625.

Neri, G. (2010). Non-Conventional Sol-Gel Routes to Nanosized Metal Oxides for Gas Sensing: From Materials to Applications. *Science of Advanced Materials* 2, 3-15.

Pengfei, Q. F., Vermesh, O., Grecu, M., Javey, A., Wang, Q., Dai, H., Peng, S., & Cho, K. J.(2003). Toward large arrays of multiplex functionalized carbon nanotube sensors for highly sensitive and selective molecular detection, *Nano Lett.* 3 347;

Roithová, J., Schröder, D., Míšek, J., Stará, I. G., & Starý, I.(2007). Chiral superbases: The proton affinities of 1- and 2-aza[6]helicene in the gas phase, *J. Mass Spectrom.* 42, 1233.

Salehi-Khojin, A., Khalili-Araghi, F., Kuroda, M. A., Lin, K. Y., Leburton, J.-Pierre, & Masel, R. I.(2011). On the Sensing Mechanism in Carbon Nanotube Chemiresistors, *ACS Nano*, 5, 153–158.

Shaffer, M.S.P., & Windle, A.H. (1999) Fabrication and Characterization of Carbon Nanotube/Poly(vinyl alcohol) Composites, *Advanced Materials,* 11, 937–941

Shimizu, Y., & Egashira, M. (1999) Basic aspects and challenges of semiconductor gas sensors, *MRS Bulletin* 24, 18-24.

Star, A.; Gabriel, J.-C. P.; Bradley, K.; Gruner, G.(2003). Electronic detection of specific protein binding using nanotube FET devices. *Nano Lett.* 3, 459-463.;

Sun, Y. P., Fu, K. F., Lin, Y., & Huang, W. J. (2002). Functionalized carbon nanotubes: properties and applications, *Acc. Chem. Res.* 35, 1096.

Valentini, L., Armentano, I., Kenny, J. M., Cantalini, C., Lozzi, L., & Santucci, S. (2003). Sensors for sub-ppm NO2 gas detection based on carbon nanotube thin films, *Appl. Phys. Lett.* 82 961.

Varghese, O. K., Gong, D., Paulose, M., Ong, K. G., & Grimes, C. A. Hydrogen sensing using titania nanotubes. *Sens. Actuators B,* 93, 338-344, 2003

Wei-De Zhang and Wen-Hui Zhang. (2009) Carbon Nanotubes as Active Components for Gas Sensors, *Journal of Sensors*, 160698.

Nanomix website, April 2011, Available from: <http://nano.com/index.html >

Applied Nanotech Holdings Inc. website, April 2011, Available from: <http://www.appliednanotech.net/>

15

Nanocomposites Based on Elastomeric Matrix Filled with Carbon Nanotubes for Biological Applications

Stefano Bianco[1], Pietro Ferrario[2], Marzia Quaglio[1]
Riccardo Castagna[2] and Candido F. Pirri[1,2]
[1]Center for Space Human Robotics, Fondazione Istituto Italiano di Tecnologia
[2] Materials and Microsystem Lab (Chilab)-Latemar Unit, Dipartimento di Scienze dei Materiali ed Ingegneria Chimica, Politecnico di Torino
Italy

1. Introduction

In recent years nanocomposite materials have prompted huge interest due to their exceptional electrical, thermal and mechanical properties. Carbon nanotubes (CNTs) have become the prevalent constituent in such nanocomposites thanks to their superior mechanical properties and unparalleled thermal and electrical transport capabilities. Carbon nanotube-based nanocomposites have mainly employed a polymer matrix, even if ceramic and metallic materials have also been used. In any case, most composite syntheses have utilized multi-walled carbon nanotubes (MWCNT).

In this chapter, a polydimethylsiloxane/multi-walled carbon nanotube composite is considered and characterized. PolyDiMethylSiloxane (PDMS) is an insulating elastomeric material whose main technological employment had been as a structural material for micro-electrical mechanical systems (MEMS) and microfluidic-based devices. Since 2005, it has been reported that by incorporating carbon nanotubes - especially MWCNTs - in polydimethylsiloxane, the resulting material possess a number of enticing properties that can be successfully harnessed in MEMS technology, both for sensing and actuation mechanisms.

We report on synthesis and morphological characterization with a field-emission scanning electron microscopy (FESEM) analysis, with the optimization of the preparation technique to asses the dispersion of nanotubes within the elastomeric matrix, as well as the affinity between the two constituent phases. The dispersion of the carbon filler and its affinity with the polymer matrix are crucial factors that determine the quality and reliability of nanocomposites. FESEM is an adequate tool for investigating such morphological features.

Owing to the promising applications of PDMS/MWCNT composites in sensing and actuation, we characterized the thermal and electrical behaviour of the material and we demonstrated its applicability for the fabrication of miniaturized devices for bio-analyses. Electrical and thermal characterization was aimed at determining the conductivities of composites with different nanotube concentrations and what changes are brought about by a rise in temperature. Concerning electrical transport, our data were analyzed by means of the percolation theory. Improvements in thermal transport characteristics present a particular interesting for their beneficial effects in the fabrication of devices that need to sustain complex thermal protocols based on fast transitions between different temperatures, like the Polymerase Chain Reaction for DNA amplification.

2. Composite and nanocomposite materials

In the continuing quest for improved performance, currently-used materials frequently reach the limits of their usefulness. Thus researchers are always striving to produce either improved traditional materials or completely new materials. Composite materials are an example of the latter category. Those are multi-phase materials obtained through the artificial combination of different materials in order to attain properties that the individual components by themselves cannot attain.

Three main features characterise composite materials:

i. the constituents are present in "reasonable" proportions, say greater than 5%;
ii. the constituent phases display different properties, and hence the composite properties are noticeably different from the properties of the constituents;
iii. they are produced by intimately mixing and combining the constituents by various means.

The constituent materials are classified into matrix and filler phases. The former is the continuous constituent that is often present in the greater quantity in the composite. The role of the filler, instead, is to enhance the properties of the matrix, including electrical, thermal, and mechanical properties [Chung, 2010]. Usually at least one of the dimensions of the filler is small, say less than 500 μm. In addition, the filler is usually described as being either particulate or fibrous. Particulate fillers have dimensions that are approximately equal in all directions, and are often spherical, cubic or platelet in shape. On the other hand, a fibrous filler is characterised by its length being much greater than its cross-sectional dimension. However, the ratio of length to the cross-sectional dimension, known as the aspect ratio (AR), can vary considerably. Long fibres with high aspect ratios give rise to what are called continuous fibre composites, whereas discontinuous fibre composites are fabricated using short fibres of low aspect ratios. The orientation of the discontinuous fibres may be random or preferred, whereas continuous fibres are usually unidirectional oriented [Matthews & Rawlings, 1999].

Recently much interest has been prompted by a novel class of composite materials, called nanocomposites. Those are composites where one of the phases, namely the filler phase, has at least one dimension of less than 100 nm. Owing to the nanometric scale of the filler, which is also termed "nanofiller", nanocomposites are endowed with huge interfacial area per volume between matrix and filler. As a result, molecular interactions between the two constituents are much stronger than in composites where the filler has a micrometric scale. Although a variety of nanoparticles, such as nanospheres, have been successfully utilised as a filler material, careful scrutiny has mainly been focused on nanofillers with high aspect ratios. This is down to the fact that fillers characterised by high AR can confer either isotropic or anisotropic properties. In particular, if the filler is randomly oriented in the host matrix the composite will display properties that are independent of the spatial direction under analysis. Otherwise, if the nanocomposite has been devised by inducing a preferred orientation to the filler, the resulting properties would be heavily dependent on the considered special direction.

Carbon nanotubes have recently emerged as an excellent nanofiller material, thanks to the combination of their small size and particular physical properties [Ajayan & Tour, 2007]. Both single-walled and multi-walled CNTs possess superior thermal and electric properties: they are thermally stable up to 2800 °C in vacuum, their thermal conductivity is about twice as high as diamond, and the electric-current carrying capacity is 1000 times higher than copper wires. Furthermore, theoretical studies carried out on the mechanical properties of CNTs indicate that the elastic modulus is greater than 1 TPa and strengths are 10÷100 times higher than the strongest steel at a fraction of their weight [Harris, 2009]. The exceptional electrical and mechanical properties of CNTs, in particular, have prompted huge interest in the production of nanotube-containing composites materials for electronic and structural applications. In many cases these nanocomposites have employed polymer matrices. By and large, polymers can be easily fabricated without damaging CNTs during processing and they represent a huge class of materials with many interesting properties. In particular, CNTs have been incorporated in polymers with weak mechanical properties and low electric transport. The resulting nanocomposite possess the properties of each component with a synergistic effect: on one hand, the nanotubes serve to provide electrical and thermal conductivity and increase stiffness, strength and toughness; on the other, the polymer matrix allows adequate transfer of loads, reducing slippage between adjacent nanotubes bound by weak Van der Waals interactions [Yakobson et al, 1996].

In particular, some efforts have been recently dedicated to the study of polydimethylsiloxane/carbon nanotubes nanocomposites and such topic will be deepened in the next paragraphs.

2.1 Properties and applications of PDMS/CNT nanocomposite

Polydimethylsiloxane is a silicon elastomer and it is a high performance material on account of its optical transparency, chemical and thermal stability, biocompatibility and mechanical flexibility [McDonalds & Whitesides, 2002]. PDMS is an organosilicon compound, commonly referred to as a silicone, whose chemical formula is $(H_3C)_3SiO[Si(CH_3)_2O]nSi(CH_3)_3$, where n indicates the number of repeating monomer units $[SiO(CH_3)_2]$ occurrences in the polymer chain. Figure 1 shows the PDMS structure and a 3D representation.

Fig. 1. Chemical formula of PDMS and its 3D structure (from Wikipedia website).

The structure exhibits low intermolecular force between the methyl groups ($-CH_3$), high flexibility of the siloxane backbone (SiO), high strength of the siloxane bond and a partial ionic nature [Hillborg & Gedde, 1999]. The linear PDMS is liquid at room temperature, because its glass transition temperature $T_g << RT$, in detail $T_g = -127$ °C. Since the methyl groups $-CH_3$ are characterised by very intense σ bonds, they offer 'protection' to the -Si-O- groups, which are reactive due to their polar nature. Among other things, the methyl groups are also able to rotate around the backbone, thus offering an isotropic protection. [Kim et al, 1999]. Rotation about siloxane bonds in PDMS is virtually free, the energy required for rotation being almost zero [Tobolski, 1960]. This free rotation is reflected in the low glass transition temperature. Its hydrophobic behaviour combined with its low T_g leads to define PDMS as an oil.

The material of interest for MEMS and microfluidic applications is obtained by adding a curing agent (CA) to the PDMS base. The polymer base (also denoted oligomer or pre-polymer) accommodates an unspecified number of vinyl groups, and the curing agent includes at least three silicon hydride bonds for each molecule and some quantity of platinum that acts as a cross-linking catalyst. The addition produces a cross-linking reaction that forms the thermosetting PDMS. Mechanically, this material is referred to as an elastomer. Its viscosity is significantly reduced after cross-linking and the material behaves as a rubber at room temperature.

The cross-linking reaction takes place as the double bond of the polymer base vinyl group ($H_2C = CH -$) is broken, leading to the formation of the $-Si-CH_2-CH_2-Si-$ bond by using one of the -Si-H- curing agent terminals.

Much interest has been prompted about the surface properties of polydimethylsiloxane. PDMS has a low surface tension, displays a moderate interfacial tension against water, has a high water repellency (i.e. it is hydrophobic), it does not present surface shear viscosity, and has a soft feel [Clarson & Semlyen, 1993]. The adjectives "high", "low", etc. are in comparison to the behaviour of hydrocarbon-based materials. Thus, PDMS has a surface tension lower than almost all organic polymers. One of the very first consequences of the PDMS low surface energy is that whenever it is interfaced with a high energy surface medium such as water a change in the polymer's organization occurs. Lavielle and Schultz [Lavielle & Schultz, 1985] demonstrated that polar groups belonging to PDMS are attracted to the surface by its contact with water. The explanation as to why PDMS is a hydrophobic material was advanced by Khorasani and co-workers [Khorasani et al, 2005], partly derived from L. Franta [Franta, 1988]. According to them, the PDMS structure is a helical one, whose inner part is made of Si-O-Si groups and the outer part consists of methyl groups. The Si-O bond angle is relatively large (120°-160°), which leads to the molecule having a certain degree of flexibility. Hydrophobicity emerges from the different electrochemical nature of the two functional groups.

PDMS is an electrical and thermal insulator, limiting its employment to structural applications that include fluidic channels, valve diaphragms, encapsulating chambers, etc. [Bokobza, 2004; Demir et al, 2005]. Carbon nanotubes can confer to it a significant enhancement of the electrical, thermal and mechanical properties. The addition of CNTs above a critical threshold value reduces the resistivity of the composite by several orders of magnitude, and it has been reported that thermal conductivity is enhanced by 65% with a 4 wt% multi-walled nanotube loading [Liu et al, 2004]. Moreover, it has been shown that the addition of 2 wt% single-walled nanotubes to PDMS produces a 203% improvement in elastic modulus and a 154% in yield strength over the pure polymer [Moniruzzaman et al,

2007]. PDMS/CNT nanocomposites show a number of notable properties that could be easily exploited in micro-electro mechanical systems. However, to date, very few works have been published regarding the employment of PDMS/CNT composites in MEMS technology. PDMS/CNTs composites have mainly been used as strain gauges and pressure nanosensors. It is known that such composites are piezoresistive materials, i.e. materials that change their nominal resistivity due to applied mechanical stress. Although the gauge factor (GF) ($\Delta R/R$) is not high compared with other materials, the sensitivity of the composite to pressure is expected to be quite large due to its large Young's modulus [Wu et al, 2009]. Under an externally applied pressure, the relative resistance measured along the same direction increases, and this effect is stronger if the concentration of carbon nanotubes is low [Hu et al, 2008]. Since most piezoresistive materials are also sensitive to temperature, it is important to characterise the temperature coefficient of resistivity (TCR) of PDMS/MWCNT. As reported by Xu and Allen [Xu et al, 2010] the measured TCRs are much smaller than their piezoresistive coefficients, less than 0.005 $°C^{-1}$. This aspect indicates that the resistivity does not vary significantly under normal working conditions. The piezoresistive properties and low TCRs have been successfully capitalised on the fabrication of strain gauges. For example, PDMS/MWCNT nanocomposites have been utilised to determine the displacement of cantilevers or micro-beams, which are important components in MEMS. The amount of displacement is transduced by a change of resistance of the composite that in turn causes a voltage imbalance in a Wheatstone bridge. Moreover, Kang and colleagues (Kang et al, 2006) have extended the composite sensor to a long strip, named "neuron sensor". They suggested that a neural system in the form of a grid could be attached to the surface of a structure to form a sensor network, enabling structural health monitoring.

On account of the electrical conductivity of PDMS/MWCNTs composites, it has been feasible to create tactile sensors through an array of intertwined capacitors [Engel et al, 2006]. The electrodes of the capacitors are made of PDMS/MWNCT composite. The network is composed of two layers of elastomer oriented orthogonal to each other. If pressure is applied, the electrodes on which the external force is addressed will get closer to their respective electrodes on the bottom matrix, thus increasing their capacitance.

Together with sensing properties, actuation characteristics were studied as well. Actuator materials have the faculty of changing their physical dimensions in response to external stimuli and transfer a variety of forms of energy into mechanical work. With regard to their corresponding energy supplies, they can be classified into electrical, thermal, pneumatic and optical actuators [Kovacs, 1998]. Various studies have revealed that when MWCNTs were embedded into a polymer matrix, including PDMS, the nanocomposites experienced either contractive or expansive actuation under light illumination determined by the nanotubes concentration and alignment. In particular, PDMS/MWCNT composites showed structural expansion in response to light under small pre-strains. Under large pre-strains, they showed instead structural contraction in response to light [Ahir & Terentjev, 2005; Ahir et al, 2006]. These properties have not been widely turned to advantage. PDMS/MWCNT composite is a good candidate material for optical actuation, which brings distinctive advantages such as wireless actuation, remote controllability, electromechanical decoupling, low noise, elimination of electrical circuits and higher-level integrity [Lu & Panchapakesan, 2007]. Those composites could therefore be useful in nanoscale devices whereby incorporating batteries is not feasible and the only form of energy is external light. Such devices are said to be working on energy harvesting (or scavenging). In addition, it has been shown that the photomechanical actuation of PDMS/MWCNT composites can be integrated with microelectromechanical technologies to develop CNT-based micro-optomechanical systems (CNTMOMS) [Lu & Panchapakesan, 2005; Lu & Panchapakesan, 2006]. Another innovative form of optical actuation and energy harvesting application has been figured out by D. Okawa and colleagues [Okawa et al, 2009]. The Zettl's group availed itself of recent studies performed on vertically aligned carbon nanotube forests (VANTs). This material exhibits high absorption of visible light (99.9%) that is quickly then turned into heat through electron relaxation, thus rendering VANTs optimum heat switches. Heating up VANTs gives rise to the formation of surface tension gradients at the interface with a liquid in contact with the nanotubes. Owing to the optical transparency of polydimethylsiloxane, to its perfect adhesion with VANTs and to its density (similar to water), VANTs were embedded in the elastomer. By focusing light in appropriate regions of VANTs, optically induced thermal surface tension gradients propelled the structures (PDMS+VANTs) either in linear or circular paths, depending on the device design.

The tailoring of the aforementioned material properties gives to the PDMS/CNT nanocomposite a multipurpose feature. Such increased properties can find a fruitful application in the fabrication of a new generation of biodevices. The improvement of thermal transport properties, in particular, can increase the efficiency and selectivity of polymer-based devices for Polymerase Chain Reaction (PCR). The efficiency and selectivity of PCR is strictly related to the execution of fast transitions among the three temperatures required by the thermal protocol. Miniaturized devices with reduced mass, as in Micro-Total Analysis Systems for DNA amplification, exhibit a lower thermal inertia than traditional bench-top instruments, increasing the amplification efficiency. The main research goal lays in giving life to a new generation of small, portable and fast devices for applications ranging from clinical medicine, to genetic disease diagnostics and forensic science.

In that scenario, the development and study of innovative materials and technological processes gained a crucial importance. On the first works silicon and glass were used, being their micro-technology well developed and robust compared to any other material [Reyes et al, 2002]. More recently, many efforts were spent in exploring the use of polymers as PDMS, PolyMethylMethAcrylate (PMMA), PolyCarbonate (PC) and Cyclic-Olefine-Copolymer (COC), which are low cost, easy to be manufactured, optically transparent and biocompatible [Zhang et al, 2006; Zhang & Xing, 2007].

Despite the promising characteristics of polymers, their low thermal conductivity is the main problem to face to design disposable PCR Lab-On-a-Chips (LOCs), being rapid temperature transitions difficult to achieve. Till now, four main strategies have been proposed to overcome this problem [Zhang et al, 2006; Zhang & Xing, 2007; Wu et al, 2009]: (i) reducing the thermal inertia of the device by reducing its dimensions; (ii) using optimized thermocyclers; (iii) performing dynamic PCR reactions; (iv) fabricating integrated heaters.

Polymeric composites, in the case of filling with thermally conducting nanomaterials, can show a dramatic improvement in thermal response with respect to traditional polymers. In that field, the improvement in reaction efficiency using PDMS/CNT nanocomposite was recently demonstrated by our group [Quaglio et al, 2011]. Moreover, combining the obtained thermal performance with the improved electrical conduction properties and some of the other material characteristics like the wettability behaviour, it is possible to design innovative all-polymeric devices with integrated heaters, opening new attractive scenarios in bioscience.

3. Nanocomposite preparation and morphological characterization

Reading through the obtainable literature on PDMS/CNT composites, it is immediately noticeable that the ways of preparing PDMS/MWCNT composites are as numerous as the research groups who have published works on the composite's properties. With the exception of a few overly complicated techniques that include demanding and multi-step functionalizations of nanotubes, all preparations require four main "ingredients": the PDMS polymer base (PB), the curing agent (CA) that activates the cross-linking reaction, CNTs and finally a solvent for nanotube dispersion within the polymer matrix. The ratio of CNTs to PDMS elastomer is chosen depending on the desired application and performance of a device. In the case of uninvolved conductors for capacitive sensors or resistive heaters, a large quota of carbon nanotubes may be added in order to increase the conductivity of the composite. On the other hand, strain and force sensitive sensors call for lower loading of nanotubes to enhance the sensitivity. In all cases, satisfactory dispersion of carbon nanotubes is an essential prerequisite. Here the adjective "satisfactory" denotes a level of dispersion whereby there are no macroscopic bundles or clusters of nanotubes. The level of homogeneity attained in the nanocomposite is usually verified with electron microscopy observations. Electron microscopy could also shed light on several other morphological features of PDMS/MWCNT composites, including a qualitative assessment of the grade of affinity between carbon nanotubes and polymer matrix.

3.1 Matrix characteristics

The chosen PDMS was the Sylgard® 184 Silicone Elastomer Kit, from Dow Corning (USA), comprising a liquid base oligomer (dimethylsiloxane with the vinyl functional group, SiH = CH_2) and a curing agent (pre-mixture of a platinum complex and copolymers of methylhydrosiloxane and dimethylsiloxane, SiH). When mixing base and curing agent at a typical ratio of ten parts base to one part curing-agent by weight, the catalyst promotes a hydrosilylation reaction between base and curing agent [Xia & Whitesides, 1998]. Its main characteristics can be listed as follows:

- it exhibits a good thermal stability, up to 190°C in air;
- its mechanical properties can be adjusted by modifying the mixing ratio between PB and CA or the polymer base molecular weight. Different stiffness levels, different elastic modulus and adhesion energy are thus obtained;
- its low viscosity makes casting into moulds down to few hundred microns feasible and it can be spinned to fill mould voids deeper than 5 µm;
- it is rather effortless to remove the cross-linked polymer from moulds;
- it is chemically inert;
- it is isotropic and homogeneous;
- it is an electrical insulator with a break-down voltage of about 2×10^7 V/m;
- it is not toxic or cancerous or in any way harmful to human beings and the environment;
- the high flexibility of the backbone allows the permeability of many gases and vapours [Hillborg & Gedde, 1999];
- hot, damp environments, such as superheated steam, are very detrimental to PDMS integrity. That is due to the partially ionic nature of the siloxane backbone that has a tendency to suffer from nucleophilic or electrophilic attack resulting in a susceptibility to hydrolysis by water [Clarson & Semlyen, 1993].

3.2 Filler characteristics

The multi-walled carbon nanotubes used in this work for nanocomposite preparation were supplied by Nano Carbon Technologies Co. Ltd (Japan).

They were synthesised through a floating reactant method. This is a particular CVD technique that allows thorough dispersion of the hydrocarbon in the catalytic particle solution. The synthesis requires an organo-metallic compound (ferrocene) as a catalyst precursor, an organic solvent (toluene) as a carbon feedstock and hydrogen as a carrier gas. Toluene and ferrocene are fed into the reactor through a microfeeding pump and temperature is raised at 1200 °C. Then a thermal treatment is accomplished for 30 min in high purity argon at 2600 °C. The thermal treatment, or thermal annealing, is used to remove residual metal catalyst particles. It has been experimentally shown that annealing catalytically-produced MWCNTs at graphitization temperatures (1600÷3000 °C) has the effect of not only removing the catalyst impurities but also of improving the structural quality of the tubes by turning them into straight "threads" [Musso et al, 2008]. Table 1 summarises the structural and physical properties of the as-prepared and annealed CNTs, as reported previously [Kim et al, 2005]. In this work, annealed MWCNTs were used.

	As-grown nanotube	Annealed nanotube
Interlayer spacing d_{002} [nm]	0.342	0.3385
Diameter [nm]	20÷70	20÷70
Aspect ratio	>100	>100
Volume density [g/cm^3]	0.005	0.005
Real density [g/cm^3]	1.89	2.1
Specific surface area [m^2/g]	28	28
Temperature of oxidation [°C]	560	600
Metal impurity	14wt%	<450 ppm

Table 1. Characteristics of the nanotube used in this study, as reported previously [Kim et al, 2005]. Interlayer spacing was estimated by XRD, diameter and length of the tube by FESEM analyses, volume density by tapping method, real density by a pycnometer, specific surface area by BET (N_2 physisorption), temperature of oxidation by TGA and metal impurity by X-ray fluorescence spectroscopy.

3.3 Nanocomposite preparation

Different preparation methodologies were suited for material preparation, especially depending on the filler content. Samples with low loading concentration were fabricated with the most straightforward technique (namely "A"). The step-by-step procedure is as follows:

1. the nanotubes and the PB quantities are weighted via digital balance and preliminary mixed with a metallic spatula;
2. the mixture is poured into an agate mortar and further mixing occurs. Blending is achieved through a pestle and in order to ensure good dispersion of nanotubes, it is advisable to apply a persistent shear force;
3. the mélange is left to degas for 24 hours;
4. the curing agent is added in a 10:1 ratio to PB, followed by long hand stirring with a pestle;
5. second degassing;

6. transfer everything into previously rinsed moulds and let two to three hours to go by, then heat up the oven at 85 °C and put the moulds inside; consider that samples with a low CNTs concentration will just demand twenty minutes to cross-link, while those with higher concentrations (>1.0 wt%) could necessitate up to two hours.

Given the low concentration of carbon nanotubes relative to the polymer matrix, the consistency of the mixture prior to casting is akin to pure PDMS's. Thence, casting into moulds is performed by simple dripping.

The second preparation method, which will be hereafter denoted by "B", is among the simplest and quickest when it comes to fabricating samples with high concentration of nanotubes (>2 wt%) and foresees a pre-dispersion of nanotubes in a solvent, as suggested for example by L. Bokobza [Bokobza, 2009]. The preparation works as follows:

1. the nanotubes were mixed with isopropanol in an open container; the amount of solvent employed rendered the mixture fluid;
2. the PDMS oligomer was mixed with an excess of isopropanol so that the resulting consistency is similar to that of the curing agent;
3. the two mixtures were sonicated together for hours;
4. solvent was let partially evaporate under chemical hood;
5. the procedure follows as from point 2 in preparation "A".

It is known from literature that the presence of nanotubes inside the matrix partially hinder the reticulation of the polymer [Xu et al, 2008]. Thus, significantly longer times are needed for the preparation "B", because both of the higher number of preparation steps and the need for a longer reticulation time.

3.4 Morphological analysis

A morphological characterization through Field Emission Scanning Electron Microscopy was performed to evaluate the dispersion degree of the filler inside the polymeric matrix. It is well known that a uniform dispersion of the filler is a big issue, since nanotubes tend to form agglomerate because of the Van der Waals forces acting on the wide exposed surface of the nanostructures. Thus, CNTs are usually in form of bundles that need to be broken by applying a strong shear force. FESEM analysis is a powerful tool to evaluate the efficiency of the above-reported preparation procedures.

It was decided to carry out the FESEM analysis on cross-sections of PDMS/MWCNT nanocomposites. After the samples had cross-linked and been extracted from moulds, a 1 mm-thick slice was cut out from each sample with a cutter blade. The cross-sections were then coated with chromium to avoid charging during electron irradiation and mounted on the instrument's sample holder.

The first set of pictures (Figure 2) is a collection of images taken from samples with increasing CNTs concentration, all obtained with the same magnification (5.00 kx). For low CNTs concentrations (<2 wt%) the samples are to all intents and purposes non-conductive, inasmuch as the nanotubes are far apart or form unconnected bundles. As the concentration is increased, the nanotubes give rise to a sort of conductive path more (4 wt%) or less (3 wt%) homogeneous. The sample with a CNTs concentration of 0.5 wt% was fabricated in line with preparation "A". The cracks present in some images do not belong to the sample but rather to the chromium coating.

The use of multiple steps of mixing, including sonication, present in preparation "B" have made it possible to achieve samples with better levels of dispersions, as it has just been shown in Figure 2 (d) and (e). In spite of the lower nanotube content, the 3 wt% sample exhibit small bundles (the bright circular spots) that are not present in the 4 wt% sample. As it has been mentioned above, the improved dispersion has the disadvantage of a very time-consuming preparation because of slow cross-linking.

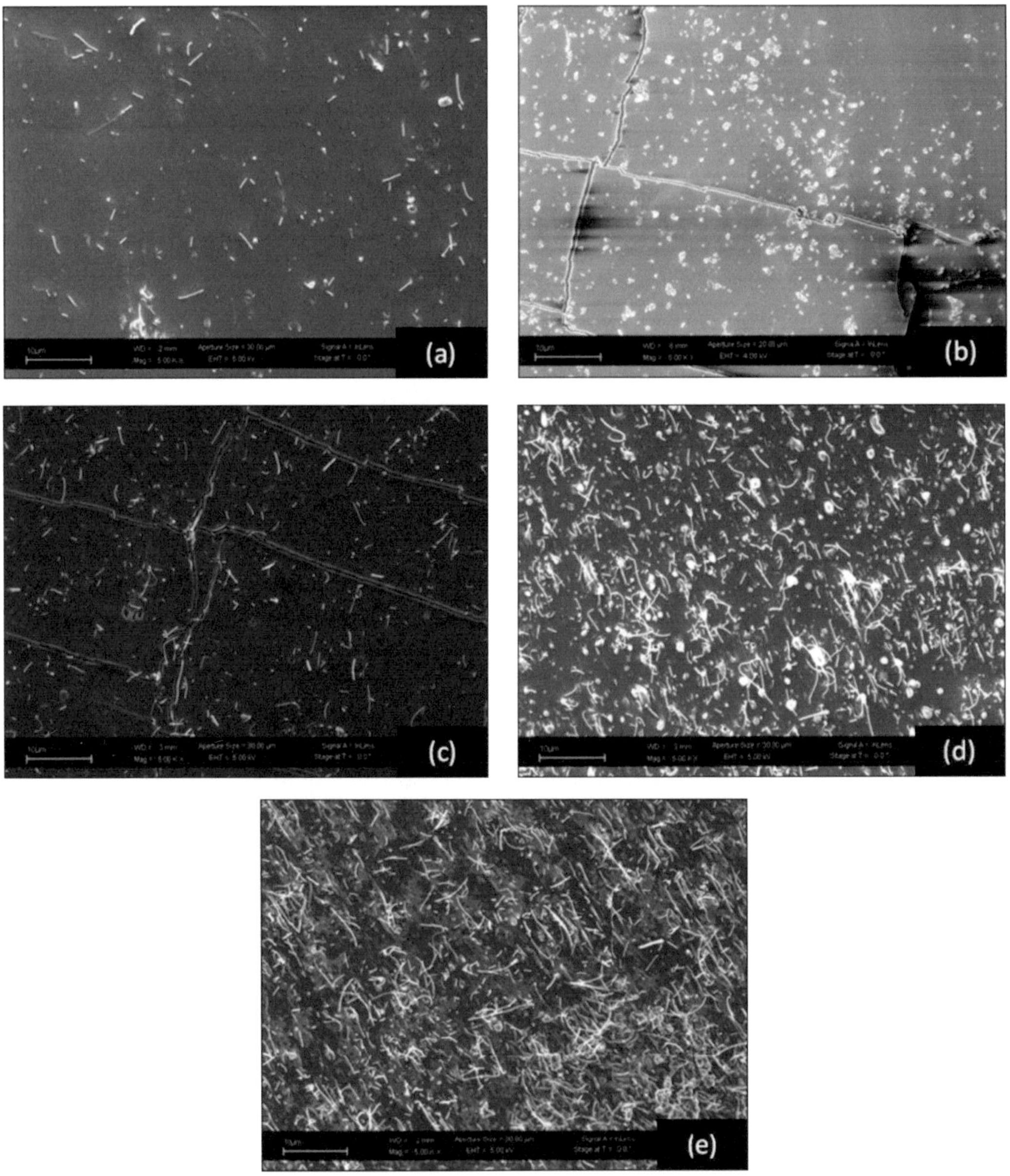

Fig. 2. FESEM images of nanocomposites with a concentration of nanotubes equal to (a) 0.5 wt% (prep. "A"), (b) 1.0 wt% (prep. "A"), (c) 2.0 wt% (prep. "B"), (d) 3.0 wt% (prep. "B") and (e) 4.0 wt% (prep. "B"). All images were shot with a magnification factor of 5000. Some of the visible cracks are not due to the composite but rather relate to the chromium layer.

Notwithstanding the use of solvents and the sonicator, the presence of nanotube clusters is ubiquitous at small scales. Indeed, as shown in the collection of pictures listed in Figure 3, nearly all the samples present regions where nanotubes are scant alongside regions where their concentration is greater. The clusters of carbon nanotubes visualised in Figure 3 are nonetheless small - of the order of just few microns - compared to the dimensions of the nanocomposites that develop at the millimetre scale; the CNT bundles exhibited by the samples fabricated with preparation "B" are the smallest achievable by hand mixing and sonication combined, and represent a limit of this technology.

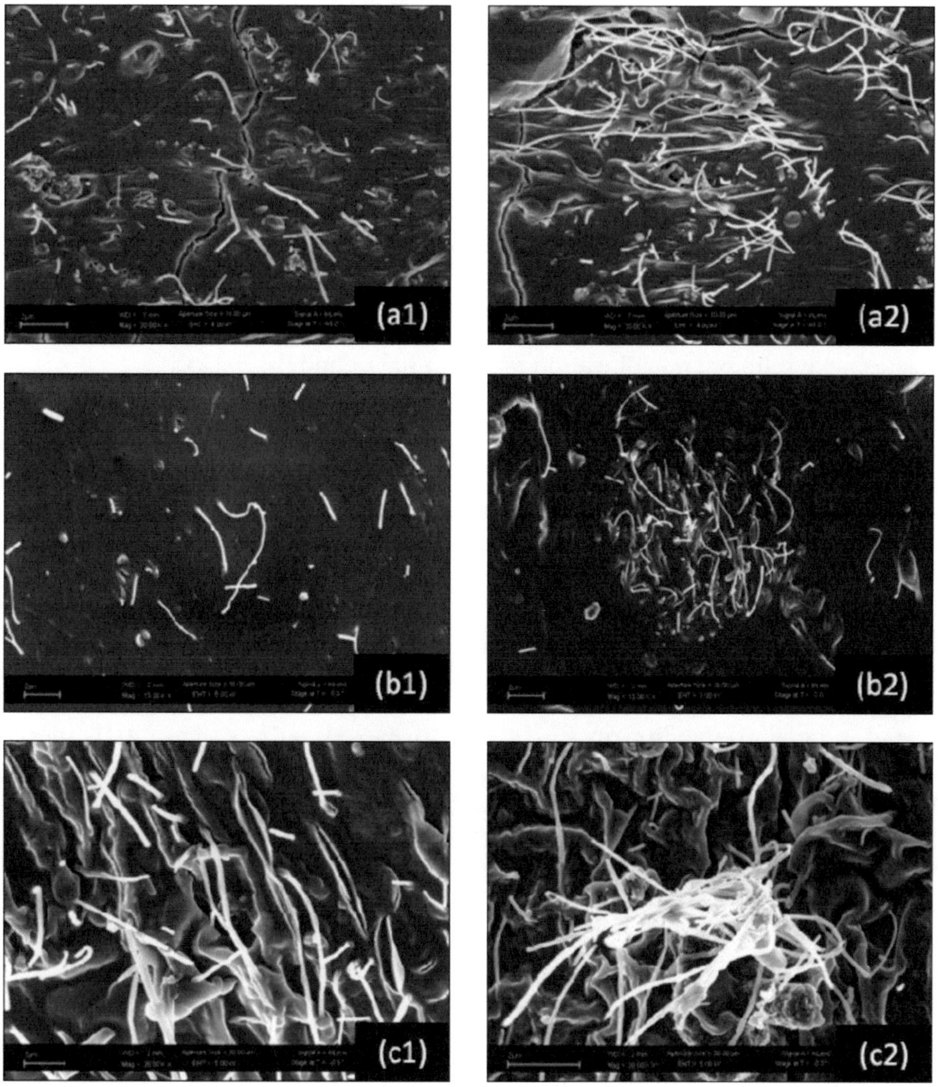

Fig. 3. The left column (1) of this matrix of images refers to regions where different samples display a relatively low concentration of carbon nanotubes. The right column (2), on the other hand, concerns regions where the CNTs concentration is higher. (a) 2 wt% (prep. "A"), 20.00 kx; (b) 0.5 wt% 15.00 kx (prep. "A"); (c) 4 wt%, 30.00 kx (prep. "B").

Furthermore, the hand mixing performed through the mortar was the essential process that ultimately ensured limited dimensions of the clusters. As a matter of fact, attempts to prepare composite using only the sonicator were tried out and Figure 4 illustrates one of those.

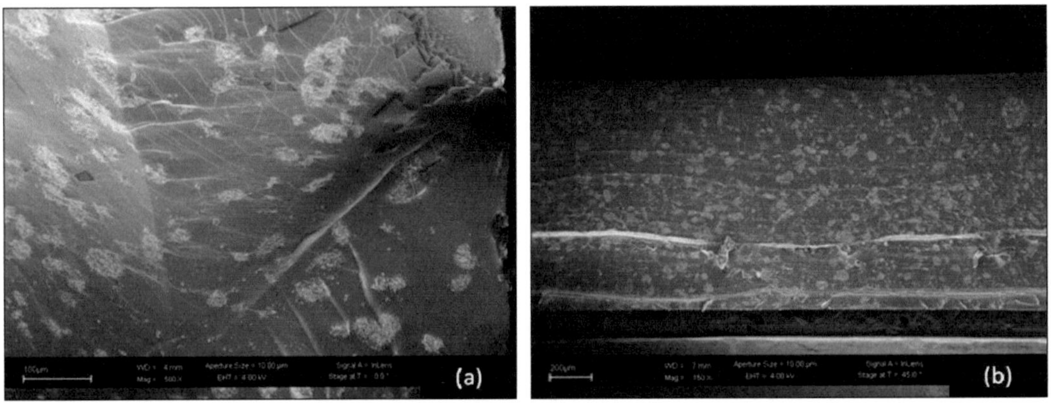

Fig. 4. Using a sonicator as an alternative to manual mixing with mortar and pestle gives rise to composites with unsatisfactory dispersion of the filler. Both images refer to 1 wt% composites, with different magnification ratios (a) 100 kx and (b) 200 kx.

The sample (1 wt%) were made similarly to preparation "A" but in lieu of using the mortar, the blending of polymer base and nanotubes first and the subsequent addition of the curing agent were achieved through sonication. The idea of using the mortar to mix the nanocomposites' constituents was basically inspired by this failed attempt.

Another advantage of carrying out the mixing through a mortar concerns the level of affinity between polymer and filler. In any nanocomposite material it is important that the constituents are intimately intertwined and closely connected, so that adequate transfer of loads and the setting up of numerous synergistic effects can take place. According to some publications, it is suggested that PDMS molecules can be absorbed on the nanotube surfaces via hydrophobic non-covalent interactions [Baskaran, 2005; Wang, 2006; Beigbeder, 2008] or PDMS has a good wettability on CNTs due to its low surface tension [Barber, 2004]. Obviously, the higher the degree of dispersion, the more pronounced the affinity between nanotubes and PDMS is. Very magnified images (200÷500 kx) of MWCNTs embedded in the nanocomposite were collected in order to estimate this important aspect. By shooting such magnified images, it was possible to measure the nanotubes' diameters (Figure 5). It was observed that more often than not it exceeded the expected value of ≤70 nm, meaning that they must be coated by the polymer which contributes to increase the tubes' diameter.

4. Electrical and thermal transport characterizations

The aim of this paragraph is to report on the electrical and thermal characterization that was performed on PDMS/MWCNT composites with different nanotube concentrations. The conductivities of the nanocomposites were obtained and fitted with a scaling law derived from percolation theory, with a good agreement between theory and experiment. Thermal transport properties were investigated by means of the laser flash technique, which is a powerful tool for the evaluation of the transport properties in solids.

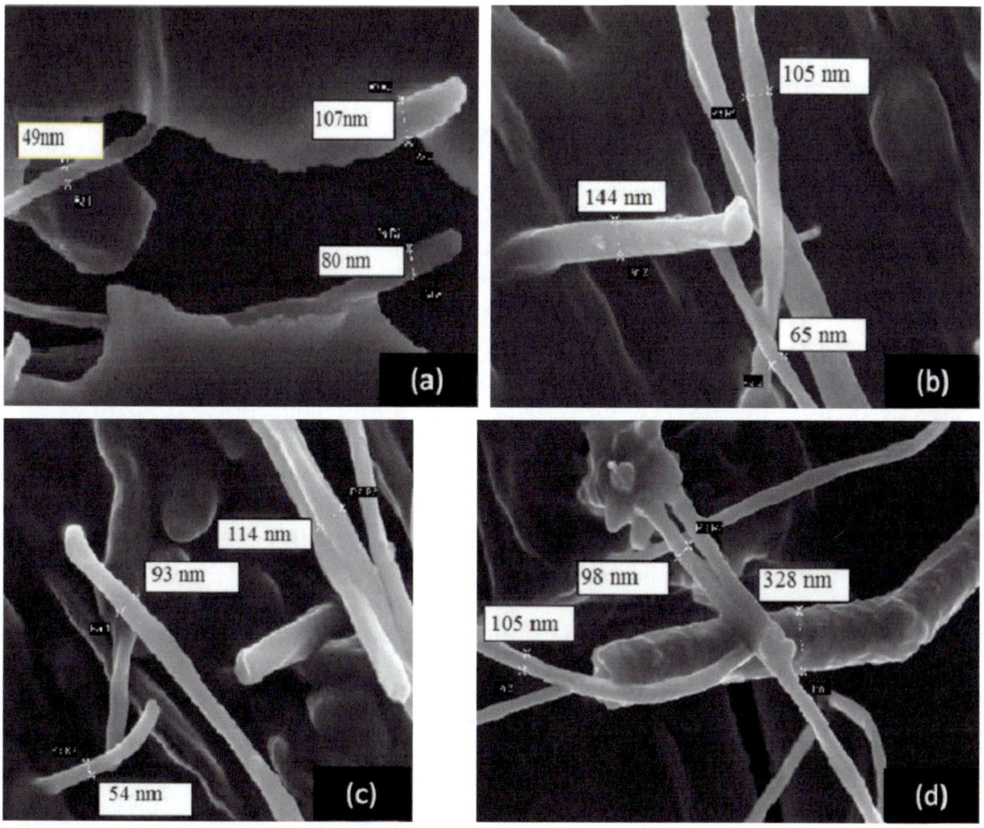

Fig. 5. FESEM images of individual carbon nanotubes belonging to (a) 2 wt% nanocomposite (prep. "A"), (b) 3 wt% nanocomposite (prep. "B"), (c) 4 wt% (prep. "B") and (d) 6 wt% (prep. "B"). According to the specifications of the nanotubes manufacturer, the clean nanotubes' diameter is around 50nm, as expounded in Table 1.

4.1 DC electrical characterization

Electrical transport properties of PDMS/CNT composite were previously reported from several research groups [Liu & Fan, 2007; Bokobza et al, 2008; Khosla & Gray, 2009; Liu & Choi, 2009], with a various range of percolation thresholds p_c's (between 0.05 wt% and 4 wt%), and this aspect is not contradictory since p_c depends principally on CNT morphological characteristics (i.e. aspect ratio). Some authors report the non-linearity of I-V curves for filler concentrations close to the percolation threshold. C.-X. Liu and J.-W. Choi [Liu & Choi, 2009] observed a parabolic rather than a linear increase of the current density at high voltages. The non-linear behaviour was more conspicuous at low filler loadings. Non-linearity of the I-V characteristic was also argued by C.-H. Liu and S.S. Fan [Liu & Fan, 2007]. The CNTs loadings were in range of 0.1÷3 wt% and the dispersion of MWCNTs was obtained through ball milling and ultra-sonication. The I-V curve for 1 wt% loading displayed a marked nonlinear behaviour. The curve is symmetric and indicates that the resistance depends upon the applied voltage. In a later article the same authors [Hu et al, 2009] showed that the non-linearity of the I-V curve is a feature mostly associated with low concentrations of MWCNTs. For a concentration of filler as low as 0.35 wt%, authors claimed that the current/voltage curve is symptomatic of the presence of a Schottky contact. The non-linearity of the curve becomes less and less noticeable with increased filler loading, like in the case of the 5 wt% sample. The reasons for the observed nonlinearity may be found in a tunnelling transport theory developed by P. Sheng in 1980 [Sheng, 1980], called fluctuations-induced tunnelling. This theory applies to random systems, such as conductor-insulator composites, disordered semiconductors, or doped organic semiconductors, where the electrical conduction is dominated by electron transfer through tunnelling between conducting segments. The theory may apply to PDMS/MWCNT composites by postulating that carbon nanotubes are coated by the elastomer and electrical transport sets in if the electrons tunnel through the insulating barrier that separated the nanotubes. In this picture of events, those barriers would be lowered by an external voltage, thus accounting for the observed resistance drop.

To correctly perform the electrical measurements (in a four point probe configuration), it is important to avoid a direct pressure on the contacts, avoiding an influence of the piezoresistive behaviour of the material on the global resistivity. In our samples, PDMS/CNT composites were spooned inside hollow parallelepiped-shaped structures formed on the surface of a pure PDMS block; the structures on the surface were sufficiently thin (<<1 mm) so that the sheet resistance approximation holds. The composites were casted inside the pre-defined geometries and let cross-linked in situ. Metallic electrodes were deposited through thermal evaporation in high vacuum condition: in particular, the electrodes consisted of high aspect ratio conductive strips transversally bridging the composites.

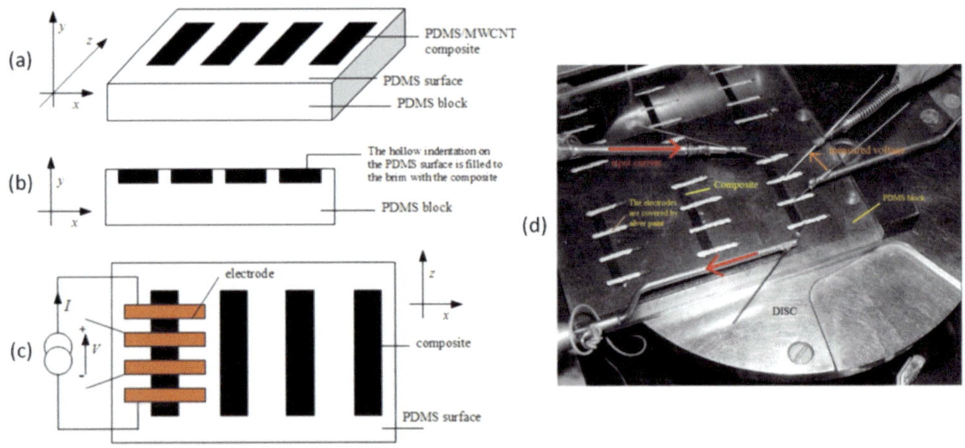

Fig. 6. A schematic diagram that illustrates the premises of the four-terminal sheet resistance measurement of PDMS/MWCNT composites. The key aspect of the new design is shown in (c): four electrodes are deposited on PDMS and on the composite and the four multimeter probes can be placed on the sections that are solely in contact with the elastomer. Picture (d) represents the final system: current is supplied by two probes and voltage drop is detected by the other two central probes. In this way a four-terminal sheet resistance measurement is made for each stripe of PDMS/MWCNT composite.

Contact with the probe is then achieved on one of the two sections that are glued to PDMS. A Ti (5 nm)/Cu (200 nm) bilayer was chosen, Ti acting as adhesion layer. The sample shape and electrical connections are represented in Figure 6.

The PDMS block offers a mere structural role therefore. Its main asset is denoted by its flexibility that confers the elastomer the ability to absorb most of the pressure applied during an I-V measurement, without transmitting it to the PDMS/MWCNT composite that is being analysed. The nanocomposites were prepared with ten different carbon nanotube concentrations: 0.5, 0.7, 1.0, 1.2, 1.5, 2.0 wt% (preparation "A"), 3.0, 4.0, 5.0 and 6.0 wt% (preparation "B"). The uncured composites were cast into the thin structures by filling them to overflowing using a small metallic spatula and then removing the excess material from the stripes by dragging a cutter blade across them. The composites with low nanotube concentrations (<1.5 wt%) were cast into the 500 µm-thick blocks for practical reasons; on the other hand, it was easier to fill the 200 µm-deep structures when dealing with higher concentrations. For each stripe two or more I-V measurements were collected depending on the concentration of carbon nanotubes. For samples with very low concentrations we were able to record voltage drops for a limited current ranges because of their high resistance. The 0.7, 1.0 and 1.2 wt% samples presented non-linear I-V curves, as shown in Figure 7 for the last two CNT concentrations. The curves attain linearity around the origin that is lost for more intense electric currents. However, for nanotubes concentrations exceeding 1.5 wt%, the I-V curves were near perfect straight lines, thus differing from what was ascertained during bulk resistance measurements.

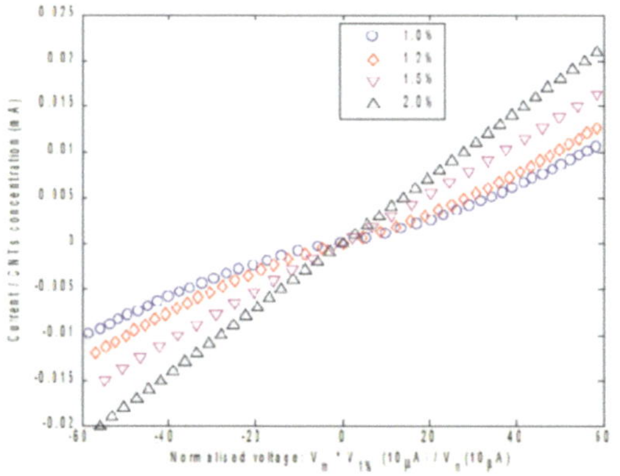

(a)

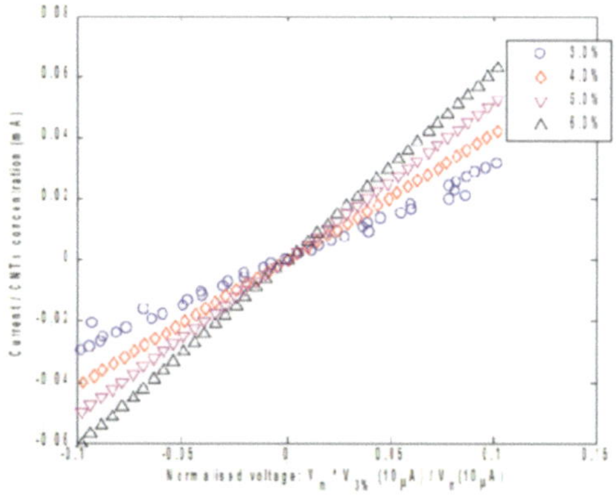

(b)

Fig. 7. (a) Normalised I-V curves of four nanocomposites (1.0, 1.2, 1.5 and 2.0 wt%, prep. "A"). The input current was set to cover the range from -10µA to +10µA during the experiment. The curves belonging to the 1.0 wt% and 1.2 wt% composites exhibit linearity around the origin and a parabolic-like trend for intense electric currents. Their differential resistance is not constant, but rather attains a maximum about the origin. (b) Normalised I-V curves for four high-CNT concentrations (prep. "B"). The curves are all linear. For all samples the input current was [-10,+10] µA.

Their trends are shown in Figure 7 (a) and (b). In order to represent their values on a common scale, the measured voltages belonging to each composition (V_m in the figures) have been normalised so that they occupy the same tension scale of the 1 wt% sample (Figure 7 (a)) and the 3 wt% sample (Figure 7 (b)). Having the same voltage range, the currents were differed from one percentage to the other by normalising the input values with respect to the CNTs content.

The fluctuations-induced tunnelling model that supposedly governs the electron transport in the composites may still be well-grounded. The fact is that the non-linear behaviour contemplated by the model is present in those nanocomposites with filler concentrations very close to the percolation threshold. But as the nanotube content is increased the insulating barriers that allegedly separate contiguous CNTs are reduced to such an extent that electrons have limited difficulty penetrating them. Our I-V curves tally quite well with what has been reported in literature. Liu and Fan [Liu & Fan, 2007] attributed the observed nonlinearity to the formation of a Schottky-like contact, but this claim seems preposterous in the light of our I-V curves. The electrical characteristic for a Schottky contact is roughly similar to that of a classic p-n junction, which is above all asymmetric about the origin. Our curves, even the non-linear ones, are instead symmetric. The single sheet resistance value was obtained for each stripe and it can then be turned into a bulk resistivity value by virtue of the following equation:

$$\rho = R_{sheet}\frac{d}{d_{el}}w\,,$$

where ρ is the bulk resistance, d and w are the width and thickness of the stripes respectively, and d_{el} is the average distance between two electrodes.

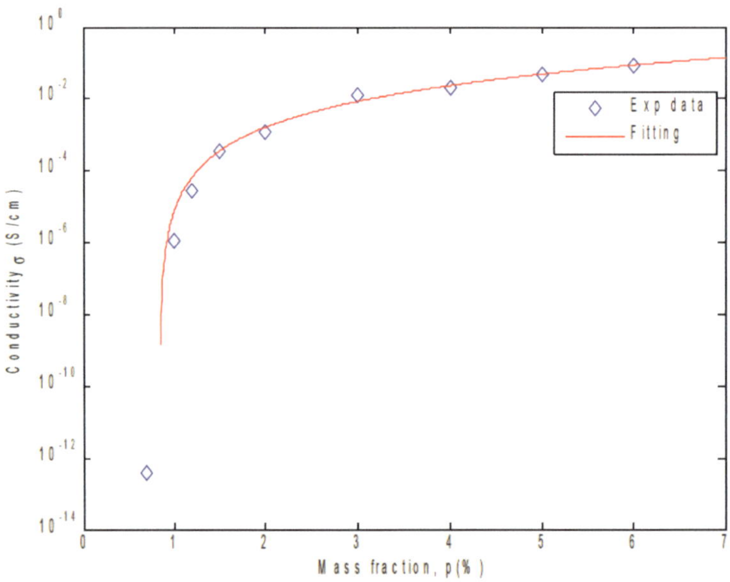

Fig. 8. Conductivity values of the PDMS/MWCNT nanocomposites for different filler concentrations (blue rhombus). Experimental data were fitted with the scaling law from percolation theory (red line). The fitting curve can be expressed as: $\sigma=1.1\cdot(p-0.85)^{2.67}$ mS/cm.

As visually represented in Figure 8, the conductivity jumps up by several order of magnitudes passing from low loadings around the percolation threshold (0.7 and 1.0 wt%) to those above that critical value. The conductivity continues to rise but seemingly its increasing trend gets sluggish for higher concentrations (5.0 and 6.0 wt%). This probably points to the existence of a plateau at higher concentrations, as indeed envisioned by percolation theory. The data shown in Figure 8 were fitted by the well-known percolation scaling law:

$$\sigma = \sigma_0 (p - p_c)^t ,$$

where σ_0 is a proportionality constant, p_c is the percolation threshold, and t is the critical exponent that characterise the percolative network [Stauffer & Aharony, 1994]. The outcome of the fitting delivered the following parameter values:

- σ_0 = 1.1 mS/cm;
- p_c = 0.85 %;
- t = 2.67.

According to the percolation theory, the power of the scaling law has the remarkable feature that it is entirely independent of the kind of system being studied, and it only depends on the dimensionality of space. The scaling law and the critical exponent are universal and it is only the percolation threshold that changes according to the percolating system under analysis. Computer simulations on "ideal systems" have all consistently shown that the critical exponent is exactly equal to 2.0. However, more realistic simulations [Stanley, 1977; Kogut & Straley, 1979] have demonstrated that the critical exponent can exceed the universal value of 2.0, with maximum values that are set by a combination of different geometrical factors. Many works in the literature regarding polymer/CNT nanocomposites report t values between 2 and 3 [McLachlan, 2005 and references therein].

4.2 Thermal transport characterization

Thermal transport properties are a crucial characteristic to face in designing disposable biodevices. For instance, the efficiency and selectivity of Polymerase Chain Reaction (PCR) is strictly related to the execution of fast thermal transitions among the three temperatures required by the protocol. Polymeric materials are very promising in the fabrication of biodevices, but their low thermal conductivity is often a bottleneck in the design of highly efficient systems. The development and study of innovative materials and technological processes gained a crucial importance to give life to a new generation of small, portable and fast devices. Polymeric composites, in the case of filling with thermally conducting nanomaterials, can show a dramatic improvement in thermal response with respect to pure polymers. Since the biocompatibility of PDMS is well known, lowering its thermal inertia can open new interesting ways for its widespread application. Some works about that topic were recently published. C.H. Liu et al. [Liu et al, 2004] reported an enhancement of 65% in thermal conductivity with 4 wt% MWCNT loading in the silicone elastomer. The increase in thermal conductivity is nowhere near as dramatic as the enhancement in electrical conductivity. T. Borca-Tasciuc [Borca-Tascuic, 2007] has argued that heat transport is inefficient owing to the high thermal resistance between neighbouring nanotubes and between nanotubes and the matrix. In order to maximise heat transport, the authors infiltrated the polymer on long, vertically-aligned carbon nanotubes. In this way, the number of tube/polymer interfaces in the direction of charge flowing is minimised. Composites grown in line with this approach will display a significant level of anisotropy as far as thermal properties are concerned in the directions parallel and perpendicular to the aligned nanotubes. Through a photo-thermoelectric technique they inferred that the thermal diffusivities are insensitive in the range of 180÷300 K. Moreover, they found that the thermal

diffusivities in the perpendicular direction are ~2÷4 times smaller than in the parallel direction and larger than effective media theory predictions using reported values for the thermal diffusivity of millimetre thick aligned multi-walled carbon nanotube arrays.
Here we report on the thermal transport properties characterization of PDMS/CNT nanocomposite with different filler content by means of the laser flash technique [Parker et al, 1961]. This methodology allows measuring the room temperature thermal diffusivity (*a*). A pulsed laser (wavelength 1064 nm, energy 3 J, spot area 0.8 mm^2) impinged on cylindrical shaped samples with a diameter of 1 cm and a thickness of ~1 mm (measured with a Mitutoyo thickness gage, resolution 0.001 mm). Since the polymeric matrix is completely transparent in the IR region, a relatively thick chromium layer was deposited by RF magnetron sputtering on the surface exposed to laser radiation, to guarantee complete adsorption of the impinging energy. Temperature was measured on the opposite side with a thermocouple.
Laser flash characterization results are reported in Figure 9. Data were analyzed following the model proposed by Parker et al. [Parker et al, 1961] that links the thermal diffusivity *a* with sample thickness *L* and time $t_{1/2}$ required for the back surface to reach half of the maximum temperature rise:

$$\alpha = 1.38 \frac{L^2}{\pi^2 t_{1/2}}$$

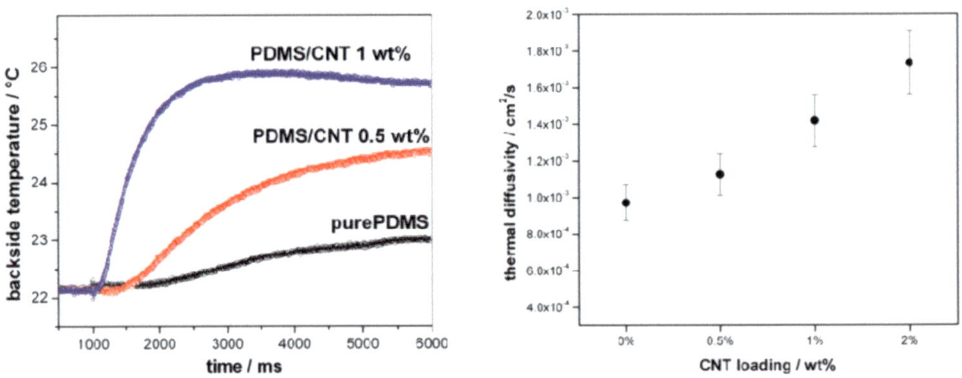

(a) (b)

Fig. 9. (a) Raw data obtained by laser flash measurements for pure PDMS and PDMS/CNT composites with 0.5 and 1 wt% CNT loading. (b) Thermal diffusivities for pure PDMS and PDMS/CNT nanocomposites with 0.5, 1 and 2 wt% filler content.

The improved thermal response is evident both in the raw curves and on the thermal diffusivity evaluations. It emerges that a ~50% enhancement in α value can be achieved with a 1 wt% nanotube charging. This result is perfectly compatible with the evaluation previously reported by Borca-Tasciuc et al. [Borca-Tascuic, 2007], since the 1.25 times increase in heat transport evaluated perpendicularly to the axis of the tubes is due to the entangled tubes, similarly to what happens in our isotropic system. A significant improvement can therefore be noticed, even with a very low filler loading.

5. Nanocomposite application for the fabrication of a stationary PCR biodevice

Recently we proposed the application of the PDMS/CNT nanocomposite for the fabrication of a miniaturized device for the Polymerase Chain Reaction for DNA amplification [Quaglio et al, 2011]. We demonstrated that the better thermal behaviour for devices made of the nanocomposite material with respect to the one made of pure PDMS allowed faster temperature transitions and a better control in steady-state temperatures. The better thermal performance of the nanocomposite-based device was mirrored by an efficiency increase in PCR reaction, demonstrating the direct enhancing effect of the CNTs content in the composite.

An example of the fabricated device is reported in Figure 10. A LOC with a non-symmetric drop-like shape was designed (volume 15 μl, including inlet and outlet cylinders), to optimize the contact between the chamber surfaces and the liquids, avoiding air bubbles formation. The reaction chamber was bonded to a silicon substrate with a thin film (approximately 100 μm thick) of the same nanocomposite, to guarantee the fluidic structure sealing.

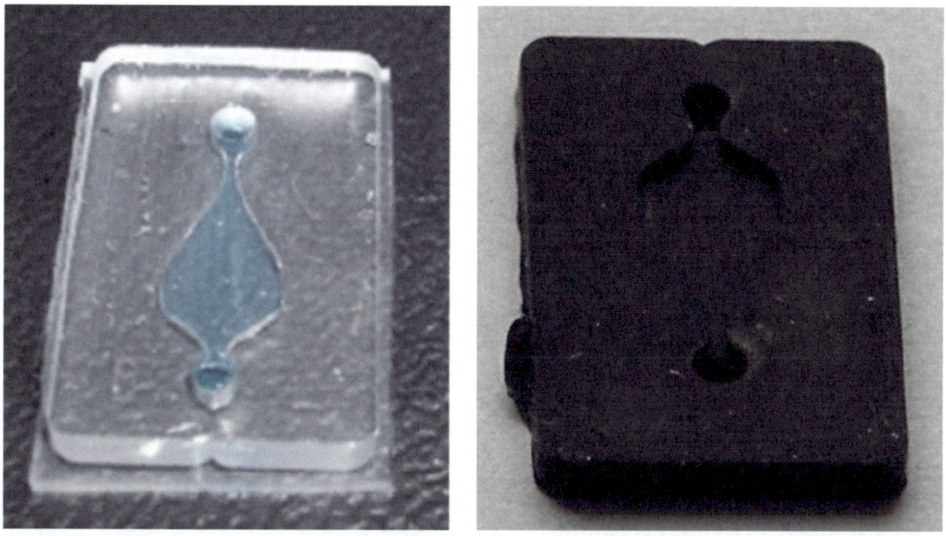

Fig. 10. Example of chips used for PCR protocols.

To estimate the thermal response on LOC device, the temperature profile during standard PCR thermocycling was evaluated directly on-chip. Temperature was measured with a thermocouple, inserted into the device close to the reaction chamber, but completely surrounded by the material, to minimize any external effect on the thermal response. On-chip thermal profiles are reported in Figure 11. We clearly observed a faster transition and a better control in steady-state temperatures when using the nanocomposite and while increasing the CNTs loading.

The performance of PDMS and PDMS/CNT LOCs was evaluated by the detection of PCR products related with the Human Actin Gene with a length of 150 base-pairs (bp) and with a different PCR protocol related with the Human β-Actin with a length of 756 bp. Details

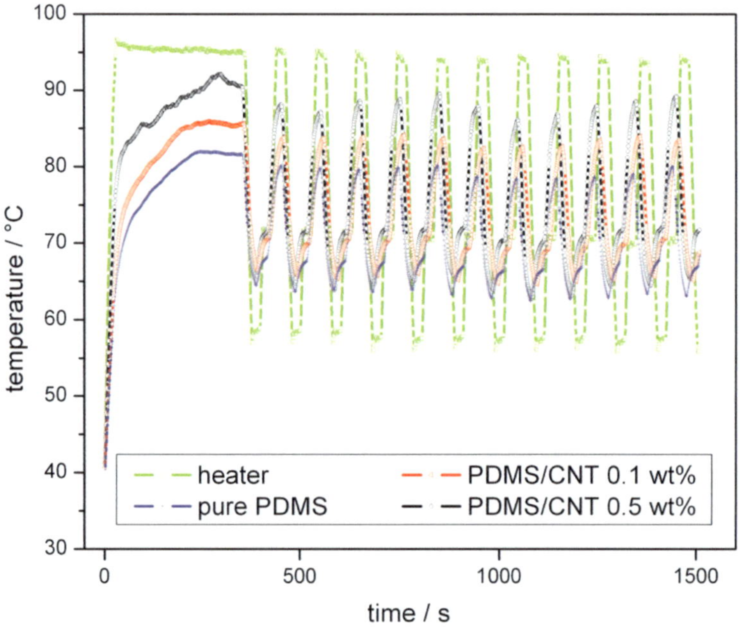

Fig. 11. Evaluation of the on-chip thermal profile during a standard PCR thermocycling, both for pure PDMS and for nanocomposites with different filler loading.

about the biological protocols are reported elsewhere [Marasso et al, 2011; Quaglio et al, 2011]. For both biological models, the PCR standard thermal protocol consisted on 5 min for pre-denaturing at 95 °C, 20 s for denaturing at 95 °C, 20 s for annealing at 55 °C, 20 s for extension at 72 °C repeated for 40 cycles, and finally a post-extension at 72 °C for 2 min. To enhance the effect of the improved thermal diffusivity of the nanocomposite, we designed a shortened protocol to allow stopping the fragment amplification during the exponential phase. Therefore, on the modified protocol the denaturing, annealing and extension steps were shortened to 10 s, the number of cycles was set to 20 and the final post-extension step was deleted.

The PCR products were analyzed by a 3% agarose gel (A5093, Sigma), electrophoresis separation stained by ethidium bromide (Sigma) and then visualized under UV light. The DNA Ladder used was the GeneRuler™ 100 base-pairs (SM0243, Fermentas). The pictures of the loaded agarose gels were analyzed with GelAnalyzer, a freeware 1D gel electrophoresis image analysis software.

Standard thermal protocol is designed for optimal PCR response in pure PDMS device. Following such protocol, PCR reaction reaches its saturation point and reaction efficiency can be evaluated from the intensity in the agarose gel picture. In other words, thermocycling is so slow to allow PDMS to reach sufficiently high temperature to permit the reaction to be efficiently performed. If the same protocol is applied to nanocomposite-based chip, exactly the same efficiency is noticed, because in both cases (on PDMS and PDMS/CNTs samples) the reactions reach the saturation plateau (and our experimental output is only related with the end-point result of the reaction). The standard protocol was used to estimate the maximum amount of amplicons concentration. To emphasize the improvement in reaction efficiency related to the improved thermal behaviour, it is mandatory to design a shortened protocol, to stop the fragment amplification during its exponential duplication and before the achievement of the plateau (the PCR phase defined as the attenuation in the rate of the exponential product accumulation) which is seen concomitantly in later PCR cycles [Kainz, 2000]. In such case, the faster transition for nanocomposite chip is expected to play a determining role in the end-point reaction output, with a significant difference with respect to what evaluated for pure PDMS chip.

Four PCR tests were performed using standard protocol, revealing a similar and reproducible concentration (Standard PCR tubes: (9.9 ± 0.2) g/l; PDMS: (9.8 ± 0.3) g/l; PDMS/CNT: (10.1 ± 0.2) g/l). Afterwards, standard PCR tubes and both PDMS and PDMS/CNT LOCs were tested in quadrupole with the shortened protocol. The agarose gel pictures reported in Figure 12 (a) indicated a DNA concentration of (9.9 ± 0.2) g/l for standard PCR tubes, (9.8 ± 0.1) g/l for the PDMS/CNT LOC and (7.6 ± 0.2) g/l for the PDMS LOC. Assuming that the concentration achieved with the standard protocol (10 g/l) was the maximum result for our bio-design, we estimated the relative efficiency of the PCR reactions. As reported in Figure 12 (c), while the DNA amplification for the shortened protocols on PDMS/CNTs LOC (97.6%) was comparable with the maximum value obtained with standard PCR tubes (98.7%), it was sensibly reduced on PDMS LOC (75.3%).

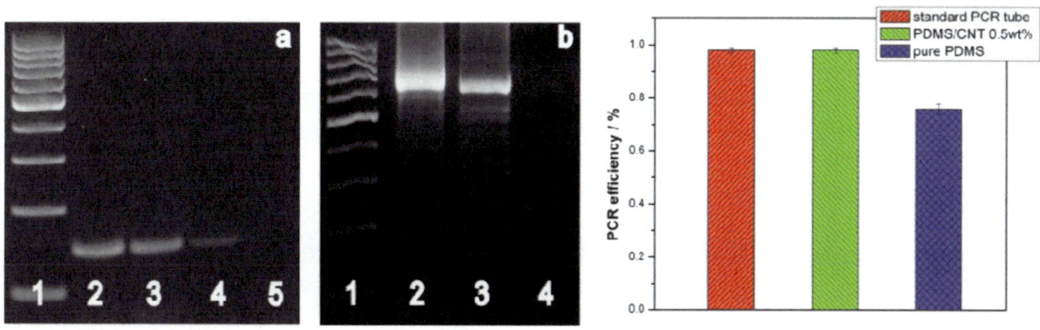

(a) (b) (c)

Fig. 12. (a) 3% agarose gel image: (1) 100 bp DNA Ladder (2) PCR amplicon obtained with standard PCR tube (3) PCR amplicon obtained with PDMS/CNT 0.5 wt% LOC (4) PCR amplicon obtained with PDMS LOC (5) negative control. (b) 3% agarose gel image: (1) 100 bp DNA Ladder (2) PCR amplicon obtained with PDMS/CNT 0.5 wt% LOC (3) PCR amplicon obtained with PDMS LOC (4) negative control. (c) Results of PCR reaction with shortened protocol. Efficiency equal to 1 is considered for PCR standard thermal protocol.

A subsequent test performed with the PCR protocol for Human β-Actin (756 bp) allowed evaluating the presence of non specific or uncompleted DNA sequences. As reported in Figure 12 (b), both LOC devices confirmed their performances concerning the amount of PCR production. Otherwise, while PDMS/CNT LOC devices showed no presence of nonspecific amplicons, the PDMS LOC revealed shorten non-specific amplicons.

With the reported experiments we were able to conclude that the PCR can be performed in nanocomposite-based chips with low filler content with a 25% improvement in reaction efficiency, without noting any non-specific duplication or enzyme segregation. In other words, a 75% reduction in reaction time can be applied without compromising the reaction efficiency.

5. Conclusions

Nanocomposite materials have generated considerable interest over the last few years and their number as well as their field of employment is predicted to rise. Among them, polydimethylsiloxane/carbon nanotube composites have only recently been investigated and the material has displayed a variety of properties that may be exploited in promising MEMS applications and biodevice fabrication. In this chapter we tried to give a general overview on the application reported recently in literature, trying to underline the possible future application and evolution of the studies. In our mind, the peculiar properties of the material were not completely fulfilled up to now, thus a huge amount of innovative applications still have to be developed. For instance, we would like to mention the possible application in energy harvesting (fully exploiting its photomechanical properties) and on the fabrication of all-polymeric biodevices based on the Electrowetting-On-Dielectrics model. These are the topics currently under evaluation in our research group and we are confident that some interesting results will be rapidly reached.

6. Acknowledgements

Authors would like to thank Nano Carbon Technologies Co Ltd (Japan) for nanotube supply. Moreover, the aid and contribution of Dr. E. Campagnoli, A. Chiodoni and S. Guastella from Politecnico di Torino (Italy) is gratefully acknowledged.

7. References

Ahir, S.V. & Terentjev, E.M. (2005). Photomechanical actuation in polymer–nanotube composites. *Nature Materials*, 4, (2005), pp. 491-495.

Ahir, S.V., Squires, A.M., Tajbakhsh, A.R. & Terentjev E. M. (2006). Infrared actuation in aligned polymer-nanotube composites. *Physical Review B*, 73, 8, (2006), pp. 085420_1-12.

Ajayan, P.M. & Tour, J.M. (2007). Nanotube composites. *Nature*, 447, (2007), pp. 1066-1068.

Barber, A.H., Cohen, S.R. & Wagner, H.D. (2004). Static and Dynamic Wetting Measurements of Single Carbon Nanotubes. *Physical Review Letters*, 92, 18, (2004), pp. 186103_1-4.

Baskaran, D., Mays, J.W., & Bratcher, M.S. (2005). Noncovalent and Nonspecific Molecular Interactions of Polymers with Multiwalled Carbon Nanotubes. *Chemistry of Materials*, 17, 13, (2005), pp. 3389-3397.

Beigbeder, A., Linares, M., Devalckenaere, M., Degée, P., Claes, M., Beljonne, D., Lazzaroni, R. & Dubois P. (2008). CH-π Interactions as the Driving Force for Silicone-Based Nanocomposites with Exceptional Properties. *Advanced Materials*, 20, 5, (2008), pp. 1003-1007.

Bokobza, L. (2004). Elastomeric Composites. I. Silicone Composites. *Journal of Applied Polymer Science*, 93, 5, (2004), pp. 2095-2104.

Bokobza, L., Rahmani, M., Belin, C., Bruneel, J.-L. & El Bounia, N.-E. (2008). Blends of Carbon Blacks and Multiwall Carbon Nanotubes as Reinforcing Fillers for Hydrocarbon Rubbers. *Journal of Polymer Science Part B: Polymer Physics*, 46, 18, (2008), pp. 1939-1951.

Bokobza, L. (2009). Some Issues in Rubber Nanocomposites: New Opportunities for Silicone Materials. *Silicon*, 1, 3, (2009), pp. 141-145.

Borca-Tascuic, T., Mazumder, M., Son, Y. Pal, S.K., Schadler, L.S. & Ajayan, P.M. (2007). Anisotropic Thermal Diffusivity Characterization of Aligned Carbon Nanotube-Polymer Composites. *Journal of Nanoscience and Nanotechnology*, 7, 4-5, (2007), pp. 1581-1588.

Chung, D.D.L. (2010). *Composite Materials, Science and Applications* (2nd edition), Springer, ISBN 9781848828308, London.

Clarson, S.J. & Semlyen, J.A. (1993). *Siloxane Polymers*, Englewood Cliffs, ISBN 0138163154, New York.

Demir, M.M., Menceloglu, Y.Z. & Erman, B. (2005). Effect of filler amount on thermoelastic properties of poly(dimethylsiloxane) networks, *Polymer*, 46, 12, (2005), pp. 4127-4134.

Engel, J., Chen, J., Chen, N., Pandya, S. & Liu, C. (2006). Multi-Walled Carbon Nanotube Filled Conductive Elastomers: Materials and Application to Micro Transducers. *Proceedings of the 19th IEEE International Conference on Micro Electro Mechanical Systems*, ISBN 0-7803-9475-5, Istambul, Turkey.

Franta, I. (1988). *Elastomers and rubber compounding materials*, Elsevier, ISBN 0444429948, Amsterdam.

Harris, P.J.F. (2009). *Carbon Nanotube Science*, Cambridge University Press, ISBN 9780521828956, Cambridge.

Hillborg, H. & Gedde, U.W. (1999). Hydrophobicity Changes in Silicone Rubbers. *IEEE Transactions on Dielectrics and Electrical Insulation*, 6, 5, (1999), pp. 703-717.

Hu, C.H., Liu, C.H., Chen, L.Z., Peng, Y.C. & Fan, S.S. (2008). Resistance-pressure sensitivity and a mechanism study of multiwall carbon nanotube networks/poly(dimethylsiloxane) composites. *Applied Physics Letters* 93, 3, (2008), pp. 033108_1-3.

Kainz, P. (2000). The PCR plateau phase - towards an understanding of its limitations. *Biochimica et Biophysica Acta (BBA) - Gene Structure and Expression*, 1494, 1-2, (2000), pp. 23-27.

Kang, I., Schulz, M.J., Kim, J.H., Shanov, V. & Shi, D. (2006). A carbon nanotube strain sensor for structural health monitoring. *Smart Materials and Structures*, 15, 3, (2006), pp. 737-748.

Kim, J., Chaudhury M.K. & Owen, M.J. (1999). Hydrophobicity Loss and Recovery of Silicone HV Insulation. *IEEE Transactions on Dielectrics and Electrical Insulation*, 6, 5, (1999), pp. 695-702.

Kim, Y.A., Hayashi, T., Endo, M., Kaburagi, Y., Tsukada, T., Shan, J., Osato, K. & Tsuruoka, S. (2005). Synthesis and structural characterization of thin multi-walled carbon nanotubes with a partially facetted cross section by a floating reactant method. *Carbon*, 43, 11, (2005), pp. 2243-2250.

Khorasani, M.T., Mirzadeh, H. & Kermani, Z. (2005). Wettability of porous polydimethylsiloxane surface: morphology study. *Applied Surface Science*, 242, 3-4, (2005), pp. 339-345.

Khosla, A. & Gray, B.L. (2009). Preparation, characterization and micromolding of multi-walled carbon nanotube polydimethylsiloxane conducting nanocomposite polymer. *Materials Letters*, 63, 13-14, (2009), pp. 1203-1206.

Kogut, P.M. & Straley, J.P. (1979). Distribution-induced non-universality of the percolation conductivity exponents. *Journal of Physics C: Solid State Physics*, 12, 11, (1979), pp. 2151-2160.

Kovacs, G.T.A. (1998). *Micromachined Transducers SourceBook*, McGraw-Hill, ISBN 0072907223, Boston.

Lavielle, L. & Schultz, J. (1985). Surface properties of graft polyethylene in contact with water. *Journal of Colloid and Interface Science*, 106, 2, (1985), pp. 438-445.

Liu, C.H., Huang, H., Wu, Y. & Fan, S.S. (2004). Thermal conductivity improvement of silicone elastomer with carbon nanotube loading. *Applied Physics Letters*, 84, 21, (2004), pp. 4248-4250.

Liu, C.H. & Fan, S.S. (2007). Nonlinear electrical conducting behavior of carbon nanotube networks in silicone elastomer. *Applied Physics Letters*, 90, 4, (2007), pp. 041905_1-3.

Liu, C.-X. & Choi, J.-W. (2009). Patterning conductive PDMS nanocomposite in an elastomer using microcontact printing. *Journal of Micromechanics and Microengineering*, 19, (2009), pp. 085019_1-7.

Lu, S. & Panchapakesan, B. (2005). Optically driven nanotube actuators. *Nanotechnology*, 16, 11, (2005), pp. 2548-2554.

Lu, S. & Panchapakesan, B. (2006). Nanotube micro-optomechanical actuators. *Applied Physics Letters*, 88, (2006), pp. 253107_1-3.

Lu, S. & Panchapakesan, B. (2007). Photomechanical responses of carbon nanotube/polymer actuators. *Nanotechnology*, 18, 30, (2007), pp. 305502_1-8.

Marasso, S.L., Giuri, E., Canavese, G., Castagna, R., Quaglio, M., Ferrante, I., Perrone D. & Cocuzza, M. (2011). A multilevel Lab on chip platform for DNA analysis. *Biomedical Microdevices*, 13, 1, (2011), pp. 19-27.

Matthews, F.L. & Rawlings, R.D. (1999). *Composite Materials: Engineering and Science*, CRC Press, ISBN 0-8493-0621-3, Cambridge.

McDonald, J.C. & Whitesides, G.M. (2002). Poly(dimethylsiloxane) as a Material for Fabricating Microfluidic Devices. *Accounts of Chemical Research*, 35, 7, (2002), pp. 491-499.

McLachlan, D.S., Chiteme, C., Park, C., Wise, K.E., Lowther, S.E., Lillehei, P.T., Siochi, E.J. & Harrison, J.S. (2005). AC and DC percolative conductivity of single wall carbon nanotube polymer composites. *Journal of Polymer Science Part B: Polymer Physics*, 43, 22, (2005), pp. 3273-3287.

Moniruzzaman, M., Chattopadhyay, J., Billups, W.E. & Winey, K.I. (2007). Tuning the Mechanical Properties of SWNT/Nylon 6,10 Composites with Flexible Spacers at the Interface. *Nano Letters*, 7, 5, (2007), pp. 1178-1185.

Musso, S., Giorcelli, M., Pavese, M., Bianco, S., Rovere, M. & Tagliaferro, A. (2008). Improving macroscopic physical and mechanical properties of thick layers of aligned multiwall carbon nanotubes by annealing treatment. *Diamond and Related Materials*, 17, 4-5, (2008), pp. 542-547.

Okawa, D., Pastine, S.J., Zettl, A. & Fréchet, J.M.J. (2009). Surface Tension Mediated Conversion of Light to Work. *Journal of the American Chemical Society*, 131, 15, (2009), pp. 5396-5398.

Parker, W.J., Jenkins, R.J., Butler, C.P. & Abbott, G.L. (1961). Flash Method of Determining Thermal Diffusivity, Heat Capacity, and Thermal Conductivity. *Journal of Applied Physics*, 32, 9, (1961), pp. 1679-1684.

Quaglio, M., Bianco, S., Castagna, R., Cocuzza, M. & Pirri, C.F. (2011). Elastomeric nanocomposite based on carbon nanotubes for Polymerase Chain Reaction device. *Microelectronic Engineering*, (2011), doi:10.1016/j.mee.2011.01.032.

Reyes, D.R., Iossifidis, D., Auroux, P.A. & Manz, A. (2002). Micro Total Analysis Systems. 1. Introduction, Theory, and Technology. *Analytical Chemistry*, 74, 12, (2002), pp. 2623–2636.

Sheng, P. (1980). Fluctuation-induced tunneling conduction in disordered materials. *Physical Review B*, 21, 6, (1980), pp. 2180-2195.

Stanley, H.E. (1977). Cluster shapes at the percolation threshold: and effective cluster dimensionality and its connection with critical-point exponents. *Journal of Physics A: Mathematical and General*, 10, 11, (1977), L211.

Stauffer, D. & Aharony, A. (1994). *Introduction to Percolation Theory*. Taylor and Francis, ISBN 0748400273, London.

Tobolsky, A.V. (1960). *Properties and Structure of Polymers*, Wiley, New York.

Wang, T., Hu, X., Qu, X. & Dong, S. (2006). Noncovalent Functionalization of Multiwalled Carbon Nanotubes: Application in Hybrid Nanostructures. *Journal of Physical Chemistry B*, 110, 13, (2006), pp. 6631-6636.

Wu, C.-L., Li, H.-C., Hsu, J.-S., Yip, M.-C. & Fang, W.. (2009). Static and dynamic mechanical properties of polydimethylsiloxane/carbon nanotube nanocomposites, *Thin Solid Films*, 517, 17, (2009), pp. 4895-4901.

Wu, J., Cao, W., Wen, W., Chang, D.C. & Sheng, P. (2009) Polydimethylsiloxane microfluidic chip with integrated microheater and thermal sensor. *Biomicrofluidics*, 3, 1, (2009), pp. 012005_1-7.

Xia, Y. & Whitesides, G.M. (1998). Soft Lithography. *Angewandte Chemie International Edition*, 37, 5, (1998), pp. 550-575.

Xu, J., Razeeb, K.M. & Roy, S. (2008). Thermal Properties of Single Walled Carbon Nanotube-Silicone Nanocomposites. *Journal of Polymer Science Part B*, 46, 17, (2008), pp. 1845-1852.

Xu, W.J., Kranz, M., Kim, S.H. & Allen, M.G. (2010). Micropatternable elastic electrets based on a PDMS/carbon nanotube composite. *Journal of Micromechanics and Microengineering*, 20, (2010), pp. 104003_1-7.

Yakobson, B.I., Brabec, C.J. & Bernholc, J. (1996). Nanomechanics of Carbon Tubes: Instabilities beyond Linear Response. *Physical Review Letters* 76, 14 (1996), pp. 2511-2514.

Zhang, C., Xing, D. (2007). Miniaturized PCR chips for nucleic acid amplification and analysis: latest advances and future trends. *Nucleic Acids Research*, 35, 13, (2007), pp. 4223–4237.

Zhang, C., Xu, J., Ma, W. & Zheng, W. PCR microfluidic devices for DNA amplification. *Biotechnology Advances*, 24, 3, (2006), pp. 243-284.

16

Investigation of the Effective Reinforcement Modulus of Carbon Nanotubes in an Epoxy Matrix

Alfonso Martone, Gabriella Faiella, Mauro Zarrelli,
Vincenza Antonucci and Michele Giordano
*IMCB-CNR Institute for Composite and
Biomedical Materials - National Research Council
Italy*

1. Introduction

Polymer nano-composite matrix could be the ideal solution for a new generation of composite materials. The continuous demand for new high performance polymer composite for various applications, in different industrial sectors, has lead many researchers to investigate the potential use of the carbon nanotubes (CNTs) as nano-reinforcements of polymer matrix for the manufacturing of traditional laminate composite.

CNTs have attracted considerable attention due to their unique mechanical, surface, multi-functional properties and strong interactions with the hosting matrix mainly associated to their nano-scale features. CNTs' impressive mechanical properties, with stiffness and strength values falling within the range of 100-1000 GPa and 2.5-3.5 GPa, respectively, make them ideal candidates to develop novel composites characterized by advanced polymer matrices (Treacy, 1996).

Despite the enormous amount of experimental data available in literature (Coleman et al. 2006, Thosterston et al., 2003), there are still controversial results concerning elastic property, strength and fracture toughness; moreover, due to inherent difficulties in processing these unconventional nanostructure as nano-fillers in polymer system, a reliable theoretical correlation of the experimental data is still in shadow. Different approaches to build an appropriate theory for predicting reinforcement efficiency of CNTs within an hosting matrix have been presented in the literature.Indeed, the reinforcement capability of carbon nanotubes in a polymeric matrix will depend on their amount, but, undoubtedly, their arrangement within the hosting medium plays a fundamental role in the load transfer mechanism. For this reason, the state and level of dispersion need to be accounted in any attempt for predicting the mechanical behaviour of the final nano-composite system.

In literature the enhancing reinforcement of CNT loading for the Young's modulus is commonly reported. However, at the same time, discrepancy among the different data is highlighted. Therefore, an important issue for modelling purpose is the lack of a reliable database for this property.

Characterization and structure-properties of nano-mechanics modelling research have shown that enhancement in mechanical properties of nano-composites are strongly

dependent upon the level of dispersion and the final morphology of the nano-fillers. CNTs dispersion process still represents a critical issue to allow the potential usage of these nano-structure as reinforcements. Dispersion and homogenization stands as a very complex phenomenon due to the natural tendency of CNTs to bundle and to aggregate mainly due to Van der Walls interactions among nanotubes (Martone et al. 2010).
In literature (Martone et al. 2010), various techniques are reported about dispersion of nanotubes in polymer resins. However, despite the enormous number of published works, most of these methods are either limited in capacity or not powerful enough to separate the agglomerates into individual nanotubes. The interested reader can refer to update survey published by Hussein et al. A good dispersion and, possibly, the alignment of CNTs within a matrix represent still a challenge and they could operate as main factors to drive the diffusion of carbon nanotubes as nano-reinforcements on industrial scale. Although these exiting opportunities, till doubts arise concerning the more efficient procedure for the uniform dispersion of the nanotube within polymer system. Thus, experimental data on mechanical property of the final nano-composite appear not always reliable and useful.
A further issue related with the current state of the art within the area of CNT/polymer nano-composite is certainly represented by the achieved level of reinforcement, with single (SWCNT) or multiwalled carbon nanotubes (MWCNT) as fillers in hosting matrix.
From the theoretical point of view, different parameters could be considered to quantify the reinforcement effect of nano-filler within the final nano-composite system. Based on classical Cox's approach, developed for paper and other fibrous materials to analyse the effect of orientation of the fibres on the stiffness and strength, a suitable expression for the effective reinforcement modulus of paper and fibrous materials can be defined as follows:

$$E_{\eta} = \frac{\left(E_C - E_m\right)}{V_{filler}} + E_m \tag{1}$$

where E_C and E_m represent, respectively, the modulus for the final composite and the hosting matrix and V_{filler} is the nanotubes volume fraction. Eq. 1 represents the equivalent modulus of reinforcement according to the measured value of the final nano-composite. An expression for the reinforcement efficiency can be properly derived by the eq.1, considering the ratio between the effective reinforcement modulus and the nominal modulus of the reinforcement, as follows:

$$\eta_{ref} = \frac{E_{\eta}}{E_{filler}} = \frac{1}{E_{filler}} \frac{\left(E_C - E_m\right)}{V_{filler}} + \frac{E_m}{E_{filler}} \tag{2}$$

where E_{filler} represents the nominal modulus of the loaded filler.
Figure 1 reports the reinforcement efficiency, computed on literature data, taken from different research works.
The data reported, in fig.1, are related with lowest nanotubes content (dilute regime) which should correspond to the minimum level of nanotubes clustering and thus highest level of dispersion for each set of data. Since the nanotubes modulus values range from 1000 GPa to 3000 GPa according to Treacy, the enhancement of elastic modulus showed by the reported data appears extremely low with a remarkable spread.
Analysis of the data reported in fig.1 reveals that the obtained efficiency of reinforcement is sensibly lower than expected, supporting great efforts to develop more standardised manufacturing processes and, hence, to gather more reliable data.

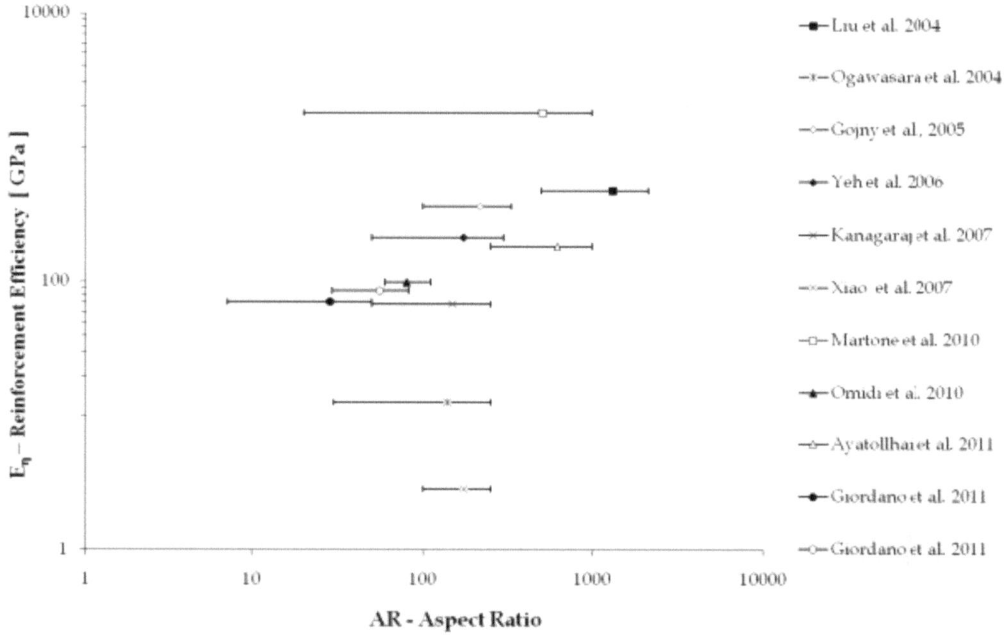

Fig. 1. Reinforcement efficiency, E_η, as function of CNT average aspect ratio for literature data.

This chapter is divided into three main sections. In the first part, the available experimental data concerning the effect on mechanical behaviour of CNTs as nano-fillers will be presented and analyzed; in the second part, a survey of modelling attempts for this property will be presented in order to highlight the major features of reinforcement mechanisms due to CNTs. The last part of the chapter will present a further development of a simple rule-of-mixture based model which accounts for different fundamental features such as contacts and waviness of nano-reinforcing elements within a polymeric system.

2. State of the art on CNTs reinforcement effects

2.1 Experimental evidences

Toward the establishment of carbon nanotubes (CNTs) as prospective mechanical reinforcement of polymers, two fundamental issues need to be addressed: the bulk mechanical properties of carbon nanotube nano-composites as function of nano-filler properties (geometry and shape, i.e. aspect ratio and curliness) and, in addition, the effect of topological distribution of nanotubes within the hosting matrix.

Currently, only the first issue has been investigated thoroughly by using different experimental techniques and theoretical models, whilst the morphology of such nano-composites have been extensively studied only by means of electrical and thermal conductivity tests. Enhancing effect of carbon nanotubes in hosting polymer still remains under shadow for two main reasons: the unavailability of reliable experimental data and the unveiling mechanisms of reinforcement associated with nanotubes.

In this paragraph, the first section will review the experimental data regarding the mechanical behaviour of CNT-polymer nano-composite, currently available in the literature;

finally, in the second part, the main developed models to predict the enhancement effect will be presented and described.

2.2 Mechanical behaviour of nano-composites

Given the potential of the CNTs as reinforcement agents, several researchers have tenaciously pursued their use in polymer nano-composites, either thermoplastics or thermosetting based system. Despite the huge quantity of available experimental data, it is widely recognised that the results are still controversial.

A paramount issue within the area of nano-composite research and development is represented by the dispersion of the filler within the matrix. In fact, the level of dispersion, in terms of uniformity and disentanglement, influences not only directly, the arrangement of tubes and, hence, the final morphology of the nano-composite, but also, indirectly, their maximum content and their reinforcement mechanism features.

Early investigation on the mechanical behaviour of CNT composites have reported enhancements in Young modulus lower than expected, revealing also negligible increasing or in some cases, significant decrease in strength and toughness.

Ogawasara et al. 2004, have manufactured composites with nanotube content up to 14% in weight, experiencing a quasi-linear behaviour of elastic modulus (see figure 2). In this case, the effective reinforcement of the nanotubes, according to the Coleman parameter (eq.1) results very low (about E_{η}=13 GPa) and constant as nanotube content increases. Moreover, beyond a certain weight fraction, a significant reduction (-20%) of strength respect to the neat polymer was measured likely due to the inherent difficulties to uniformly disperse the high amount of nano-filler within the hosting matrix.

An early work on mechanical behaviour of polymer loaded with nanotubes was published by Shaffler and Windle in 1999, considering multi-walled carbon nanotubes in PVA system. The reported amount of nano-filler, used in this work, appears incredible high (up to 60%) whereas the gaining in mechanical modulus, measured by dynamic mechanical analyser (DMA), results extremely low (η_E=4.7 GPa). A complete review up to 2006 for the nominal reinforcement efficiency on Young modulus for various hosting systems, different carbon nanotubes types and dispersive technique is reported in Coleman 2006.

While scientists have studied the CNT induced reinforcement focusing on systems characterised by an "enormous" nano-filler content, other researchers have pointed out the importance of the percolative behaviour of carbon nanotubes (Kovacs et al. 2007, 2009) and have particularly analysed the enhancement contribution of electrical property. It is well know, in fact, that onset transition between insulator conductor behaviour falls at CNT content around 0.10%w, according to both nanotubes aspect ratio and final morphology.

Studies on low volume fraction nano-composites have highlighted an higher mechanical efficiency associated to carbon nanotubes. De Zhang et al., 2004 reported a modulus variation from 0.396 GPa to 0.852 GPa by using only 1 %wt of MWCNTs in Polyamide 6 corresponding to a reinforcement of 91 GPa. They also observed a significant increase in yield strength from 18 to 40 MPa. These results were mainly associated to the achieved good level of dispersion as revealed by the performed microscopy analysis and the reinforcement efficiency was varying according to the CNT content.

Liu et al, 2004 studied the mechanical performances of high aspect ratio nanotubes in a thermoplastic matrix at low volume fraction (<1%). The experimental data of bending modulus and yield strength, presented in this work, showed that the incorporation of a small amount of MWNTs into a polymer matrix can significantly improve those mechanical

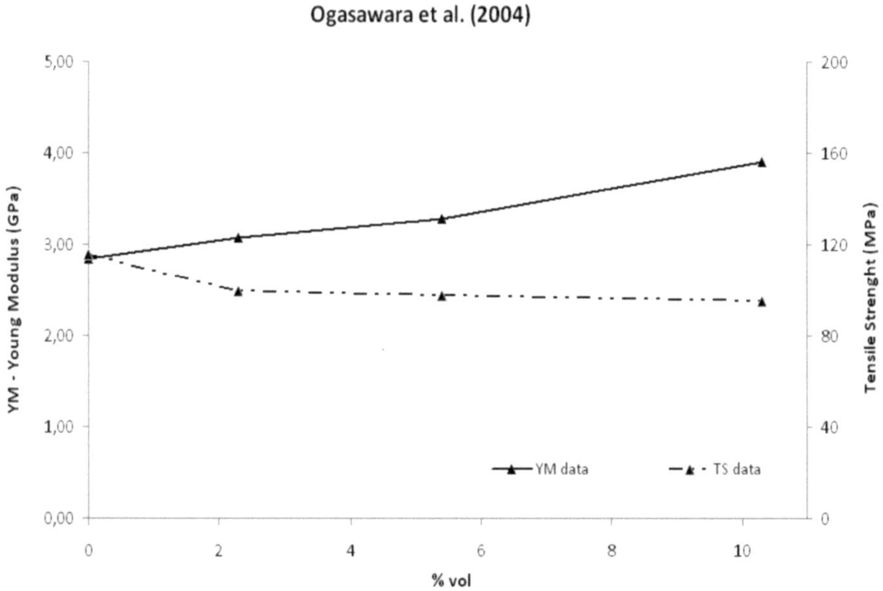

Fig. 2. Young Modulus (YM) and Tensile Strength (TS) data vs nanotube content. Ogawasara et al., 2004;

properties (see figure 3). Microscopy observations indicated the successful achievement of a uniform and fine dispersion of MWNTs throughout PA6 matrix and a strong interfacial adhesion between nanotubes and the matrix, which are responsible for the remarkable enhancements in overall mechanical properties.

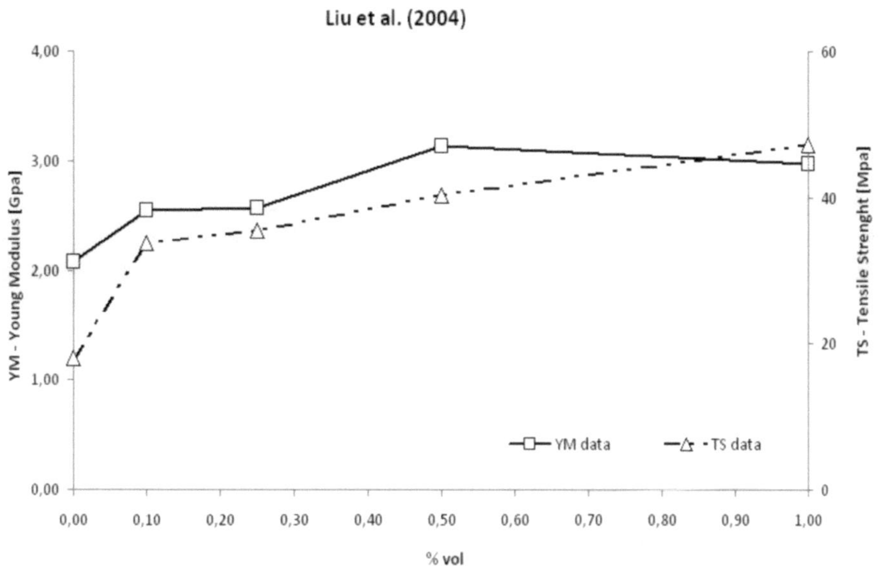

Fig. 3. Young Modulus (YM) and Tensile Strenght (TS) data vs nanotube content. Liu et al., 2004;

Data published by Ayatollahi et al. 2011 report a direct proportionality of both Young's modulus and tensile strength for a epoxy polymer filled with different nanotube contents (see figure 4). A limited reinforcement efficiency for the Young modulus and slight increase of the tensile strengths is noticeable.

Very recent data published by Martone et al. (2010) and Martone et al. (2011) revealed that the Young modulus of carbon nanotubes within an hosting polymer system strongly depend upon nano-filler aspect ratio which also affects the critical nanotube content corresponding to a fully connected path (see figures 5). The averaged aspect ratios of the nano-fillers were, respectively, equal to 29 and 55 and 505. It has been verified that, for all cases, the traditional rule of mixture cannot predict correctly the experimental values. In fact, a highly non linear and non monotonic trend, as function of the MWCNT content, was found and a maximum value was observed for each nano-composites modulus curve.

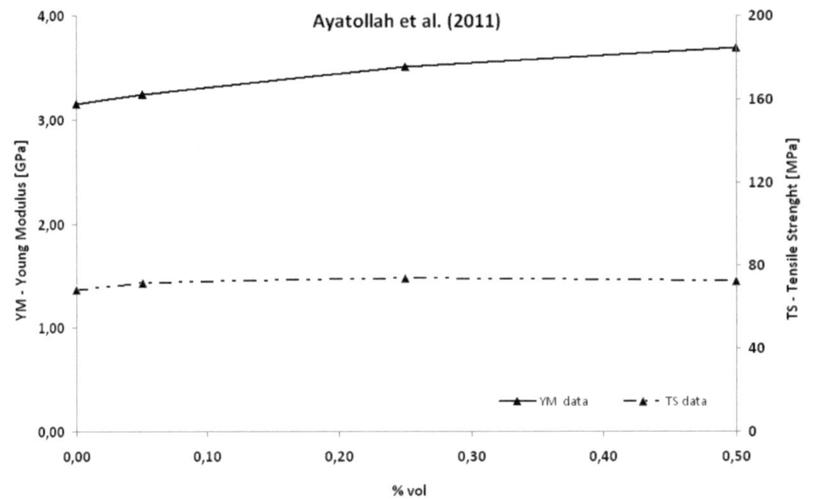

Fig. 4. Young Modulus (YM) and Tensile Strenght (TS) data vs nanotube content. Ayatollahi et al. 2011.

The effective reinforcement modulus, E_{η}, results a monotonic decreasing function of the volume nanotubes content for each considered aspect ratios. CNTs contribute to the composite mechanical stiffness with an effective modulus that decreases with arising of the aspect ratio values in the low concentration region and it continuously decreases as the filler content increases. It can be noticed that while the curves trend is independent on the CNT aspect ratio, the CNT reinforcement in very dilute regime ($V_{NT} \rightarrow 0$) is severely a function of nanotube aspect ratio. It worth to note also that the transition region do not overlay with the experimental percolation threshold leading to the conclusion that the progressive development of the nanotubes network within the matrix do affect the mechanical efficiency of the filler even below the formation of the percolative path.

It is worth to highlight the non-linear behaviour of Young's modulus as function of CNT content reported by the curves in fig.5. In fact, the relative enhancement in composite modulus decreases progressively with filler fraction according to Ogasawara, Liu and Ayatollahi, in turn the tensile strength increases linearly with the nanotube content for Ogasawara and Liu, while decreases for higher filler percentage in case of Ayatollahi.

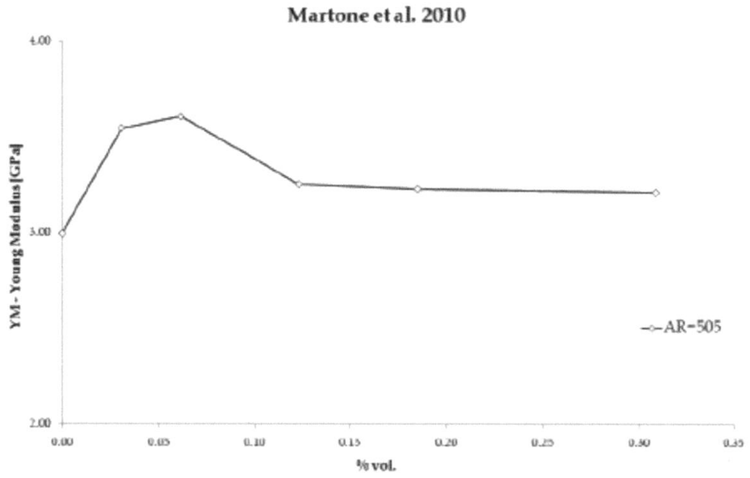

a)

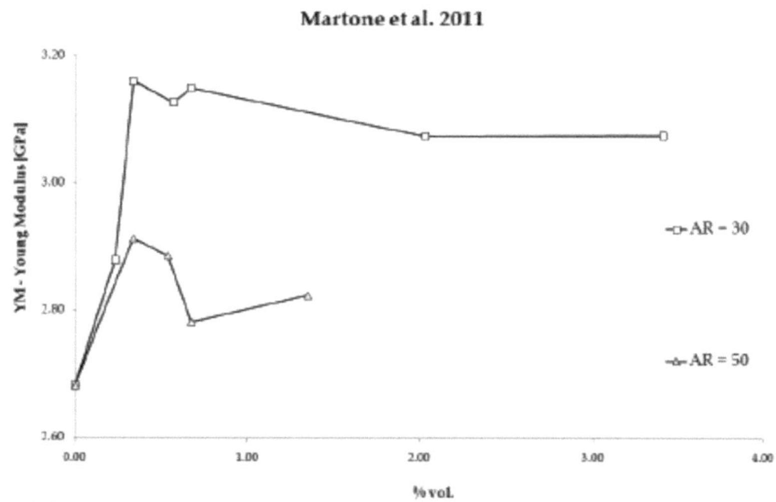

b)

Fig. 5. Young Modulus (YM) data vs. CNT content from a) Martone et al. (2010) and b) Martone et al. (2011)

Figure 6 shows data sets published among the 2004 and the 2011 for the Young's modulus of CNT-polymer nano-composite at different CNT contents revealing a controversial dependency and very different trends. The reported data sets have been grouped in three distinct categories, according to their functionality with CNT content: inversely linear behaviour in blue colour , non linear behaviour in red colour and almost constant curves in purple.

A linear trend of curve can be recognised for some of these data sets (*blue sets*) whereas some other sets highlight a non-linear functionality with CNT content showing a enhancement of the modulus at certain percentage (*red sets*). In some other cases a negligible proportionality with CNT content (*purple sets*) is recorded conversely with previous data.

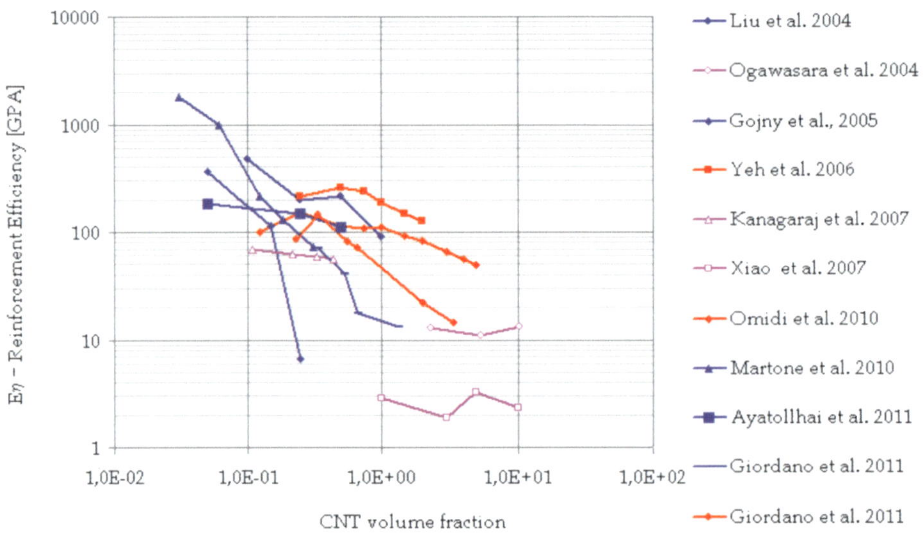

Fig. 6. Available literature data of reinforcement efficiency, E_η , vs. CNT content.

Further contradictory findings on experimental data relative to mode I fracture toughness (K_{Ic}) are reported in literature, thus driving the actual effort of many researchers to attain better and more reliable experimental information on effective mechanism of reinforcement and enhancement. The geometrical features of carbon nanotubes and processing techniques strongly affect the final results as also confirmed by limited amount of fracture toughness data.

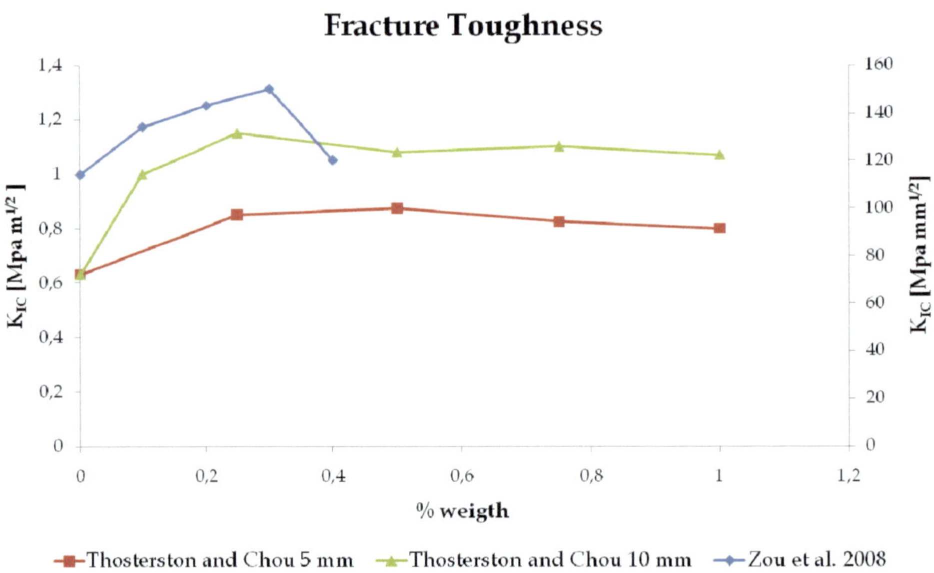

Fig. 7. Fracture toughness of CNTs composites. a) Thostenson and Chou, 2006. b) Zhou et al. 2008.

Thostenson and Chou, 2006 used the calendering technique to disperse MWCNTs in epoxy and used single-edge notch bend (SENB) specimen to measure K_{Ic}. In this work, nano-composites at different loading contents were manufactured by tuning the dispersion parameters (distance between calendar cylinders, temperature, cylinder rates etc…). Both the investigated nano-composites exhibits a maximum enhancement in fracture toughness at low CNTs content (comparable with statistical percolation threshold), followed by a decreasing behaviour (figure 7). The measured maximum K_{Ic} was obtained for 0.2 wt.% of MWCNTs when the gap between the rolls of calendering machine was 10 μm.
Zhou et al., 2008 investigated the fracture toughness behaviour at low weight content of CNTs recording similar enhancements (figure 7). In this latter case CNTs were dispersed in the matrix by ultra-sonication technique.
Experimental data suggest that CNTs would be able to exploit their potential as structural reinforcement when are nearly isolated in the hosting medium. This latter condition requires not only to manufacture the nano-composites at low CNT volume content, but also to employ high energy dispersion techniques capable to de-bundle the pristine nanotube ropes and aggregates.
A potential approach to enhance the load transfer mechanism matrix could be attained by strengthening the interfacial bonding among nanotubes and hosting system. The functionalization of CNT-surfaces by tailored chemical groups, such as carboxyl- or glycidyl-groups could enable covalent bonding between CNTs and hosting system, improving the interfacial stress transfer. Gojny et al. 2005 have investigated the effect of nanotube functionalization on the ultimate strength and fracture toughness of composite showing the mechanical performances at low nanotube content could be improved. A study of strengthen mechanism due chemical functionalization of the nanotubes is out of the scope of this work.

2.3 Modelling approaches

In spite of the outstanding mechanical properties of single tubes, long fibre reinforced nanopolymer composites exhibit a very limited improvement of mechanical performances, if compared to conventional advanced composites. This contradictory behaviour can be explained by considering that the reinforcing contribution of carbon nanotubes is yielded not only by their amount within the material, but also by the state of dispersion, orientation, shape and number of contacts within the hosting system. All these features play a critical role on the final reinforcement enhancement, and they should be taken into account to develop reliable models for prediction of nano-composite effective properties.
Since CNTs could be assimilated to "*tiny*" short particles, the classical micromechanics approaches for short fibre reinforced composites (SFRC) were employed as first attempts to develop predictive models of CNT/polymer composites (Coleman et al.,2006, Thostenson and Chou).

Voight and Reuss bounding models

A preliminary analysis to quantify the composite modulus can be performed by estimating the upper and the lower bounds. Bounding methods approximate both the stress and the strain field within the composite allowing the estimation of the modulus by a minimization of the overall strain energy. These methods cannot predict, directly, the composite stiffness, whereas they provide a proper range for the real composite stiffness.
The Voight and Reuss method represents the most simple bounding model. The lower and upper bounds are evaluated imposing, that fibre and reinforcement are subjected to the

same stress and strain field. The main assumption of this model concerns to the isotropic behaviour of the final composite system and, hence, when reinforcement and matrix have very different stiffness, the predicted bounds are quite far apart. As rule of thumb, the elastic modulus of the composite could be evaluated averaging the bounding values according to the following expression:

$$E_c = \lambda E_{up} + (1-\lambda) E_{low} \tag{3}$$

The λ coefficient accounts for the reinforcement arrangement in the matrix, for random planar distribution it is assumed to be 3/8 (Tucker). The latter equation does not take into account any geometry of the fibre and it uses only three independent variables.

Halpin-Tsai model

A popular and widely adopted model to predict the stiffness of SFRC is the Halpin-Tsai equation (HT) that has been originally developed for continuous unidirectional composite (Halpin et al. 1976). This model is undoubtedly the most simple. The HT equations correlate empirically the property of composite material with specific characteristics of matrix and reinforcing phases together with their proportions and geometries. The model was derived by the work of Hermans, 1967 and Hill, 1964 by noting that three Hermans's equations for the stiffness could be re-written with a single expression according to the following equations:

$$\mathrm{E}_{\mathrm{NC}} = \left[\frac{3}{8} \frac{1+\varsigma \eta_{\mathrm{L}} \mathrm{v}_{\mathrm{NT}}}{1-\eta_{\mathrm{L}} \mathrm{v}_{\mathrm{NT}}} + \frac{5}{8} \frac{1+2\eta_{\mathrm{T}} \mathrm{v}_{\mathrm{NT}}}{1-\eta_{\mathrm{T}} \mathrm{v}_{\mathrm{NT}}} \right] \tag{4}$$

$$\eta_{\mathrm{L}} = \frac{\left(\mathrm{E}_{\mathrm{NT}} / \mathrm{E}_{\mathrm{m}} \right) - 1}{\left(\mathrm{E}_{\mathrm{NT}} / \mathrm{E}_{\mathrm{m}} \right) + \varsigma} \tag{5}$$

$$\eta_T = \frac{\left(E_{NT} / E_m \right) - 1}{\left(E_{NT} / E_m \right) + 2} \tag{6}$$

Halpin and Tsai, in adapting their approach to short-fibre composites, noted that the ζ parameter must lie between 0 and +∞ and they suggested also that this parameter could be correlated with the geometry of the reinforcement. Moreover, when calculating the longitudinal Young modulus, it should vary as a function of the fibre aspect ratio (figure 8), whereas for the other engineering constants, an independency upon the reinforcement shape of the filler, could be suitably assumed.

The efficiency of the Halpin-Tsai equations has been valuably assessed for low concentration of filler. However it has been verified that at high volume fractions, discrepant predictions are computed. Modification of the HT model was published during the years mainly trying to correlate the ζ parameter with volume fraction. Despite the encouraging agreement of the data, the corrections are still based on fitting parameters without any theoretical supports (Nielsen et al., 1970, Lewis et al., 1970 and Hewitt et al., 1970)

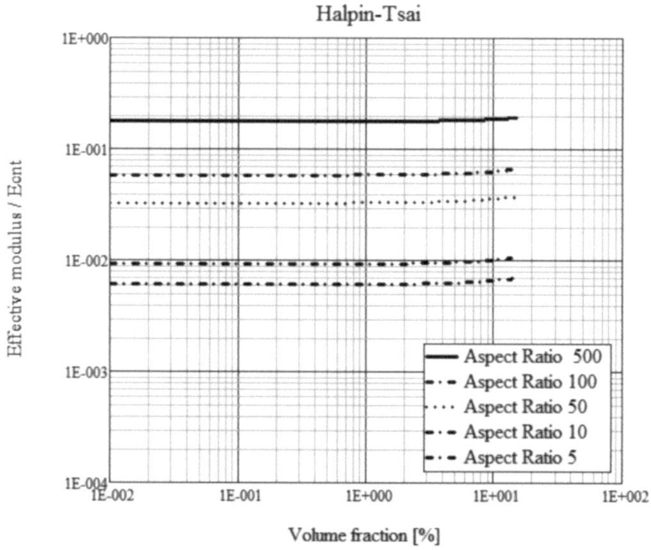

Fig. 8. Reinforcement efficiency vs aspect ratio by H-T model

Thostenson and Chou approach of H-T models

Assuming the Halpin-Tsai model, Thostenson and Chou considered that, in the case of MWCNT, only the outer shell would carry the load as logical assumption of the relatively low bonding with inner layers. According to this assumption, the effective MWCNT elastic modulus was evaluated by considering the application of all loads only to the outer cross section, according to the following equation:

$$E_{eff} = \frac{4t}{d} E_{NT} \tag{7}$$

where t and d represent, respectively, the graphite layer thickness (0.34 nm), and, the nanotube outer diameter; while the ζ parameter, for low volume fractions, is assumed as:

$$\varsigma = 2\frac{l}{d} \tag{8}$$

The authors have added to the simple aspect ratio, the nanotube diameter as reasonable parameter influencing the reinforcement efficiency. The following equation reports the longitudinal elastic modulus as function of fibre aspect ratio and fibre diameter:

$$E_L = E_m \frac{\left(1 + 2\left(\frac{l}{d}\right)\left(\frac{\left(\frac{E_{NT}}{E_m}\right) - \left(\frac{d}{4t}\right)}{\left(\frac{E_{NT}}{E_m}\right) - \left(\frac{l}{2t}\right)}\right) \upsilon_{NT}\right)}{\left(1 - \left(\frac{\left(\frac{E_{NT}}{E_m}\right) - \left(\frac{d}{4t}\right)}{\left(\frac{E_{NT}}{E_m}\right) - \left(\frac{l}{2t}\right)}\right) \upsilon_{NT}\right)} \tag{9}$$

Predictions computed by using Thostenson correction show a reduced level of efficiency at constant aspect ratio for each curve (figure 9) but trends substantially similar to the Halpin-Tsai predictive curves. Shokrieh and Rafiee, 2010 have demonstrated through FE analysis that Halpin-Tsai based methods overestimates the composite modulus and they have suggested to adjust the model predictions by introducing an equivalent solid reinforcement with interphase region as filler.

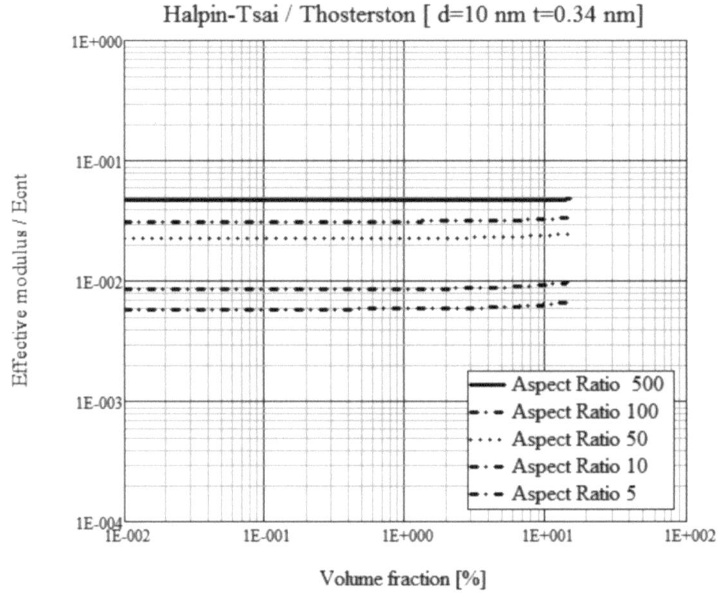

Fig. 9. Reinforcement efficiency prediction according to H-T/Thostenson modifications for different aspect ratios, AR=5, 10, 50, 100, 500

Yeh correction of H-T equation

Methods based on the previous correction for the H-T approach have revealed a good agreement with experimental data only in the dilute regime whilst they are unable to predict the non-linear behaviour observed by many authors at higher filler content (Yeh et al., 2006; Martone et al., 2010). Further modifications were introduced by Yeh et al.,2006 in order to account for the aggregation effect of carbon nanotube on mechanical response of the final nano-composite. They suggested to modify the shape factor ζ by adding an exponential term to model the possible nanotube aggregation, according to the following equation:

$$\varsigma = 2\frac{l}{d}e^{-(a\,v_{NT}+b)} \tag{10}$$

where a and b represent adjustable parameters.
Higher levels of aggregation and networking of CNT within the final system necessarily suggest an higher aggregation-associated values (a and b).
Although this correction leads up to a good correlation with experimental data, this approach is rather sophisticated since it needs the earlier determination of three constants (a,b and c). Moreover the physical meaning of the two aggregation factors is missed.

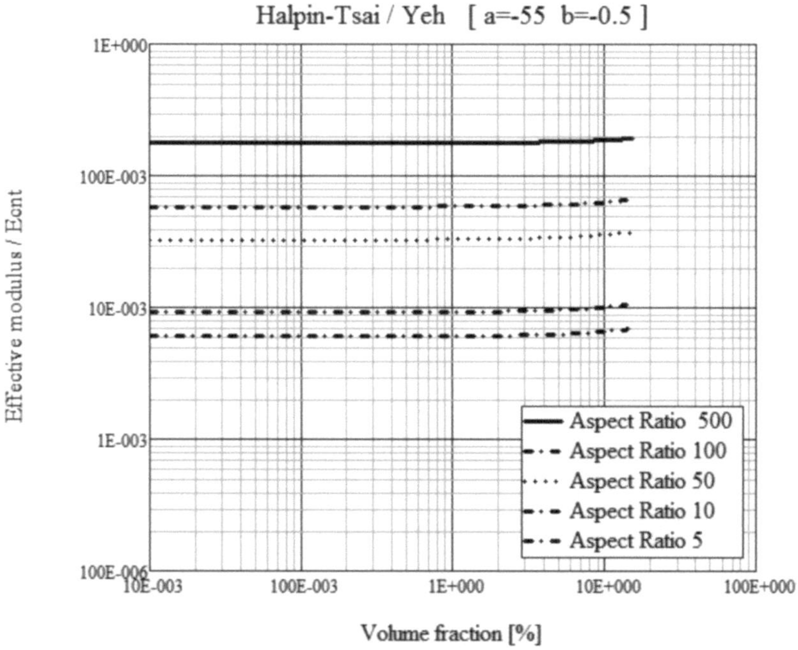

Fig. 10. Reinforcement efficiency prediction according to H-T/ Yeh correction for different aspect ratios, AR=5, 10, 50, 100, 500

Figure 10 reports the predictive curves of reinforcement efficiency as function of nanofillers volume content for different aspect ratio.

Shear Lag based approach

Among the different methods based on the mixtures rule, the approach based on "*Shear Lag*" theory indeed represents the most common, intuitive and reliable. In fact, despite some serious speculative flaws, this model probably has gained popularity, due to the lack of algebraic complexity and the physical appeal of the original scheme. The main limitation of this model is represented by the capability to predict only longitudinal stiffness of the system.

Cox's model

Cox (1952) introduced the concept of the effective tensile modulus of a short fibre embedded into a matrix by defining the efficiency, η, as the reduction ratio of its intrinsic tensile modulus, E_f. This approach, in the case of aligned fibres, is described by the following equations:

$$E = \eta_\ell v_f E_f + \left(1 - v_f\right) E_m \tag{11}$$

$$\eta_\ell = 1 - \frac{\tanh\left(a \, l/d\right)}{\left(a \, l/d\right)} \tag{12}$$

where η_l is the length efficiency factor and the parameter a, accounts for the fibre packing and the Young moduli ratio. Developments of the Cox model were conducted mainly upon the way of choosing suitably the radius of the surrounding matrix cylinder (hexagonal, circular, square etc.).

Readers interested in short fibre modelling approach for the case of unidirectional composite are remanded to an awesome overview by Tucker et Liang,1999.

Although the mentioned models were developed for micro-sized reinforcing elements and for considerable volume content rather than nano-metric structures and dilute regime, many researchers have attempted to model CNTs reinforcement effect by using these approaches. Results are not encouraging, unless some different issues, related mainly with the nano-metric structure of these unconventional fillers are accounted for.

The direct implementation of the reported models to the case of nano-composites reveals critical discrepancy and inefficiency due to the main assumption that only the specific content of CNTs drives the reinforcement efficiency. Actually, conventional modelling approaches for short or particulate composites neglect issues such as, topology, dispersion morphology, contacts among the nano-fillers and possible sticky potential which indeed represent critical features in the case of nano-filler for predictive models.

The topological configuration of the fillers, as example, has been demonstrated to significantly act in the case of the electrical conductivity of the final nano-composite. In fact, according to the available literature, it is well known, that percolative phenomenon drives the conductive-non conductive behaviour of nano-composites (Gojny et al., 2005, Kovacs et al., 2007, 2009).

Despite conductive properties of nano-composites have been well-described in the framework of percolation theory and an appreciable agreement with a huge amount of experimental data is reported, a lack of theoretical understanding and modelling capability for mechanical behaviour is still missing.

In the last decade, many researchers and scientists have proposed analytical and numerical models to predict the mechanical behaviour of polymer matrix filled with carbon nanotubes. Most of the works were essentially based on modification of classical approaches originally developed for short fibre and particulate composites, by considering, in some cases, specific features such as waviness or topological distribution of nano-fillers.

Carman-Reifsnider correction

Among the different variants of the rule of mixtures, the most prominent model has been developed by Carman-Reifsnider, 1992. Assuming a coaxial fibre-matrix scheme, i.e. cylindrical fibres surrounded by a concentric matrix shell, the fibre volume fraction results a square function of the radii ratio, thereby the stress transfer coefficient becomes:

$$a = \sqrt{\frac{-3E_m}{2E_{NT}\ln(v_{NT})}} \tag{13}$$

The length efficiency factor (η_l) approaches to 1 as the aspect ratio increases, highlighting that higher aspect ratio fillers are preferred to improve the matrix stiffness at low reinforcement content. The main advantages of this model concerns with the possibility to overcome the non-linear behaviour (figure 11) reported by experimental data expressing the reinforcement efficiency as follows:

$$\frac{E_c - E_m}{E_m} = \left(\eta_\ell \frac{E_{NT}}{E_m} - 1\right) v_{NT} \tag{14}$$

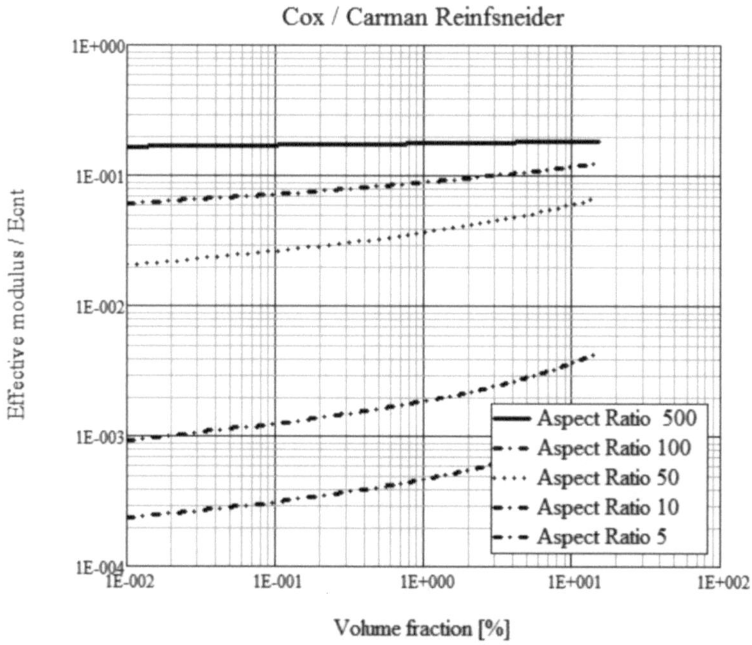

Fig. 11. Cox's model prediction of reinforcement efficiency according Carman-Reifsnider correction for different aspect ratios, AR=5, 10, 50, 100, 500

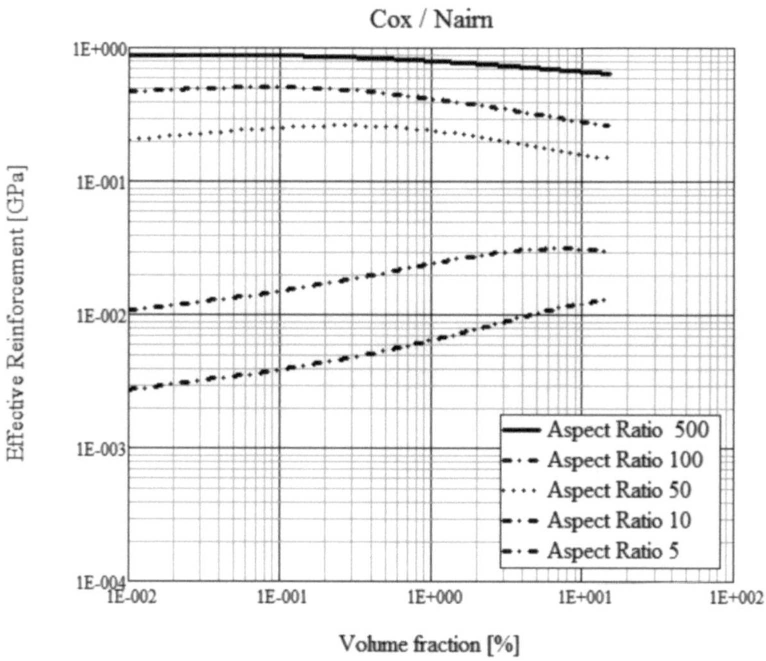

Fig. 12. Reinforcement efficiency predictions according to shear-lag approach including Nairn correction for different aspect ratios, AR=5, 10, 50, 100, 500

Energy-based model by Nairn

A severe flaw of the shear lag based approach is the prediction of composite stiffness at very low fibre volume fraction. Nairn, 1997 published an accurate study on the use of shear lag method. In this work a modified definition of the stress transfer parameter, based on energy concerns, is proposed. Equations 15 and 16 report the expression of shear-lag based model with Nairn modifications to evaluate the composite modulus

$$\beta = \sqrt{\frac{2}{E_f E_m} \cdot \frac{E_f V_f + E_m\left(1-V_f\right)}{\frac{\left(1-V_f\right)}{\left(4\cdot\frac{E_f}{2\left(1+\nu_f\right)}\right)} + \frac{1}{\left(2\cdot\frac{E_m}{\left(1+\nu_m\right)}\cdot\left(\frac{1}{1-V_f}\ln\left(\frac{1}{V_f}\right)-1-\frac{1-V_f}{2}\right)\right)}}} \tag{15}$$

$$E_c = \frac{V_f E_f + \left(1-V_f\right)E_m}{1+\frac{E_f}{E_m}\frac{V_f}{1-V_f}\frac{\tanh\left(\beta \ell/2\right)}{\beta \ell/2}} \tag{16}$$

The predictions computed by the Nairn suggested shear lag model are reported in figure 12 for different aspect ratio values up to 10% percentage of filler volume fraction.

Shear-lag modification by Omidi et al.

The relationship between mechanical properties of the final nano-composite system and CNT loading could be assumed linear only at low volume fraction. Omidi et al.,2010 have suggested a modified version of eq. 11 by adding an exponential factor as reported in eq. 11:

$$\frac{E_c - E_m}{E_m} = \left(\eta\frac{E_{NT}}{E_m}-1\right)\upsilon_{NT}\, e^{\alpha\upsilon NT} \tag{17}$$

Even though the fitting is much more congruent with experimental data, the exponential factor is not correlated with any physical justification or theoretical assumption weakening the final modelling validity. A further modification to include nanotube waviness and spatial orientation distribution, is reported in eq. 18, according to the following expression:

$$\eta = \eta_0 \eta_\ell \eta_w \tag{18}$$

The distribution of oriented nanotubes, included in the model by the η_0 factor, results 1/3 and 1/5 for randomly oriented CNT, in 2D (in plane) and in 3D respectively. The α coefficient in equation (17) is defined as:

$$\alpha = \frac{\ln\beta}{\upsilon_{NT}}, \; \beta = \frac{\left(\left(E_c - E_m\right)/E_m\right)}{\left(\eta\frac{E_{NT}}{E_m}-1\right)\upsilon_{NT}} \tag{19}$$

The above method provides good predictions of CNT-Polymer composite mechanical behaviour. However, it needs as input not only the experimental parameter, α, obtained by

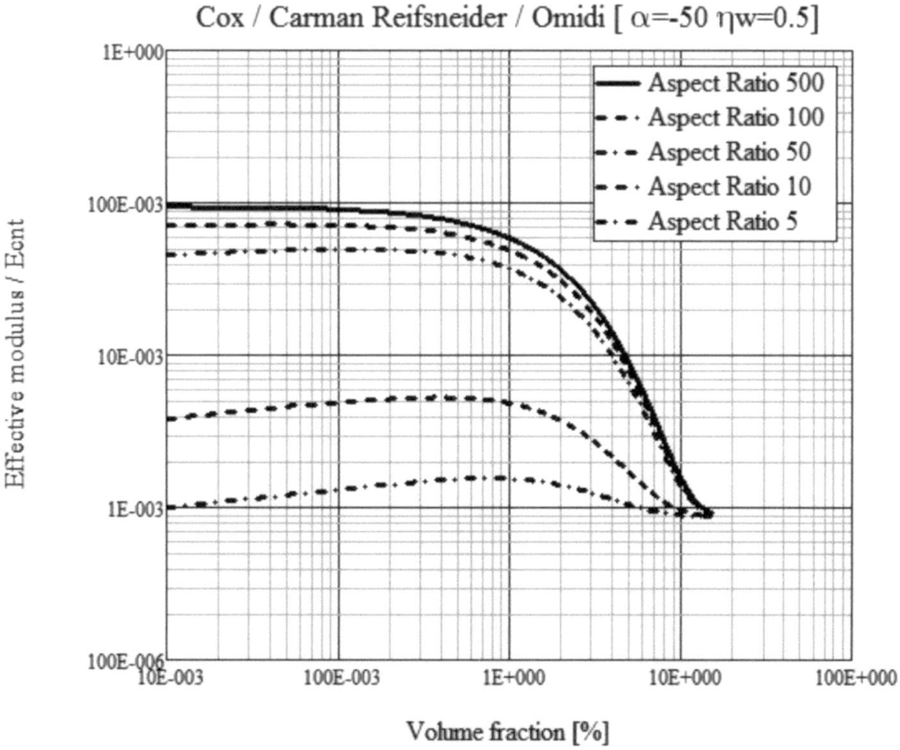

Fig. 13. Reinforcement efficiency predictions for shear lag model according to Cox-Carman-Reifsnider-Omidi corrections for different aspect ratios, AR=5, 10, 50, 100, 500

testing the nanocomposite at the high CNT volume fraction, but also the fitting parameter, η_w, associated with CNT waviness (figure 13).

Attempts have been proposed to account for the suspension of cluster within the material (Guzma´n de Villoria et al., 2007, Chatterjee, 2008). To the best author knowledge, a proper developed theory able to take into account nano-composite topology is still missed. Moreover, all the above modifications were proposed as rule of thumb to predict the elastic behaviour of such composites through the introduction of empirical correction factors characterised by weak physical meaning.

All analysed models report as main assumption the uniform dispersion of the filler content. Indeed, this hypotheses seems not realistic to represent CNT-Polymer systems, as widely demonstrated by the available literature. Dispersion technique, processing parameters and typology of the hosting matrix will modify the final nano-composite, strongly affecting its final mechanical performances. Contacts among nanotubes can be assumed to occur with a major probability in the case of higher nano-filler content. Moreover, the assumption of interconnected nanostructure leads to the consideration that excluded volume and effective aspect ratio of nano-reinforcements differ significantly from the nominal value.

In the following paragraph, the rule of mixture will be introduced highlighting a recent modification of the original model to account for the nanotubes connectivity published by Martone et al., 2011. Recent experimental data will be also analysed and compared with model predictions to support the validity and the effectiveness of this approach.

3. Recent advanced in CNT reinforcement effect

3.1 Tube contact effects in the Cox's model

To the best knowledge of the authors, all the proposed methods to predict mechanical performances of nano-composites are essentially based on semi-empirical modification of short fibre micromechanics. The available models do not account for the progressive formation and growth of connected paths among loaded nano-elements, which may represent the most relevant features for the enhancement effect of mechanical property. The main models (Halpin-Tsai and Shear-lag) presented in the previous paragraph and various modifications (Cox, Nairn, Reifsneider, Yeh) show a negligible dependency of the reinforcement efficiency on the nano-filler aspect ratio. The only approach, which considers a strong relation between the efficiency and the aspect ratio, is shown by the Omidi et al. correction, which in turn, requires experimental data at very high CNT content and a curve fitting parameter as input for the computation of the model. The connectedness and the waviness of the nanofiller within the hosting matrix indeed affect the mechanical enhancement of the final nano-composite. For this reason, a predictive model for the mechanical reinforcement efficiency of CNT should take necessarily into account such features.

In this paragraph an attempt to include the connectedness of tubes and the waviness feature of the carbon nanotubes within the hosting medium is discussed and described. The main achieved results, experienced on CNT/nano-composites by many researchers, could be correctly summarised as follows:

1. the dependence of the effective reinforcement modulus upon the tubes aspect ratio. In the very dilute regime the effective elastic modulus of CNT depends directly on the averaged aspect ratio of the filler.
2. the reduction of mechanical efficiency at increasing volume content. As the nanotube content increases due to the progressive increment of tube-to-tube contacts the effectiveness of transferring load decreases.

Shear-lag-based approaches have been widely used to analyse the effect of the aspect ratio on mechanical of carbon nanotubes composite. The main assumption of these models is the perfect adhesion between the phases where the stress transfer occurs via a shear mechanism. By applying the shear lag models, the composite modulus results an average value of the specific components modulus that are reduced by a numerical factor depending on the aspect ratio of the reinforcing filler. In this case, the most prominent expression for the prediction of the nano-composite Young's modulus is given by equation 20:

$$
\begin{aligned}
E_c &= E_\eta \cdot \phi + E_m \cdot (1-\phi) \\
E_\eta &= \eta \cdot E_{NT} \\
\eta &= 1 - \frac{\tanh(K \cdot AR)}{(K \cdot AR)} \\
K &= \sqrt{\frac{-\frac{2}{1+\nu m}}{\frac{E_{NT}}{E_m} \cdot \ln(\phi)}}
\end{aligned}
\qquad (20)
$$

The stress transfer parameter, K, is computed, according to the Carman-Reifsnider assumption, considering, as fibres arrangement, a cylindrical packing surrounded by the matrix. While the progressive formation of contacts among tubes creates a fully connected web which improves the materials conductive (thermal and electrical) behaviour, the mechanical performances are characterized by an opposite tendency. Actually, contacts among the "tubes" create points at which negligible load transfer could be assumed. Therefore, a predictive model for such nano-composites needs to account for arising number of contact points with the CNT content. With the aim to achieve this latter target, the study of particle packing is paramount; nevertheless, study of the density of random rods or fibre packing has not received great attention until the work published by Philipse in 1996, on the stacking configuration of high aspect ratio rods.

The Random Contact Model is based on the assumption that two particles of any shape can contact with a probability which is averagely independent on other contact formations, in other terms, if contacts are uncorrelated, their number for each particle varies linearly with the particle concentration. The theory states that the average number of contacts per particle $<c>$ is dependent on the normalized average excluded volume, V_{ex}, and the particles volume fraction, φ, according to the following expression:

$$\langle c \rangle = \varphi \cdot \frac{V_{ex}}{V_p} \tag{21}$$

where V_p is the particle volume.

The excluded volume of an object is defined as the volume around an object into which the centre of another similar object is not allowed to enter if overlapping of the two objects is to be avoided. According to Onsager, 1949, the average excluded volume of a pair of random rods, modelled by cylinders of length L with two hemispherical caps of diameter D, is given by

$$V_{ex} = \frac{\pi}{2} \cdot L^2 D + 2\pi \cdot D^2 L + \frac{4}{3}\pi \cdot D^3 \tag{22}$$

A further feature to account for the correct computation of the excluded volume of "*tube*" elements, is related to the achieved waviness in the final nano-composite system. Waved objects, in fact, will necessarily induce a further contribution to the excluded volume.

Thus the average number of contacts per particle $<c>$ for tubes is dependent on "*tube*" aspect ratio (AR) and their final waviness level (w), according to the following expression:

$$\langle c \rangle = w \cdot \phi \cdot \left(4 + \frac{3AR^2}{3AR + 2}\right) \tag{23}$$

where the w represents the "*tube*" waviness level. This parameter has been introduced as the ratio between the effective average excluded volume and that evaluated for a straight rod to account for the CNT waved arrangement within the final nano-composite.

A number of theoretical studies are available in literature to evaluate the effect of nanotube waviness on the excluded volume. An exhaustive discussion on the calculation of excluded volume for curved rods is beyond the scope of this work. Remarkable efforts have been reported by Behran and Sastry, 2007, Li and Chou, 2007, Ma and Gao, 2008, nevertheless, these studies are limited by the main assumption of rod shapes. As stated previously, contacts between nanotubes imply a reduction of the stress transfer efficiency of the reinforcement, afterwards the rule of mixtures could be modified by reducing the stress transfer coefficient by the average number of contacts

$$\eta = 1 - \frac{\tanh\left(\dfrac{K \cdot AR}{1+<c>}\right)}{\left(\dfrac{K \cdot AR}{1+<c>}\right)} \tag{24}$$

Below the hypotheses of contacting nanotubes, the effective aspect ratio of filler could be defined as

$$AR_e = \frac{AR}{1+<c>} \tag{25}$$

Figure 14 and 15 report the predictive curves of the present contact tube model based on rule of mixture in the case of straight and waved tubes respectively over almost four filler volume fraction percentage decades.

In particular, predictions have been computed considering (fig.14) a constant waviness level (w=1) and, hence, schematizing the "*tube*" elements as rigid (i.e. neglected contribution of the excluded volume associated to the waviness level) for different average aspect ratio, AR=5, 10, 50, 100, 500. It is worth to observe that the original shear lag approach is not able to describe the progressive reduction of reinforcement efficiency, whilst modified expression including random contact model, excluded volume and waviness may very accurately reproduce the experimental values at very dilute regime. Waviness accounts the augmentation of contacts between nanotubes due to the curliness of fillers.

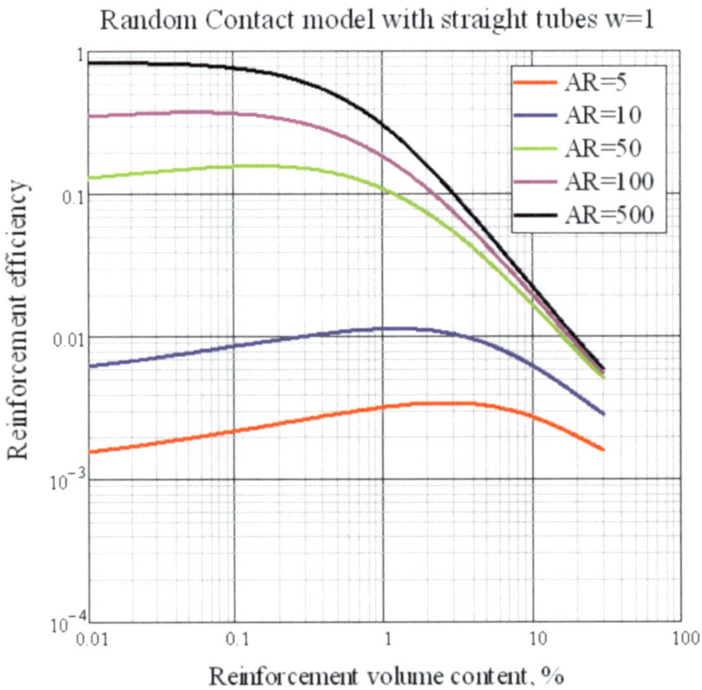

Fig. 14. Predictions of reinforcement efficiency for "straight tubes" (w=1) at different aspect ratio AR= 5, 10, 50, 100, 500 according to the developed model

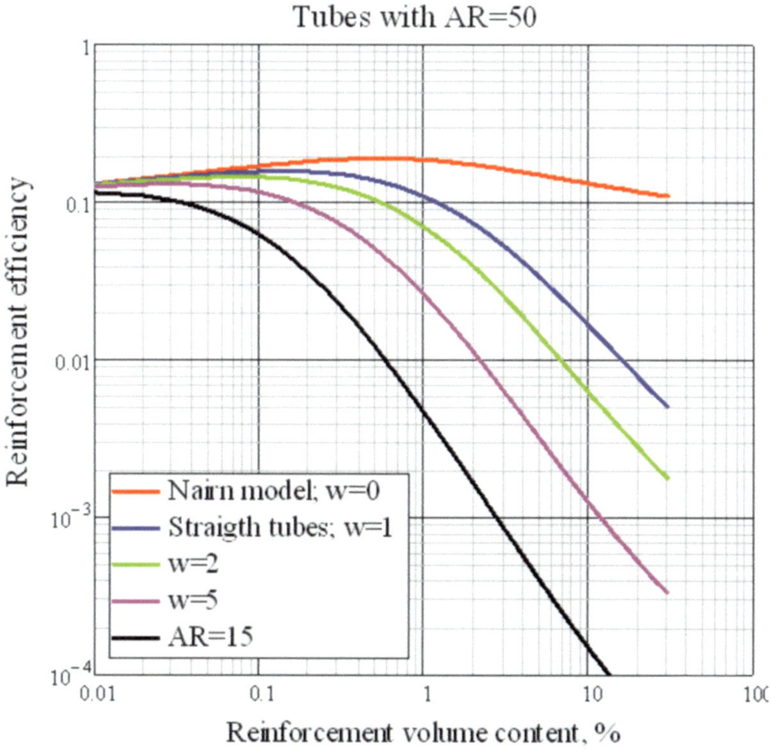

Fig. 15. Predictions of reinforcement efficiency according to the developed models for the case of AR=50 at different magnitudes of waviness, w=0,1,2,5

3.2 Comparison: literature data vs modified Cox's model

In this section, the proposed modification of Cox's model presented in paragraph 3.1, will be applied to available literature experimental data. Topological parameters (AR_e and w) will be evaluated by curve fitting for each set of data.

Table 1 reports the details of the considered CNT nano-composite taken from the scientific literature of the last 10 years. It can be noticed that the data set are related to different typology of carbon nanotubes and hosting matrix. The aspect ratio is spread over three decades and, moreover, experimental results are associated to nano-composite samples that have been manufactured by different techniques: high energy mixing such as sonication or low or moderate mixing such as mechanical stirring by magnetical agitation.

Figures 16 a) and b) show the modelling curves and corresponding experimental data sets. Model parameters are the CNTs elastic modulus, kept constant at the value of 3 TPa for all simulations, CNT aspect ratio, and the CNT waviness parameter, evaluated by curve fitting. In figure 16, for each data set the fitting parameters are overlaid. It is worth to note that the waviness parameter could be correlated to the aspect ratio, namely the higher the aspect ratio the higher the waviness parameter. Table 2 reports the best fitting parameters of the presented model, evaluating the effective filler aspect ratio, AR_f, and waviness parameter, w for each set of data. It is important to notice that the fitting aspect ratio falls in the nominal range for nanocomposites mixed by ultrasonication, whilst nanocomposites mixed by mechanical stirring led to an effective aspect ratio lower than the nominal one.

Reference	Matrix	MWNT	Manufacturing	Diameter [nm]	AR	E_η [Gpa]
Omidi et al., 2010	Epoxy LY-5052	RIPI CVD	Sonication	10-50	20-300	148
Ogasawara et al., 2004	Polyimide Tri-A-Pi	CNRI CVD	Mechanical Stirring	20-100	10-200	13
Martone et al. 2010	Epoxy RTM6	CVD	Sonication	9.5	20-1000	1784
Martone et al. 2011	Epoxy RTM6	CVD	Sonication	110-170	22-52	144
Martone et al. 2011	Epoxy RTM6	CVD	Sonication	40-70	7-50	71
Xiao et al., 2007	Polyethylene LDPE	NanoLab CVD	Mechanical Stirring	10-20	50-500	6
Yeh et al., 2006	Phenolic PF-650	Own CVD	Sonication	15-40	50-300	221
Ayhatollai et al. 2011	Epoxy ML-506	CVD	Stirring + Sonication	10-30	250-1000	184
Liu et al. 2004	Nylon6	CVD	Melt Compounding	7-20	500-2000	472
Kanagaraj et al., 2007	Polyethylene HDPE	Nanotech CVD	Magnetical Stirring	60-100	50-250	68

Table 1. Physical properties of nanocomposites considered from available experimental data.

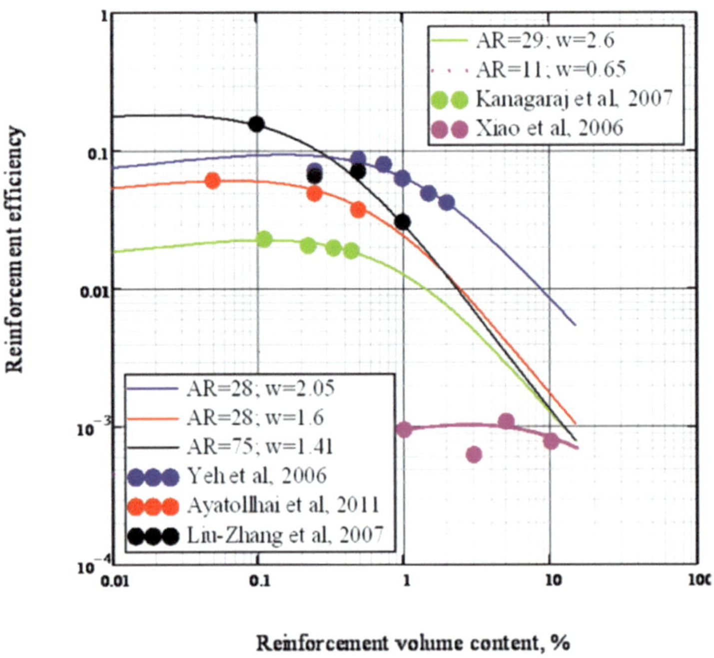

a)

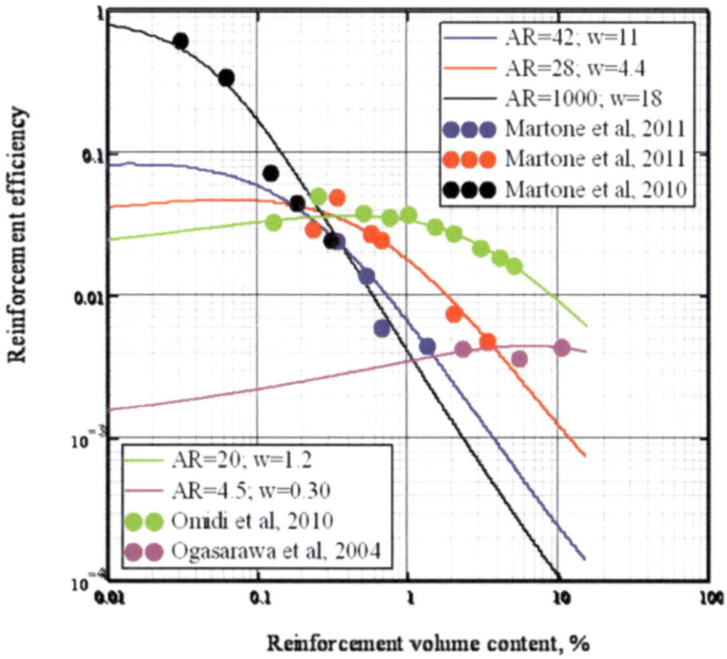

b)

Fig. 16. Comparison between experimental data and model predictions. The fitting parameters are superimposed to each curve.

	AR	*w*
Omidi et al., 2010	20	1.2
Ogasawara et al., 2004	4.5	0.3
Martone et al. 2011 AR505	1000	18.0
Martone et al. 2011 AR30	42	11
Martone et al. 2011 AR55	28	4.4
Xiao et al., 2007	11	0.65
Yeh et al., 2006	28	2.0
Ayhatollai et al. 2011	28	1.6
Liu et al. 2004	75	1.41
Kanagaraj et al., 2007	29	2.6

Table 2. Fitting parameters (AR-aspect ratio and w-waviness) for different data sets.

4. Conclusion

The potential usage of carbon nanotubes to enhance the mechanical behaviour of matrix polymer indeed represents a potential breakthrough for a new generation of advanced polymer composites. The mechanical reinforcement of a polymer by carbon nanotubes is highly attracting the effort of the scientific community mainly for two reasons. Firstly, the wide amount of available experimental data in the scientific literature results controversial

and uncertainties arise regarding the effects induced by the different carbon nanotubes geometry, loading fraction, network formation and processing conditions.
Secondly, reliable models, for the prediction of the reinforcement feature with increasing filler content, are still not available as currently used models do not correctly represent the various issues associated with content, morphology and type of nano-fillers. Some of these issues have been considered in the proposed reinforcement efficiency model which the chapter presents and describes. Experimental findings support the evidence that in carbon nanotube reinforced polymers, contacting "tube" fillers implies a significant decrease of stress transfer efficiency between reinforcing nanotubes and hosting matrix via a reduction of the effective filler length. Philipse's model of random contacting nanotubes has been considered to allow the evaluation of the average reduction of the effective tube aspect ratio by means of reinforcing particle excluded volume computation. Finally, the shear lag theory of the short fibre micromechanics provides the necessary framework for the evaluation of the composite modulus. The nanotubes aspect ratio stands as the controlling parameter for the determination of the mechanical properties of CNTs reinforced polymers. A physical parameter accounting for the waviness of the nanotubes within the polymer matrix is provided in order to describe the deviation of the excluded volume evaluated for curved nanotubes from the straight rod assumption. The model is able to reproduce the non linear reinforcement behaviours of different aspect ratios carbon nanotubes in different matrices reported in the literature. A topological explanation of the reduction of the reinforcement efficiency of carbon nanotubes within polymer matrices upon their contacting and clustering has been provided addressing the major phenomenological behaviours. Further investigation is needed to better clarify the role of nanotubes curvature within a hosting polymer matrices on elastic modulus of the final nano-composite system.

5. References

Ayatollahi, M.R.; Shadlou, S.; Shokrieh, M.M. (2011). Fracture toughness of epoxy/multi-walled carbon nanotube nano-composites under bending and shear loading conditions. *Materials and Design*, Vol. 32, pp.2115–2124. doi:10.1016/j.matdes.2010.11.034.

Behran, L.; Sastry, A. M. (2007). Modeling percolation in high aspect-ratio fibre systems. II. The effect of waviness on the percolation onset. *Physical Review E*, Vol. 75, N.041121.

Carman, G.P.; Reifsnider K.L. (1992). Micromechanics of short-fibre composites. *Composites Science and Technology*, Vol. 43, N. 2, pp. 137-146.

Chatterjee, A.P. (2008). Modeling the elastic moduli of fibre networks and nanocomposites: Transversely isotropic filler particles. *Journal of Applied Physics*, Vol. 103, pp. 064316.

Coleman, J.N.; Khan, U.; Blau, W.J.; Gun'ko, Y.K. (2006). Small but strong: A review of the mechanical properties of carbon nanotube-based composites. *Carbon*, Vol. 44, pp. 1624-1652. doi: 10.106/j.carbon.02.038.

Cox, H.L. (1952). The elasticity and strength of paper and other fibrous materials. *British Journal of Applied Physics*, Vol. 3, N. 3, pp. 72-76. doi: 10.1088/0508-3443/3/3/302.

De Zhang, W.; Shen, L.; Phang, I.Y.; Liu T.X. (2004). Carbon nanotubes reinforced nylon-6 composite prepared by simple melt-compounding. *Macromolecules*, Vol. 37, N. 2, pp. 256–259.

Gojny, F.H.; Wichmann, M.H.G.; Fiedler, B.; Schulte, K. (2005). Influence of different carbon nanotubes on the mechanical properties of epoxy matrix composites - A comparative study. Composites Science and Technology, Vol. 65, pp.2300–2313.

Guzma´n de Villoria, R.; Miravete, A. (2007). Mechanical model to evaluate the effect of the dispersion in nanocomposites. *Acta Materialia* , Vol. 55, pp. 3025–3031.

Halpin J. C. Kardos J. L. (1976) The Halpin±Tsai Equations: A Review. Polym Eng Sci, Vol. 16, pp. 344-352.

Hermans, J.J. (1967) The elastic properties of fiber reinforced materials when the fibers are aligned. *Proc. Kon Ned Akad v Wetensch B*, Vol. 65, pp. 1-9

Hewitt RL, de Malherbe MC. (1970) An Approximation for the Longitudinal Shear Modulus of Continuous Fibre Composites. *Journal of Composite Materials* ,Vol. 4, pp. 280-282.

Hill, R. (1964) Theory of Mechanical Properties of Fibre-Strengthened Materials: I Elastic Behaviour. *Journal of Mechanical Physics of Solids* Vol. 12, pp. 199-212.

Hussain, F.; Hojjati,M.; Okamoto, M.; Gorga, R.E. (2006). Review article: Polymer-matrix Nanocomposites, Processing, Manufacturing, and Application: An Overview. *Journal of Composite Materials*, Vol. 40, N. 17, pp. 1511-1575.

Kanagaraj, S.; Varanda, F.R.; Zhiltsova, T.V.; Oliveira, M.S.A.; Simoes, J.A.O. (2007). Mechanical properties of high density polyethylene/carbon nanotube composites. *Composites Science and Technology*, Vol. 67, pp. 3071–3077.

Kovacs, J.Z.; Mandjarov, R.E.; Blisnjuk, T.; Prehn, K.; Sussiek, M.; Müller, J.; Schulte, K.; Bauhofer, W. (2009). On the influence of nanotube properties, processing conditions and shear forces on the electrical conductivity of carbon nanotube epoxy composites. *Nanotechnology*,Vol. 20.

Kovacs, J.Z.; Velagala, B.S. ; Schulte, K.; Bauhofer W. (2007). Two percolation thresholds in carbon nanotube epoxy composites. *Composite Science and Technology*, Vol. 67, pp. 922-926.

Lewis TB, Nielsen LE. (1970) Dynamic Mechanical Properties of Particulate-Filled Composites. *Journal of Applied Polymer Science* Vol. 14, pp. 1449-1471.

Li, C.; Chou, T.W. (2004). Modeling of elastic buckling of carbon nanotubes by molecular structural mechanics approach. *Mechanics of Materials*, Vol. 36, pp.1047–1050.

Li, C.; Chou, T.W. (2007). Continuum percolation of nanocomposites with fillers of arbitrary shapes. *Applied Physics Letters*, Vol. 90, N.174108.

Liu, T.X.; Phang, I.Y.; Shen, L.; Chow, S.Y.; Zhang, W. D. (2004). Morphology and mechanical properties of multiwalled carbon nanotubes reinforced nylon-6 composites. *Macromolecules*, Vol. 37, N. 19, pp. 7214–7222.

Ma, P.C.; Kim, J-K; Tang, B.Z. (2007). Effects of silane functionalization on the properties of carbon nanotube/epoxy nanocomposites. Composites Science and Technology, Vol. 67, pp. 2965–2972.

Ma,H.M.; Gao X.L. (2008). A three-dimensional Monte Carlo model for electrically conductive polymer matrix composites filled with curved fibres. *Polymer*, Vol. 49, pp. 4230–4238. doi:10.1016/j.polymer.2008.07.034.

Martone, A.; Faiella, G.; Antonucci, V.; Giordano, M.; Zarrelli, M. (2011). The effect of the aspect ratio of carbon nanotubes on their effective reinforcement modulus in an epoxy matrix. *Composites Science and Technology*, doi: 10.1016/j.compscitech. 2011.04.002.

Martone, A.; Formicola, C.; Giordano, M.; Zarrelli, M. (2010). Reinforcement efficiency of multi-walled carbon nanotube/epoxy nano composites. *Composites Science and Technology*, Vol. 70, Issue 7, pp. 1154-1160, doi:10.1016/j.compscitech.2010.03.001.

Nairn, J.A. (1997). On the use of shear-lag methods for analysis of stress transfer in unidirectional composites. *Mechanics of Materials*, Vol. 26, pp. 66-80.

Nielsen LE. 1970. Generalized Equation for the Elastic Moduli of Composite Materials. J Appl Phys, vol. 41, pp. 4626-4627.

Ogasawaraa, T.; Ishidaa, Y.; Ishikawaa, T.; Yokota, R. (2004). Characterization of multi-walled carbon nanotube/phenylethynyl terminated polyimide composites. *Composites: Part A*, Vol. 35, pp. 67–74.

Omidi, M.; Hossein Rokni, D. T.; Milani, A.S.; Seethaler R.J.; Arasteh, R. (2010). Prediction of the mechanical characteristics of multi-walled carbon nanotube/epoxy composites using a new form of the rule of mixtures. *Carbon*, Vol. 48, pp. 3218-3222. doi:10.106/j.carbon.2010.05.007.

Onsager, L. (1949). The effects of shape on interaction of colloidal particles. *Annals of the New York Academy of Science*, Vol. 51, pp. 627-659.

Philipse, A.P. (1996). The Random Contact Equation and Its Implications for (colloidal) Rods in Packings, Suspensions, and Anisotropic Powders. *ACS publications Langmuir*, Vol. 12, N. 5, pp. 1127-1133.

Shaffler,M.S.P and A.H.Windle, (1999) Fabrication and characterization of carbon Nanotube/Poly(vinyl alcohol) Composites. Advanced Materials, Vol.11, N.11, pp. 937-941.

Shokrieh, M.M.; Rafiee, R. (2010). Investigation of nanotube length effect on the reinforcement efficiency in carbon nanotube based composites. *Composite Structures*, Vol. 92, pp. 2415–2420. doi:10.1016/j.compstruct.2010.02.018.

Sun, L.; Warren, G.L.; O'Reilly, J.Y.; Everett, W.N.; Lee, S.M.; Davis, D. et al. (2008). Mechanical properties of surface-functionalized SWCNT/epoxy composites. Carbon, Vol. 46, pp. 320–328.

Thostenson, E.T.; Chou, T.W. (2003). On the elastic properties of carbon nanotube-based composites: modeling and characterization. *Journal of Physics D: Applied Physics*, Vol. 36, pp. 573-582.

Thostenson, E.T.; Chou, T.W. (2006). Processing–structure–multi-functional property relationship in carbon nanotube/epoxy composites. Carbon, Vol. 44, pp. 3022–3029.

Treacy, M.M.J.; Ebbesen, T.W.; Gibson, J.M. (1996). Exceptional high Young's modulus observed for individual carbon nanotubes. *Nature*, Vol. 381, pp.678-680.

Tucker III, C.L.; Liang, E. (1999). Stiffness predictions for unidirectional short-fibre composites: Review and evaluation. *Composite Science and Technology*, Vol. 59, pp. 655-671.

Valavala, P.K.; Odegard, G.M. (2005). Modeling techniques for determination of mechanical properties of polymer composites. *Reviews on Advanced Material Science*, Vol. 9, N. 1, pp. 34-44.

Wang, Q.; Duan, W. H.; Liew, K. M.; He, X. Q. (2007). Inelastic buckling of carbon nanotubes. *Applied Physics Letters*, Vol. 90, pp. 033110.

Xiao, K.Q.; Zhang, L.C., Zarudi, I. (2007). Mechanical and rheological properties of carbon nanotube-reinforced polyethylene composites. *Composites Science and Technology* Vol. 67, pp. 177–182.

Yeh, M.K.; Tai, N.H.; Liu, J.H. (2006). Mechanical behavior of phenolic-based composite materials. *Carbon*, Vol. 44, pp. 1-9. doi:10.1016/j.carbon.2005.07.005.

Zhou, Y.; Pervin, F.; Lewis, L.; Jeelani, S. (2008). Fabrication and characterization of carbon/epoxy composites mixed with multi-walled carbon nanotubes. Material Science and Engineering A , Vol. 475, pp. 157–65.

17

Synthesis of Carbon Nanotube-Metal Oxides Composites; Adsorption and Photo-Degradation

Vinod K. Gupta[1,2] and Tawfik A. Saleh[2]
[1]*Department of Chemistry, Indian Institute of Technology Roorkee*
[2]*Chemistry Department, King Fahd University of Petroleum & Minerals*
[1]*India*
[2]*Saudi Arabia*

1. Introduction

The unique chemical, physical and mechanical properties of carbon nanotubes (CNTs) have stimulated extensive investigation since their discovery in the early 1990s (Iijima, 1991). The primary research interests focus on the development of simple and cost effective methods for the synthesis of CNTs. To synthesize single wall carbon nanotubes (SWCNT) and multi wall carbon nanotubes (MWCNT), various techniques have been developed. Examples are laser ablation, arc discharge, and chemical vapor deposition (CVD). However, CVD is considered to be more productive method to produce CNTs with acceptable price. Therefore, the extraordinary properties of CNTs and the continuous decrease in their price make them potentially useful in a wide range of applications like in catalysis, photocatalysis and adsorption. The adsorption properties of CNTs depend on different factors such as their large specific surface area, highly porous and hollow structure, light mass density and strong physical interaction ability.

Recently, the applicability of CNTs as adsorbent has been extensively investigated. The applications of CNTs for the removal of pollutants from aqueous solutions have been studied. Numerous experimental studies have already been carried out on the adsorption of heavy metals and hazardous organic materials (Li, et al., 2002; Li, et al., 2003; Chen, et al., 2006; Hyung and Kim, 2008; Goering, et al., 2008).

For further enhancement the adsorption ability of CNTs, purification and activation or functionalization of CNTs is essential and can be performed via oxidation treatment (Cho, et al., 2010; Shao, et al., 2010). Purification is necessary to remove the impurities, like catalyst particles, soot, amorphous carbon and other forms, which are often found, mixed with CNTs. Impurities can reduce the adsorption by their coating on the surfaces of CNTs, and as a result influencing the adsorption on the external surface of the nanotubes (Agnihotri et al., 2006). The enhancement in the adsorption ability of CNTs via oxidation treatment with oxidizing agent, such as nitric acid, hydrogen peroxide, potassium permanganate, or a mixture of nitric acid and sulfuric acid, is most probably ascribed due to an increase in hydrophilicity of CNTs. Oxidation treatment of CNTs results in creating new functional groups on the surface of nanotubes.

In recent years, metal oxides nanoparticles supported on carbon nanotubes have been extensively studied and found to be effective adsorbent for the removal of heavy metal ions and hazardous organic chemicals from water; and also as catalysts. The synthesis methodology of such materials will be discussed in the followed section. The discussion will highlight the synthesis process starting with activation or functionalization of CNTs as a preliminary step for the synthesis of such materials.

2. Synthesis strategy

The formation of carbon nanotubes/metal oxide (CNT/MO) composites is both of fundamental and technological interest since the composite combines the unique properties of CNTs and metal oxide nanoparticles, and in addition exhibit some new extraordinary properties caused by the interaction between them (Eder, D. , 2010); Chu, et al., 2010). Thus, a new class of nanocomposites can be made with extraordinary properties meeting a wide range of applications in different disciplines. For example, alumina is one of the most widely used ceramic materials because of its high hardness, good oxidation resistance and chemical stability. However, its applications are limited because of its low fracture toughness. CNTs have been widely used to reinforce and increase the fracture toughness of alumina owing to their unique one dimensional structure with robust mechanical and thermal properties (Chan, et al., 2005).

Various approaches are used to synthesize CNT/MOs composites. This includes hot pressing of composite powder, pressureless sintering technique, direct in-situ growth, in situ CVD synthesis route, high-intensity ultrasonic radiation method, assembling pre-synthesized metal oxide nanoparticels as building blocks on CNTs, spontaneous formation of metal oxide nanoparticels on CNTs, thermal decomposition of metal oxides precursor directly onto the surface of carbon nanotube, hydrothermal crystallization, sol-gel followed by spark plasma sintering process, surfactant wrapping sol-gel method, chemical precipitation and controlled heteroaggregation method (Liu et al., 2009; Zhang et al., 2010; Keshri et al., 2010; Datye et al., 2010; Zhen et al., 2007; Flahaut et al., 2000; Lupo et al., 2004; Gupta et al., 2011; Zhang et al., 2006; Gao et al., 2008; Chan et al., 2005; Yang et al., 2009; Estili and Kawasaki, 2008).

Over all, the methods based on what is called wet chemistry are perhaps the most widely utilized methods in everyday chemistry. One of the most important reasons of their popularity is simplicity and cost comparing to others. However, in these methods, activation or functionalization of CNTs is a primary step that should be conducted before the synthesis and preparation of CNT/MO nanocomposite materials. Functionalization of CNTs plays an important role in facilitating the binding, embedding or loading nanoparticels on the surface of nanotubes since it introduces chemical functional groups to the surface of nanotubes. The successful in functionalization of the nanotubes is the first step to properly prepare the nanocomposite.

2.1 Functionalization of CNTs

Activation or functionalization of CNTs by oxidation treatment introduces chemical functional groups. It can be performed using oxidizing agents such as nitric acid, sulfuric acid, a mixture of sulfuric acid and nitric acid, potassium permanganate, sulfuric acid in presence of potassium permanganate, hydrogen peroxide in presence of nitric acid, hydrogen peroxide, ozone, an oxygen-based atmosphere by an inductively coupled plasma or microwave energy and water (Cuentas-Gallegos et al., 2006; Wang et al., 2009; Smith et al., 2009; Lu et al., 2008; Otvos et al., 2006; Aviles et al., 2009; Haydar et al., 2000; Peng et al., 2011; Ionescu et al., 2006; Hojati-Talemi & Simon, 2009).

Treatment with different oxidizing agents might introduce different functional groups on the CNTs surface including alcoholic, carboxylic, aldehydic, ketonic, and esteric oxygenated functional groups, see Fig.1. Hydroxyl groups (b) on the edge of nanotube could be of phenolic character. Oxygen could be substituted for edge carbon atoms in xanthene- or ether-type (d). Carboxyl groups (a) might give carboxylic anhydrides (e) if they are close together. Also, carbonyl groups might condense to lactone groups (f) or form lactols (g) if they are in close to hydroxyl groups or carboxyl groups. The existence of carbonyl groups could be isolated (c) or arranged in quinone-like structure (h). The groups (a, b, e f, and g) have weak acidic character. The acidity of o-CNTs can be determined by acid-base titration method or Boehm titration (Boehm, 1994).

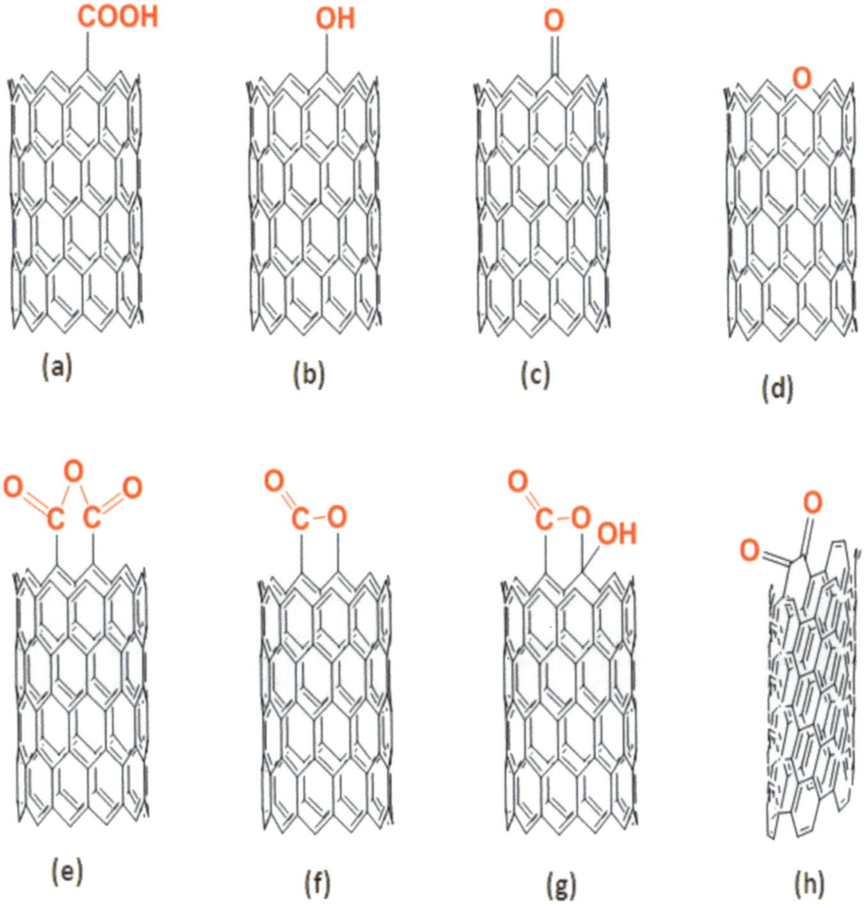

Fig. 1. Possible structures of oxygen-containing groups on the carbon nanotube surface

Synthesis of CNT/MO requires the nanotubes' surface to have mainly alcoholic and carboxylic functional groups which can facilitate the binding of the nanoparticles into the nanotubes. Over other methods, treatment of CNTs by nitric acid or mixtures of sulfuric acid and nitric acid have been expected to mainly graft carboxylic and alcoholic functional groups onto the CNTs surface (Martinez et al., 2003; Chiang et al., 2011). It is also worth mentioning that the contents of carboxylic and alcoholic groups increase with increasing oxidation time.

The characterization of the presence of functional groups on the surface of nanotube and the efficiency of functionalization on nanotubes can be professionally carried out by various instruments like transmission electron microscope (TEM), field emission scanning electron microscope (FESEM), energy dispersive X-ray (EDX), Raman spectroscope, thermogravimeter (TGA), and fourier transformed infrared spectroscope (FT-IR) and X-ray photoelectron spectroscope (XPS).

Morphology and structure of CNTs is characterized by TEM and FESEM. The impurities and defects such as carbonaceous materials, amorphous carbon coatings and catalyst particles are observed as black dots inside the body of nanotubes or on its surface. TEM provides sufficient resolution for the nanotubes' length and diameter distribution; and opening of the ends of nanotubes. TEM and SEM can allow observing if damage to the surface of CNTs after chemical treatment has occurred and observing the presence of different materials around nanotubes, if any. EDX measurements are used for quantitative representation of the components in the CNTs sample. After oxidation process of CNTs, it is expected that EDX reflect more oxygen on the CNTs sample as a result of grafting new oxygen-containing functional groups.

FT-IR is used to analyze the chemical bonding and type of functional groups grafted onto the nanotubes. The application of FTIR in the study of the surface chemistry of nanotubes can provide direct means to observe the interactions occurring at the surface during adsorption and to determine the structure of the adsorbed species. The FTIR spectra of oxidized CNTs (o-CNTs) exhibit various characteristic bands. Example of FTIR spectra of o-MWCNTs is shown in Fig. 2.

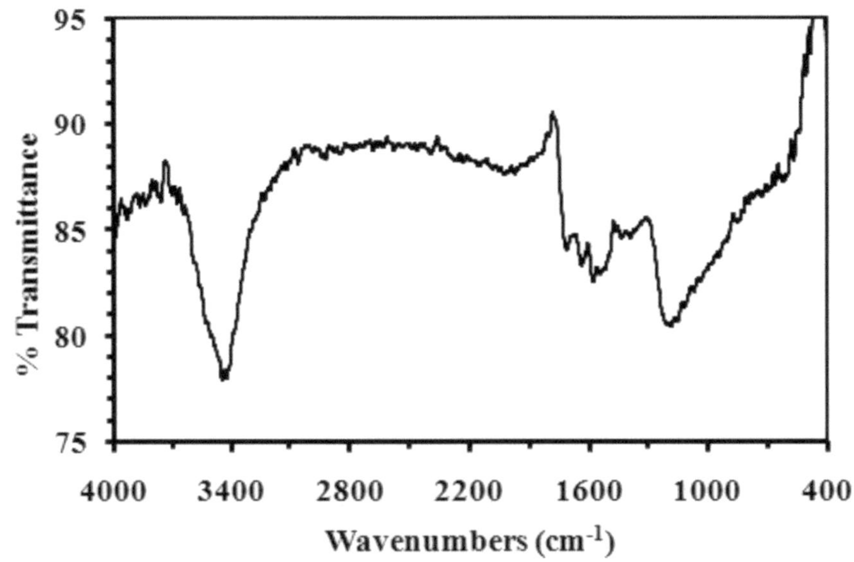

Fig. 2. FTIR spectra of MWCNT oxidized with mixtures of H_2SO_4/HNO_3 at 100 ^0C.

The IR spectrum of o-MWCNTs, Fig. 2, shows absorption peak at 1715 cm^{-1} corresponding to the stretching vibration of C=O from the carboxylic groups (-COOH). The carbonyl characteristic peak is also observed at about 1636 cm^{-1} and can be assigned to the carbonyl group from quinine or ring structure. The broad peaks at 1180 cm^{-1} could be assigned to C–O stretch from phenol or lactone groups; and also to C–C bonds. The band at 2910 cm-1 and 2850 cm^{-1} corresponds to CH stretching. The peak at around 3400 cm^{-1} corresponds to OH stretching. This peak can be assigned to the hydroxylic group of moisture, alcohol, or carboxylic groups. The aromatic C=C stretch is observed at around 1580 cm^{-1} in spectra of both pristine CNTs and o-CNTs.

XRD is used in order to ascertain the quality and crystalline nature of nanotubes as opposed to amorphous carbon materials. Fig. 3 shows the X-ray diffraction pattern of the o-MWCNTs. The intensity counts versus two theta. The angle is between the incident and scattered beams. The pattern shows intense peak at $2\theta = 25.9^{o}$ corresponding to the (002) reflection. Compared to the normal graphite, $2\theta = 26.5^{o}$, this peak shows a downward shift; which is attributed to an increase in the sp^2, C=C layers spacing (Zhang et al., 2002). The other diffraction peaks, are at the angles 2θ of 42.7^{o}, 43.9^{o}, 53.5^{o} and 77.5^{o} and indexed to the (100), (101), (004) and (110) reflections.

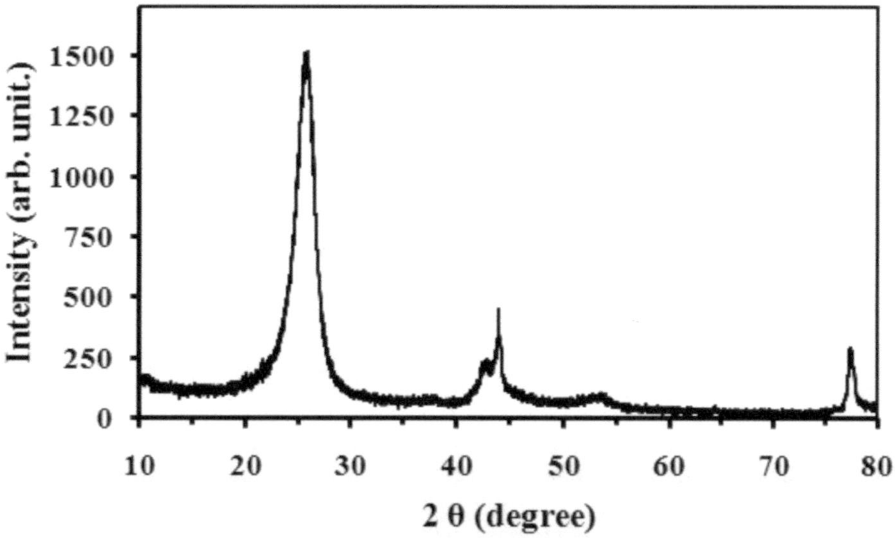

Fig. 3. XRD patterns of MWCNT oxidized with HNO3 at 120 ^{o}C

Raman spectroscopy is frequently used to characterize CNTs. It reflects different characteristic spectra for sp^3, sp^2, and sp carbons, as well as for disordered sp^2. Two bands is usually observed showing the characteristics of CNTs. The bands point the graphite band (G band), at about 1580 cm^{-1}, and the disorder and defects of the structure, named D band, at around 1330 cm^{-1}. The ratio between the intensity of D band and G bands ($I_{D/G}$), is related to the degree of disorder of the CNTs. When oxidation treatment of CNTs is performed, these characteristic peaks should still be identified to indicate that the treatment process does not damage the nanotube structure. An increase of the ratio values ($I_{D/G}$), of CNTs after oxidation treatment, ensures that the functional groups are covalently grafted with the CNTs. It indicates clearly that the oxidation treatments of CNTs breaks some of its bonds and inserts new chemical bonds, interpreted as defects on the nanotube structure (Osorio et al., 2008).

The TGA analysis of CNTs presents good indication of the functionality. o-CNTs starts decomposing earlier than the pristine-CNTs. The thermal stability of the pristine CNTs decreases with increasing oxidative functionalisation. The decrease of the initial burning temperature of the o-CNTs is closely related to the introduction of the defects and functional groups upon oxidation treatments. This is attributed to the earlier decomposition of the grafted carboxylic groups on the nanotubes. For example, the decomposition of the pristine-MWCNTs occur (TGA under flowing nitrogen, temperature range of 90–860 °C, at rate of 15 °C/min) at the onset temperature of 639 °C while of o-MWCNTs, it occurs at 627 °C (Okpalugo et al., 2005).

The XPS survey scan of CNTs is used to reveal the chemical compositions of the uppermost surface of the nanotubes. The major peaks due to the C1s and O1s photoelectrons can be observed in the scan. For example, the change in the C1s and O1s can be monitored and the O/C atomic ratios can be calculated to indicate the degree of surface oxidation. General conclusion is that the higher the O/C ratio, the more the oxygen atoms are bound to the nanotubes in the oxidation treatment. In addition, XPS O1s peaks demonstrate additional information on the nature of the surface oxygen-containing groups including the contribution of chemisorbed H_2O.

2.2 Synthesis and characterization of carbon nanotube/metal oxides composites

Carbon nanotubes/metal oxide (CNT/MO) composites can be prepared by various methods as mentioned earlier in this section. Here, the focus is on the wet chemistry methods where the CNTs act as nucleation sites for the metal oxides. In the first section, the synthesis of CNT/MO will be discussed. In the second section, the characterization of the CNT/MO nanocomposites using different characterization tools will be described with highlight on the interpretation of some experimental data.

2.2.1 Synthesis of CNT/MO composites

Generally, the o-CNTs are dispersed in an appropriate solvent like acetone, ethanol, propanol or n-methyl-2-pyrrolidone; untile well homogeneous suspension is formed. The dispersion can be achieved via sonication; some surfactants are used for dispersion. At the same, the precursor of the metal oxide is dissolved in a suitable solvent. Then, the latter is drop-wise added into the dispersed o-CNTs; and the mixture is sonicated for a while, then, magnetically stirred for some time. After that, the suspensions' mixture is transferred into round-bottomed flask and refluxed at high temperature; say 120- 200 ^{0}C, in oil bath or hot plate, with magnetic stirring, for specific time depends on the type of metal oxide. After that, the system is allowed to cool naturally to room temperature. Then, the mixture is filtered and washed with suitable solvent like distilled water and ethanol several times. After that, the resulted composite is dried at 100 ^{0}C. Final step in the strategy of synthesis is the calcinations process of the composite at various temperatures and for different times. The end product consists of CNTs coated, or embedded with nano-sized metal oxide; see the visualization illustration in Fig. 4.

The ratio of MO nanoparticels on CNTs is controlled by the initial concentrations and the type of the precursors of the metal ions. It is worth mentioning that the high ratio of MO would result in the agglomerating of the nanoparticles on the nanotubes. Additionally, the

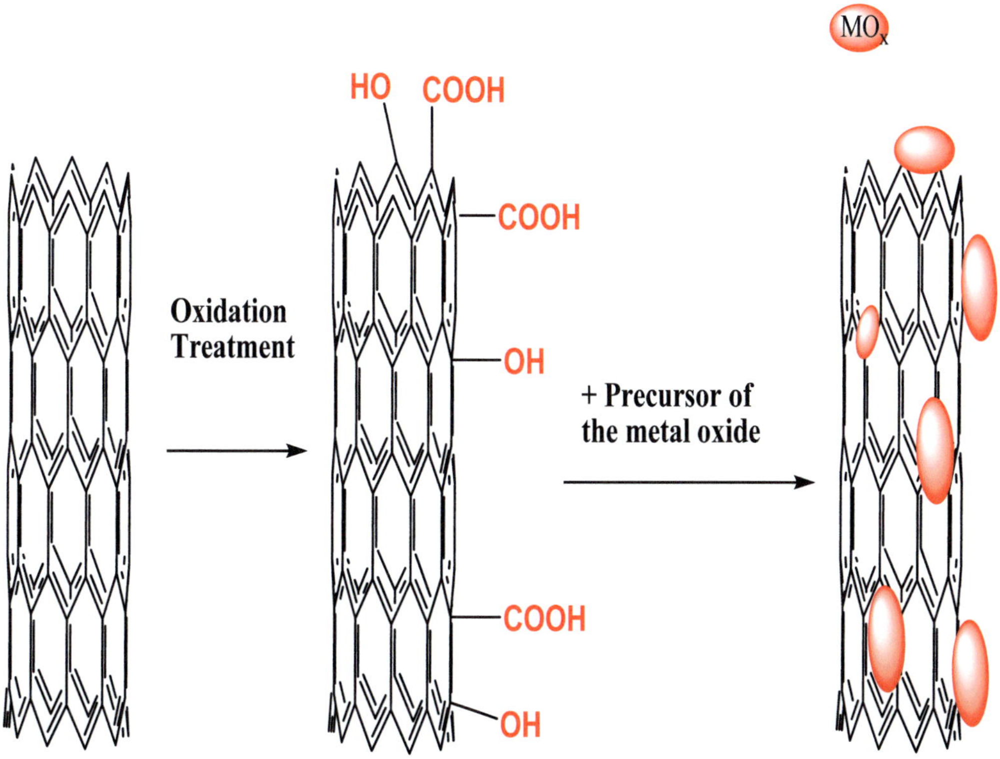

Fig. 4. Illustration representation of the synthesis of Carbon nanotubes/metal oxide (CNT/MO) composites

distribution of nanoparticels on the surface of the nanotubes is controlled by the synthesis treatment including the addition of the precursors into well dispersed o-CNTs, the extent of treatment oxidation of pristine CNTs. The nanoparticle size is controlled by the calcinations temperature and time. The rate of increasing the calcinations temperatures can also have an effect on the nanoparticle size. Controlling these parameters can yield a composite of uniformly dispersed metal oxide nanoparticels on the surface of the nanotubes.

As illustrated in Fig. 5, the nanotube surface presents polar groups like hydroxyl or carboxyl groups which are able to interact with the oxygen of the metal oxide, like aluminum oxide. The interaction could be through hydrogen bonding, or the oxygen atoms of hydroxyl or carboxyl groups interact with aluminum atoms through the pair of electrons on the oxygen atoms. The formation of chemical bond is also possible.

2.2.2 Characterization of CNT/MO composites

Characterization of the as-prepared carbon nanotube/metal oxides (CNT/MO) nanocomposite is performed by means of different characterization tools like high resolution transmission electron microscope (HRTEM), field emission scanning electron microscope (FESEM), energy dispersive X-ray spectroscopy (EDX), thermogravimetric analysis (TGA), Raman spectroscope, UV-vis spectrometer and fourier transformed infrared spectroscope (FT-IR), X-ray photoelectron spectroscope (XPS) and photoluminescence (PL) technique; in addition to electrochemical techniques. The techniques are complementary.

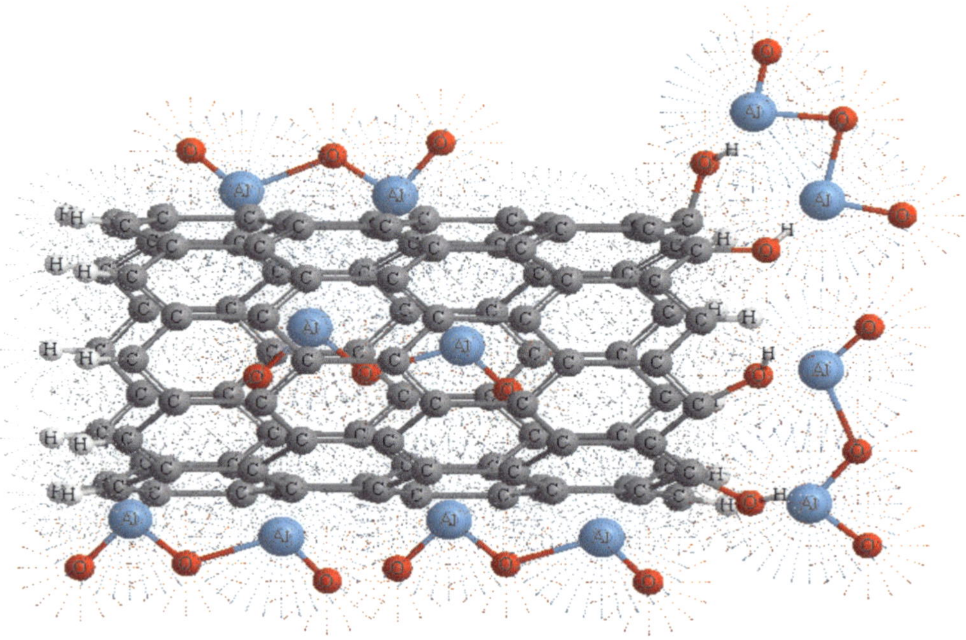

Fig. 5. Structural representative of bossible ineraction between Al_2O_3 and carbon nanotube in MWCNT/Al_2O_3 nanocomposite.

They can be used to investigate the chemical and physical properties of the nanocomposites. In the following paragraphs, the importance of each technique for the characterization of the composites will be highlighted with detailed examples of some experimental results.
HRTEM image indicates the shape and distribution of nanparticles on the surface of the nanotubes. The average size and diameters of the nanparticles can also be determined from the TEM images. Particularly, TEM is employed to observe the loading state of the nanocomposites. Specifically, TEM images show the surface of the nanotubes coated, decorated or embedded with the nanoparticles. It might be also used to compare the surfaces o-CNTs with that of CNT/MO nanocomposite in term of the roughness and compact.
SEM micrograph is used to investigate the surface morphology and the general morphological features of the prepared composites nanomaterials. It also reveals the uniformity of dispersion of the nanoparticles on the nanotubes. The average particle size can also be calculated. SEM image can be taken with low or high magnification. As an example, Fig. 6 presents the low magnification SEM to reflect the morphology of the CNT/alumina nanocomposite in which CNTs seem to be decorated with aluminum oxide nanomaterial.
The EDX or EDS measurements are usually used to confirm the percentage, the atomic ratio, of components on the surface of the nanocomposites. It should be noticed that EDX spectrum is better to be measured for large surface area of the sample for better representation, in other words, lower errors. That means the SEM image from which EDX is measured could be in μm scale, as shown in the inset of Fig. 7. However, it is not expected to get the exact atomic ratio of the metal oxides since some oxygen contribution comes from the oxygen-containing functional groups, such as carbonyl or carboxyl groups, on the surface of CNTs.

Fig. 6. Field emission scanning electron microscope (FESEM) of MWCNT/Al_2O_3 nanocomposite

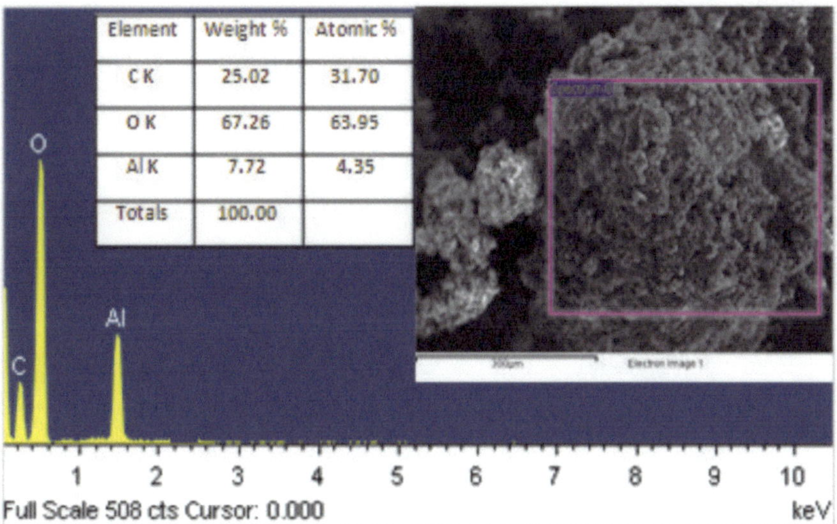

Fig. 7. Energy dispersive X-ray (EDX) spectrum of the area indicated in the Inset: low magnification scanning electron microscope (SEM) of MWCNT/Al_2O_3 nanocomposite; Inset: Table presents the atomic ratio, percentage of the components in the nanocomposite.

Thermogravimetric analysis is used to confirm the presence of metal oxides particles in the nanocomposite. From TGA of CNT/MO nanocomposites, it is expected to have initial weight loss between 250 to 500 °C and can be attributed to condensation reactions of metal-hydroxyl residual groups on the surface of the prepared nanocomposites materials (Amais et al., 2007).

XPS analysis is also of important. In the C1s XPS spectra, the main peak is usually observed at 284.6 eV which is attributed to the C–C bonds. Additional peaks are observed at 285.8 eV, 286.8 eV and 289.3 eV; and are assigned to C–O and C=O bonds; and COO groups (Jitianu et al., 2004). The shift of these peaks to higher binding energy could be indicative of binding of metal oxide to the nanotubes.

The XRD patterns are used to characterize the crystalline structure of the CNT/MO nanocomposites. Comparing the XRD patterns of o-CNTs and CNT/MO nanocomposite, the characteristic peaks for o-CNTs at the positions of 25.9° and 43.9°, see Fig. 3, might disappear or become thinner in the XRD pattern of the nanocomposites. The reason is that these peaks in the nanocomposites are overlapped by the peaks of the metal oxide nanoparticles. For example, in the XRD pattern of MWCN/TiO_2 composites, only anatase and brookite phases attributed to TiO_2 have been observed. The possible explanation is that the main peak of nanotube at 25.9° is masked with the main peak of anatase TiO_2 at 25.4° since their positions are so close. Additional is that the crystalline extent of carbon nanotube is lower than the crystalline extent of TiO_2, leading to the shielding of the peaks of CNTs by those of TiO_2 (Yu et al., 2005). Example is the XRD pattern, Fig. 8, of MWCNT/WO_3 nanocomposite. In this pattern the diffraction peaks revealed the presence of hexagonal and orthorhombic tungsten oxide nanoparticles. Therefore, the presence of tungsten oxide nanoparticles peaks might lead to the mask the characteristic peak of the nanotubes at 2 θ equal to 25.8 degree.

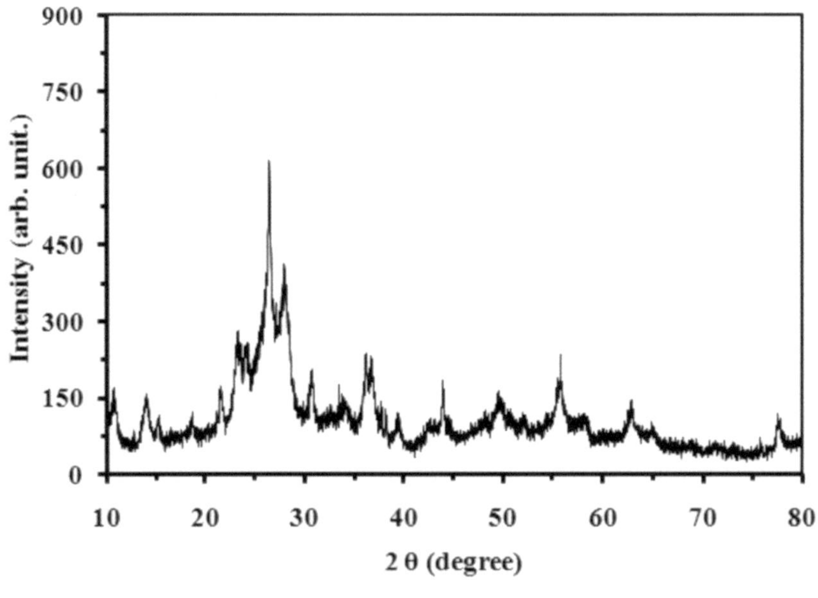

Fig. 8. XRD patterns of MWCNT/WO_3 nanocomposite.

Raman spectroscopy is a powerful characteristic technique that is employed to distinguish between the characteristic peaks of TiO_2 and those of CNTs since peaks from each components; nanotube and metal oxide are in most cases separated in frequency, and thus the phases are distinguishable. It is worth mentioning that to run samples on Raman, the energy power should be as low as possible to prevent the effect on the nanotube absorption and get good results. The energy power is preferably to be selected between 10 and 100 mW.

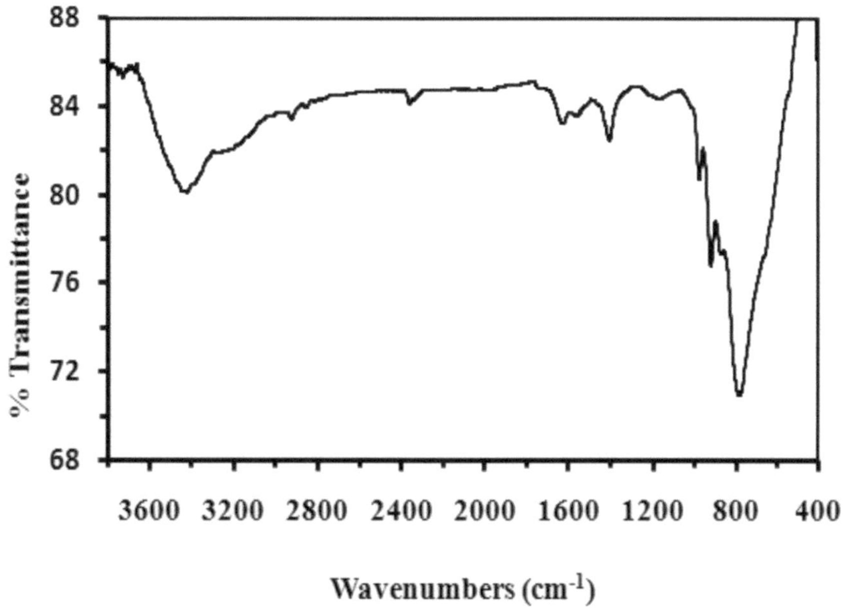

Fig. 9. FTIR spectra of MWCNT/WO_3 nanocomposite.

Comparing the IR spectra of o-CNTs and the nanocomposite is of interest. The comparison should focus on two characteristic peaks. The first peak is the characteristic peak of the pure metal oxide nanoparticels, which is usually appear between around 400 and 700 cm^{-1} depends on the type of metal oxide. For example this peak appear at 700 cm^{-1} in the IR spectrum of WO_3 nanoparticles. The shift of this peak in the spectra of nanocomposite could be attributed to strong bond between the metal and the nanotube through oxygen-containing group. The second peak is the band assigned to C=O stretching vibration. The shift of this peak toward lower wavenumber is an indication of the carboxylic ion binding between nanoparticles and the nanotube. Example is the IR spectrum of MWCNT/WO_3 nanocomposite , depicted in Fig. 9. This spectrum shows the shift of the peak of W-O covalent bond from 701cm^{-1} to 770cm^{-1}. It also shows the shift of the band assigned to C=O stretching vibration from 1670 to 1606 cm^{-1}, reader can compare Fig. 9 with Fig. 2.

3. Applications of CNT/MO nanocomposites

CNT/MO nanocomposites are new generation with interesting properties that not only combine the properties CNTs and MOx but also hold new properties caused by the interaction between them. In addition, combine these materials in a composite can overcome some disadvantages. For example, metal oxide nanoparticles have a tendency to agglomerate because of their dangling bonds. In the composite, dispersion of the nanoparticles on the surface of the nanotube prevents their agglomeration. It should be noticed that the aggregation of nanotubes in aqueous media is solved by functionalization. As a result, the composites attractive wide applications compared with the isolated nanoparticles because CNTs act as carrier to stabilize the nanoparticles, maintaining their integrity. Examples of the fields of applications are water treatment, environment, catalysis and others. Examples of the nanocomposites that can be prepared are Ti_2O_3, ZnO, WO_3, Fe_2O_3, SnO_2, CeO_2, Al_2O_3, ZrO_2, SiO_2, V_2O_4 and Er_2O_3 on CNTs. In the next section, we will discuss the application of the composites as adsorbents followed by the discussion of their applications to enhance catalytic activity in the second section. A mechanism will be discussed to explain the enhancement of the photocatalytic properties of the nanocomposite.

3.1 Nanocomposites as sorbents

The sorption applications of CNT/MO nanocomposites to tackle environmental pollution problems have received considerable attention. The nanocomposites provide chemically inert surfaces with high specific surface areas for physical adsorption. They have relatively uniform structure, providing more adsorption sites.

For the application of the nanocomposite as sorbents in batch sorption experiments, different parameters need to be optimized, such as agitation speed, contact time and adsorbent dosage, pH of the aqueous media, initial concentration of the pollutant, effect of presence of other cations and anions in the media. Also, for the application of the nano-composite as adsorbents in fixed bed column experiments, different parameters need to be optimized, such as flow rate, the bed thickness and diameters, pH of the aqueous media, initial concentration of the pollutant, effect of the presence of other cations and anions in the media.

For example, MWCNT/alumina nanocomposite has been reported to be an effective sorbent for the removal of lead ions from aqueous solution with pH range of 3-7. The efficiency of the nanocomposite has been tested in batch mode and fixed bed, column mode. It has been reported that by increasing agitation speed, contact time and adsorbent dosage the amount of the removal is increased. The percentage of lead removal increases by decreasing the flow rate and increasing the bed thickness (Gupta et al., 2011). CNT-iron oxides magnetic composites has been successfully applied as adsorbent for removal of different targets from water such as Pb(II), Cu(II) (Peng et al., 2005), Ni(II), Sr(II) (Chen et al., 2009b) and Cr(III) (Gupta et al., 2011). The incorporation of iron oxide magnetite with CNTs has been used for europium adsorption. This composite is potentially a promising to facilitate the separation and recovery of CNTs from solution with magnetic separation technique. The beauty of this method is that CNTs do not become another source of contamination (Chen et al., 2009a). Also, the magnetic MMWCNT nanocomposite has successfully utilized as an adsorbent for removal of cationic dyes from aqueous solutions (Gong et al., 2009).

The adsorption ability of manganese oxide-coated carbon nanotubes has been utilized for the removal of lead(II) from aqueous solution. The effect of metal oxide nanoparticels loading level has been investigated. It has been reported that the adsorption capacity of Pb(II) increased with increasing MnO_2 load because of the availability of more binding sites. The optimum reported load is 30 wt% (Wang et al., 2007).

3.2 Nanocomposites as catalysts

CNTs have been widely used to support and enhance the catalytic activity of metal oxides. CNT/WO_3 nanocomposite has been utilized as catalysts for C6 olefin skeletal isomerization in different activation conditions. Comparing with tungstated zirconia, the nanocomposite shows higher skeletal isomerization selectivity at 200 °C with higher conversion level without deactivation. The optimum load of WO_3 on CNTs was reported to be 5% (Pietruszka et al., 2005). Such composite shows also higher photocatalytic activity comparing with WO_3 nanoparticels. The composite has been applied for the degradation of dye. It could be concluded that doping WO_3 into CNTs enhances the photocatalytic activity (Wang et al., 2008).

Another comparison of the photocatalytic activity has been conducted between TiO_2 and TiO_2/CNTs composites for acetone degradation. It has been noticed that the presence of a small amount of CNTs can enhance photocatalytic activity of TiO_2 (Yu et al., 2005). This is supported by another study in which photocatalytic activity of TiO_2/SWCNT composite has been investigated degradation of organic dyes and 2,6-dinitro-p-cresol. High photocatalytic activity of the TiO_2/SWCNT composite has been observed comparing with that of TiO_2 nanoparticles. This could be attributed to the synthetic effect of larger surface area, smaller crystalline size, and the formation of the chemical bond between TiO_2 and the nanotube which improve interface contact (Zhou et al., 2010; Wang et al., 2009). It has been reported that the bond of carbon–oxygen–titanium can extend the light absorption to longer wavelengths and therefore potentially leading to the improvement of the photocatalytic activity (Woan et al., 2009).

MWCNTs has been used to enhance the catalytic activity of zinc oxide for the acetaldehyde removal (Saleh et al., 2011) and cyanide degradation in aqueous media (Saleh et al., 2010) under UV irradiation. Zinc oxide nanoparticels are embedded on multiwall carbon nanotubes via thermal hydrolysis method. Great enhancement in photocatalytic activity has been recorded by the MWCNT/ZnO nanocomposite.

3.2.1 Mechanism

The enhancement in photocatalytic performance of the CNT/MO nanocomposites can be rationally explained in a mechanism based on two criteria. The first is adsorption ability of the nanotubes since the adsorption is a key process in the catalytic destruction of organic pollutants. While the second is the electron transfer from nanoparticels into the nanotubes, in other words the electron injection (Xu et al., 2010 & Martinez et al., 2011). In general, irradiation of metal oxide semiconductor with a photon of sufficient energy, greater or equal to band gap energy could promote an electron from valence band (VB) to conduction band (CB). The promoted electron leaves a vacancy in the valence band creating positive hole which can react with hydroxyl forming a powerful oxidant or a hydroxyl radical. The promoted electrons reduce dissolved oxygen to produce superoxide anion radical. The hydroxyl radical ($^{\bullet}OH$) and superoxide anion radical ($O_2^{\bullet -}$) oxidize the target.

CNTs can adsorb oxygen molecule on its surface thus the promoted electrons in CB might directly react with the adsorbed oxygen molecule and form superoxide anion radical. This will increase the number of radicals ready for reacting with the target. Also CNTs can adsorb the target which is considered to be the first step in the degradation since the pre-adsorption of the target on the surface of catalyst particles is a prerequisite for efficient photo-degradation (Zhao et al., 1998).

On the other hand, the strong interaction and formation of ester bonds between the metal and the nanotube leading in a close contact which offers an effective route of electron transfer from the CB in to the nanotube. This can reduce the recombination of photo-generated electron-hole pairs. This conclusion is supported by the higher activity of the

CNT/MO composites and by being CNTs relatively good electron acceptors while semiconductors as good electron donor under illumination (Subramanian et al., 2004; Geng et al., 2008).

4. Conclusion

In this chapter we discuss the preparation of the new class of CNT/MO nanomaterials that offer great potential applications in different fields. The chapter has analyzed the methodology of functionalization and activation of the nanotubes as a first step for preparation of CNT/MO nanocomposites. The details of functional groups characterization have also been discussed. The synthesis of CNT/MO nanocomposites is detailed with the characterization interpretation. Some case-studies have been carried out referring to a realistic use of CNT/MO nanocomposites as adsorbents that have excellent adsorption ability for many kinds of pollutants like inorganic and organic compounds.

The advantage is that CNTs provide large number of nucleation centers available on its surface for receiving the nanoparticles and preventing their agglomeration. This makes such material not only good in adsorption but also in enhancement of the catalytic activity. The reason given for the high catalytic performance is the optimum dispersion of MO on the CNTs. The mechanism of enhancing the photo-catalytic activity is discussed based on adsorption ability of the nanotube and the transfer of promoted electrons from metal oxide to CNT. The most important processing issue is the interfacial strength between the CNTs and the MO nanoparticles.

Over the last years, much progress has been made in adsorption applications of CNT/MO nanocomposites. In spite of high costs, using CNTs as adsorbents may be advantageous in future owing to their high adsorption capacities. However, there are still much to do to enhance CNT/MO nanocomposites' adsorption properties in future. Efforts in synthesis methods to develop a cost effective way of CNTs production are required. Investigations on CNT/MO nanocomposites on applications in adsorption and catalysis are recommended.

5. Acknowledgment

Authors acknowledge the support of King Fahd University of Petroleum and Minerals, (KFUPM) Dhahran, Saudi Arabia, for this work.

6. References

Agnihotri, S.; Mota, J. P. B.; Rostam-Abadi, M. & Rood, M. J. (2006). Adsorptionsiteanalysis of impurity embedded single-walled carbon nanotube bundles, Carbon 44: 2376–2383.

Amais, R. S.; Ribeiro, J. S.; Segatelli, M. G.; Yoshida, I. V.P.; Luccas, P. O. & Tarley C. R. T. (2007). Assessment of nanocomposite alumina supported on multi-wall carbon nanotubes as sorbent for on-line nickel preconcentration in water samples, Separation and Purification Technology 58: 122–128

Aviles, F.; Cauich-Rodríguez, J. V.; Moo-Tah, L.; May-Pat, A. & Vargas-Coronado, R. (2009). Evaluation of mild acid oxidation treatments for MWCNT functionalization, Carbon 47: 2970–2975.

Boehm, H. P. (1994). Some aspects of the surface chemistry of carbon blacks and other carbons. Carbon 32: 759–69.

Chan, B. M., Cha, S. I.; Kim, K. T., Lee, K. H. & Hong, S. H. (2005). Fabrication of carbon nanotube reinforced alumina matrix nanocomposite by sol–gel process, Materials Science and Engineering A 395: 124–128

Chen, C. L.; Wang, X. K. & Nagatsu, M., (2009a). Europium Adsorption on Multiwall Carbon Nanotube/Iron Oxide Magnetic Composite in the Presence of Polyacrylic Acid, Environ. Sci. Technol. 2009, 43, 2362–2367

Chen, C., Jun Hu, Dadong Shao, Jiaxing Li, XiangkeWang, (2009b). Adsorption behavior of multiwall carbon nanotube/iron oxide magnetic composites for Ni(II) and Sr(II), Journal of Hazardous Materials 164 923–928

Chen, C.L.; Wang, X.K. (2006). Adsorption of Ni(II) from aqueous solution using oxidized multiwall carbon nanotubes, Ind. Eng. Chem. Res. 45: 9144–9149.

Chiang, Y.; Lin, W. H. & Chang, Y. C. (2011). The influence of treatment duration on multi-walled carbon nanotubes functionalized by H2SO4/HNO3 oxidation, Applied Surface Science 257 2401–2410

Cho, H. H.; Wepasnick, K.; Smith, B.A.; Bangash, F.K.; Fairbrother, D.H. & Ball W.P. (2010). Sorption of aqueous Zn[II] and Cd[II] by multiwall carbon nanotubes: the relative roles of oxygen-containing functional groups and graphenic carbon, Langmuir 26: 967–981.

Chu, H.; Wei, L.; Cui, R.; Wang, J. & Li, Y. (2010). Carbon nanotubes combined with inorganic nanomaterials: Preparations and applications. Coord Chem Rev 2010, 254: 1117-1120.

Cuentas-Gallegos, A.K.; Martínez-Rosales, R.; Rincón, M.E.; Hirata, G.A. & Orozco, G. (2006). Design of hybrid materials based on carbon nanotubes and polyoxometalates, Opt. Mater. 29: 126–133.

Datye, A.; Wu, K. H.; Gomes, G.; Monroy, V.; Lin, H.T.; Jozef, V. & Vanmeensel, K. (2010). Synthesis, microstructure and mechanical properties of Yttria Stabilized Zirconia (3YTZP) – Multi-Walled Nanotube (MWNTs) nanocomposite by direct in-situ growth of MWNTs on Zirconia particles, Composites Science and Technology 70: 2086–2092

Eder, D. (2010). Carbon Nanotube-Inorganic Hybrids. Chem Rev. 110: 1348-1352.

Estili, M. & Kawasaki, A. (2008). An approach to mass-producing individually alumina-decorated multi-walled carbon nanotubes with optimized and controlled compositions, Scripta Materialia 58: 906–909

Flahaut, E.; Peigney, A.; Laurent, C.; MarlieRe, C.; Chastel, F. & Rousset, A. (2000). Carbon nanotube–metal–oxide nanocomposites: microstructure, electrical conductivity and mechanical properties, Acta Mater. 48: 3803–3812

Gao, B.; Peng, C.; Chen, G. Z. & Puma, G.L. (2008). Photo-electro-catalysis enhancement on carbon nanotubes/titanium dioxide (CNTs/TiO_2) composite prepared by a novel surfactant wrapping sol–gel method, Applied Catalysis B: Environmental, 85: 17–23

Geng, Q.; Guo, Q.; Cao, C. & Wang, L. (2008). Investigation into NanoTiO_2/ACSPCR for Decomposition of Aqueous Hydroquinone, Ind. Eng. Chem. Res. 47: 2561-2568

Goering, J.; Kadossov, E. & Burghaus, U. (2008). Adsorption kinetics of alcohols on single wall carbon nanotubes: an ultra high vacuum surface chemistry study, J. Phys. Chem. C 112: 10114–10124.

Gong, J. L.; Wang, B.; Zeng, G.; Yang, C.; Niu, C.; Niu, C.; Zhou, W. & Liang, Y. (2009). Removal of cationic dyes from aqueous solution using magnetic multi-wall carbon nanotube nanocomposite as adsorbent, Journal of Hazardous Materials 164: 1517–1522

Gupta, V.K.; Agarwal, S. & Saleh, T. A. (2011). Synthesis and characterization of alumina-coated carbon nanotubes and their application for lead removal, Journal of Hazardous Materials 185: 17–23

Gupta, V.K.; Agarwal, S. & Saleh, T. A. (2011). Chromium removal by combining the magnetic properties of iron oxide with adsorption properties of carbon nanotubes, Water Research 45: 2207-2212

Haydar, S.; Moreno-Castilla, C.; Ferro-García, M.A.; Carrasco-Marín, F.; Rivera-Utrilla, J.; Perrard, A. & Joly, J.P. (2009). Regularities in the temperature-programmed desorption spectra of CO2 and CO from activated carbons, Carbon 38 (2000) 1297–1308.

Hojati-Talemi, P.; Simon, G. (2009) Microwave-based treatments for multi-walled carbon nanotubes, , Phys. Status Solidi C 6, 10: 2170–2173

Hyung, H. & Kim, J.H., (2008). Natural organic matter (NOM) adsorption to multi-walled carbon nanotubes: effect of NOM characteristics and water quality parameters, Environ. Sci. Technol. 42: 4416–4421.

Iijima, S. (1991). Helical Microtubules of Graphitic Carbon, Nature 354: 56-58

Ionescu, R.; Espinosa, E.H.; Sotter, E.; Llobet, E.; Vilanova, X.; Correig, X.; Felten, A.; Bittencourt, C.; Van Lier, G.; Charlier, J. C. & Pireaux J. J. (2006). Oxygen functionalisation of MWNT and their use as gas sensitive thick-film layers, Sensors and Actuators B 113 36–46

Jitianu, A.; Cacciaguerra, T.; Benoit, R.; Delpeux, S.; Beguin, F. & Bonnamy, S., (2004). Synthesis and characterization of carbon nanotubes–TiO_2 nanocomposites, Carbon 42: 1147–1151

Keshri, A. K.; Huang, J.; Singh, V.; Choi, W.; Seal, S. & Agarwal, A. (2010). Synthesis of aluminum oxide coating with carbon nanotube reinforcement produced by chemical vapor deposition for improved fracture and wear resistance, Carbon 48: 431–442.

Li, Y. H.; Ding, J.; Luan, Z. K.; Di, Z.C.; Zhu, Y.F.; Xu, C. L.; Wu, D.H. & Wei, B.Q. (2003) Competitive adsorption of Pb2+, Cu2+ and Cd2+ ions from aqueous solutions by multiwalled carbon nanotubes, Carbon 41 2787–2792.

Li, Y. H.; Wang, S. G.; Wei, J.Q.; Zhang, X. F.; Xu, C. L.; Luan, Z. K.; Wu, D. H. & Wei, B. Q. (2002) Lead adsorption on carbon nanotubes, Chem. Phys. Lett. 357 263–266.

Liu, X.; Huber, T. A.; Kopac, M. C. & Pickup, P. G. (2009). Ru oxide/carbon nanotube composites for supercapacitors prepared by spontaneous reduction of Ru(VI) and Ru(VII), Electrochimica Acta 54 (2009) 7141–7147

Lu, C. & Chiu, H. (2008). Chemical modification of multiwalled carbon nanotubes for sorption of Zn2+ from aqueous solution, Chem. Eng. J. 139 (2008) 462–468.

Lupo, F.; Kamalakaran, R.; Scheu, C.; Grobert, N. & Ruhle, M. (2004). Microstructural investigations on zirconium oxide–carbon nanotube composites synthesized by hydrothermal crystallization, Carbon 42 (2004) 1995–1999

Martinez, M. T.; Callejas, M. A.; Benito, A.M.; Cochet, M.; Seeger, T.; Anson, A.; Schreiber, J.; Gordon, C.; Marhic, C.; Chauvet, O.; Fierro, J.L.G. & Maser, W.K. (2003). Sensitivity of single wall carbon nanotubes to oxidative processing: structural modification, intercalation and functionalisation, Carbon 41: 2247–2256.

Martinez, C.; Canle, M. L.; Fernandez, M. I.; Santaballa, J. A. & Faria, J.; (2011). Kinetics and mechanism of aqueous degradation of carbamazepine by heterogeneous photocatalysis using nanocrystalline TiO_2, ZnO and multi-walled carbon nanotubes - anatase composites, Applied Catalysis B: Environmental 102: 563–571

Okpalugo, T. I. T.; Papakonstantinou, P.; Murphy, H.; Mclaughlin, J. & Brown, N. M. D. (2005). Oxidative functionalization of carbon nanotubes in atmospheric pressure filamentary dielectric barrier discharge (APDBD), Carbon 43: 2951–2959

Osorio, A. G.; Silveira, I. C. L.; Bueno, V. L. & Bergmann, C.P. (2008). $H_2SO_4/HNO_3/HCl$ Functionalization and its effect on dispersion of carbon nanotubes in aqueous media, Applied Surface Science 255 2485–2489

Otvos, Z.; Onyestyák, G.; Hance, A.; Kiricsi, I. & Rees, L.V.C. (2006). Surface oxygen complexes as governors of neopentane sorption in multiwalled carbon nanotubes, Carbon 44: 1665–1672.

Peng, K.; Liu, L.Q.; Li, H.; Meyer, H. & Zhang, Z. (2011). Room temperature functionalization of carbon nanotubes using an ozone/water vapor mixture, Carbon 49: 70–76

Peng, X.; Luan, Z.; Di, Z.; Zhang, Z. & Zhu, C. (2005). Carbon nanotubes-iron oxides magnetic composites as adsorbent for removal of Pb(II) and Cu(II) from water, Carbon 43: 855–894

Pietruszka, B.; Gregorio, F. D.; Keller, N. & Keller, V. (2005). High-efficiency WO_3/carbon nanotubes for olefin skeletal isomerization, Catalysis Today 102–103: 94–100

Saleh, T. A.; Gondal, M.A. & Drmosh, Q.A., (2010). Preparation of a MWCNT/ZnO nanocomposite and its photocatalytic activity for the removal of cyanide from water using a laser, Nanotechnology 21 (49), 8: doi:10.1088/0957-4484/21/49/495705

Saleh, T. A.; Gondal, M. A.; Drmosh, Q.A.; Z Yamani, .H. A. & AL-yamani, A. (2011) Enhancement in photocatalytic activity for acetaldehyde removal by embedding ZnO nano particles on multiwall carbon nanotubes, Chemical Engineering Journal, 166(1), 407-412

Shao, D. D.; Ren, X. M.; Hu, J.; Chen, Y.X. & Wang, X. K. (2010). Preconcentration of Pb2+ from aqueous solution using poly(acrylamide) and poly(N,N-dimethylacrylamide) grafted multi walled carbon nanotubes, *Colloids Surf. A* 360: 74–84.

Smith, B.; Wepasnick, K.; Schrote, K. E.; Cho, H. H.; Ball, W. P.; Howard Fairbrother, D. (2009). Influence of Surface Oxides on the Colloidal Stability of Multi-Walled Carbon Nanotubes: A Structure-Property Relationship, Langmuir 25: 9767–9776

Subramanian, V.; Wolf, E.; Kamat, P. V. (2004). Catalysis with TiO_2/Gold Nanocomposites. Effect of Metal Particle Size on the Fermi Level Equilibration, J. AM. CHEM. SOC. 126: 4943-4950

Wang, H.; Wang, H. & Jiang, W. (2009). Solar photocatalytic degradation of 2,6-dinitro-p-cresol (DNPC) using multi-walled carbon nanotubes (MWCNTs)–TiO_2 composite photocatalysts, Chemosphere 75: 1105–1111

Wang, S.; Shi, X.; Shao, G.; Duan, X.; Yang, X. & Wang, X. (2008). Preparation, characterization and photocatalytic activity of multi-walled carbon nanotube-supported tungsten trioxide composites, Journal of Physics and Chemistry of Solids 69: 2396- 2400

Wang, S.G.; Gong, W.; Liu, W.; Yao, Y.; Gao, B. & Yue, Q. (2007). Removal of lead(II) from aqueous solution by adsorption onto manganese oxide-coated carbon nanotubes, Separation and Purification Technology 58: 17-23

Woan, K.; Pyrgiotakis, G. & Sigmund, W. (2009). Photocatalytic Carbon-Nanotube-TiO_2 Composites, Adv. Mater. 21: 2233-2239

Xu, Y. J.; Zhuang, Y. & Fu, X. (2010). New Insight for Enhanced Photocatalytic Activity of TiO2 by Doping Carbon Nanotubes: ACase Study on Degradation of Benzene and Methyl Orange, J. Phys. Chem. C 114: 2669-2676

Yang, Q.; Deng, Y. & Hu, W. (2009). Preparation of alumina/carbon nanotubes composites by chemical precipitation, Ceramics International, 35: 1305-1310

Yu, Y.; Yu, J. C.; Yu, J. G.; Kwok, Y.C.; Che, Y.K.; Zhao, J.C.; Ding, L; Ge, W.K. & Wong, P.K. (2005). Enhancement of photocatalytic activity of mesoporous TiO_2 by using carbon nanotubes, Applied Catalysis A: General 289: 186-196

Zhang, D.; Shi, L.; Fu, H. & Fang, J. (2006). Ultrasonic-assisted preparation of carbon nanotube/cerium oxide composites, Carbon 44: 2849-2867

Zhang, H. B.; Lin, G. D.; Zhou, Z. H.; Dong, X.; Chen, T. (2002). Raman spectra of MWCNTs and MWCNT-based H2 adsorbing system, Carbon 40: 2429-2436.

Zhang, S. C.; Fahrenholtz, W. G.; Hilmas, G.E. & Yadlowsky, E. J. (2010). Pressureless sintering of carbon nanotube-Al_2O_3 composites, Journal of the European Ceramic Society 30 1373-1380

Zhao, J.; Wu, T.; Wu, K.; Oikawa, K.; Hidaka, H. &Serpone, N. (1998). Photoassisted Degradation of DyePollutants. 3. Degradation of the Cationic Dye Rhodamine B in Aqueous Anionic Surfactant/TiO_2 Dispersions under Visible Light Irradiation: Evidence for the Need of Substrate Adsorption on TiO_2 Particles, Environ. Sci. Technol., 32: 2394-2400

Zhen, F.; Kunzai, J.C.; Cui, F. S.; Xu, Y. & Kuang, Y. (2007). Preparation and capacitive properties of cobalt-nickel oxides/carbon nanotube composites, Electrochimica Acta 52: 2959-2965

Zhou, W.; Pan, K.; Qu, Y.; Sun, F.; Tian, C.; Ren, Z.; Tian, G. & Fu, H. (2010). Photodegradation of organic contamination in wastewaters by bonding TiO_2/single-walled carbon nanotube composites with enhanced photocatalytic activity Chemosphere 81: 555-561

18

Foam Materials Made from Carbon Nanotubes

Kyuya Nakagawa
Research Centre for Nano-Micro Structure Science and Engineering, University of Hyogo, Japan

1. Introduction

1.1 What is CNT foam?

Foam is a macroporous material and comes in many forms, such as polystyrene shock absorbers, dish-washing sponges, and foam board insulation. This chapter describes a technique to prepare foam materials from carbon nanotubes (CNTs). The use of a CNT-made foam as a kitchen sponge may not be a good idea, however. We of course expect the nano-carbons to have a specific functionality on the prepared bulk material. A challenge would be to place a function derived from nanomaterials on a bulk structuralized material. It would be highly useful to make up a material with a specific function that is attained only by nano- and micro-structuralization. A major feature of a foam material is that it has a function derived from its microstructures (macropores). The present chapter discusses a preparation strategy to produce a material with controlled microstructures from CNTs.

Fig. 1. Foam materials made from carbon nanotubes. Samples in grey and white (pictured bottom left) were prepared by mixing titanium nano-powders.

1.2 How is it made?

Recent studies explain how to make from CNTs highly porous foam materials, which are composite materials made from polymers and CNTs. CNT foams can be prepared using several methods, as illustrated in Fig. 2. One is a bottom-up method using phase separations induced by thermal phase transition, polymerization reaction, solubility shift, and so on. When a phase separation is induced in a CNT suspension, a phase that cannot include dispersed materials (such as CNTs) generates a counter phase in which the dispersed materials are concentrated (concentrated phase). Frequent phase separation produces a microstructure in the resultant solution after the pure phases are removed by post-treatment. As shown in Fig. 2A, this phase separation route is mainly realized by a gas-liquid phase separation process (foaming) and solid-liquid phase separation process (freezing). CNT foam preparation using a poly(vinyl alcohol) foaming process was first reported by Shaffer and Windle (1999). Subsequent researchers found that foaming processes with polystyrene, polyurethane, poly(methyl methacrylate), and so on showed potential in preparing CNT foam with controlled pore structures (Shen et al., 2005; Park & Kim, 2008; Xiang et al., 2009; Chen et al., 2010; Hermant et al., 2009; Zeng et al., 2010).

It has long been known that freeze-drying a solution results in porous materials, where the pore structures are replicas of the ice crystals formed during freezing. However, it was only recently that an attempt was made to prepare porous CNT foams via freeze-drying. Nabeta and Sano (2005) conducted a trial to prepare freeze-dried CNT foams from a CNT-gelatin aqueous suspension. This would make it the first study to employ a solid-liquid phase separation process for producing macroporous CNT foams (in their report, they stated that the microstructures were created by gelatin gel formation; however, this may be only partly true: gel formation contributes to producing nanostructures, whereas ice creates microstructures). In any case, it has clearly been shown by related studies that ice is an interesting tool for creating interconnected 3D microstructures in a freeze-dried CNT foam (Lau et al., 2008; Thongprachan et al., 2008; Kwon et al., 2009; Olivas-Armendariz et al., 2010; Nakagawa et al., 2010). Fairly well-aligned beautiful macropore formations in the freeze-dried foams have been demonstrated by a Spanish group during intensive practical studies (Gutiérrez et al., 2007a; 2007b; Abarrategi et al., 2008).

Phase separation techniques would allow further processing variations, as described in a report by Leroy et al. (2007). They demonstrated CNT foam preparation via a bubbling and subsequent freeze-drying process that enabled unique microstructure formation. Sol-gel processing would also be useful in producing CNT foams with nano-structured CNT arrays (Worsley et al., 2009). In order to create a controlled microstructure by using a phase separation process, we must control the kinetics of the phase transition (e.g. evaporation for foaming process, solidification for freezing process). In this regard, freezing would be advantageous for planning an engineering strategy because freezing can simply be controlled by a thermal system. The following sections deal with the processing strategy for obtaining desirable microstructures on a CNT foam prepared via a phase separation process induced by freezing.

As depicted in Fig. 2B, CNT deposition onto a ready-made foam material is an alternative CNT foam preparation method (Boccaccini et al., 2007; Ji et al., 2009; Zawadzak et al., 2009; Meng et al., 2009). It is one type of coating technique for CNTs on a porous media. Pore structures of the media can be appropriately designed using a number of other methods.

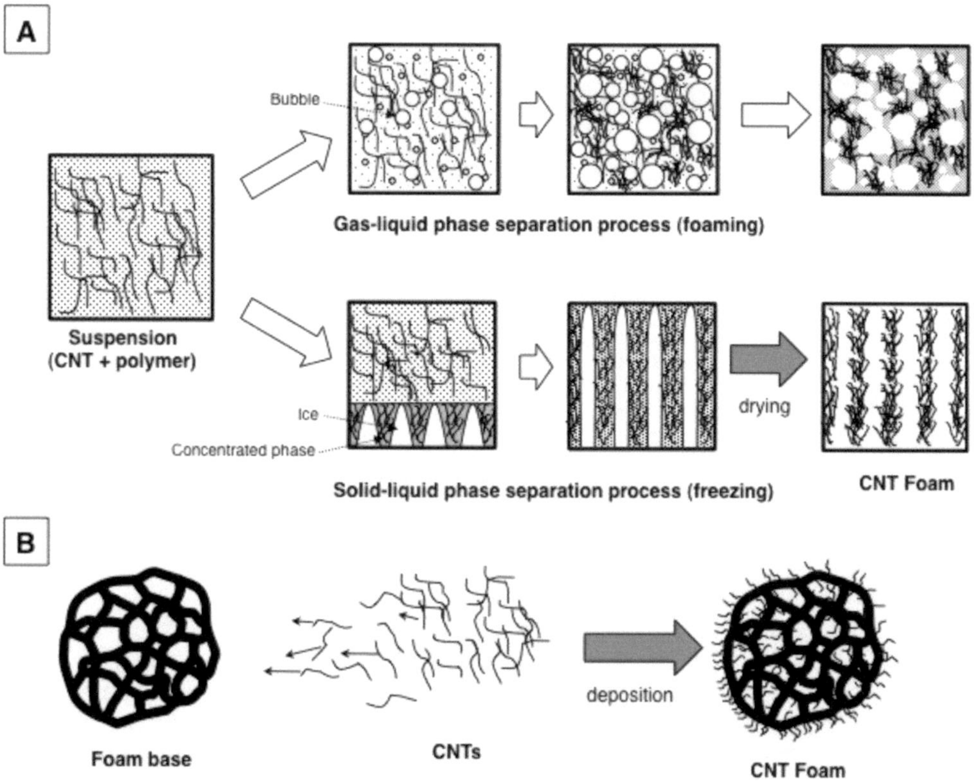

Fig. 2. Foam material preparation from carbon nanotubes.

2. Freezing as a tool to control microstructure

2.1 Solidification of colloidal suspension

Freezing induces a pattern formation in a colloidal suspension when the velocity of freezing front progress is maintained at a certain value. The microstructure of a frozen colloidal suspension cannot be obtained by simply replacing the water in unfrozen solution by ice. The freezing front rejects suspended particles when the front velocity is sufficiently low. However, it engulfs particles when the velocity is too high. The physics of this particle rejection phenomenon at the solidification front have been investigated based on several theories (Wilde & Perepezko, 2000). It has recently been identified that interfacial premelting phenomena virtually control the behaviour of a foreign particle at the solidification interface (Dash et al., 1995, 2006; Wettlaufer & Worster, 2006). When a solidification interface meets a particle, repulsive intermolecular interaction between the solid and the particle creates a thin liquid layer for reducing total free energy at the interface (reducing instability caused by the repulsive intermolecular interaction). This liquid layer formation, due to interfacial premelting, allows a particle to move away from the freezing front by repulsive interaction with the ice (Fig. 3). The repulsive force is a dominating factor that controls the particle transfer kinetics. Put simply, a particle is rejected by the ice front when the particle mass-transfer velocity is higher than the ice front velocity; otherwise it is engulfed. Following this theory, suspended particles are sometimes rejected and engulfed in a frozen bulk to form a complex microstructure. Further theoretical approaches predict a periodicity of microstructures made by ice in a colloidal suspension (Peppin et al., 2007, 2008, 2010). However, as yet, no complete theory has yet been put forward for designing a micrometre-scale 3D architecture constructed by ice.

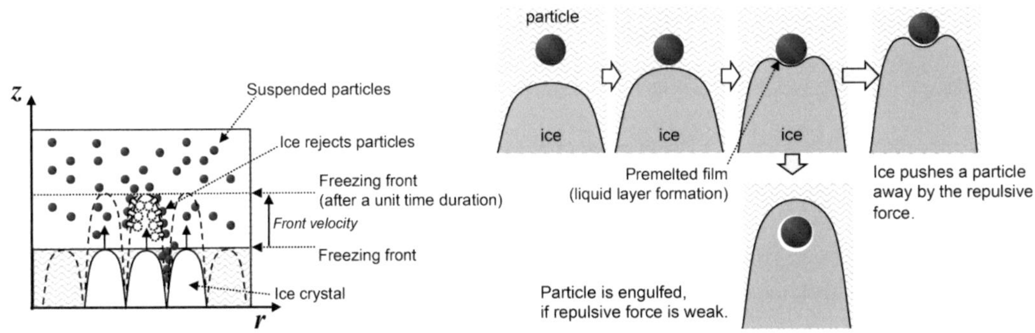

Fig. 3. Schematic illustration of the solidification of a colloidal suspension.

2.2 Engineering approach

This section describes an engineering approach to assessing microstructure formations in freeze-dried CNT foams prepared via several processing routes. As shown in Fig. 4, freezing was performed via either contact freezing with a heat exchanger or immersion freezing in a cryo-bath. Details of these experiments can be found in a report by Nakagawa et al. (2010). These two freezing methods make it possible to prepare several types of microstructures, as explained in the following.

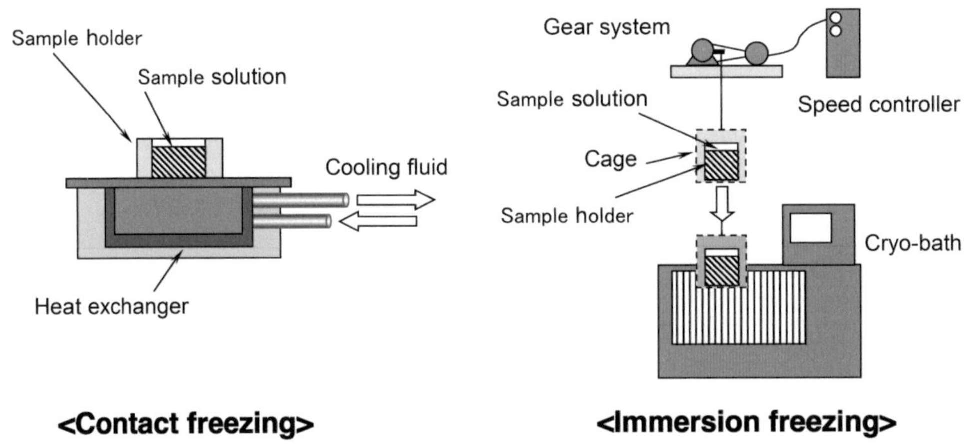

Fig. 4. Freezing methods.

2.2.1 Materials and methods

Materials

Commercial multi-walled CNTs (MWCNTs) (Bayer Materials Science Ltd., 13–16 nm mean outer diameter, 1 to >10 μm length, purity >95%, synthesized by chemical vapour decomposition) were used in this work. Carboxymethyl cellulose (CMC) sodium salt surfactant (Fluka, Sweden) was used as a dispersing agent. Some 1 wt% of MWCNT was dispersed in 10 mL of an aqueous solution of CMC sodium salt (1 wt%) with the aid of an ultrasonic homogenizer (UH-300, SMT company, Japan) for 30 minutes at 15 W output. The suspension thus obtained was placed in a sample holder equipped with a cooling device.

Contact freezing

A 0.25-mL acrylic sample holder (diameter D = 8 mm, height H = 5 mm) was fixed on a heat exchanger, through which a liquid coolant (methanol) was continuously circulated to control the temperature of the sample. With this external arrangement, the freezing phenomena would not be affected by artificial obstacles. In the experiment, the sample solution temperature was initially stabilized at an ambient temperature (25 °C) for 15 minutes, then cooled to -40 °C at a selected cooling rate, namely -3.0 or -0.1 °C/minute. The frozen sample thus obtained was consecutively freeze-dried on the same cooling shelf. It should be noted here that the observed cooling rate was the value used by the cryostat to control its circulating coolant temperature, which may differ slightly from the temperature of the heat exchanger surface that contacted the sample solution.

Immersion freezing

An immersion system equipped with an electrical gear system was designed to lower a sample cage at a fixed speed into a cryo-bath, whose temperature was maintained at a selected value (-40 °C with methanol, or -196 °C with liquid nitrogen). A thin square copper plate was attached to the bottom edge of a glass tube (inner diameter D = 8 mm, height H = 15 mm, wall thickness d = 1 mm), and the setup was placed in a cage. The glass tube was filled with the sample solution and left to equilibrate to room temperature. It was then arranged in the immersion system in such a manner that the sample bottom lay just above the pre-cooled coolant surface. The cage was lowered into the cryo-bath at a selected immersion speed (67 or 20 µm/s) until the coolant level almost reached the top edge of the sample tube. The obtained frozen sample was then immediately transferred from the cage to the cooling shelf of the freeze-dryer.

Freeze-drying

The frozen sample was placed on the shelf of a heat exchanger pre-cooled at -40 °C. After evacuating the system (chamber absolute pressure of 10–20 Pa), the shelf temperature was reset at -20 °C, and the samples were left under these conditions for four days to complete the sublimation. The drying curves of various CNT solid foams have been reported in previous work (Nakagawa et al., 2008). Typically, it took approximately 60 hours to finish the drying process under the chamber pressure of 10 Pa and shelf temperature of -20 °C.

Model calculation

Temperature profiles during the freezing steps were simulated via a mathematical freezing model developed by Nakagawa et al. (2007) in order to estimate freezing front geometry, freezing front velocity, and temperature gradient in a frozen zone. The mathematic model was based on classical Fourier-type thermal conduction equations and solved by commercial finite-element analysis software, COMSOL 3.6. The 2D axisymmetric model took into account the real geometry of the sample container and its surroundings, drawn using the software. The key details of this modelling approach (e.g. basic equations, model geometry, boundary conditions) are explained in previous publications (Nakagawa et al., 2007, 2010).

2.2.2 Microstructures of the prepared foams made from CNTs

A foam material possesses porous microstructures, as noted in the Introduction. The microstructure of the prepared CNT foam were observed with SEM, as depicted in Fig. 5. These images confirm that fine macropores were formed with a particular regularity. A typical microstructure obtained via contact freezing was a bundle of cylindrical pores (Fig. 5a). A layered monolith was prepared via immersion freezing (Fig. 5b). By enlarging the matrix of the pore wall, it was found that it was made from CNTs. However, it must be borne in mind that these architectures are maintained by the polymers that help CNTs to disperse in the original solution. The sponge-like elastic feature of a typical CNT foam sample could be clearly observed via this polymer (Fig. 1). The foam materials prepared in the presented methods were highly porous, with a porosity of about 95–98% and a bulk density of 0.05–0.06 g/cm^3.

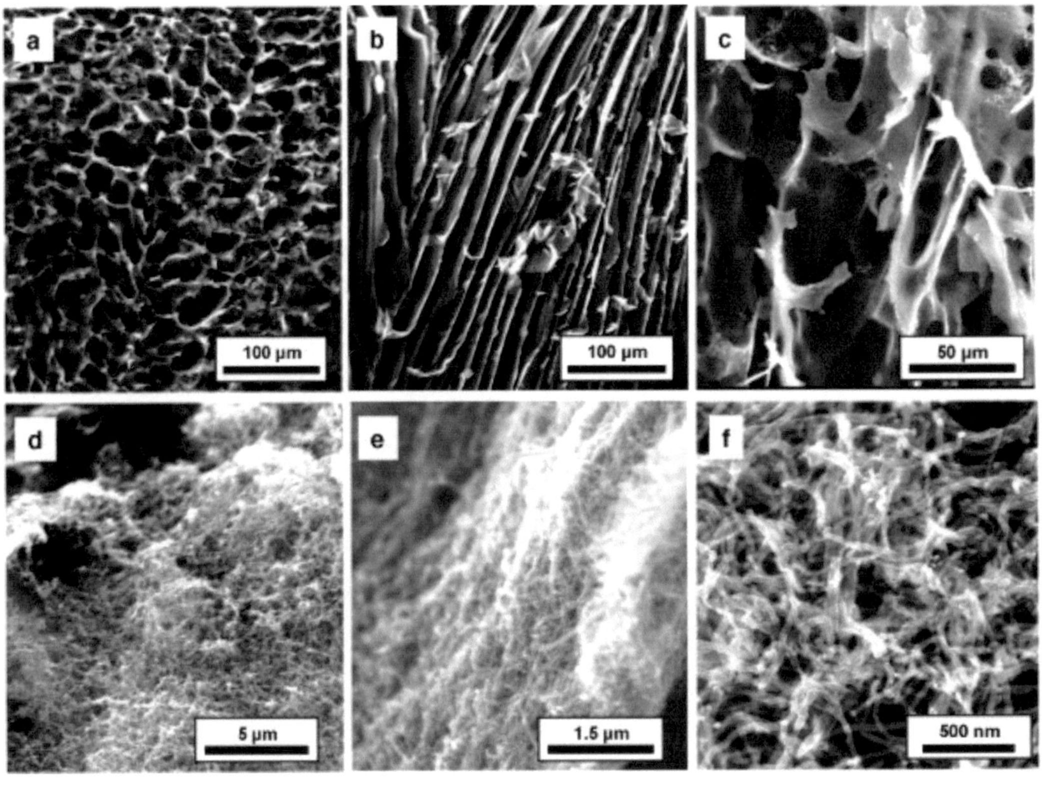

Fig. 5. SEM images of CNT foams: (a) a sample prepared via contact freezing, (b) a sample prepared via immersion freezing, (c)–(f) foam matrix images taken at various magnifications.

Microscopic images of the prepared CNT foams are shown in Fig. 6. The black domain consists of the CNTs, and the brighter portions represent the pores of the freeze-dried samples. The 2D images were obtained after embedding the samples in epoxy resin. First, it is evident that the domain of the CNTs exhibits a fairly well-interconnected network. It is also clear from these figures that the observed pore sizes were influenced by the freezing condition, while the microstructures were dependent on the freezing method. Comparison of the horizontal and the corresponding vertical cross-sectional images of a sample prepared via contact freezing at a fast cooling rate reveals that both were assembled of mostly columnar pores (Fig. 6A-1, -2). In fact, it is difficult to find clear differences or tendencies in their morphology, which suggests that the growth directions of ice crystals along the *z* and *r* axes were quite random. In contrast, lamellar structures were observed on the horizontal cross-sectional images obtained via a slow cooling rate ((Fig. 6B-1) while columnar structures appeared on the vertical cross-sections (Fig. 6B-2). This fascinating contrast suggests that ice crystal growth along the *z*-direction was constrained by the slow cooling condition. The cooling rate was a key parameter that decided the mean pore size, if the sample was frozen with a contact cooling plate. In short, a faster cooling rate led to smaller macropores. Generally, smaller pores appeared at the bottom of the sample. This is because the ice crystal growth rate was extremely rapid just after the breakpoint of supercooling. After the onset of supercooling, ice crystals grew at a moderate velocity controlled by the thermal flux through the system, thereby resulting in larger pore sizes.

Interestingly, when samples were frozen via immersion freezing, we could observe an obvious morphological difference. Continuous lamellar structures commonly appeared on the vertical cross-sectional images (Fig. 6C-2, 6D-2), and columnar structures appeared on the horizontal cross-sectional images although their aspect ratio was sometimes very large (Fig. 6C-1, 6D-1). Ice crystals predominantly grew in the *z*-direction, and bundles of uniformly arranged dendritic ice crystals contributed to construct a honeycomb monolith. Even though slightly smaller pore sizes were observed at the bottom of the sample owing to the influence of a supercooled layer, the lamellar structures along the *z*-direction appeared rather uniformly throughout the entire sample height. In fact, the immersion rate influenced the aspect ratio of the columnar cross section. In short, the aspect ratio became larger when the samples were immersed more slowly, suggesting that the aspect ratio depends on a kinetic balance between the ice crystal growth rates along the *z*- and *r*-directions.

As previously explained, the ice crystal growth rate was found to correspond roughly to the immersion rate, although the mean pore size did not show a linear relation to the latter. Incidentally, it is worth noting that, although the mean freezing front velocity of sample A1, which was experimentally detected as ~25 μm/s, was essentially equivalent to the immersion rate of B3 (20 μm/s), the microstructures of these two samples are by no means similar. In other words, it is necessary to discuss how these microstructures were produced based on rational data that essentially govern ice crystal appearance and growth during the freezing stages.

2.2.3 How does freezing control microstructure?

A model calculation is a useful tool to estimate property data taking into account the geometry of the sample undergoing a transient process. The freezing process is a transient process that is controlled by thermal flux in a sample solution. The phase transition of water (solidification) provides latent heat to the system. Therefore, the heat must be removed from the system for the continuance of freezing. Thermal flux strongly relates to sample geometry, the thermal conductivity of each component (e.g. solution, solid layer, sample holder, ambience), and cooling protocols. A mathematical model has been developed by the

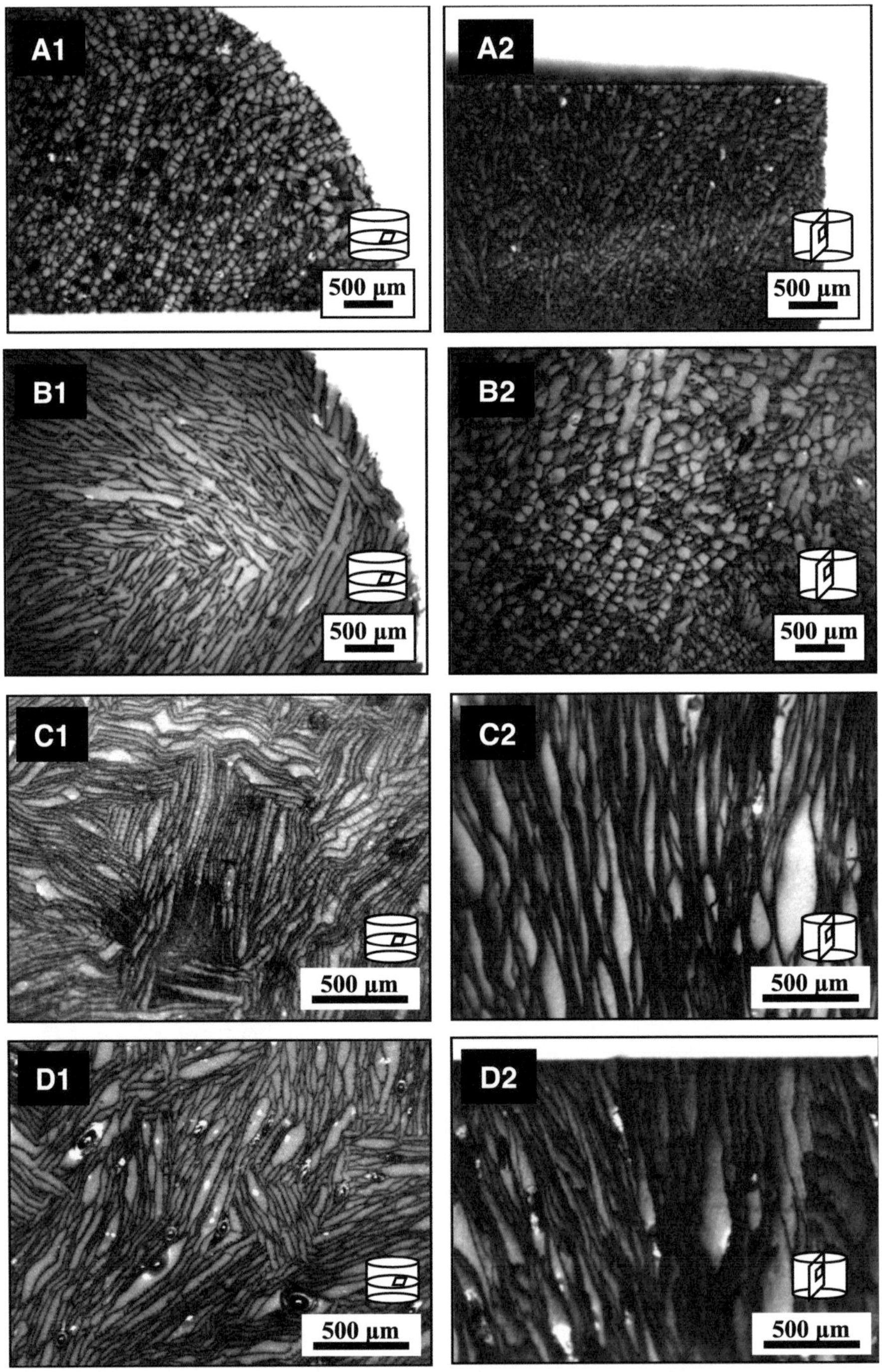

Fig. 6. Microscopic images of CNT foams: (A1, 2) sample prepared via contact freezing (−3 °C/min), (B1, 2) sample prepared via contact freezing (−0.1 °C/min), (C1, 2) sample prepared via immersion freezing (67 µm/s), (D1, 2) sample prepared via immersion freezing (20 µm/s).

author, as reported in previous publications (Nakagawa et al., 2007, 2010). By using the mathematical model simulation, we can obtain thermal histories of the sample solutions during contact freezing and immersion freezing. These thermal profiles are useful for predicting microstructure formation in a freeze-dried foam material.

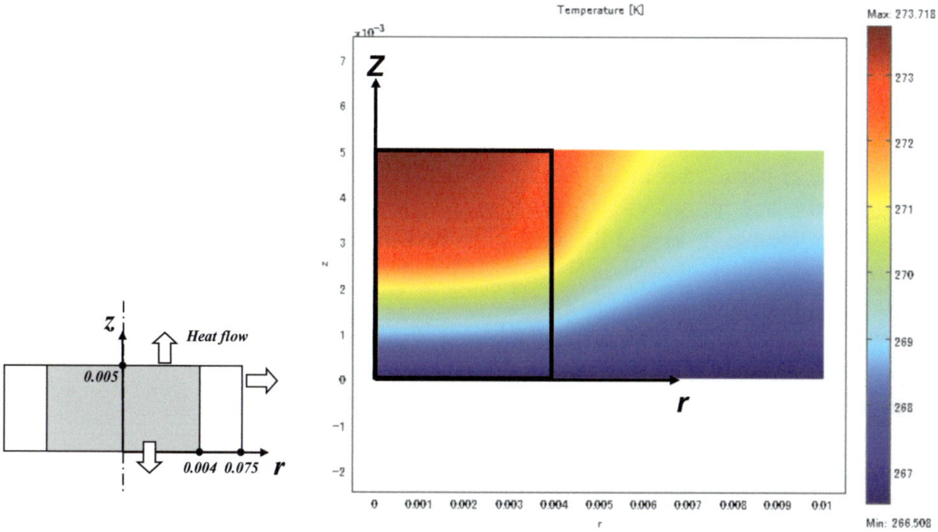

Fig. 7. Temperature profile during freezing (contact freezing).

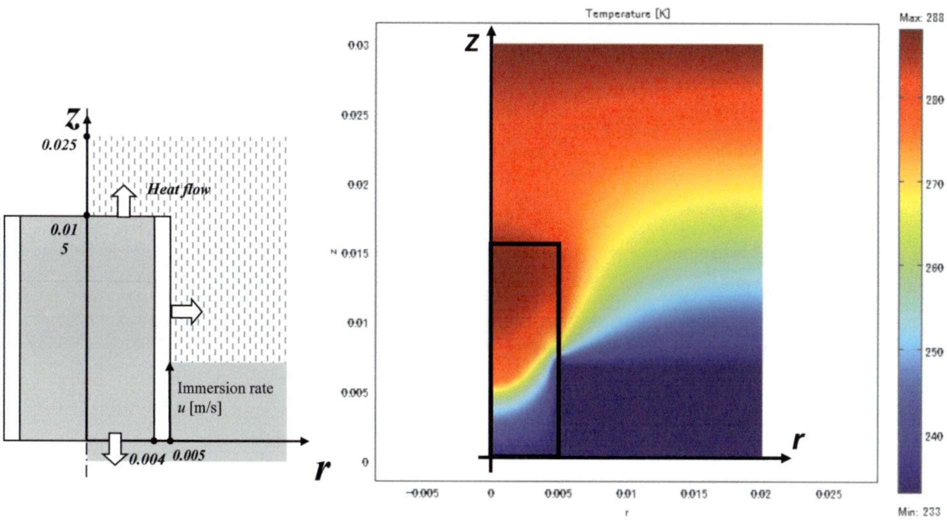

Fig. 8. Temperature profile during freezing (immersion freezing).

Through microscopic observation it was found that the present two freezing protocols lead to unique microstructures. The most fascinating feature of the immersion freezing is the appearance of a uniformly arranged monolithic layered microstructure in the sample, which was not observed in the samples prepared via contact freezing. Both freezing methods can

be referred to as unidirectional freezing systems. However, even if the freezing rates were adjusted to achieve equivalent values for both contact and immersion freezing, we cannot produce samples with equivalent microstructures. This suggests that the ice formation must be identified by the geometry of the system. The term "unidirectional freezing" risks misidentification of a freezing system if it is used without identifying the thermal flow in the freezing system. Figures 7 and 8 are snapshots of the thermal profiles during contact freezing and immersion freezing, respectively. For both types of freezing, the freezing fronts move on vector $(r, z) = (0, 1)$, so that we may call this "unidirectional freezing". However, the shapes of the freezing fronts differ greatly from one another. We can confirm the differences more clearly with the concrete calculated values in Fig. 9. It is strongly believed that this is a critical parameter that determines the 3D microstructure made by ice.
Let us focus on the overall growth direction of the growing ice crystals in a newly appearing freezing layer (thickness R, in unit time), as sketched in Fig. 10 and Fig. 11.

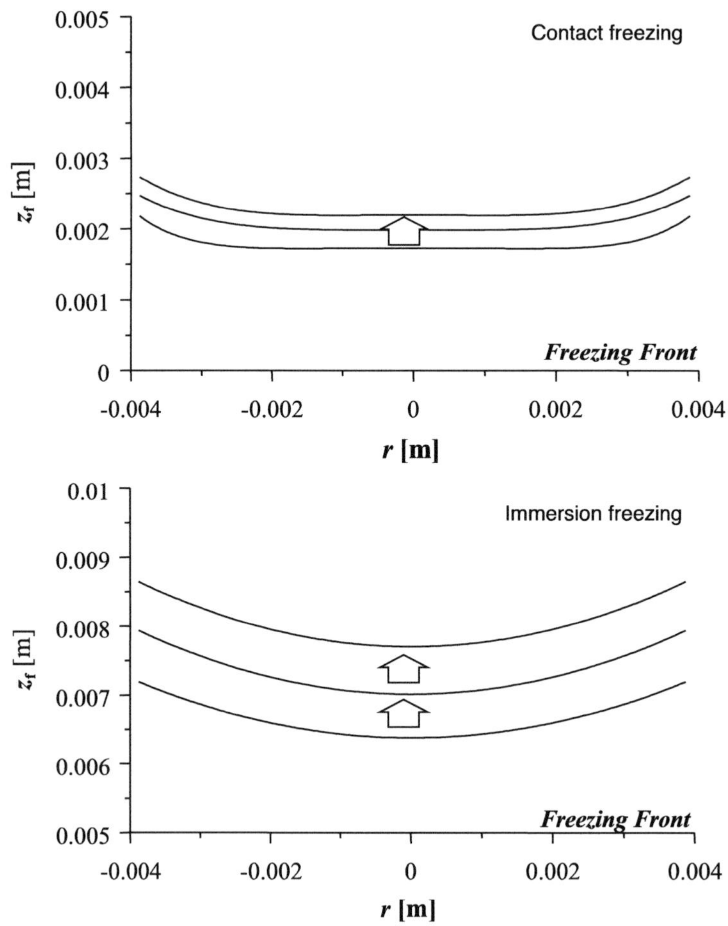

Fig. 9. Freezing front geometries.

If a freezing front is flat as simulated during contact freezing, the temperature distribution in the layer would allow a growing ice crystal to choose an arbitrary growth direction on the interface unless there is some obstructing ice. In addition, the typical microstructures obtained via contact freezing were randomly assembled columnar ice crystals (Fig. 6-A), while the lamellar structures only appeared in the samples prepared at a slower cooling rate (Fig. 6-B). Specifically, the relatively slow freezing front velocity (small R values) mean that ice crystal growth was permitted only in a layer of limited thickness per unit of time (Fig. 10), thereby resulting in the rather unique microstructure shown in Fig. 6-B.

When we employ immersion freezing, we should bear in mind that the freezing front is curved, as shown in Fig. 9. In this case, several scenarios may lead to the creation of the unique microstructure observed in the samples (Fig. 6-C, D). First, an existing ice crystal at an adjacent higher freezing front level would present a barrier for the growing ice crystal, and as a consequence, there would be less opportunity for growth in the $+r$-direction. In the second scenario, no ice crystals can grow into a domain where the temperature is higher than the equilibrium freezing point. By this reasoning, the zone in which ice grows toward the $-r$-direction in unit time is controlled by the vector **R** (Fig. 11). Therefore, it may be that isotropic ice crystal growth (toward the z-direction) is encouraged during immersion freezing, while anisotropic growth is permitted under contact freezing. The temperature gradient in an already frozen zone should influence the choice of the direction because a larger temperature gradient would encourage z-direction crystal growth whereas a smaller gradient would lead to tilted ice growth (Deville et al., 2007). The simulation results predict that the temperature gradient values during immersion freezing are larger than their contact-freezing counterparts. However, this is not a parameter that exclusively governs the isotropy of ice crystal growth.

These insights suggest that, although the basic ice crystallization phenomena (and subsequent particle rejection phenomena) were comparable in both freezing methods, the observed temperature distribution and the movement of the freezing front in the entire freezing bulk essentially controlled the ice crystal morphology in the frozen matrix. Thus, a practical design of material microstructures for the freeze-drying process could be achieved via controlling thermal flow through the freezing system.

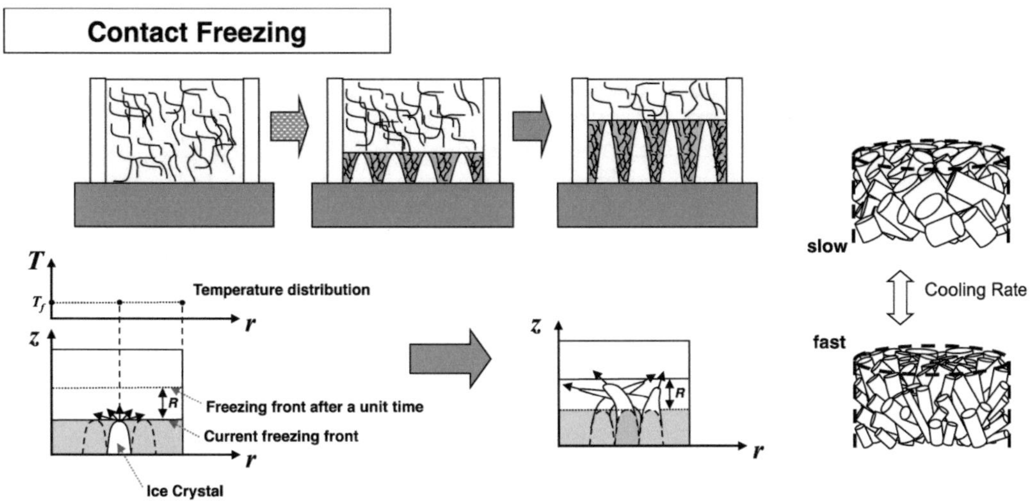

Fig. 10. Sketches of ice formation during contact freezing.

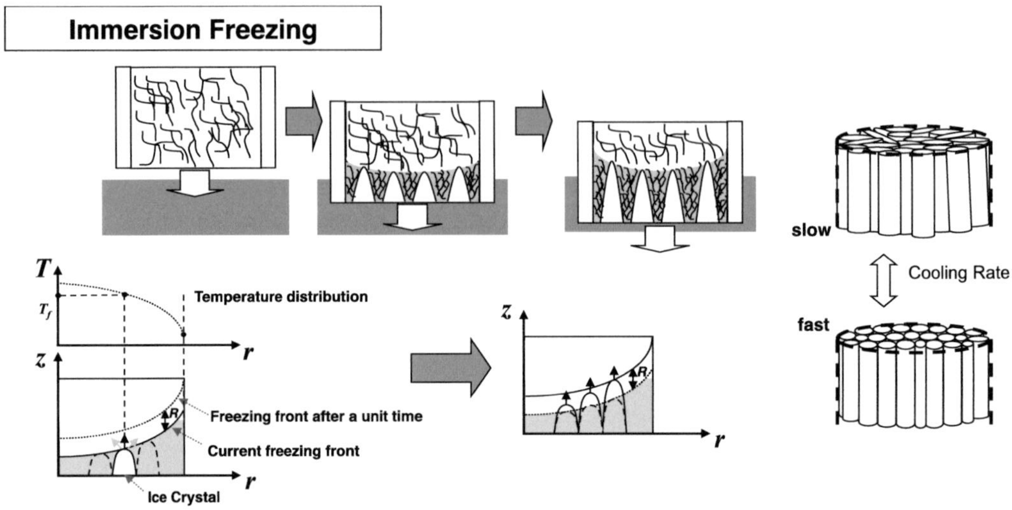

Fig. 11. Sketches of ice formation during immersion freezing.

3. Insights into the functional properties

3.1 A bulk material derived from nanomaterials

As is widely known, some nanomaterials have specific functions that are primarily evoked at the nanometre scale, so that much research has been carried out for the nano-tech industries. CNTs are one of the most intensively studied nanomaterials. However, no promising commercial application based on CNTs has yet been found. CNT foam is a material with a porous structure (microstructure) constructed with nano-porous carbon, that is, CNTs. It would be reasonable to expect that CNT foam attains functions derived from both its nanostructure and its microstructure. As discussed above, we can produce CNT foam whose microstructure is controlled with a particular regularity. The following sections describe several attempts that corroborate the characteristics of CNT foams derived from CNTs and their microstructures.

3.2 Electrical conductivities of CNT foams

3.2.1 Response to the ambient pressure

In this section, we discuss the electrical conductivity of the prepared CNT foams. The foam samples prepared in this work showed electrical resistivity in the range of 1–10 kΩ/m. This is of course derived from interconnected CNTs in a foam material. It is known that the CNT is an excellent adsorbent for gas molecules owing to its huge surface-to-volume ratio, hollow geometry, and unique electrical properties (Sun et al., 2007; Chang et al., 2008). In addition, a detectable change in electrical conductivity of the CNT is connected to the process of gas adsorption–desorption on the nanotube surface (Romanenko et al., 2007). Electron transfer through the CNTs is promoted when certain gas molecules are adsorbed onto the CNT walls (Chang et al., 2008). In other words, the electrical conductivity of the CNT is dependent on the interactions between the adsorbed gas molecules and the CNT wall. Similarly, it has been reported that the electrical resistance of the CNTs respond sensitively to the adsorption of various gases (Suehiro et al., 2003; Quang et al., 2006). In this sense, the CNT has recently become a good candidate as a sensitive material for the detection of gases such as H_2, O_2, NO_x, NH_3, and VOCs with improved real-time monitoring (Sayago et al., 2008). It would therefore be reasonable to expect these features in CNT foams and to anticipate a link between the gas detection feature and microstructure of CNT foam.

We connected probes to the prepared CNT foam with conductive adhesive. The sample with the probes was placed in a vacuum chamber, and the pressure in the chamber and electrical resistivity were monitored with a digital multimeter. Figure 12 shows typical examples of the relationship between the electrical resistivity of the CNT foam and the chamber pressure. It is clear that the resistivity responds rapidly to the rise and drop in the pressure. It is straightforward to attribute this change in electrical resistivity to the adsorption and desorption of air on the foam walls composed of CNTs. It was found by the author's research group that the absolute resistivity value decreased with increasing CNT content, and the specific value of resistivity difference (corresponding to 0.1 MPa pressure difference) decreased with increasing CNT content (Thongprachan et al., 2008). The contacts between the CNTs in the bulk walls were influenced by the CNT concentration, and the difference in the contacts appeared as a difference in the electrical resistance of the sample.

What would happen if we introduce a gas other than air into the evacuated chamber? Attempts were carried out with He, ethane, and iso-butane. These three inert gases differ in molecular size. We expected that the microstructure formation in CNT foam may influence the diffusivity of these gases. Samples prepared via contact freezing and immersion freezing were employed for this dynamic electrical response test, with the results shown in Fig. 13.

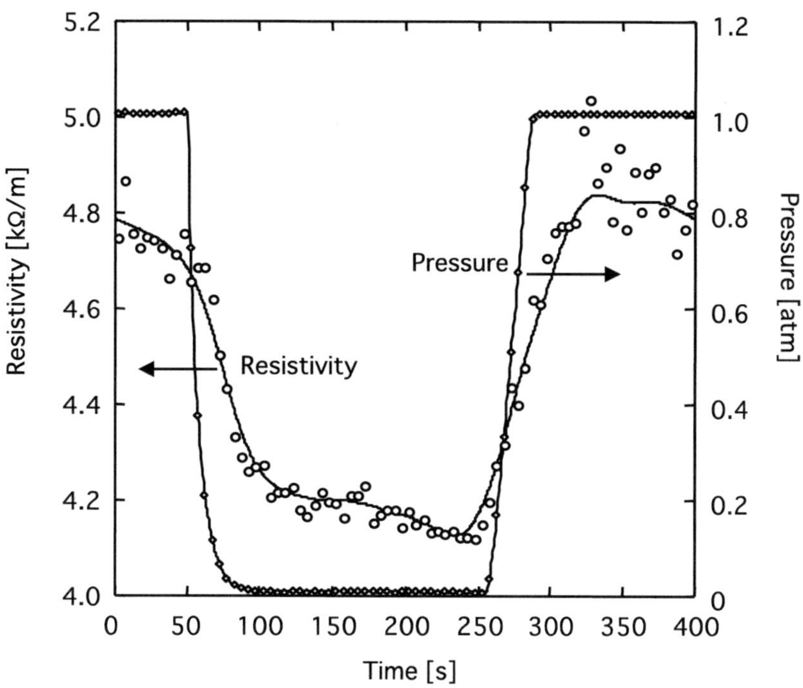

Fig. 12. Relationship between electrical resistivity and gas pressure for CNT foam.

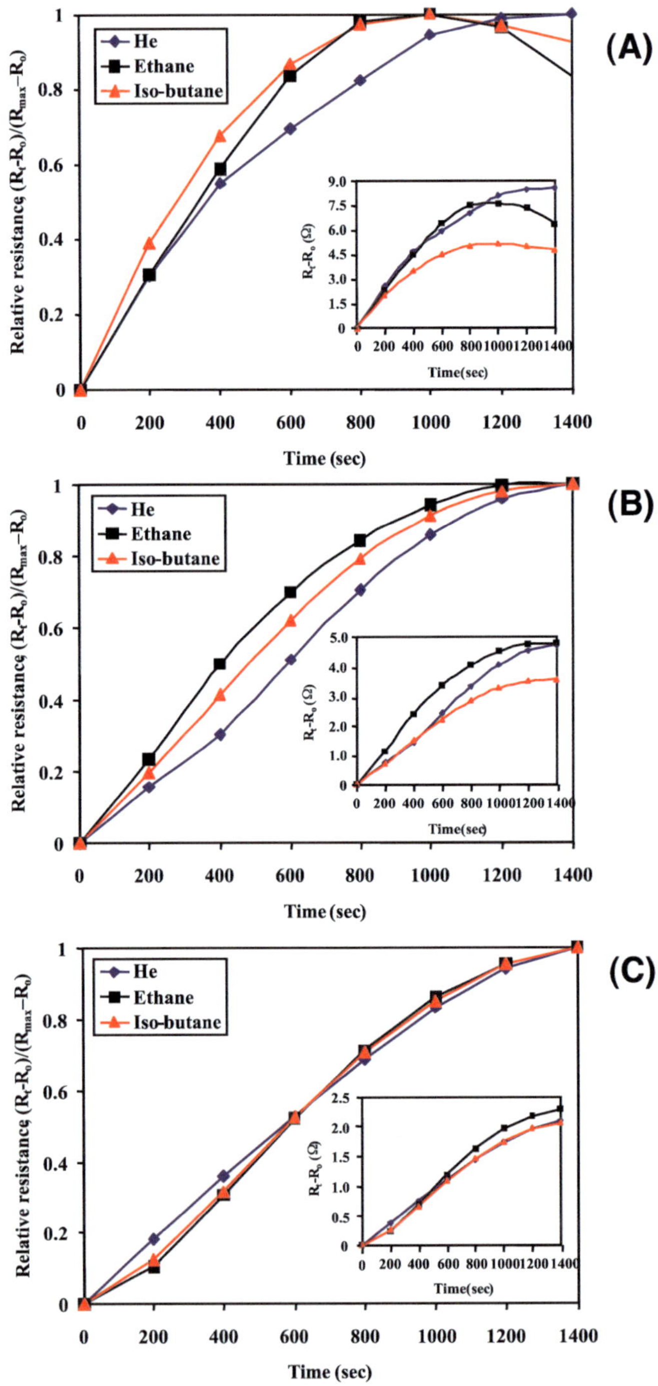

Fig. 13. Dynamic electrical response of CNT foams against uptake time for He, ethane, and iso-butane. Sample prepared: (A) via contact freezing (slow cooling), (B) via contact freezing (rapid cooling), and (C) via immersion freezing.

As expected, the gas type and microstructure appear to be related to the electrically conductive characteristics of the CNT foams. It was clear that different patterns of normalized gas-uptake electrical response appeared for each CNT foam specimen after coming into contact with each of the three test gases. The pattern difference corresponds to a difference in the freezing conditions. As described in the previous section, the CNT foams produced via contact freezing possessed a pore structure with randomly oriented columnar pores, whereas the immersion freezing method gave uniformly oriented pore structures. The freezing conditions (i.e. cooling rate or immersion rate) significantly affected the pore size. In short, CNT foam produced with a faster freezing rate usually had a smaller macropore size. For the sample prepared via contact freezing, among the three tested gases, it was found that He attained an equilibrium state most slowly (Fig. 13-A). In fact, the required gas adsorption equilibrium time increased approximately in the following order: iso-butane = ethane < He. Surprisingly, no significant differences in the normalized responses were observed between the three gases in the case of CNT foam obtained via immersion freezing.

It may be logical to attribute the resulting differences in the normalized gas-uptake electrical response for each gas species to the gas diffusion process in CNT foam. It was ensured that each CNT foam specimen used in this experiment possessed significantly different microstructures. The macropores, however, are not sufficiently important in controlling the mass transfer of gases such as He (they consist of almost free space in terms of angstrom-scale molecules). Therefore, it would be reasonable to conclude that the nano-space made by CNT assembly is the major factor in controlling gas diffusion regimes, and that the manner of ice formation controls the nanostructures made by CNT arrays. Direct evidence for shedding light on whether ice creates such a structure is still lacking. However, this may be possible because ice pushes CNTs away in creating a microstructure during freezing. Considering the repulsive force balance between ice and a suspended tubular-shaped nanomaterial (CNT), a directional force provided by an ice crystal would allow it to take on ordered alignment. This point should be investigated further to examine the potential for material with specific functions derived from both nano- and micro-structuralization. As revealed in earlier research, CNT arrays show promise in the creation of advanced material functionalities (Ding et al., 2007; Velasquez-Garcia et al., 2010).

In any case, if we wish, for instance, to employ the CNT foam as a gas-sensing material, we can control its sensitivity and pattern of response by selecting suitable preparation conditions, which in turn influence the microstructure and nanostructure in the sample. Recently, sensor arrays consisting of different composite materials, a so-called electronic nose, have been fabricated by several researchers (Hu et al., 2004; Lu et al., 2006; Star et al., 2006). Since each conductive composite has its own response characteristics as determined by specific interactions between the polymer matrix and the gas of interest, the response pattern given by the detector array contains hidden information about odour species and concentration, which can be extracted through signal processing. A CNT foam may be an interesting candidate as an electronic nose because it has the potential to provide various responses by changes in its structure. Foam preparations from various CNTs are also of interest in investigations into a wide range of applications. As reported in the literature, CNT foams can be applied, among other things, to bio-scaffolds, electronic parts, and shielding material from electromagnetic radiation (Zawadzak et al., 2009; Gutiérrez et al., 2007a, 2007b; Nakagawa et al., 2011; Thomassin et al., 2008). Further in-depth research into CNT foam should provide tantalizing applications of CNTs.

3.2.2 Application to gas diffusion layers of a fuel cell (PEMFC)

This section describes another attempt at exploring CNT foam applications, which is to apply CNT foams to gas diffusion layers (GDLs) for a proton exchange membrane fuel cell (PEFMC). A PEMFC is constructed from a proton exchange layer, catalytic layers, and GDLs. Pt is usually applied to the catalyst on the proton exchange membrane surface or on the GDLs. The proton exchange layer and the catalytic layers are inserted between the GDLs. The GDL allows uniform access of the fuel gas (e.g. hydrogen) and oxidant (e.g. oxygen) to the catalyst layer (Fig. 14-E).

The performance of PEMFC is determined by properties relating to electrochemical reactions. The particular importance of GDL lies in it being a medium that allows gas diffusion, electron conduction, and the mass-transfer control of water produced during an electrochemical reaction at the catalytic layer. The water transfers through the cathode side of the GDL and is subsequently exhausted from the system or used as a moisture source for the proton exchange membrane. The water flow in the GDL is known as liquid-vapour two-phase flow and is considered to be a critical parameter in fuel cell performance (Chen et al., 2004). The temperature gradient in the GDL is crucial to the water vapour pressure that influences the water mass transfer (Kawase et al., 2008). The thermal conductivity and surface nature are thus important characteristics in the GDL material. Ziegler and Gerteisen (2009) have shown that the surface nature and microstructure of GDLs almost completely determine the liquid water distribution and, as a consequence, affect the power-generating performance of the fuel cell. Therefore, the challenge in preparing potential GDL materials is to satisfy all of these requirements (e.g. good electrical conductivity, low mass-transfer resistance, high thermal conductivity, controlled surface characteristics) with a sensible material design strategy. One possibility may be to employ CNT foams as GDLs.

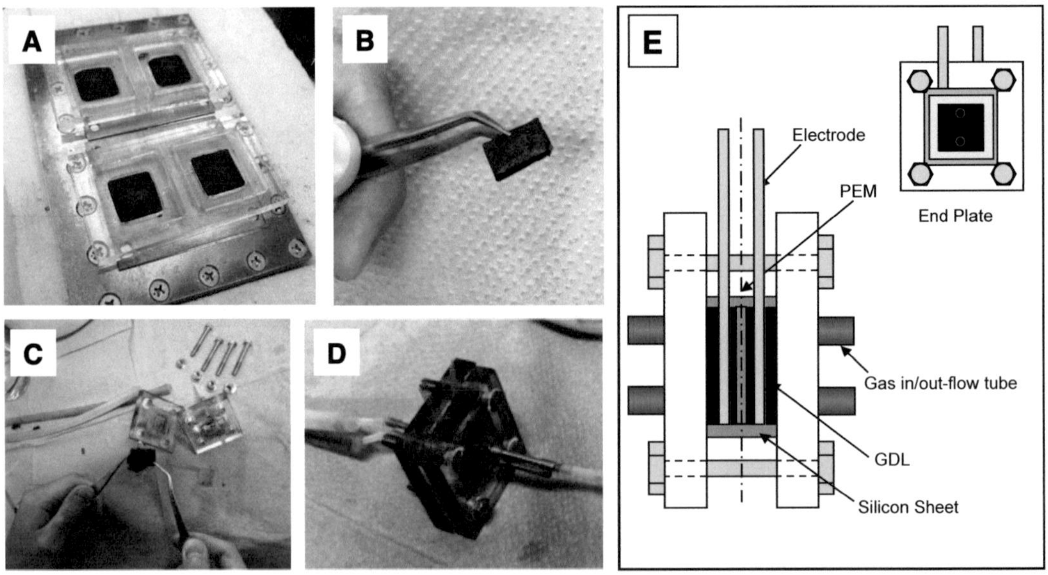

Fig. 14. Setup of PEMFC; (A) CNT foam preparation for PEMFC (contact freezing system), (B) prepared CNT foam, (C, D) assembling a lab-scale PEMFC, (E) assembly instructions.

CNT foams were prepared from an aqueous suspension of CNTs dispersed by chitosan. Thin film-shaped CNT solid foams were prepared (15×15×1 mm), and a lab-scale PEMFC was assembled as shown in Fig. 14. The microstructures of the prepared solid foams can be seen in Fig. 15. As described above, the preparation conditions essentially controlled the microstructures of the prepared CNT foams. The macropores of the prepared carbon foams were heterogeneously distributed in the dried bulk materials, and in general, pores at the top layer were larger than those at the bottom. Increasing the CNT content in the original suspension led to a decrease in mean pore sizes. Furthermore, the amount of CNTs was found to be related to the carbon networks. The carbon foam that contained an excess amount of CNT lost its interconnected carbon networks after preparation. The cooling rate during the freezing step clearly affected the pore sizes. Mean pore sizes decreased with increasing cooling rate.

It was confirmed that the PEMFC assembled with the CNT foams prepared in this study displayed fairly good fuel cell performance. The performance was by no means inferior to that of a fuel cell prepared with conventional carbon paper as GDL material. This suggests that both the CNT content of a freeze-dried bulk and the freezing conditions influenced the performance of prepared fuel cells. It is believed that the carbon networks formed during the freezing step are closely linked to cell performance. Hence, the carbon foam with the defection in its interconnected microstructure did not exhibit good-quality GDL.

Impedance measurement is a useful method of characterizing a battery cell that makes it possible to estimate film resistance and kinetic resistance separately by the adoption of an equivalent circuit model to PEMFC (Fig. 16).

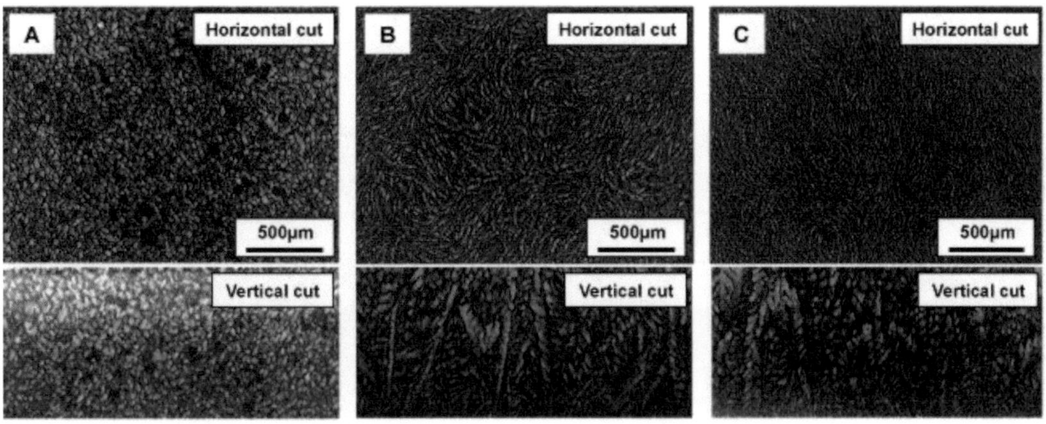

Fig. 15. Microstructures of the prepared CNT foams; (A) Sample-A *(CNT content 15 mg/sheet, prepared via rapid freezing)*, (B) Sample-B *(CNT content 9 mg/sheet, prepared via slow freezing)*, (C) Sample-C *(CNT content 9 mg/sheet, prepared via rapid freezing)*.

It is approximately considered that the film resistance (R_s) corresponds to ohmic resistance such as resistance of the materials and conduction loss at the boundaries of the GDL or membrane. In contrast, the kinetic resistance (R_s+R_c) relates to the identical resistance of the device while the circuit is running, which can be regarded as an index of the electrochemical property. The Nyquist diagrams obtained from the impedance measurement on the prepared fuel cells are shown in Fig. 17. We can see that the prepared carbon foams were useful in reducing the film resistances of the PEMFCs, probably owing to the elasticity and

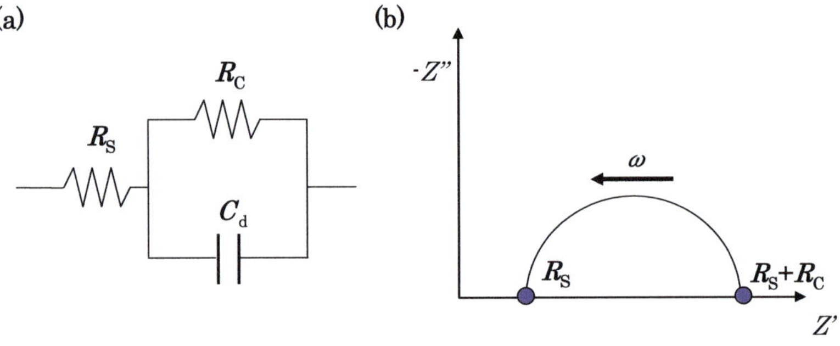

Fig. 16. Equivalent circuit model of PEMFC (a) and Nyquist diagram (b).

specific microstructure of the prepared CNT foams. The kinetic resistances were found to be dependent on the processing conditions of GDLs, that is, the freezing conditions. It is believed that the freezing process controls the degree of overlap among those CNTs that influence the electrochemical properties in a fuel cell. As addressed in the section 3.2.1, it may be that the ice formation controls the nanostructures created by CNT arrays. This idea is further supported by the fact that the thermal conductive characteristics of the prepared solid foams depended on the freezing conditions. Before closing this chapter, it would be interesting to look at the thermal conductivities of the CNT foams that were used in the PEMFC experiment (Table 1). The apparent thermal conductivities of the prepared CNT foams were measured as described in a previous publication (Nakagawa et al., 2011). It was confirmed that the value increased with increasing freezing rate. Although a simple correlation of the values could not be given, it would provide indirect evidence that the nanostructures made by CNT arrays are controlled by ice formation during the freezing step.

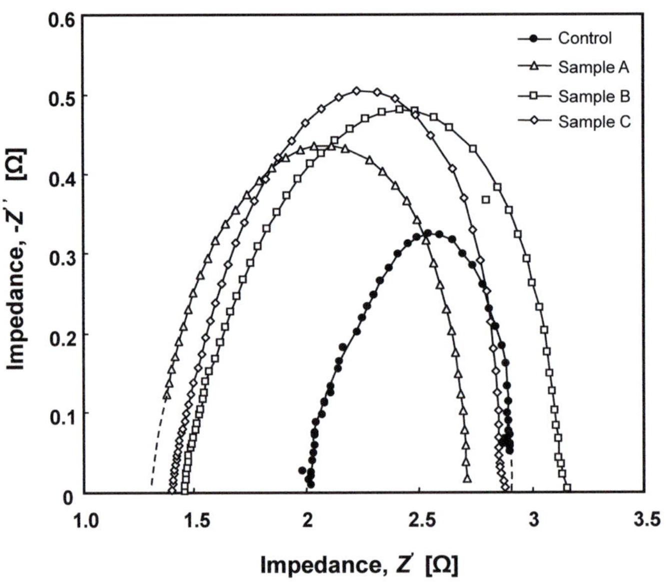

Fig. 17. Nyquist diagrams of the prepared PEMFCs (*samples are as described in Fig. 15: commercial carbon paper was used to obtain control data*).

	Apparent thermal conductivity [W/mK]	Apparent thermal diffusivity [m^2/s]
Sample A	0.71	7.4×10^{-6}
Sample B	0.16	2.7×10^{-6}
Sample C	0.42	4.7×10^{-6}
Control (commercial carbon paper)	0.21	0.75×10^{-6}

Table 1. Apparent thermal conductivity of the prepared CNT foams.

4. Concluding remarks

This chapter has described a method of preparing CNT foam preparation using freeze-drying in which the ice formation process could strategically be applied for microstructure control and design. Following solidification theory of a colloidal suspension, ice formation from a CNT aqueous suspension, as expected, created a microstructure in the resultant frozen solution. It was found that the thermal flow during freezing dominated the pattern formation of ice crystals. Therefore, macropores produced in the freeze-dried material (as a replica of ice crystals) may be designed by carefully planning the freezing operation with an external chilling device. Recent literature has shown that unidirectional freezing is key to controlling the microstructure and to creating unique pore structures. However, as pointed out in this chapter, ice microstructure formation cannot be correctly identified without consideration of thermal flow (more specifically, the geometry of the solidification front) in the freezing solution as a whole. From an engineering point of view, sample geometry and thermal flux need to be taken into account in microstructure design. It may be an interesting finding that freezing controls not only the microstructures of freeze-dried CNT foam but also the arrangement of the interconnected CNT arrays (or the degree of overlap). These structural modifications on the nanoscale perhaps determine the subsequent characteristics and functionality of the CNT foams, such as their electronic and thermal properties. It is hoped that the attempts described in this chapter may provide hints for developing applications of CNT foams. The CNT foams demonstrated here remain at the level of prototype and provide scope for future work. For example, it would be interesting to create a composite of an electro-conducting polymer with CNTs for producing a CNT foam (Sweetman et al., 2008).

5. Acknowledgements

The author benefited from contributions by Dr Thongprachan Napawon and is grateful for her sustained efforts and intensive experimental work on CNT foam analysis. Thanks are also due to Mr Yoshinaga Yasumura for his critical assistance in investigations into the feasibility of CNT foams for fuel cell application.

6. References

Boccaccini, A.R.; Chicatun, F.; Cho, J.; Bretcanu, O.; Roether, J.A.; Novak, S. & Chen, Q. (2007). Carbon Nanotube Coatings on Bioglass-Based Tissue Engineering Scaffolds. *Advanced Functional Materials*, 17: 2815–2822

Chang, Q.; Zhao, K.; Chen, X.; Li, M. & Liu, J. (2008). Preparation of gold/polyaniline/multiwall carbon nanotube nanocomposites an application in ammonia gas detection. *Journal of Materials Science*, 43: 5861-5866

Chen, J.; Matsuura, T. & Hori, M. (2004). Novel gas diffusion layer with water management function for PEMFC. *Journal of Power Sources*, 131: 155-61.

Chen, L.; Linda, R.O. & Schadler, S. (2010). The influence of carbon nanotube aspect ratio on the foam morphology of MWNT/PMMA nanocomposite foams. *Polymer* 51: 2368-2375

Dash, J.G.; Fu, H. & Wettkaufer, J.S. (1995) The premelting of ice and its environmental consequences. *Reports on Progress in Physics.*, 58: 115–167

Dash, J.G.; Rempel, A.W. & Wettlaufer, J.S. (2006) The physics of premelted ice and its geophysical consequences. *Review of Modern Physics*, 78: 695–741

Deville, S.; Saiz, E. & Tomsia, A. (2007) Ice-templated porous alumina structures. *Acta Materialia*, 55: 1965-1974.

Ding, F.; Lin, Y.; Krasnov, P.O. & Yakobson, B.I. (2007) Nanotube-derived carbon foam for hydrogen sorption. *Journal of Chemical Physics*, 127: 164703

Gutiérrez, M.C.; Garcia-Carvajal, Z.Y.; Hortigüela, M.J.; Yuste, L.; Rojo, F.; Ferrer, M.L. & del Monte, F. (2007a). Biocompatible MWCNT scaffolds for immobilization and proliferation of E. coli. *Journal of Material Chemistry* 17: 2992-2995

Gutiérrez, M.C.; Hortigüela, M.J.; Amarilla, J.M.; Jimenez, R.; Ferrer, M.L. & del Monte, F. (2007b). Macroporous 3D architectures of self-assembled MWCNTs surface decorated with Pt nanoparticles as anodes for direct methanol fuel cell. *Journal of Physical Chemistry C*, 111: 5557–5560

Hermant, M.C.; Verhulst, M.; Kyrylyuk, A.V.; Klumperman, B. & Koning C.E. (2009). The incorporation of single-walled carbon nanotubes into polymerized high internal phase emulsions to create conductive foams with a low percolation threshold. *Composites Science and Technology* 69: 656–662

Hu, J.W.; Chen, S.G.; Zhang, M.Q.; Li, M.W. & Rong, M.Z. (2004). Low carbon black filled polyurethane composite as candidate for wide spectrum gas-sensing element. *Materials Letters*, 58: 3606-3609

Jia, L.; Stevens, M.M.; Zhu, Y.; Gong, Q.; Wu, J. & Lian, J. (2009). Preparation and properties of multi-walled carbon nanotube/carbon/polystyrene composites. *CARBON* 47: 2733–2741

Kawase, M.; Inagaki, T. & Miura, K. (2008). Nonisothermal through-plane transport model of PEMFC with local VLE assumption. *ECS Transaction*, 16: 563-573.

Kwon, S.M.; Kim, H.S. & Jin, H.J. (2009). Multiwalled carbon nanotube cryogels with aligned and non-aligned porous structures, *Polymer* 50: 2786–2792

Lau, C. & Cooney, M.J. (2008). Conductive Macroporous Composite Chitosan-Carbon Nanotube Scaffolds. *Langmuir*, 24: 7004-7010

Leroy, C.M.; Carn, F.; Backov, R.; Trinquecoste, M. & Delhaes, P. (2007). Multiwalled-carbon-nanotube-based carbon foams. *Carbon* 45: 2307–2320

Lu, Y.; Partridge, C.; Meyyappan, M. & Li, J. (2006). A carbon nanotube array for sensitive gas discrimination using principal component analysis. *Journal of Electroanalytical Chemistry*, 593: 105-110

Luis Fernando Velásquez-Garcia, L.F.; Gassend B.L.P. & Akinwande, A.I. (2010) CNT-Based MEMS/NEMS Gas Ionizers for Portable Mass Spectrometry Applications. *Journal of Microelectromechanical Systems*, 19: 484-493

Meng, D.; Ioannou,J. & Boccaccini, A.R. (2009). Bioglass-based scaffolds with carbon nanotube coating for bone tissue engineering. *Journal of Materials Science; Materials in Medicine*, 20: 2139-2144

Nabeta, M. & Sano, M. (2005). Nanotube Foam Prepared by Gelatin Gel as a Template. *Langmuir*, 21: 1706-1708

Nakagawa, K.; Thongprachan, N.; Charinpanitkul, T. & Tanthapanichakoon, W. (2010). Ice Crystal Formation in the Carbon Nanotube Suspension: A Modelling Approach. *Chemical Engineering Science*, 65: 1438-1451

Nakagawa, K.; Yasumura, Y.; Thongprachan, N. & Sano, N. (2011). Freeze-dried Solid Foams Prepared from Carbon Nanotube Aqueous Suspension: Application to gas diffusion layers of a proton exchange membrane fuel cell. *Chemical Engineering and Processing*, 50: 22-30

Nakagawa, K; Hottot, A.; Vessot, S. & Andrieu, J. (2007). Modelling of Freezing Steps during Freeze-Drying of Drugs in Vials. *AIChE Journal*, 53: 1362-1372

Olivas-Armendáriz, I.; Garcia-Casillasa, P.; Martiinez-Sánchez, R.; Martiinez-Villafane, A. & Martínez-Pérez C.A. (2010). Chitosan/MWCNT composites prepared by thermal induced phase separation. *Journal of Alloys and Compounds* 495: 592–595

Park, K.W. & Kim, G.H. (2009). Ethylene Vinyl Acetate Copolymer (EVA)/Multiwalled Carbon Nanotube (MWCNT) Nanocomposite Foams. *Journal of Applied Polymer Science*, 112: 1845–1849

Peppin, S.S.L., Majumdar, A. & Wettlaufer, J.S. (2010) Morphological instability of a non-equilibrium ice-colloid interface. *Proceedings of the Royal Society A*, 466: 177–194.

Peppin, S.S.L., Worster, M.G. & Wettlaufer, J.S. (2007) Morphological instability in freezing colloidal suspensions. *Proceedings of the Royal Society A*, 463, 723–733.

Peppin, S.S.L.; Wettlaufer, J.S. & Worster, M.G. (2008) Experimental Verification of Morphological Instability in Freezing Aqueous Colloidal Suspensions. *Physical Review Letters*, 100: 238301

Quang, N.H.; Trinh, M.V.; Lee, B.H. & Huh, J.S. (2006). Effect of NH_3 gas on the electrical properties of single-walled carbon nanotube bundles. *Sensors and Actuators B*, 113: 341-346

Romanenko, A.I.; Anikeeva, O.B.; Kuznetsov, V.L.; Buryakov, T.I.; Tkachev, E.N. & Usoltseva, A.N. (2007). Influence of helium, hydrogen, oxygen, air and methane on conductivity of multiwalled carbon nanotubes. *Sensors and Actuators A*, 138: 350-354

Sayago, I.; Santos, H.; Horrillo, M.C.; Aleixandre, M.; Fernandez, M.J.; Terrado, E., et al. (2008). Carbon nanotube networks as gas sensors for NO2 detection. *Talanta, 77*: 758-764

Shaffer, S.P.M. & Windle, A.H. (1999). Fabrication and characterization of carbon nanotube/poly(vinyl alcohol) composites. *Advanced Materials*, 11: 937-941

Shen, J.; Zeng, C. & Lee L.J. (2005). Synthesis of polystyrene–carbon nanofibers nanocomposite foams. *Polymer* 46: 5218–5224.

Star, A.; Joshi, V.; Skarupo, S.; Thomas, D. & Gabriel, J.C.P. (2006). Gas sensor array based on metal-decorated carbon nanotubes. *The Journal of Physical Chemistry B*, 110: 21014-21020.

Suehiro, J.; Zhou, G. & Hara, M. (2003). Fabrication of a carbon nanotube-based gas sensor using dielectrophoresis and its application for ammonia detection by impedance spectroscopy. *Journal of Physics D: Applied Physics*, 36: L109-14

Sun, G.; Liu, S.; Hua, K.; Lv, X.; Huang, L. & Wang, Y. (2007). Electrochemical chlorine sensor with multi-walled carbon nanotubes as electrocatalysts. *Electrochemistry Communications*, 9: 2436–2440

Sweetman, L.J.; Moulton, S.E. & Wallace, G.G. (2008). Characterisation of porous freeze dried conducting carbon nanotube-chitosan scaffolds. *Journal of Materials Chemistry*, 18: 5417-5422

Thomassin, J.M.; Pagnoulle, C.; Bednarz, L.; Huynen, I.; Jerome, R. & Detrembleur, C. (2008). Foams of polycaprolactone/MWNT nanocomposites for efficient EMI reduction. *Journal of Materials Chemistry*, 18: 792-796

Thomassin, J.M.; Pagnoulle, C.; Bednarz, L.; Huynen, I.; Jerome, R. & Detrembleur, C. (2008) Foams of polycaprolactone/MWNT nanocomposites for efficient EMI reduction. Journal *of Materials Chemistry*, 18: 792-796

Thongprachan, N.; Nakagawa, K.; Sano, N.; Charinpanitkul, T. & Tanthapanichakoon, W. (2008). Preparation of macroporous solid foam from muti-walled carbon nanotubes by freeze-drying technique. *Materials Chemistry and Physics*, 112: 262-269

Wettlaufer. J.S. & Worster, M.G. (2006). Premelting Dynamics. *Annual Review of Fluid Mechanics*, 38: 427–452

Wilde, G. & Perepezko, J.H. (2000) Experimental study of particle incorporation during dendritic solidification. *Materials Science and Engineering A*, 283: 25–37

Worsley, M.A.; Kucheyev, S.O.; Satcher, J.H.; Hamza, A.V. & Baumann, T.F. (2009). Mechanically robust and electrically conductive carbon nanotube foams. *Applied Physics Letters* 94, 073115

Xiang, Z.; Chen, T.; Li, Z. & Bian, X. (2009). Negative Temperature Coefficient of Resistivity in Lightweight Conductive Carbon Nanotube/ Polymer Composites. *Macromolecular Material and Engineering*, 294: 91–95

Zawadzak, E.; Bil, M.; Ryszkowska, J.; Nazhat, S.N.; Cho, J.; Bretcanu, O.; J.A. & Boccaccini, A.R. (2009). Polyurethane foams electrophoretically coated with carbon nanotubes for tissue engineering scaffolds. *Biomedical Materials* 4: 015008 (9pp)

Zeng, C.; Hossieny, N.; Zhang, C & Wang, B. (2010). Synthesis and processing of PMMA carbon nanotube nanocomposite foams. *Polymer* 51: 655–664

Zhang, S.; Zhang, N.; Huang, C.; Ren, K. & Zhang, Q. (2005). Microstructure and Electromechanical Properties of Carbon Nanotube/ Poly(vinylidene fluoride-trifluoroethylene-chlorofluoro- ethylene) Composites. *Advanced Materials*, 17: 1897-1901

Ziegler, C. & Gerteisen, D. (2009). Validity of two-phase polymer electrolyte membrane fuel cell models with respect to the gas diffusion layer, *Journal of Power Sources*, 188: 184-191

19

Wear Properties of Cu-CNT Nanocomposites

Rajendrakumar G Patil[1], Vishnukanth Chatpalli[2] and Ramesh C S[3]
[1]*Dept. of Mechanical Engg, Atria Institute of Technology*
[2]*KVT & SDC, Govt. of Karnataka*
[3]*Dept. of Mechanical Engineering, PES Institute of Technology*
India

1. Introduction

Currently, Carbon Nano Tubes (CNTs) are gaining wide spread applications in several high technological areas owing to its remarkable mechanical and electronic properties. Carbon nano tubes were discovered to be the fourth form of carbon by the classic experiments of Ijima in 1991. After that paper there was a flurry of activity worldwide on studies and emphasis on synthesis of CNTs (Ebbesen et al.). Worldwide the major emphasis has been on to develop quality CNTs, several routes of processing CNTs do exist which are Laser Ablation, Electric Arc Discharge and Chemical Vapor Deposition. The properties, both electrical and mechanical of the CNTs have been studied extensively all over the world by several researchers (Tans et al.).It is reported that the CNTs do processes a very high modulus of one Tera Pascal (Harris) and electrical conductivity (1013 A/cm^2). Because of these superior properties CNTs are a potential candidate as reinforcement material to produce composites. Copper and its alloys are used commercially because of its high electrical and thermal conductivity coupled with strength.

The hard and refractory ceramic reinforcements such as silicon carbide, Alumina, titanium oxide have commonly been used as reinforcements in the ductile copper matrix. Although they posses higher strength, better wear resistance they exhibit lower ductility and thermal conductivity when compared with copper. Further the enhancement in the elasticity modulus of copper based particulate composites is not significant. It is also reported that CNTs can be an ideal reinforcements to develop high-quality composites owing to their low weight and exceptional properties. Kashyap et al., have reported that use of CNT as reinforcements in Aluminum matrix has resulted in drastic improvements in strength which has been attributed to Orowan looping. It is reported that use of CNTs in Ni-matrix has resulted in lowering of Coefficient of friction and wear rates of the composites when compared with nickel matrix. However meager information is available as regards to development and study on tribological behavior of Cu/CNT composites. Dong et al., developed the composites using HIP process and reported. Copper with its low stacking fault energy enables one to observe free dislocations on a Transmission Electron Microscope (TEM) since dynamic recovery will not take place in copper during processing. This aspect has not been studied so far in the literature. In the light of the above, this research focuses on development and characterization of tribological properties of Cu/CNT (MWCNTs and SWCNTs). There is only one paper where Dong et al. studied the sliding wear behavior of

Carbon Nanotube/Copper composite. Hence, this aspect is the thrust of the present research i.e., to reinforce copper with Multi-Walled Carbon Nanotubes (MWCNTs) and Single-Walled Carbon Nanotubes (SWCNTs) to enhance its wear and friction behavior. Friction and wear tests have been conducted on both sintered pure copper and the developed composites.

2. Arc discharge chamber – production of CNTs

The Carbon arc method of synthesizing CNTs and Fullerenes consists of a stainless steel chamber, a DC power unit and inert gas supply.The arc reaction chamber fabricated is a hollow cylinder of internal of internal diameter 140 mm and wall thickness of 14 mm. Large internal diameter was provided for easy mounting, removal of electrodes and carbon soot from the inside surface of the chamber. Material selected for the reactor being stainless steel mainly due to their corrosion resistant property and it was ensured that the diameter of reactor were large enough to prevent excess heating of the wall due to arcing of electrodes (>30000 C).Water circulation through the walls of the chamber, to the electrode holders and also through the pipes surrounding the electrodes ensures proper cooling. The thickness of the chamber wall was made sufficient enough to hold a rough vacuum at high temperature. A view port is provided on the chamber to visualize the plasma generation and arcing of electrodes. Flanges are mounted at the reactor ends and to guarantee proper vacuum, o-rings were provided. The top flange of the reactor has a cooling system arrangement for the cathode (copper) to prevent the melting of electrodes due to high temperature developed during arcing process. During the early stages of the arcing process the anode is to be fed regularly to obtain a stable arc, which is being subsequently sustained by maintaining a constant gap of 1mm between the electrodes.

2.1 Power source

A saturated core reactor controlled DC power supply was used with a rating of 80V and 300A. For generation of carbon nanostructures, ripples are maintained within 5% by providing capacitor filter circuit to the power unit and an oscilloscope to monitor the arcing process of the electrodes. The voltage applied for the experiment varies depending on the inert medium used and electrode separation distance. With Helium gas at 500 torr, at a voltage of 20V and separation of 1mm or less between electrodes, gives best results.
Current does not directly influence CNT growth but rather, the heating effect of current is responsible for CNT growth. Heating to temperatures above 3000°C ensures nanotubes to remain open ended without closure there by allowing CNTs to grow. However, the same heating effect can result in the sintering of CNTs if sufficient cooling is not provided. The magnitude of current should ideally be maintained at a minimum to achieve a self-sustained stable arc.

2.2 Inert atmosphere

The inert atmosphere in the chamber is critical to the growth of CNTs. Helium is by far the most popular inert medium. However, lot of successful experiment has been conducted with other gases and even liquids. The temperature/annealing growth mechanism for CNTs by arc discharge process requires cooling of the CNT samples grown at the cathode by inert media and hence the thermal conductivity and pressure of the gas becomes an important factor in the synthesis methodology. High thermal conductivity media provide better cooling for the CNT sample and prevent sintering.

2.3 Inert gas pressure

If the inert medium is a gas, then the pressure of the gas plays a critical role in the yield and quality of CNTs produced. At 500 torr there was a considerable concentration of CNTs. Beyond this pressure the yield of CNTs reduced. The increase in pressure of helium from 20 torr to 500 torr is beneficial to CNT growth due to better cooling of the sample by the inert medium. For generation of CNTs both SWNT and MWNT, Helium gas was vaccumised at 500 torr in the reaction chamber using a direct drive rotary vacuum pump.

2.4 Electrodes

Another very important part of the arc discharge process is electrode material and its geometry. Cooling of electrodes plays an important role in CNT synthesis. CNTs are deposited on the cathode, hence the use of copper cathodes provide better cooling than graphite, which was originally used. The anode is composed of graphite; it may include catalyst for SWNT synthesis. Both anode and cathode can be water-cooled. This reduces chances of sintering of CNT. For the preparation of single walled nanotubes or filled nanotubes, catalysts/material must be inserted into a hole drilled in the center of the graphite anode. As can be expected, Fe, Co, and Ni have been used in conjunction with lanthanide, actinide and transition metals as effective catalysts for SWCNT synthesis. The most favored configuration is a 6mm anode and a 12mm cathode.The experimental setup consisting of arc discharge chamber, inert gas, power source and cooling system is as shown in Figure 1.

Fig. 1. Arc Discharge Apparatus

2.5 Synthesis and purification of carbon nanostructures

A stainless steel arc evaporation reactor with a viewing port as discussed was used for synthesis of these Nano structures. The chamber is connected to a rotary vacuum pump and a helium supply.

The graphite (anode) and copper (cathode) electrodes are mounted on the flanges of the chamber. Cooling being the most important requirement for production of these nano structures and hence water-cooling was provided for cathode, anode and the reactor. Position of anode can be adjusted from outside the chamber so that a constant gap can be maintained during arc evaporation. A DC power supply was used and arc-discharge is usually carried out below 20V. The current depends upon the diameter of graphite rod and also on helium pressure.

2.5.1 Multi-Walled Carbon Nanotubes (MWCNTs)

MWCNTs are synthesized by using Arc Discharge setup. An electric arc is struck between the graphite anode and copper cathode of 6mm and 12mm diameter respectively in a helium atmosphere of 500 torr. The current density is approximately 150 A/cm^2 DC, current and voltages are maintained at 75A and 20V respectively. Helium gas atmosphere can be static or dynamic during the processes.

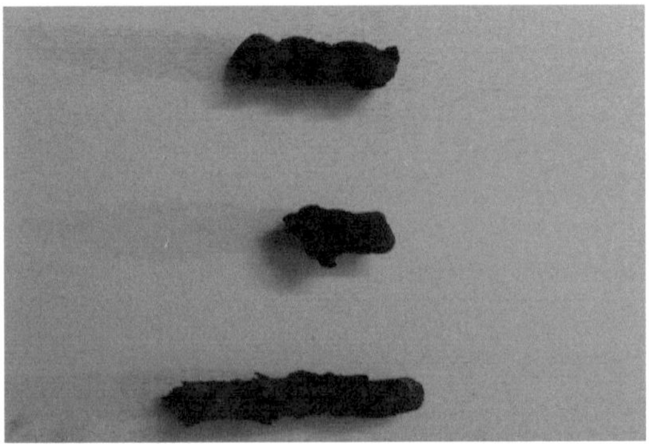

Fig. 2. Cigar like structure deposited on the Copper cathode during Arc evaporation process

The soft-core material contains MWCNTs, polyhedral particles and various kinds of graphitic particles. The soft core thus obtained is dispersed in ethanol and is sonicated for 20 minutes. This solution is then decanted and the residue is sonicated again to recover any remaining nanotubes. The decanted solution is then heated to evaporate ethanol leaving behind nanotubes.

Optimum operating parameters for production of MWCNTs:

Potential drop across electrodes = 20 V
Helium pressure = 500 torr
Anode= Ø 6 mm and Cathode= Ø 12 mm
Current density = (150-200) Amp/cm^2
Inter-electrode distance during stationary period of discharge = 1mm
Average inter-electrode temperature = 4000 K
Deposit rate at cathode surface = 1mm/min
Good cooling of electrode & chamber

2.5.2 Single - Walled Carbon Nanotubes (SWCNTs)

SWCNTs are produced in the same arc evaporation chamber as shown in figure 1 by co-vaporization of graphite and bimetal catalyst in a composite anode. The graphite rod is

6mm in diameter and 150mm in length. A hole of 4mm is drilled axially and densely packed with mixture of Cobalt and Nickel in the ratio 3:1 and 96% graphite powder. The helium pressure is maintained at 500 torr. A stable arc is formed at 20V and 75A. The gap between electrodes is maintained around 2mm. SWCNTs are formed as a web like structure in the chamber and also as a collaret around the cathode. It should be noted that cylindrical deposit also grows at the cathode, consisting of a hard gray shell and a soft core. The soft core has poorly developed columnar structure and contains MWCNTs and polyhedral particles.

The product obtained can be divided in to three structural types; a spongy soft belt called collaret is formed around the cylindrical deposit while relatively a strong cloth like soot is found on the chamber walls and finally a web like structure is suspended in the chamber volume between cathode and walls. The above three type contains varying amount of SWCNTs, Fullerenes and amorphous carbon. To extract SWCNTs, the soot is heated at 450°C for 20 minutes to vaporize volatile impurities. The residue is then treated with $HN0_3$ to dissolve metal catalyst particles. The soot, free from volatile impurities and catalyst particles is filtered to separate it from $HN0_3$ and the residue thus obtained is dissolved in ethanol followed by sonication for 20 minutes. The decanted solution is finally heated to vaporize ethanol to obtain pure SWCNTs.

Optimum operating parameters for production of SWCNTs:

Potential drop across electrodes = 20 V
Helium pressure = 500 torr
Composite Anode = 7 mm outside diameter & inside diameter 4mm
Cathode = Ø 12 mm
Current density = 150 - 200 Amp/cm^2
Inter-electrode distance during stationary period of discharge = 1 mm
Average inter-electrode temperature = 4000 K
Deposit rate at cathode surface = 1.5 mm/min
Good cooling of electrode & chamber

3. Methodology / approach

The route to produce CNTs is electric arc-discharge, process parameters such as voltage, current, graphite electrode purity and the gap between electrodes and the pressure of the carrier gas (Helium) was studied. The multi-walled CNTs obtained from Arc Discharge process was sonicated in an ultrasonic sonicator. These nanostructures were purified by solvent method and their purity was confirmed by Thermal Gravimetric Analysis (TGA). CNTs were characterized by Raman spectroscopy and Electron microscopy to validate their identity and quality. Then the CNTs will be coated by Nickel. The MWCNTs will be given a electroless coating of Nickel as followed by Q Q Li et al., and also by Dong et al., for obtaining good wettability of CNTs with Copper. The coating of Nickel will be confirmed by X-ray diffraction and then Nickel coated CNTs and copper powder will be ball milled for (30 to 60) minutes and compacted in a hydraulic press of 100 ton capacity to obtain green compacts and sintered in Vacuum to increase density of the compacts .These compacts will be extruded to remove voids and tested for wear properties. Thin slices will be cut and Jet polished and the thin foil samples will be studied in a TEM. CNTs dispersion in the nano composite will be ascertained and also study of dislocations distributed with CNTs will be carried out. These green compacts will be sintered at suitable temperature to ensure excellent bond between each grain and particle. The sintered compacts will be subjected to metallurgical studies like SEM and TEM to investigate the homogeneity of the distribution of the CNTs in the Copper Matrix. The powdered composite compacts will be subjected to hot extrusion with a suitable extrusion ratio to obtain a dense sample of nano composite. The post sintering densities of SWCNT/Cu and MWCNT/Cu composites were found to be 98% of the theoretical density. However, TEM images indicate that the distribution of CNTs in metal matrix homogeneously is the most critical issue to enhance the mechanical properties of CNT/metal nanocomposites. The volume fraction of CNTs was determined by analyzing the carbon contents using an Elemental Analyzer. SWCNTs and MWCNTs were used in varying weight percentages to reinforce commercial purity Copper matrix. Pin-on-disc machine has been used to evaluate the friction and wear behavior of the studied materials under dry sliding conditions.

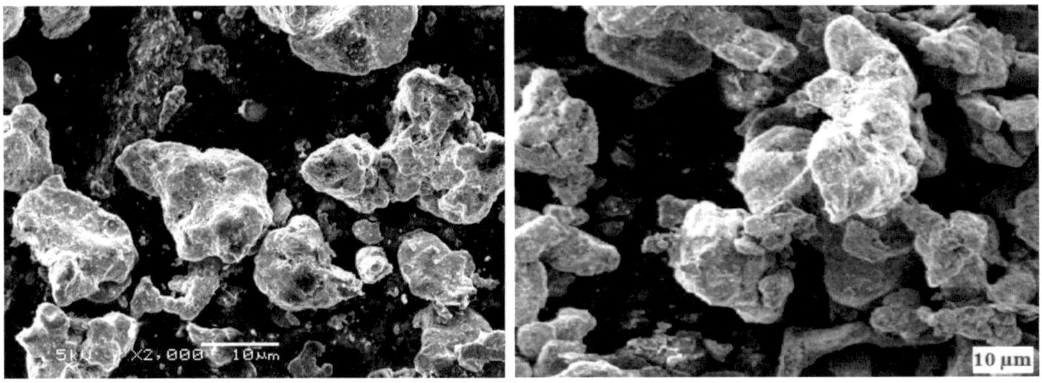

Fig. 3. SEM Images of Copper

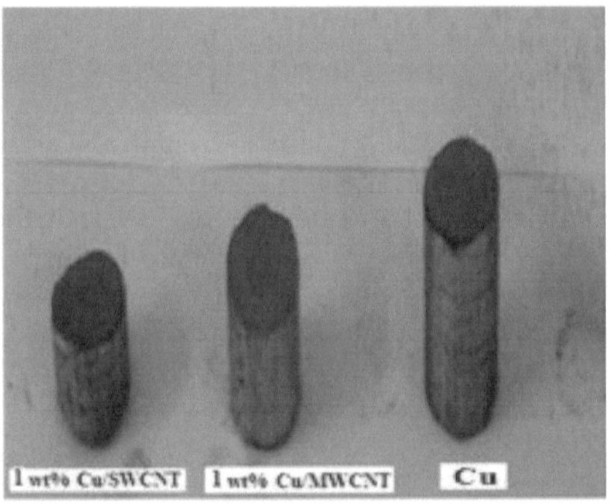

Fig. 4. Powder Compacted (Pre-sintered) Wear Test Specimens

The specimen diameter is 10mm and length is 22mm and sintering temperature is 700^{0}C for 1 hour. The hardness of the counter disc is 60RC. Load was varied from (20-150) N for various sliding velocities. SEM properties are compared with those of pure copper

Fig. 5. SEM Image of Wear Tested Pure Cu sample

specimens. SEM of worn surfaces has been investigated. The incorporation of carbon nanotubes and nickel-coated carbon nanotubes in the copper matrix composites improved tribological properties compared with those of pure copper specimens. The tests were conducted for CNT wt % 1, 3 & 5 and found 3 wt % (12 to 15 vol. %) to be optimum and tabulated.

4. Wear mechanism under dry sliding condition

Compared with carbon fiber, the reinforced effect of nanotube is greater owing to the high strength and aspect ratio of the nanotubes. It is also the evidence, that the working-hardening layer of the subsurface of the worn surface become shallower and the worn chips become smaller. Owing to Carbon nanotubes and the higher toughness of the Nanotube/Cu composite, fracture cracks in the Nanotube/Cu composite have more difficulty propagation and flaking. Therefore, the nanotubes reinforced composite has better wear resistance properties.

In previous researches, when the CNT/metal or CNT/ceramic nanocomposites are fabricated by molecular level process, the chemical bonding formed between the CNTs and the matrix ions provides homogeneous distribution of CNTs as well as high interfacial strength. Therefore, it is confirmed that such remarkable enhancement of hardness by CNT reinforcement is originated from the high interfacial strength at CNT/Cu interface, the homogeneous distribution of CNTs within Cu matrix and attained high relative densities. Thus, based on this result, it can be shown that the improvement of mechanical properties of CNT reinforced nanocomposites is expected when the external load can be shared by homogeneously distributed CNTs through the load transfer from matrix to CNTs by sound interfacial strength at CNT/matrix. Under dry sliding wear condition, the wear loss of CNT/Cu nanocomposites is reduced to 1/3 compared to those of pure Cu matrix. This result means that this nanocomposite shows three times higher wear resistance by addition of CNTs.

When the surface of pure Cu flakes away during the wear process, the worn chips are formed by peeling of Cu grains near the worn surface. However, in case of CNT/Cu nanocomposite, the Cu grains are not easily peeled from the worn surface by the pinning of homogeneously implanted CNTs across Cu grains in CNT/Cu nanocomposites. At the same time, the CNTs exposed to the worn surface during wear process can act as a lubricating carbon film owing to its low friction coefficient. Therefore, the wear loss of CNT/Cu nanocomposites is remarkably decreased with increasing volume fraction of CNTs due to the effect of homogeneous distribution of CNTs in Cu matrix and low friction coefficient of exposed CNTs on the worn surface. The remarkable enhancement of hardness is originated from the homogeneously distributed CNTs in Cu matrix, high interfacial strength at CNT/Cu interfaces and high relative density of nanocomposites. The dispersed CNTs in Cu-matrix nanocomposite provide considerably enhanced wear resistance by retarding the peeling of Cu grains during sliding wear process. Thus, the homogeneous distribution of CNTs with sound interface in Cu matrix is an important technological issue to enhance the mechanical behavior and wear resistance of CNT/Cu nanocomposite.

4.1 Wear test results of pure Cu specimen

The following are the tabulated wear test results obtained under varying load and speed conditions.

1. Copper: 99%
2. Sintering: 750 ^{0}C and 1 hour
3. Specimen: Dia. = 10mm and Length = 22mm

Load (kg)	Speed (rpm)	Time	Wear (μm)	Frictional force (N)	Temp. (^{0}C)	Weight Loss (gm)	
5	100	5 min.	14	6.6	30	Before	13.6314
		10 min.	20	8.8	32	After	13.6174
		15 min.	25	9.5	34	Diff.	00.0140
		20 min.	30	11.1	37		
5	200	5 min.	16	6.3	33	Before	13.6174
		10 min.	23	8.5	36	After	13.5879
		15 min.	29	9.2	39	Diff.	00.0295
		20 min.	34	10.8	42		
5	300	5 min.	19	6.1	37	Before	13.5879
		10 min.	25	8.2	41	After	13.5467
		15 min.	33	8.8	46	Diff.	00.0412
		20 min.	37	10.4	49		
8	100	5 min.	18	8.1	33	Before	13.6314
		10 min.	26	10.9	37	After	13.6039
		15 min.	32	12.5	40	Diff.	00.0275
		20 min.	39	13.8	44		
8	200	5 min.	19	7.7	35	Before	13.6039
		10 min.	31	10.2	39	After	13.5526

		15 min.	37	11.8	42	Diff.	00.0513
		20 min.	41	13.2	47		
8	300	5 min.	23	7.4	39	Before	13.5526
		10 min.	35	9.9	44	After	13.4712
		15 min.	41	11.2	48	Diff.	00.0814
		20 min.	47	12.1	51		
12	100	5 min.	21	8.6	38	Before	13.6314
		10 min.	34	10.9	43	After	13.6003
		15 min.	39	12.5	48	Diff.	00.0311
		20 min.	44	14.7	56		
12	200	5 min.	24	8.2	42	Before	13.6003
		10 min.	37	10.7	46	After	13.5549
		15 min.	43	12.3	53	Diff.	00.0454
		20 min.	48	13.9	59		
12	300	5 min.	26	7.7	46	Before	13.5549
		10 min.	42	10.6	49	After	13.4698
		15 min.	48	11.7	57	Diff.	00.0851
		20 min.	53	13.3	63		

4.2 Wear test results of MWCNT/Cu composite

The following are the tabulated wear test results obtained under varying load and speed conditions.

i. Cu + MWCNT: 1 Wt %
ii. Sintering: 750 ^{0}C and 1 hour
iii. Specimen: dia. = 10mm and length = 22mm

Load (kg)	Speed (rpm)	Time	Wear (μm)	Frictional force (N)	Temp. (^{0}C)	Weight Loss (gm)	
5	100	5 min.	07	3.6	28	Before	11.5401
		10 min.	09	3.7	29	After	11.5381
		15 min.	13	3.9	31	Diff.	00.0020
		20 min.	16	4.2	33		
5	200	5 min.	06	3.4	31	Before	11.5381
		10 min.	07	3.5	33	After	11.5363
		15 min.	10	3.7	34	Diff.	00.0018
		20 min.	12	4.0	36		
5	300	5 min.	05	3.2	32	Before	11.5363
		10 min.	06	3.3	35	After	11.5348
		15 min.	08	3.5	37	Diff.	00.0015
		20 min.	11	3.8	39		

8	100	5 min.	12	4.1	28	Before	11.5401
		10 min.	16	4.4	29	After	11.5374
		15 min.	19	4.9	31	Diff.	00.0027
		20 min.	21	5.4	33		
8	200	5 min.	07	3.6	30	Before	11.5374
		10 min.	09	3.7	31	After	11.5352
		15 min.	13	4.5	33	Diff.	00.0022
		20 min.	16	4.9	34		
8	300	5 min.	05	3.2	32	Before	11.5352
		10 min.	06	3.3	35	After	11.5333
		15 min.	08	3.9	37	Diff.	00.0019
		20 min.	11	4.4	39		
12	100	5 min.	17	4.4	28	Before	11.5401
		10 min.	22	4.7	29	After	11.5374
		15 min.	25	4.9	31	Diff.	00.0032
		20 min.	28	5.4	33		
12	200	5 min.	12	4.2	30	Before	11.5346
		10 min.	16	4.4	31	After	11.5331
		15 min.	19	4.6	33	Diff.	00.0028
		20 min.	21	5.1	34		
12	300	5 min.	07	3.8	32	Before	11.5352
		10 min.	09	4.1	35	After	11.5333
		15 min.	13	4.3	37	Diff.	00.0025
		20 min.	16	4.8	39		

Cu + MWCNT: 3 Wt %; Sintering: 750 ^{0}C and 1 hour;
Specimen: dia. = 10mm and length = 22mm

Load (kg)	**Speed (rpm)**	**Time**	**Wear (μm)**	**Frictional force (N)**	**Temp. (^{0}C)**	**Weight Loss (gm)**	
5	100	5 min.	04	2.9	26	Before	11.5401
		10 min.	05	3.1	28	After	11.5386
		15 min.	09	3.3	30	Diff.	00.0015
		20 min.	11	3.7	33		
5	200	5 min.	03	2.8	29	Before	11.5386
		10 min.	05	3.1	30	After	11.5373
		15 min.	09	3.2	33	Diff.	00.0013
		20 min.	10	3.6	37		
5	300	5 min.	2	2.7	30	Before	11.5373
		10 min.	4	3.0	31	After	11.5361

		15 min.	8	3.1	33	Diff.	00.0012
		20 min.	9	3.2	36		
8	100	5 min.	11	4.0	27	Before	11.5401
		10 min.	13	4.2	29	After	11.5379
		15 min.	17	4.7	30	Diff.	00.0022
		20 min.	19	5.1	31		
8	200	5 min.	06	3.3	30	Before	11.5379
		10 min.	08	3.5	31	After	11.5360
		15 min.	11	4.2	32	Diff.	00.0019
		20 min.	17	4.6	34		
8	300	5 min.	04	3.1	31	Before	11.5360
		10 min.	05	3.2	34	After	11.5343
		15 min.	07	3.7	36	Diff.	00.0017
		20 min.	14	4.2	37		
12	100	5 min.	15	4.2	27	Before	11.5401
		10 min.	20	4.5	29	After	11.5373
		15 min.	23	4.7	31	Diff.	00.0028
		20 min.	25	5.1	32		
12	200	5 min.	10	4.0	31	Before	11.5373
		10 min.	14	4.1	32	After	11.5349
		15 min.	17	4.3	34	Diff.	00.0024
		20 min.	19	4.9	35		
12	300	5 min.	06	3.7	32	Before	11.5349
		10 min.	07	3.9	33	After	11.5327
		15 min.	11	4.1	35	Diff.	00.0022
		20 min.	14	4.5	37		

Cu + MWCNT: 5 Wt %; Sintering: 750 ^{0}C and 1 hour;
Specimen: dia. = 10mm and length = 22mm

Load (kg)	**Speed (rpm)**	**Time**	**Wear (µm)**	**Frictional force (N)**	**Temp. (^{0}C)**	**Weight Loss (gm)**	
5	100	5 min.	02	2.7	25	Before	11.5401
		10 min.	04	2.8	27	After	11.5387
		15 min.	07	3.0	30	Diff.	00.0014
		20 min.	09	3.3	32		
5	200	5 min.	02	2.6	28	Before	11.5387
		10 min.	03	2.9	29	After	11.5375
		15 min.	07	3.0	31	Diff.	00.0012
		20 min.	08	3.2	35		

5	300	5 min.	02	2.5	30	Before	11.5375
		10 min.	03	2.7	32	After	11.5364
		15 min.	06	2.9	33	Diff.	00.0011
		20 min.	07	3.1	35		
8	100	5 min.	09	3.9	29	Before	11.5401
		10 min.	11	4.0	30	After	11.5381
		15 min.	14	4.4	32	Diff.	00.0020
		20 min.	16	4.9	33		
8	200	5 min.	05	3.1	31	Before	11.5381
		10 min.	07	3.3	32	After	11.5363
		15 min.	09	3.9	34	Diff.	00.0018
		20 min.	11	4.2	35		
8	300	5 min.	03	3.0	32	Before	11.5363
		10 min.	04	3.1	35	After	11.5348
		15 min.	06	3.5	37	Diff.	00.0015
		20 min.	09	4.1	39		
12	100	5 min.	13	3.9	29	Before	11.5401
		10 min.	17	4.0	31	After	11.5375
		15 min.	20	4.2	33	Diff.	00.0026
		20 min.	22	4.8	34		
12	200	5 min.	08	3.8	32	Before	11.5375
		10 min.	12	4.0	34	After	11.5352
		15 min.	15	4.1	37	Diff.	00.0023
		20 min.	16	4.7	39		
12	300	5 min.	05	3.4	34	Before	11.5352
		10 min.	06	3.6	35	After	11.5333
		15 min.	09	3.8	37	Diff.	00.0019
		20 min.	12	4.3	40		

4.3 Wear test results of SWCNT/Cu composite

The following are the tabulated wear test results obtained under varying load and speed conditions.

i. Cu + SWCNT: 1 Wt%
ii. Sintering: 750 ^{0}C and 1 hour
iii. Specimen: dia. = 10mm and length = 22mm

Load (kg)	Speed (rpm)	Time	Wear (μm)	Frictional force (N)	Temp. (^{0}C)	Weight Loss (gm)	
5	100	5 min.	07	4.2	30	Before	10.5401
		10 min.	10	4.5	32	After	10.5378

		15 min.	11	5.2	34	Diff.	00.0023
		20 min.	13	5.5	36		
5	200	5 min.	06	4.1	29	Before	10.5378
		10 min.	09	4.4	31	After	10.5356
		15 min.	10	5.1	33	Diff.	00.0022
		20 min.	12	5.4	35		
5	300	5 min.	05	4.0	28	Before	10.5356
		10 min.	08	4.3	30	After	10.5336
		15 min.	09	5.0	32	Diff.	00.0020
		20 min.	11	5.3	33		
8	100	5 min.	14	5.3	30	Before	10.5401
		10 min.	19	5.8	31	After	10.5381
		15 min.	23	6.3	32	Diff.	00.0020
		20 min.	31	6.9	33		
8	200	5 min.	16	5.2	40	Before	10.5381
		10 min.	22	5.5	43	After	10.5352
		15 min.	26	6.1	46	Diff.	00.0029
		20 min.	29	6.4	48		
8	300	5 min.	18	5.0	42	Before	10.5352
		10 min.	23	5.3	44	After	10.5319
		15 min.	27	5.9	48	Diff.	00.0035
		20 min.	31	6.2	50		
12	100	5 min.	18	5.8	48	Before	10.5401
		10 min.	23	6.1	52	After	10.5378
		15 min.	30	6.5	55	Diff.	00.0023
		20 min.	34	6.8	58		
12	200	5 min.	20	5.5	51	Before	10.5378
		10 min.	27	5.9	55	After	10.5353
		15 min.	33	6.2	58	Diff.	00.0025
		20 min.	36	6.6	61		
12	300	5 min.	22	5.3	53	Before	10.5353
		10 min.	30	5.5	57	After	10.5326
		15 min.	35	6.0	60	Diff.	00.0027
		20 min.	39	6.3	63		

Cu + SWCNT: 3 Wt %; Sintering: 750 ^{0}C and 1 hour;
Specimen: dia. = 10mm and length = 22mm

Load (kg)	Speed (rpm)	Time	Wear (μm)	Frictional force (N)	Temp. (^0C)	Weight Loss (gm)	
5	100	5 min.	06	4.0	29	Before	9.3312
		10 min.	08	4.3	31	After	9.3292
		15 min.	10	4.9	33	Diff.	0.0020
		20 min.	12	5.1	35		
5	200	5 min.	08	3.9	30	Before	9.3292
		10 min.	10	4.1	32	After	9.3274
		15 min.	11	4.7	35	Diff.	0.0018
		20 min.	14	4.9	37		
5	300	5 min.	09	3.6	31	Before	9.3274
		10 min.	11	3.9	34	After	9.3257
		15 min.	13	4.3	37	Diff.	0.0017
		20 min.	16	4.7	39		
8	100	5 min.	10	5.1	30	Before	9.5332
		10 min.	15	5.6	32	After	9.4952
		15 min.	20	5.9	34	Diff.	0.0380
		20 min.	28	6.5	35		
8	200	5 min.	14	4.9	32	Before	9.4952
		10 min.	19	5.4	34	After	9.4587
		15 min.	23	5.7	35	Diff.	0.0365
		20 min.	25	6.1	37		
8	300	5 min.	15	4.8	33	Before	9.4587
		10 min.	21	5.2	34	After	9.4235
		15 min.	25	5.3	36	Diff.	0.0352
		20 min.	28	5.8	38		
12	100	5 min.	14	5.9	53	Before	9.5332
		10 min.	20	7.2	58	After	9.4852
		15 min.	27	8.6	61	Diff.	0.0480
		20 min.	30	9.0	63		
12	200	5 min.	12	5.6	54	Before	9.4852
		10 min.	17	6.8	60	After	9.4392
		15 min.	24	8.2	62	Diff.	0.0460
		20 min.	27	8.9	65		
12	300	5 min.	11	5.4	56	Before	9.4392
		10 min.	15	6.3	62	After	9.3947
		15 min.	23	7.8	63	Diff.	0.0445
		20 min.	25	8.5	66		

Cu + SWCNT: 5 Wt %; Sintering: 750 ^{0}C and 1 hour;
Specimen: dia. = 10mm and length = 22mm

Load (kg)	**Speed (rpm)**	**Time**	**Wear (μm)**	**Frictional force (N)**	**Temp. (^{0}C)**	**Weight Loss (gm)**	
5	100	5 min.	04	3.8	30	Before	9.3220
		10 min.	06	4.0	32	After	9.3201
		15 min.	07	4.5	34	Diff.	0.0019
		20 min.	09	4.9	36		
5	200	5 min.	07	3.6	31	Before	9.3201
		10 min.	08	3.8	33	After	9.3184
		15 min.	10	4.1	36	Diff.	0.0017
		20 min.	12	4.7	39		
5	300	5 min.	05	3.4	33	Before	9.3184
		10 min.	06	3.6	35	After	9.3169
		15 min.	08	3.9	37	Diff.	0.0015
		20 min.	10	4.4	40		
8	100	5 min.	08	4.8	32	Before	9.3220
		10 min.	13	5.2	33	After	9.2950
		15 min.	17	5.5	36	Diff.	0.0270
		20 min.	24	6.0	38		
8	200	5 min.	12	4.6	33	Before	9.2950
		10 min.	17	4.9	35	After	9.2730
		15 min.	19	5.1	37	Diff.	0.0220
		20 min.	22	5.8	40		
8	300	5 min.	10	4.4	35	Before	9.2730
		10 min.	15	4.6	36	After	9.2535
		15 min.	16	4.9	39	Diff.	0.0195
		20 min.	19	5.4	42		
12	100	5 min.	17	5.4	55	Before	9.3220
		10 min.	23	6.8	59	After	9.2885
		15 min.	30	8.2	64	Diff.	0.0365
		20 min.	33	8.7	68		
12	200	5 min.	14	5.2	56	Before	9.2885
		10 min.	21	6.5	61	After	9.2543
		15 min.	26	7.9	65	Diff.	0.0342
		20 min.	29	8.4	70		
12	300	5 min.	12	4.9	58	Before	9.2543
		10 min.	19	6.2	61	After	9.2211
		15 min.	23	7.4	67	Diff.	0.0332
		20 min.	26	8.0	72		

SEM observations (Figure 6) show that there are some crates and flake-like wear scars on the worn surface of the Cu/CNT nanocomposites, which appears to be a typical characteristic of adhesive wear, and many carbon chips also can be seen.

Fig. 6. SEM Images of Wear Tested MWCNT/Cu composite sample

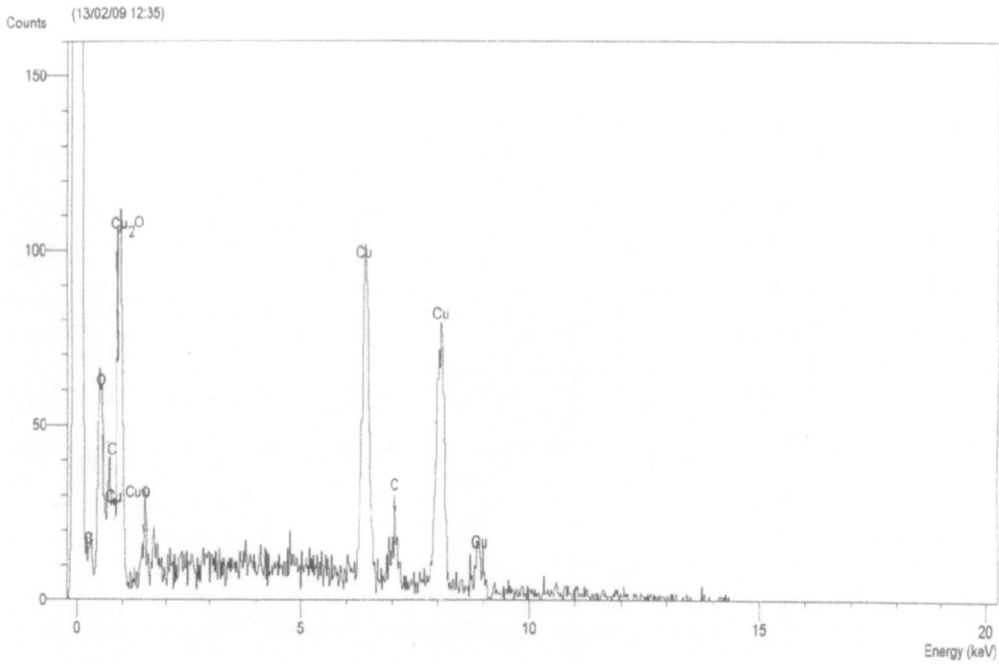

Fig. 7. XRD result of the worn chips of the Cu/MWCNT composite showing formation of Carbon and Oxides of Copper(Cu2O and CuO)

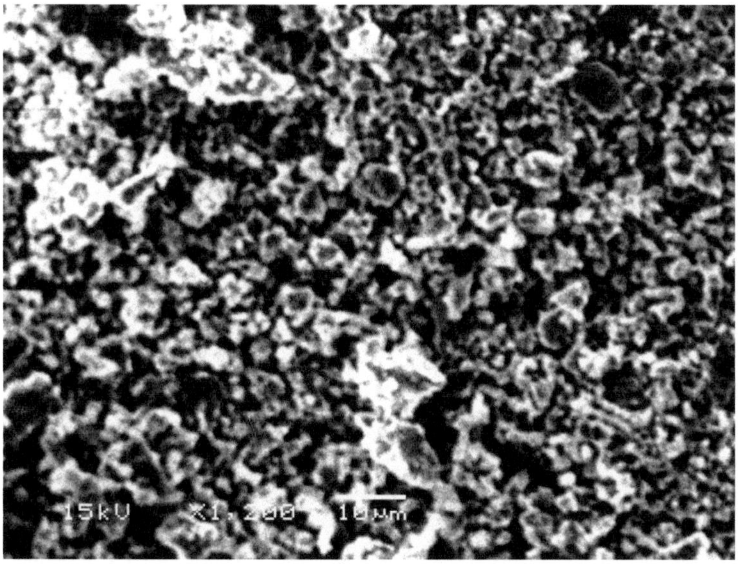

Fig. 8. SEM Image of worn chips of the Cu/MWCNT composite

Fig. 9. SEM Images of Wear Tested SWCNT/Cu composite sample

XRD results (Figure 7) show that there exist some oxides of copper (Cu_2O and CuO) in the worn chips of the composite (Figure 8), suggesting that oxidation wear appears to be the main wear mechanism of the composite. The coefficient of friction and the weight loss for the Cu/CNTs composite is less than the pure copper specimen made by the same method. As the Cu-matrix got worn out gradually, carbon nanotubes in the matrix near the surface were exposed and became the working film on the worn surface. Now the worn

contacting surfaces changed from the original metal surfaces into metals with a lubricating carbon film.
Based on oxidation theory, the carbon film not only provides lubrication in the wearing process and reduces the wear surface exposed in the air, but also prevents the copper from oxidizing. Therefore, the weight loss of the composite is reduced, which also been supported by the experimental results that both the coefficient of friction and weight loss for the Cu/CNT composite are much lower than those of the pure copper specimen made by PM.

4.4 Effect of the nanotubes volume fraction

Increasing the carbon volume fraction significantly decreases the coefficient of friction and weight loss for both the Cu/CNT composite. When carbon content exceed 8%, the graph of the weight loss for the Cu/CNT composite becomes flat and when carbon content is beyond 12%, the weight loss raises. This phenomenon is contributed to the increase of porosity of the composite when the nanotube content is beyond 12%.When nanotube content is low (< 8%), a carbon film can not cover wear surface. When the nanotubes volume fraction increases, the carbon film become larger and cover the wear surface, while the effect of lubrication and impedance to oxidization increases. The optimum nanotubes content is between 12 and 15%.

4.5 Effect of load

The effects of the applied load on the weight loss for Cu/CNT composite are shown below. It is found that the increasing the load had a minimal effect on Cu/CNT composites as compared to pure copper specimen. In another observation, the temperature of the pure copper specimen increase linearly as the load increases, whereas in case of Cu/CNT composite it increases marginally as the volume fraction of CNTs increase, providing the indirect evidence of good thermal conductivity of nanotubes.

Constants: Load=5Kgs, Speed=200 rpm, Duration=20 min			
Composition of CNT (Wt%)	**Coefficient of Friction**		
	Pure Copper (Cu)	Cu + MWCNT	Cu + SWCNT
0	0.22	0.22	0.22
1	NA	0.082	0.11
3	NA	0.073	0.099
5	NA	0.065	0.096

Table 1. Values of Coeff. Of friction for different Wt% of CNTs

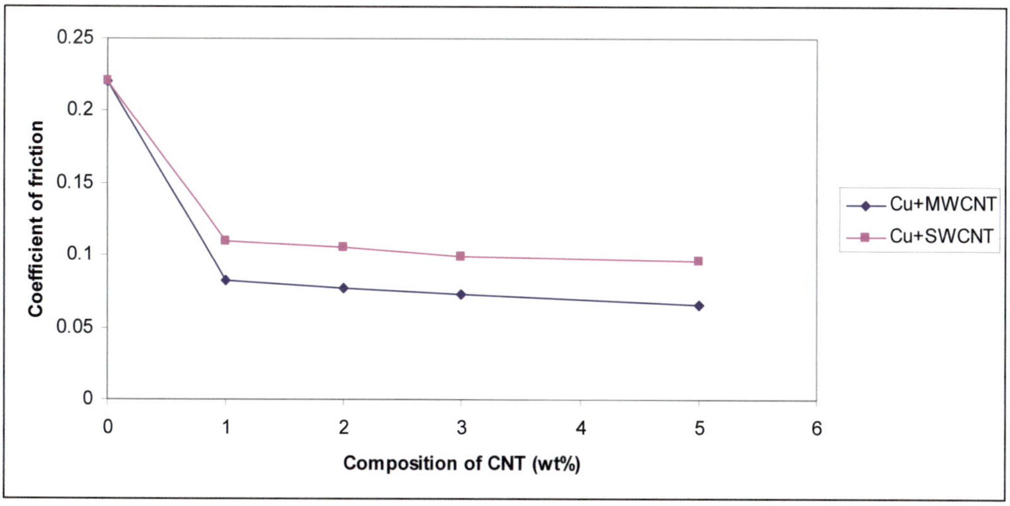

Fig. 10. CNT Composition Vs Coefficient of Friction of Cu, Cu/MWCNT & Cu/SWCNT

Constants: Load=5Kgs, Speed=200 rpm, Duration=20min			
Composition of CNT (Wt%)	Wear Rate (mm^3/Nm)		
	Pure Copper (Cu)	Cu+MWCNT	Cu + SWCNT
0	11.24	11.24	11.24
1	NA	0.0065	0.0092
3	NA	0.002	0.0031
5	NA	0.0030	0.0042

Table 2. Values of Wear rate for different Wt% of CNTs

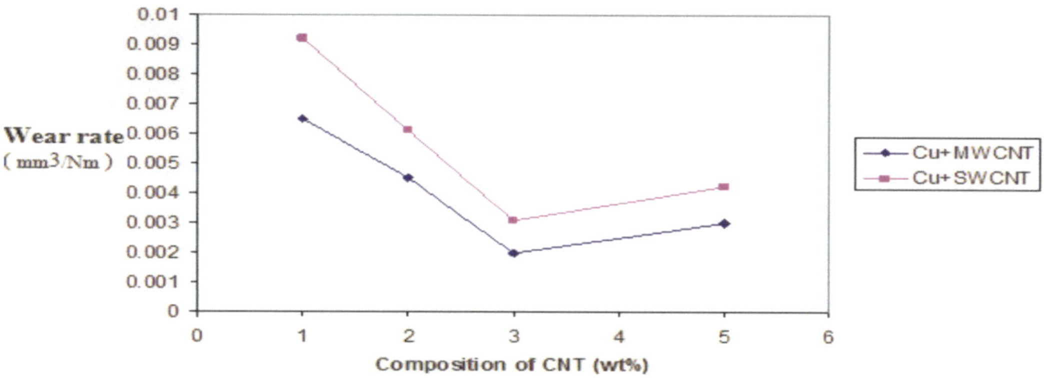

Fig. 11. CNT Composition Vs Wear rate

Constant: Composition of CNT = 1 wt%, Speed = 200 rpm		
Load (N)	Weight Loss (gms)	
	Cu + MWCNT	Cu + SWCNT
50	0.0018	0.0022
80	0.0022	0.0029
120	0.0028	0.0035

Table 3. Load Vs Weight Loss at 1 Wt% of MWCNT & SWCNT

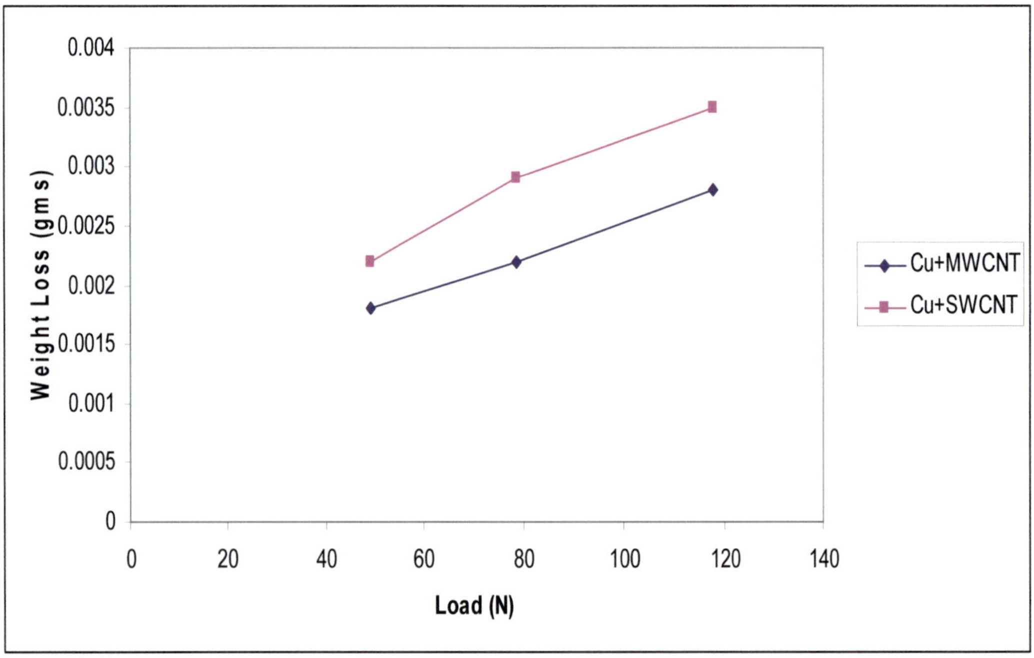

Fig. 12. Load Vs Weight loss

Constant: Composition of CNT = 3 wt%, Speed = 200 rpm		
Load (N)	Weight Loss (gms)	
	Cu + MWCNT	Cu + SWCNT
50	0.0013	0.0018
80	0.0019	0.0037
120	0.0024	0.0046

Table 4. Load Vs Weight Loss at 3 Wt% of MWCNT & SWCNT

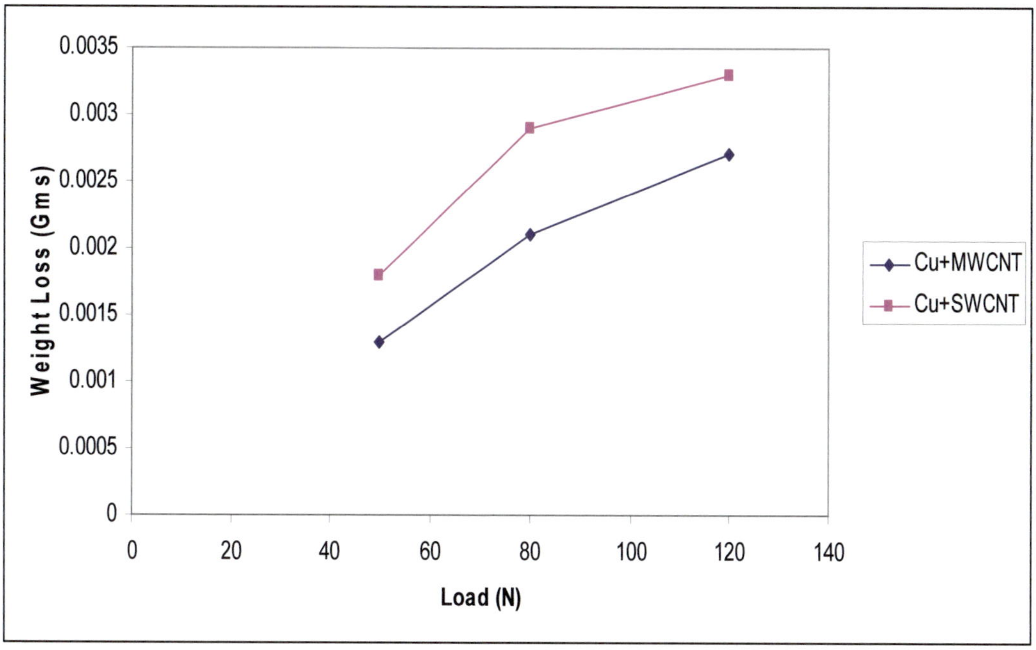

Fig. 13. Load Vs Weight loss

Constant: Composition of CNT = 5 wt%, Speed = 200 rpm		
Load (N)	Weight Loss (gms)	
	Cu + MWCNT	Cu + SWCNT
50	0.0012	0.0017
80	0.0018	0.0022
120	0.0023	0.0034

Table 5. Load Vs Weight Loss at 5 Wt% of MWCNT & SWCNT

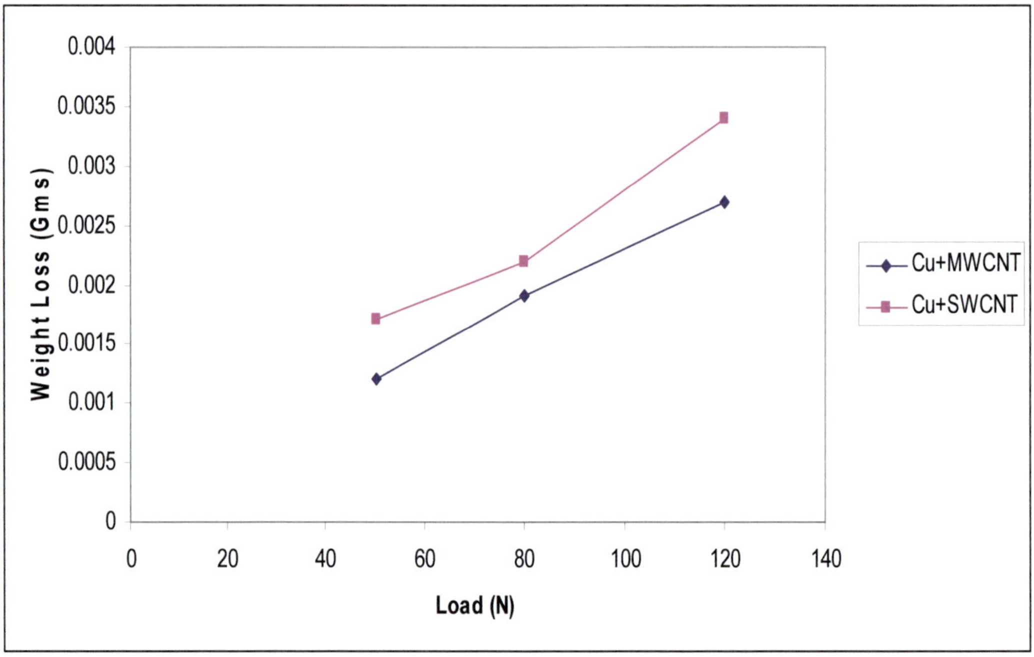

Fig. 14. Load Vs Weight loss

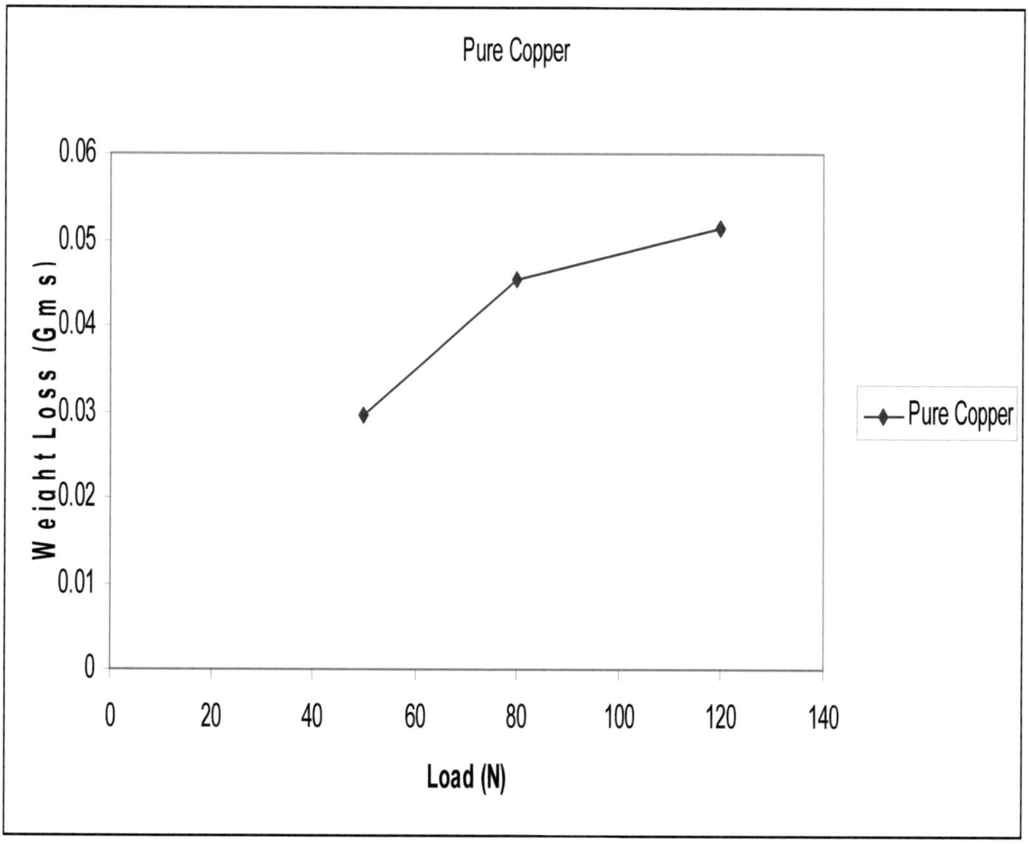

Fig. 15. Load Vs Weight loss

Load (N)	Weight Loss (gms)
50	0.0295
80	0.0454
120	0.0513

Table 6. Load Vs Weight Loss for Pure Copper at 200 rpm

5. Conclusion

CNTs both Multi-walled and Single-walled were synthesized by Arc Discharge method. This method was adopted for its simplicity and cost effectiveness. Arc discharge setup was successfully fabricated to generate good quality CNTs. The yield of MWCNTs was optimum for a helium pressure of 500 torr and a DC current of 75 amps at 20 volts. Adequate cooling of the chamber and graphite electrodes is a pre-requisite for their synthesis.

It is concluded that the homogeneous distribution of CNTs with sound interface in Cu matrix is an important technological issue to enhance the mechanical behavior and wear resistance of CNT/Cu nano composite. Oxidation wear is the main wear mechanism for the CNT/Cu composite under dry sliding conditions. The formation of carbon film can reduce the friction and wear rate. Compared with pure Cu composite, the CNT/Cu nanocomposite has a lower coefficient of friction and reduced weight loss. Increasing the nanotube volume fraction can significantly decrease both the coefficient of friction and wear rate of the composite. The optimum nanotubes content is 3 wt % i.e., between (12 and 15) Volume %.

6. Acknowledgment

The author wish to thank Sri. A.S.Chinnaswamy Raju, MD, Atria Group, Dr. M.S.Shivakumar, Principal and Management of AIT, Bangalore, for encouragement and support.

7. References

Ijima (1991) , *Helical microtubules of graphitic carbon,* Nature, Vol 354, P 51-58.

Kashyap et al., (2005)*Strengthening in carbon nanotube/aluminium (CNT/Al) composites,* Scripta Materialia, USA 53, 10, P 1105 - 1212.

Q-Q Li et al. (1997), *Structure of the Global Nanoscience and Nanotechnology Research Literature,* Japanese journal Applied Physics, 3613, P 501-503.

R Martel et al. (1998), *Single and multi-wall carbon nanotube field-effect transistors,* Applied Phy. letters ,73, 2447.

S J Tans et al.,(1998), *Room-temperature transistor based on a single carbon nanotube,* Nature 393, 49.

S R Dong et al. (2001), *An Investigation of the sliding wear behavior of Cu-Matrix composite reinforced by Carbon Nanotubes,* Materials Science and Engg. A 313(2001) P 83-87.

T W Ebbesen and P M Ajayan. (1992), *Wondrous World of Carbon Nano Tubes,* Nature, 358, 220.

T W Ebbesen et al. (1995), *Long wavelength optical response of incipient fullerene nanotubes,* Chemical Physics letters, 209, 83.

W J Harris (2000), *Carbon Nanotubes and related structures,* Cambridge University Press.